AF837864

Eukaryome Impact on Human Intestine Homeostasis and Mucosal Immunology

Nancy Guillen
Editor

Eukaryome Impact on Human Intestine Homeostasis and Mucosal Immunology

Overview of the First Eukaryome Congress at Institut Pasteur. Paris, October 16–18, 2019.

Springer

Editor
Nancy Guillen
Institute Pasteur
Paris, France

ISBN 978-3-030-44825-7 ISBN 978-3-030-44826-4 (eBook)
https://doi.org/10.1007/978-3-030-44826-4

© Springer Nature Switzerland AG 2020
This work is subject to copyright. All rights are reserved by the Publisher, whether the whole or part of the material is concerned, specifically the rights of translation, reprinting, reuse of illustrations, recitation, broadcasting, reproduction on microfilms or in any other physical way, and transmission or information storage and retrieval, electronic adaptation, computer software, or by similar or dissimilar methodology now known or hereafter developed.
The use of general descriptive names, registered names, trademarks, service marks, etc. in this publication does not imply, even in the absence of a specific statement, that such names are exempt from the relevant protective laws and regulations and therefore free for general use.
The publisher, the authors and the editors are safe to assume that the advice and information in this book are believed to be true and accurate at the date of publication. Neither the publisher nor the authors or the editors give a warranty, expressed or implied, with respect to the material contained herein or for any errors or omissions that may have been made. The publisher remains neutral with regard to jurisdictional claims in published maps and institutional affiliations.

This Springer imprint is published by the registered company Springer Nature Switzerland AG
The registered company address is: Gewerbestrasse 11, 6330 Cham, Switzerland

Eukaryome Foreword

I am so pleased to see we have finally come to the point where we start to understand that parasites and us are not alone in infections. We and parasites are not fighting (or parasites are simply utilizing us) one-on-one, but microbial community (including bacteria, archaea, and eukaryotes) plays, as a whole, a significant role in modulating gene expression, metabolism, and virulence of the parasites, and also, on the contrary, physiology, immunity, and tolerance of the host. Such mechanisms have started to unveil. This book is a showcase of such orchestration of human and microbial community including eukaryotic microorganisms, which, eukaryotic microorganisms, had been almost neglected. This book also gives the most updated overview by experts in the fields on genetic diversity and evolution of microbial eukaryotes, immune response to and immune modulation by parasitic eukaryotes, genome organization, gene regulation, pathogenesis, and drug resistance, as well as cutting-edge models and analytical systems to investigate crosstalk of eukaryome-microbiome-host in commensal and symptomatic infections.

Tomo Nozaki
The University of Tokyo
Tokyo, Japan

Preface

The first international congress on *Eukaryome Impact on Human Intestine Homeostasis and Mucosal Immunology* was held in October 2019 at Pasteur Institute, Paris, France. This congress brought together experts in the fields of genomics, infection, pathogenicity and new technologies to enrich concepts, share methodologies, and advance interdisciplinary research programs.

Our knowledge of the influence of microbes on the fate of intestinal diseases is scanty, particularly with regard to the major parasitic diseases. Multiple demographic or economic parameters of our societies could contribute to the origin of emerging infections; for example, poverty, urbanization, climate change, conflict, and political instability, as well as population migration. These features challenge us to know the impact (present and future) of parasitic diseases on public health. The intestine is a nutrient-rich environment harboring a complex and dynamic population of 100 trillion microbes, the microbiome. Until now, the scientific attention has been mainly focused on the bacterial part of the microbiome, which constitutes the most abundant and diverse fraction of intestinal ecosystem. Moreover, bacteria are sharing the gut ecosystem with a population of uni- and multicellular eukaryote organisms. Our interest focuses on the **eukaryotes microbes** inhabiting the intestine named **intestinal eukaryome** (including fungi, protists, and helminths). The knowledge about the reciprocal influence of the microbiome on the eukaryome and their independent or combined impact on homeostasis and intestinal diseases is scanty and can be considered as an important emerging field.

The objectives of EUKARYOME congress were in three areas of research:

1. Phylogenetic, prevalence, and diversity of intestinal eukaryotic microbes, their enigmatic historical evolution and their potential contributions to mucosal immune homeostasis.
2. Integrative biology to study the molecular cell biology of parasites–host interactions and the multiple parameters underlining the infectious process.

3. Forefront technologies to develop and exploit tissue engineering and microfluidics to establish three-dimensional (3D) systems helping to understand homeostasis and pathological processes in the human intestine and for drug discovery.

These topics were organized in four parts according to the following interplay:

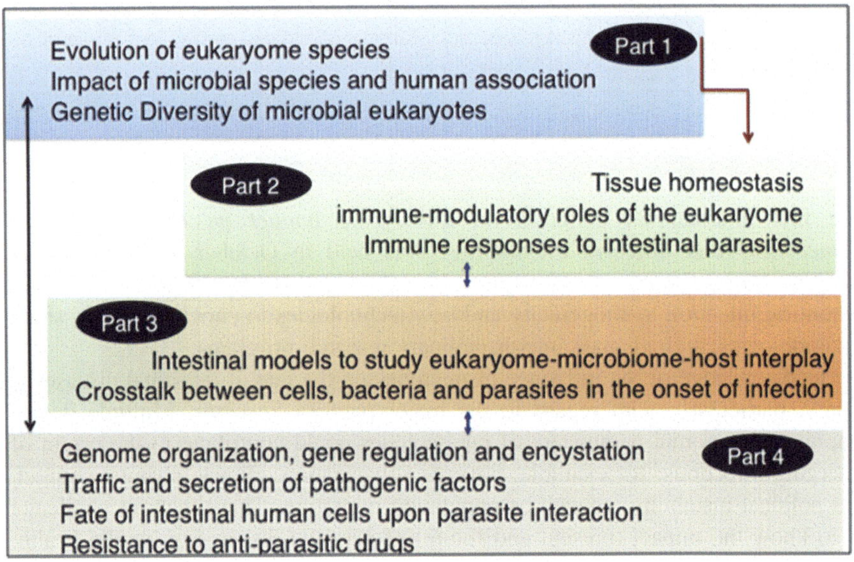

Part 1: Evolution of Eukaryome Species. Impact of Microbial Species and Human Association. Genetic Diversity of Microbial Eukaryotes

A direct association between the presence of eukaryotic microbes and changes in the bacterial intestinal microbiome has been observed in the absence of gastrointestinal disorders, disease, or inflammation. Thus, the eukaryome should be considered as a member of the intestinal ecosystem with contributors that regulate the diversity and composition of the community. The presentations and discussions were on the coevolution of humans and parasites and cellular processes, such as encystation, as important evolutionary features of unicellular organisms (Session 1 and 2). Then session 2 ended with a comparison between the eukaryome of healthy and infected people. We have also discussed genetic diversity and focused on commensal and opportunistic protists. The objective was to specify the impact of the presence of parasites on the fate of intestinal diseases.

Session 1: Evolution and impact of parasite species in humans

Laure Segurel	Human lifestyles, protozoa and the gut microbiota
Jan Tachezy	*Mastigamoeba balamuthi* at the cross of evolutionary roads
Anastosios Tsaousis	Exploring the biology and evolution of *Blastocystis* and its role in the microbiome
Cecilia Ximenez	Bacterial intestinal microbiota changes associated with *Blastocystis* infection: consequence of protist colonization? or the result of local inflammation
Robert Hirt	The molecular basis of Trichomonads' capacity to jump between hosts and mucosal sites

Session 2: Cell differentiation and species diversity of unicellular eukaryotes

Upinder Singh	Regulation of gene expression during conversion in *Entamoeba*
Pauline Schaap	Resolving Amoebozoan encystation from *Dictyostelium* evo-devo, comparative genomics and transcriptomics
David Carmena	The current molecular epidemiological scenario of *Cryptosporidium*, *Giardia* and *Blastocystis* in Implication for Public Health
Carol Gilchrist	*Cryptosporidium* infection in malnourished children

Part 2: Tissue Homeostasis. Immune-Modulatory Roles of the Eukaryome. Immune Responses to Intestinal Parasites

There is an emerging paradigm that the human microbiome is central to many aspects of health and may have a role in preventing enteric infections, but very little is known on the role of the eukaryome in these aspects. The mechanisms underlying the immune response to parasite embedded in the bacterial–host ecosystem and the consequences due to environmental changes, as alimentary diet, are unexplored to date. The possible immuno-modulatory role of eukaryote microbes was discussed (Session 3). The potential mechanism of action of these microbes supporting the immune response and the stimulation of mucus production, which alleviates the symptoms of colitis improving intestinal health was of interest. Internationally recognized specialists in eukaryome, microbiome, and immunity (Session 4) gave new insights on these topics.

Session 3: Tissue homeostasis, immune-modulatory roles of the eukaryome

William A. Petri	Role of Innate Lymphoid Cells and Th17 in regulation of intestinal infection and immunity
Arthur Mortha	Commensal protozoa shape mucosal immune homeostasis

Philippe Jay	Epithelial sentinels linking luminal danger to immune responses
Kateřina Jirků Pomajbíková	Helminth therapy—nowadays and tomorrow
Marie Claire Arrieta	Contribution of fungi to gut microbiome ecology, early-life immune development and airway inflammation
Stacey Burgess	Innate trained immunity in Giardia and Entamoeba infection

Session 4: Tissue protection and immune response to infections

Malin Johansson	Gut mucus- protection and interaction
Eric Martens	Combined low dietary fiber and mucus-degrading gut bacteria cause lethal colitis in genetically susceptible mice
Kris Chadee	Roles of the mucus layer and microbiota in innate host defense against *Entamoeba histolytica*
Hannelore Lotter	Immune responses underlying hepatic amebiasis
Alfonso Olivos	Redox, oxidative stress and heat stress in experimental amoebic liver abscess

Part 3: Intestinal Models to Study Eukaryome-Microbiome-Host Interplay. Crosstalk Between Cells, Bacteria and Parasites in the Onset of Infection

The relationship between eukaryotic commensal or invasive parasite activities may be conditioned (at least in part) by differences in the host's gut microbiota. However, the mechanisms underlying the molecular switch from a commensal phenotype to a pathogenic phenotype have not been characterized in detail; in particular, because of the lack of experimental models to study human diseases and intestine homeostasis. In addition, pathogenesis depends on the interplay between tissue renewal, microbes, and the immune system. The objective was to explore the three components "parasites-bacteria-host" interplay in the intestine. Firstly, the functional interplay between the parasites and the human host during invasion of the tissue was discussed (Session 5 and 6) and then a focus was made on the contribution of tissue engineering and microfluidics as forefront technologies to build intestinal models allowing the study of the human complex intestinal microbial ecosystem (Session 7).

Session 5: Host response and cell damage upon parasite interaction

Alok Bhattacharya	Phagocytosis by *Entamoeba histolytica.*
Serge Ankri	Parasite-bacteria responses to host immune stresses
Kumiko Nakada-Tsukui	What determines trogocytosis and phagocytosis

Katherine S. Ralston	Acquisition of host cell membrane proteins during *Entamoeba histolytica* trogocytosis
Sunando Datta	Role of EhC2B, a C2 domain containing actin modulator in amoebic erythrophagocytosis

Session 6: Host modulation on infection outcome

Tomoyoshi Nozaki	Host modulation by *Entamoeba histolytica* via secreted lysine glutamic acid rich protein 2
Julio Cesar Carrero	Role of Extracellular Traps on intestinal parasites. Relationship with virulence
Elisabeth Labruyère	Impact of tissue components on *Entamoeba histolytica* migration
Chelsea Marie	Host factors involved in *Cryptosporidium* invasion
Staffan G. Svärd	The Giardia secretome and its role during host-parasite interactions

Session 7: Intestinal models to study eukaryome-microbiome-host

Nancy Allbritton	Large intestine on chip for mucus and microbiome studies
Arturo Aguilar	An intestinal 3D model gives insight on amoebiasis
Magdalena Kasendra	Synergistic engineering of human intestine: organoids meet organs-on-chips
Nathalie Sauvonnet	Mechanical forces and 3D topology of the colonic epithelium are critical for enteric pathogens invasion using a biomimetic human gut on a chip interplay
Jean Christophe Olivo Marin	Cells, images and numbers

Part 4: Genome Organization, Gene Regulation and Encystation, Traffic and Secretion of Pathogenic factors, Fate of Intestinal Human Cells Upon Parasite Interaction. Anti-parasitic Drugs and Therapies

The molecular mechanisms underlying the virulent phenotype of intestinal parasites are mostly unknown. The sessions 8, 9, and 10 were devoted to parasite biology focusing on (i) genome organization and gene regulation; (ii) the processes of traffic and secretion of pathogenic factors allowing to feed on bacteria, kill human cells and to invade the tissue; (iii) the parasite organelles involved in the survival in anaerobic environments such as the human colon and (iv) the persistence of protozoan species due to both, encystation in the intestinal lumen and resistance to anti-parasitic drugs.

Session 8: Traffic and secretion of pathogenic factors

Mark Field	Evolution, traffic and membrane transport.
Esther Orozco	Arming the puzzle of Entamoeba histolytica phagocytosis: New pieces.
Som Lata	PIP3 binding protein screening reveals unique molecules involved in endocytic processes of Entamoeba histolytica.
Carmen Faso	The ins and outs of *Giardia lamblia* endomembrane trafficking.

Session 9: Gene expression, regulation sustaining pathogenicity

Iris Brucchaus	Gene silencing and overexpression to study pathogenicity factors of *Entamoeba histolytica*
Nancy Guillen	The transcriptome from a compact genome
Laurence Marchat	Polyadenylation machineries in intestinal parasites: latest advances in *Entamoeba histolytica*
Samie Amidou	Molecular characterization of *Entamoeba histolytica* tRNA genes
Michael Duchene	*Entamoeba* glycan biosynthesis genes, how the lipopeptidophosphoglycan (LPPG) might be made
Janna Alazzaz	Probiotics as anti-Giardia defenders: determination of the control mechanism
Gretchen Ehrenkaufer	Identification of drugs with activity against multiple pathogenic amoebae

In addition to the 46 presentations, 20 posters were selected and three of them were awarded by the Springer Scholarship:

- EhRho6 mediated actin degradation in *Entamoeba histolytica* by A. Narooka, A. Apte, P. Yadav, S. Datta
- Quinoxaline T-001 affects pathogenicity mechanisms of *Entamoeba histolytica* associated with cytoskeleton functions by J. Soto Sanchez, A. Garcia, O. Medina Contreras, L. Caro Gómez, A. Paz Gonzáles, L. Marchat, G. Rivera, R. Moo Puc, E. Ramírez Moreno
- CXXC-rich protein involves in RBC phagocytosis and cell-monolayer degradation abilities of the *E. histolytica* parasite by N. Khomkhum, S. Leetachewa, T. Nozaki, F. Moreau, A. L. Coria, C. Chadee, S. Moonsom.

Overall, this first edition of the EUKARYOME congress (92 participants) brought together international experts to discuss the latest developments and the latest procedures to understand the impact of eukaryotic microbes on intestinal homeostasis and infectious disorders. This research is in its infancy but appears as one of the most promising avenues of discovery in infectious diseases. The meeting ended with a discussion of the areas of interest that permeated the meeting, particularly the exciting new models of 3D-intestinal models as systems of interest for moving toward translational applications.

Paris, France Nancy Guillen

Contents

Chapter 1
Eukaryome: Emerging Field with Profound Translational Potential

Nancy Guillen

Abstract The human intestinal eukaryome comprises a diverse set of eukaryotic organisms living in the intestinal lumen. These are in permanent or transient interaction with commensal organisms such as bacteria. This interaction has a direct impact on human well-being. The eukaryome remains one of the least understood components of the gut microbiota despite its permanent association with the host during the natural selection of species. The emerging work hypothesis is that eukaryome and bacteria in synergy influence the complex mechanism underlying the microbial crosstalk with the human gut during health or disease. New microbiota studies should therefore include the characterization of the components and function of the eukaryome to discover its role in homeostasis and intestinal diseases.

Keywords Eukaryome · Microbiome · Human populations · Immune response · Tissue models

Introduction

The human gastrointestinal tract (from the mouth to the rectum) is home to a complex and dynamic population of microorganisms, the microbiota, which exerts a strong influence on the host during homeostasis and disease.[1] The role of bacteria, as the most abundant microbial population component of the microbiome, has been a topic of interest in recent decades (Proctor et al. 2019; Alivisatos et al. 2015). Typical commensal bacterial species present in the intestine are normally beneficial and fulfill useful and necessary functions such as aiding the digestion of food. In addition, intestinal bacteria play a crucial role in maintaining homeostasis as well as in the

[1]Microbiota is the complex community of microorganisms including bacteria but also viruses (including bacteriophages), Archaea, eukaryotes such as fungi and protozoa living in consortia in sites such as the gastrointestinal tract (Lokmer et al. 2019).

N. Guillen (✉)
Institut Pasteur, Centre National de la Recherche Scientifique-ERL9195, 25 Rue du Dr Roux, 75015 Paris, France
e-mail: nguillen@pasteur.fr

© Springer Nature Switzerland AG 2020

N. Guillen (ed.), *Eukaryome Impact on Human Intestine Homeostasis and Mucosal Immunology*, https://doi.org/10.1007/978-3-030-44826-4_1

protection against pathogens. According to abundant literature, the microbiota plays a fundamental role in the induction, education and function of the immune system of the host. However, the microbiota is very versatile and one of the main factors that determine the composition of the intestinal microbiota over the course of life is the alimentary diet, which is diverse. The microbiome is thus considered as a complex ecosystem reaching the equilibrium during homeostasis by the interaction of microbial species with human tissues but also according to the chemical nature of their environment. Altered intestinal bacterial composition, named dysbiosis, has been associated with effects of inflammatory diseases and infections (Thursby and Juge 2017).

Bacteria are not the only microorganisms living in the intestine. Other complex interactions exist between different microbial populations and the intestinal tract; they must also be considered when analysing homeostasis and diseases (Parfrey et al. 2011). These organisms include archaea, eukaryotic microorganisms and multicellular parasites colonizing the luminal part of the gastrointestinal tract. Particularly, the eukaryotic microbial population has evolved with the host over thousands of years and in combination with bacteria, it forms a complex ecosystem for which there is very little information on their mutually beneficial relationship; this is an important fact because the success of pathogenic processes that can be generated by eukaryotic microorganisms is undoubtedly influenced by their relationship with the intestinal flora. This review has the intention to comment on some rare but interesting results linking the presence of eukaryotic microbes to the modulation of the immune response during homeostasis and intestinal infection. The evolutionary traits of eukaryotic microbes and their relationship with bacteria and the host leading to their survival are also of interest in this summary. All these features have been the core of the EUKARYOME congress. In the next paragraphs there is a summary of the central points discussed during the eukaryome meeting. In addition, the articles in this proceeding book make the point on numerous data that have been presented in the meeting.

The Diversity of Species Within the Human Intestinal Eukaryome

Culture-independent molecular techniques (e.g. DNA extraction from faeces and high throughput sequencing) have been used to estimate the occurrence of eukaryotes in the human intestine, thereby revealing the diversity of these ubiquitous organisms. These studies, reviewed in (Hamad et al. 2016), indicate the presence of four major groups of eukarya colonizing the intestine of humans. The first group, the Excavata Phyla Metamonad, comprises two classes: Parabasalides (*Dientamoeba fragilis*, *Pentatrichomonas hominis* and *Trichomonas* spp.) and Diplomonadida (*Giardia lamblia*). The second group, RAS, includes three of the most diverse eukaryotic divisions (Alveolates, Rhizaria and Stramenopiles). Alveolates are parasites distributed in three principal lineages: (i) Apicomplexans, which cause severe parasitic intestinal infection in humans (*Cryptosporidium*, *Cyclospora*, *Cystoisospora* and *Sarcocystis*); (ii) non-pathogenic Ciliates (*Balantidium coli*, *Stentor roeseli* and *Vorticella campanula*); and (iii) Stramenopiles, which have morphological characteristics similar to fungi, such as *Blastocystis* which is the sole member of Stramenopiles that inhabits the human gut and eventually may cause enteritis. The third group consists of Amoebozoa with the Conosa phylum (*Entamoeba*, *Iodamoeba* and *Endolimax*) which can cause diseases in humans. The fourth group, Opisthokonta, includes all uniflagellate eukaryotes as well as metazoans and fungi. Platyhelminths and nematode species belonging to the metazoan kingdom correspond to the genera of helminthic parasites (including cestodes, trematodes and nematodes) causing infection in millions of people worldwide. More than 140 fungal genera (including ascomycetous, basidiomycetous and zygomycetous) have been discovered as permanent or transient biota in the intestinal tract; many of them are either beneficial or commensal (e.g. *Saccharomyces*) and are considered as probiotics with a role in the treatment of diarrhoea (Gouba and Drancourt 2015). Recent data, summarized below, show the impact of several of these species in the intestinal ecosystem and therefore on public health.

Evolution of Microbial Eukaryotes in the Human Intestine

In the literature, it is common to assume that parasitic species (and their pathogenic factors) derive from an ancestral organism capable of adapting to the parasitic lifestyle (Parfrey et al. 2014). This hypothesis is difficult to assess because of the lack of information concerning the nature of specific factors of the ancestral organism predisposing to parasitism. Comparative genomics and subsequent studies in cell biology provide two related functions that should help us figure out the transition from a free lifestyle to the parasitic life mode of protozoan organisms; these are respiration and encystation. Both functions are essential to the survival of ancient protozoa without mitochondria, including *Entamoeba, Giardia*, and *Trichomonas*, which all live in low-oxygen environments, or *Cryptosporidium* and *Microsporidia*, which have reduced their metabolic abilities as intracellular parasites (Liu et al. 2016).

A good example to address the issue of the transition from the free life style to parasitism lies in the genomic comparisons between free-living *Mastigamoeba* and parasitic *Entamoeba*, both from the Amoebozoa group. These studies revealed in *Entamoebae* a remarkable loss of genes related to adaptation to low oxygen niches (e.g. those involved in the tricarboxylic acid cycle and oxidative phosphorylation), a mitochondrion-derived organelle called mitosome (Embley and Martin 2006) was instead preserved. Mitosomes are mitochondrion-related organelles that have largely lost typical mitochondrial functions; these also exist in *Giardia*, *Cryptosporidium* and *Microsporidia* (Shiflett and Johnson 2010). In *E. histolytica*, the major function of mitosomes is sulfate activation leading exclusively to the synthesis of sulfolipids (e.g. cholesteryl sulfate) through the activity of sulfotransferases as cholesteryl sulfotransferase, which is also necessary for encystation (Mi-ichi et al. 2015a, b). Sulfatases and sulfotransferases encoding genes have been acquired by lateral gene transfer in the *Entamoeba* lineage. Consequently, the metabolism of sulfolipids is not conserved in *Mastigamoeba* indicating specialization of this pathway (Mi-ichii et al. 2017). These findings suggest (i) that sulfolipid metabolism has a causal relationship with parasitism (Mi-ichi et al. 2017), (ii) LGT from bacteria play a role in some functions that modifies the free-living style of ameobazoan, and (iii) *Mastigamoeba* is at the cross road of evolution of Amoebazoan.

Encystation is highly relevant for human health because encysted pathogens are resistant to treatment and the spreading of cysts in the environment is a major cause of prevalence of infectious diseases such as amoebiasis, giardiasis and thrichomoniasis. Due to the impact of these infectious diseases in humans, notable efforts are made to understand the molecular bases of encystation. For instance, a recently identified transcription factor (ERM-BP) regulates *Entamoeba* encystation (Manna et al. 2018). ERM-BP binds to NAD+ and promotes gene transcription. Remarkably, increased levels of NAD+ enhance encystation consistently with the role of carbon source depletion in triggering *Entamoeba* encystation. Other pathogenic free-living Amoebozoa, such as *Acanthamoeba,* survive environmental starvation stress by encystment and non-pathogenic free-living amoebae such as *Dictyostelia* develop spores (Kawabe et al. 2015), although many *Dictyostelia* can also encyst. Sporulation is triggered by cAMP activation of the protein kinase PKA, with cAMP levels being controlled by the adenylate cyclases ACG and ACR and the phosphodiesterase RegA. In Amoebozoa, RegA, AcrA and PKA are deeply conserved (Du et al. 2014). Examples of encystation also exist in other intestinal eukaryotic microbes including *Giardia*, *Cyptosporidium*, *Blastocystis*, and *Blantidium*. There are exceptions, however, as in the case of *Entamoeba gingivalis* found in the dental cavities of humans or *Diantemoeba fragilis*, which is widespread in the colon. For these two last Amoebozoa clear evidences showing encystation in their life cycle are so far non-existent.

Impact of Eukaryome Species in Human Populations

To examine how the gut microbiome differs among human populations, several investigations have been conducted to identify bacterial species in fecal samples. The loss of bacterial diversity among people living in the United States was found when compared to residents of the Amazonas of Venezuela or from rural areas in Malawi (Yatsunenko et al. 2012). These distinctive features were evident in early childhood and adulthood. Other explorations have shown that the decrease in bacterial diversity is systematic in the intestinal microbiome of individuals from industrialized countries compared to that of non-industrialized countries (Chabe et al. 2017). These studies, conducted in countries located at latitudes distant from each other, have concluded that human diet, urbanization levels, hygiene practices and the use of antibiotics of the industrialized lifestyle affect the diversity of bacteria in the intestinal microbiome. However, further studies suggest that the geographical distance of latitudes could have a significant effect on the interpretation of microbial diversity data in the human gut (Dikongue and Segurel 2017). Indeed, studies on the microbiota diversity at the local level with populations differing only in a limited number of factors have been more informative than studies comparing very remote populations. For example, small-scale comparisons of rural populations in Africa with contrasting livelihoods but similar local environmental and urban levels, and shared or distinct genetic ancestry, have shown an increase in bacterial diversity of the microbiome. Interestingly, increased bacterial diversity is more common in human populations infected with *Entamoeba* (Morton et al. 2015). In addition, many bacterial taxa for which abundance was significantly correlated with *Entamoeba* infection have opposite abundance profiles to those correlated with various autoimmune disorders (Morton et al. 2015). The impact of intestinal colonization by a protozoan on the composition of the bacterial microbiota was also demonstrated during infection with *Blastocystis* (Audebert et al. 2016; Nieves-Ramirez et al. 2018), a component of the microbiota detected with a high prevalence in healthy people. The question of whether *Blastocystis* is a pathogen or a commensal of the human gut still has no definitive answer. However, analysis of bacterial microbiota composition in individuals with *Blastocystis* showed higher bacterial diversity with greater abundance of Clostridia and a lower level of Enterobacteriaceae. These findings indicate that colonization by this protist could be associated to a healthy intestinal microbiota. In addition, in healthy patients, *Blastocystis* has also been found in association with other protozoa such as non-pathogenic species of *Entamoebae* (Lokmer et al. 2019). In addition, co-colonization by *Entamoeba* spp. and *Blastocystis hominis* has been linked to a healthy condition in patients from the Ivory Coast, whereas samples from patients infected by *Giardia duodenalis* were related to a dysbiotic condition (Iebba et al. 2016). These findings indicate that multiple protozoa are common residents of the healthy human gut worldwide and the key open question is to know whether the increase in bacterial diversity associated to the presence of protozoa (in mono or multicolonization) is a fact concerning only non-pathogenic protozoa.

Overall, the data suggest a specific role of intestinal eukaryotes in increasing intestinal microbial diversity. Deciphering the mechanisms governing the interaction of protozoa with the intestinal flora is the matter of emerging research.

Protozoan Feed on Bacteria from the Microbiome

In addition to encystation occurring in the colon after growth and starvation of trophozoites, the survival of enteric parasites such as *Entamoebae* is dependent on nutritional sources in part derived from the phagocytosis of bacteria. Thus, infection by *E. histolytica* is significantly correlated with intestinal microbiome composition and diversity. For instance, in some individuals, the symptomatic infection due to *E. histolytica* correlates with changes in the level of bacteria members of the *Prevotellaceae* family. Elevated levels of *P. copri* were found in patients with diarrheagenic *E. histolytica* infections suggesting the influence of the microbiome on the state of the gut (Gilchrist et al. 2016). Furthermore, significant changes in the indigenous gut flora of amoeba infected individuals show an increase in abundance of *Bifidobacterium* and at the same time, depletion of *Bacteroides*, *Lactobacillus*, *Clostridium leptum*, *Clostridium coccoides* subgroup, *Campylobacter* and *Eubacterium* (Verma et al. 2012). Using rRNA based metagenomic analysis it has been shown that *E. histolytica* preferentially phagocytoses some of the beneficial bacteria that are required for the maintenance of a healthy gut amongst which are *Lactobacillales, Erysipelotrichales, Clostridales, and Bifidobacteriales.* Phagocytosis of these bacteria may cause dysbiosis in the gut and create conditions for parasite proliferation in the human intestinal lumen (Iyer et al. 2019). Further, phagocytosis of pathogenic *Escherichia coli* by amoebae increases epithelial cell damage (Galvan-Moroyoqui et al. 2008). Recently, it has been shown that live bacteria of the family *Entereobacteriaceae* including commensal components of the microbiota help *E. histolytica* to survive oxidative stress (Varet et al. 2018) and to establish itself in the intestinal mucosa (Shaulov et al. 2018). In a global view, it is possible to conclude that the presence of intestinal protists such as *E. histolytica*, *G. duodenalis*, and *B. hominis* can induce significant changes in the bacterial diversity of the microbiome.

The Eukaryome, Public Health and Zoonosis

The intestinal eukaryotic microbes vary from parasitic to opportunistic, to commensal or mutualistic. Historically, parasites are considered to have a pathogenic effect on the human host with high medical incidence in malnourished or immuno suppressed hosts. This is the case of *Cryptosporidium* spp., *Giardia intestinalis* and *Entamoeba histolytica*, as well as the helminths *Ascaris lumbricoides* and *Strongyloides stercoralis*. Nevertheless, some intestinal eukaryotic microbes associated with humans do not cause damage (e.g. *Entamoeba dispar, Entamoeba coli, Dientamoeba fragilis*). An interesting example is *Blastocystis*, which has been linked to intestinal disorders but whose pathogenicity still remains controversial due to its high prevalence in asymptomatic carriers. Epidemiological studies have indicated that *Blastocystis* spp. have high predominance in underdeveloped countries ranging to over 50% of the population in some developing countries (Alfellani et al. 2013; Ramirez et al. 2014). Furthermore, colonization rates for *Blastocystis* are close to 20% in Europe (El Safadi et al. 2016; Bart et al. 2013). Gastrointestinal dwelling nematodes infect approximately 1 billion people worldwide causing significant ill health (Pullan et al. 2014), *Cryptosporidium* accounts for more than 60,000 deaths (Troeger et al. 2017). In addition, *Cryptosporidium* spp., *Giardia* and *Blastocystis* sp. are able to infect a wide range of animal species other than humans, indicating that the study of these pathogens should be ideally approached under the One Health concept (Robertson et al. 2019). Accurate diagnosis of intestinal parasites is important to properly manage infected populations, in particular of severely malnourished children (Mmbaga and Houpt 2017).

Tissue Homeostasis and Immune-Modulatory Roles of the Eukaryome

The gastrointestinal tract is colonized by a large number of commensal microbes and is constantly exposed to ingested antigens and potential pathogens. The regulation of intestinal tolerance therefore represents the main task of the immune system of the intestinal mucosa. The bacterial component of the microbiota has been extensively studied for its contribution to the immune system during homeostasis and subsequent infection (Reinoso Webb et al. 2016; Hooper et al. 2012). Nowadays, thanks to research models based on mice, it is well accepted that commensal bacteria play a vital role in the development of a fully functional immune system in the intestine. For example, germ free animals remove their deficiencies in the immune system upon colonization by microorganisms. Thus, interactions between commensal bacteria and intestinal mucosa induce highly regulated, innate and adaptive immune responses (Maynard et al. 2012). Some non-exclusive examples follow. The segmented filamentous bacterium (SFB) is a commensal that induces T helper 17 (Th17) cells differentiation in the small intestine (Ivanov et al. 2009). Activated Th17 cells produce IL-17A, IL-17F, IL-6, IL-22, TNF-α and GM-CSF and protect the host against fungal and bacterial infections. IL-22 promotes the formation of tight junctions in epithelial cells and increases the production of mucin and antimicrobial proteins (Zindl et al. 2013), all of which limit bacterial invasion of the lamina propria. Bacteria such as *Clostridia* and *Bacteroidetes* induce immunoregulatory T cells (Tregs), which are known to suppress immune responses to self and to bacterial antigens promoting epithelial repair and microbe tolerance (Atarashi et al. 2011). The key question is: how can this solid panorama of bacterial contribution to the immune system be influenced (or modified) by eukaryotic microbes?

An important recent finding is that dysbiose of the intestinal microbiome (after antibiotic treatment) increases susceptibility to amoeba colitis in humans and in a mouse model of intestinal infection. The mechanism of this increased susceptibility is based on the reduced recruitment of neutrophils in the intestine (Watanabe et al. 2017), which is in part due to a decrease in the surface expression of the CXCR2 chemokine receptor. Other interesting data show that bone marrow-derived dendritic cells (BMDC) from SFB-colonized mice have higher levels of IL-23 production in response to stimulation with *E. histolytica*. IL-23 is involved in the differentiation of Th17 cells in a pro-inflammatory context and, in particular, in the presence of cytokines as TGF-β and IL-6. The adoptive transfer of BMDC from an SFB (+) mouse to an SFB (–) mouse is therefore sufficient to provide protection against *E. histolytica,* and induction of IL-17A is necessary to provide this protection. These data indicate that colonization of intestinal mice with a specific commensal bacterium may provide protection during amoebiasis (Burgess et al. 2014). In addition, BMDCs isolated from mice (fed with sufficient protein diet) infected with *Giardia* produce more IL-23 compared to uninfected controls; while BMDCs from mice fed with a low protein diet and infected does not. These results suggest that *Giardia* intestinal infection and the nutritional status of the host may have a role on inflammatory cytokine production (Burgess et al. 2019).

Cytokines as Major Players of Intestinal Immunity During Parasite Infection

Interleukin-33 is a cytokine of the IL-1 family produced by endothelial cells, epithelial cells and fibroblasts, it acts as an alarmin released after cellular necrosis, altering the immune system in the event of tissue injury or stress (Miller 2011). Amoebic infection significantly increased the level of IL-33 protein in the murine model of amoebic colitis, and the IL-33 mRNA level was significantly higher in intestinal biopsies of patients infected with *E. histolytica*. Indeed, IL-33 seems to protect against amoebic infection. An increase in the expression of IL-5 and IL-13, as well as an increase in the number of M2 macrophages, at the site of infection, are correlated with this protection. IL-33 is also an important guardian of the gut barrier during *Clostridium difficile* infection via activation of group 2 innate lymphoid cells (Frisbee et al. 2019). In contrast, helminths neutralize alarm in signals of infection by suppressing IL-33 release as one of the multiple mechanisms of immunomodulation exerted by helminths, which are considered master manipulators of host immunity (Hotez et al. 2008) because the host accommodates and tolerates these parasites that release multiple immunomodulatory factors (Maizels et al. 2018). This feature leads to the development of a therapeutic strategy called helminthic therapy, which relies on the use of controlled exposure to non-pathogenic or slightly pathogenic worms in order to beneficially regulate host immunity (Maizels et al. 2018).

Microbiome studies from humans naturally infected with intestinal nematodes including whipworm have been carried out with varied outcomes regarding the influence of helminth infection (Lee et al. 2014). Potentially, it should depend on factors including co-infection, diet, age and the level of worm burden (Cooper et al. 2013). Protists, such as Trichomonads, which colonize the intestinal tract, are parasites that also manipulate immune responses. For instance, *Tritrichomonas musculis* activates epithelial inflammasome to induce the release of IL-18 (Escalante et al. 2016), which in turn promotes dendritic cell-driven Th1 and Th17 immunity, thereby altering the ability of the tissue to respond to bacterial infections (Chudnovskiy et al. 2016). Expansion of a Th1 immune response is observed in the large intestine and cecum. The consequences of *T. musculis* immunomodulation have been observed upon infection of the caecum by *Salmonella*, which causes massive inflammation in mice lacking *T. musculis*; on the contrary, mice colonized with *T. musculis* remained unaffected by the bacterial pathogen. These data suggest that *Trichomonas* species, as permanent members of the healthy mouse microbiome, are regulators of immune homeostasis and immune responses. *T. musculis* has orthologs in humans (Chudnovskiy et al. 2016), it belongs to the parabasalides class of the Excavata group, which includes other species infecting humans such as *Pentatrichomonas hominis* and *Dientamoeba fragilis*, but it is still unclear whether these protists are commensal or pathogen (Barratt et al. 2011). *Dientamoeba fragilis* exhibits a morphology similar to *Histomonas meleagridis* (except for the lack of flagella) and a close phylogenetic relationship between *Dientamoeba fragilis* and *Histomonas* has been described. *H. meleagridis* is a microaerophilic flagellated extracellular protozoan parasite of the Tritrichomonadida order infecting gallinaceous birds. This parasite lacks a cyst stage and is transmitted via the eggs of a nematode, but in vitro *H. meleagridis* can be propagated only in the presence of bacteria (Hess et al. 2015). Remarkably, the sole presence of *H. meleagridis* induces histomoniasis in turkeys and chickens but it depends on the presence of intestinal bacteria (Doll and Franker 1963).

Tissue Protection Against Microbial Eukaryotes

The mucosal surfaces of the gastrointestinal tract are the first barrier protecting the tissue from pathogens and are the support of the intestinal microbial ecosystem. A major component of the intestinal lumen is mucus that acts as a physical barrier, a trap for pathogens and a nutritional source. It is rich in the net-forming heavily glycosylated MUC2 mucin (Johansson et al. 2011; Pelaseyed et al. 2014). The intestine epithelium is composed of proliferative crypts, which contain intestinal stem cells and villi. Specialized cells present in the villi are: absorptive enterocytes, mucus-secreting goblet cells, hormone-secreting enteroendocrine cells and Paneth cells. Blood flow and peristaltic motion waves provide a dynamic environment for the epithelial cells. Fundamental to the intestinal epithelium is the extracellular matrix that provides structural and biochemical support to cells and contributes to the transfer of the biochemical signals required for tissue morphogenesis, differentiation and homeostasis. The population of luminal commensal bacteria generates chemical gradients that probably affect the physiology of the epithelial cells (Goto 2019). Pathogenic organisms have however evolved mechanisms to penetrate the mucus barrier and to gain access into the epithelium. Goblet cells are known to play an important role in intestinal immunity by producing mucin (Mahapatro et al. 2016). These cells are also involved in the transit of luminal antigens to lamina propria dendritic cells suggesting that goblet cells couple mucus to the immune system (McDole et al. 2012). Infection of mice with the helminth *Trichuris muris* indicates that goblet cell number and size increases throughout the course of the infection until immune resistance is established (Glover et al. 2019). *E. histolytica* overcomes the protective mucus layer using a combination of mucinase/glycosidase and the potent mucus secretagogue activity (Ohshima et al. 2014) of globet cells (Cornick et al. 2017). *E. histolytica* binds colonic MUC2, which they degrade using the cysteine proteinase A5. Commensal microbiota colonizing the outer mucus layer increases innate host defense against *E. histolytica* and dysbiosis (antibiotic and/or alterations in the mucus layer) renders the colonic epithelium susceptible to *E. histolytica,* which suppress the activity of the goblet cell transcription factor Math1 (Leon-Coria et al. 2018). All these data indicate that an important effect on goblet cells biology is exerted by *E. histolytica*.

Environmental factors such as diet also play a critical role in intestinal diseases. The bacterial microbiota is essential to ensure important metabolic functions that cannot be accomplished by the host. Certain intestinal bacteria (obligate anaerobes) are able to degrade complex non-digestible carbohydrates (fibers) to produce short chain fatty acids (SCFA) such as acetate, butyrate and propionate. These metabolites, rather than glucose, are the preferred energetic substrate of colon epithelial cells and, as we discussed before, the nutritional status of the host is important for parasitic infections (Burgess et al. 2019). SCFA prevent encystation in *Entamoebae* and hyperacetylation of histone 4, indicating unusual sensitivity of this parasite's histone modifying enzymes to SCFA (Byers et al. 2005; Byers and Eichinger 2008). Furthermore MUC2 and butyrate (an SCFA) contribute to the synthesis of cathelicidin (an antimicrobial peptide) in response to *E. histolytica* (Cobo et al. 2017). Apart from dietary fiber, an alternative energy source for the microbiota is the glycoprotein-rich mucus layer. For instance, it has been found that fiber deprivation leads the gut microbiota to degrade the colonic mucus barrier (Desai et al. 2016) and promotes severe colitis in mice lacking IL-10, a cytokine for which loss of function is associated with intestinal diseases.

Close below the mucus layer there is the epithelial barrier. Intestinal epithelial cells (IECs) are central players in homeostasis as they integrate microbial and immunological cues to orchestrate the complex network of microbes-host interactions. IECs constitute a physical and immune barrier to pathogens, the presence of butyrate increases barrier integrity through an increase in the synthesis of tight junction proteins (Peng et al. 2009). Among IECs, Tuft cells are critical for protection during enteric infections and inflammatory responses (Steele et al. 2016). As sensory cells these are able to detect luminal signals including nutrients and harmful components. Tuft cells have a role in Type 2 immune responses (Th2) typically initiated by parasites (Howitt et al. 2016), including commensal protozoan or helminths, and represent the primary source of interleukin-25, which induces their expansion (Howitt et al. 2016; Gerbe et al. 2016; von Moltke et al. 2016). In mice colonized with *Tritrichomonas* species (undergoing Tuft cell differentiation via the IL-13-STAT6 pathway), there is an increase of type II innate lymphoid cells (ILC2) numbers. ILC2 are cells that produce high levels of Th2 cytokines and in particular epithelial cytokines IL-25 and IL-33 activate ILC2, as in the case of helminths infection (Huang et al. 2018; Burrows et al. 2019).

Metabolites such as succinate, derived from rich diet or as a fermentative end-product of protist, engage the succinate receptor 1-GPR91- on Tuft cells and trigger ILC2 cytokine production. During infection with *Tritrichomonas*, activation of Tuft cells by succinate triggers a signalling pathway leading to the crosstalk between Tuft and ILC2 cells necessary for anti-parasite response (Schneider et al. 2018); this response is also induced by helminths (Schneider et al. 2018).

Intestinal Models to Study Eukaryome-Microbiome-Host Interplay

The understanding of parasite behaviors during intestinal infection is limited by the lack of experimental models similar to human in vivo conditions. Despite the great complexity shown by the human intestine, several approaches have been adopted to diminish the gap between the conventional cell culture models and human intestine. Tissue substitutes are emerging as promising tools for understanding human intestinal diseases (Dutton et al. 2019). Several 3D-tissue scaffolds and microfluidic systems mimic the natural microenvironment of the small intestine (Costello et al. 2014; Kim et al. 2012; Shin and Kim 2018; Jalili-Firoozinezhad et al. 2019). The structure of colon crypts has recently been recapitulated by the cultivation of human intestinal epithelium on a collagen scaffold microfabricated with an array of crypt-like invaginations (Wang et al. 2014), and a thick layer of mucus has been produced in these models (Wang et al. 2019). Although all of these substitutes for intestinal tissue are emerging as promising tools for understanding human tissues behavior in health and disease conditions, their implementation in daily laboratory work is complex.

A human gut-on-a-chip microdevice has been used to analyse the interaction of commensal microbes with human intestinal epithelial cells. The data showed that probiotic bacteria and antibiotic treatments can suppress villous lesions induced by pathogenic bacteria (Kim et al. 2016). A gut-on-a-chip microdevice recently allowed the identification of metabolites secreted by the human microbiome, capable of regulating infection with enterohaemorrhagic *Escherichia coli* (Tovaglieri et al. 2019). A complex human gut microbiome has been cultured in an anaerobic intestine-on-a-chip (Jalili-Firoozinezhad et al. 2019).

The first attempt to use 3D-intestinal models for parasitic infection has been done with *Cryptosporidium*. A bioengineered three-dimensional (3D) human intestinal tissue model was build using silk protein as the scaffolding biomaterial. This model supports long-term *Cryptosporidium* infection (DeCicco RePass et al. 2017; Cardenas et al. 2020), with the achievement of intracellular stages, including those involved in asexual and sexual replication. In addition, *Cryptosporidium* can infect, propagate and complete its complex life cycle in epithelial organoids derived from human small intestine and lung. Temporal analysis of the *Cryptosporidium* transcriptome during organoid infection reveals dynamic regulation of transcripts related to its life cycle (Heo et al. 2018). Another 3D-intestinal model mimics human amoebic intestinal infection (Rojas et al. 2020). This 3D-intestine model, derived from co-cultured cell lines, has important features for the disease such as the presence of a mucus barrier and an epithelial barrier. Tissue imaging, proteomics and transcriptomics highlight the importance of several virulence markers occurring in patients or in experimental models, but they also demonstrate the involvement of under-described molecules and regulatory factors in the amoebic invasive process. Certainly, the advances in the field of tissue engineering provide clues on how the construction of a human colon model could help us understand the host's intestinal physiology and its changes following parasite infection.

The following parts of the EUKARYOME meeting are not discussed here because they were extensive. Several sections featured presentations on how intestinal parasites hijack the cellular machineries of human cells to proceed with infection. In particular, injecting factors to control gene expression, using secreted cytokines to further invade the intestinal tissue, degrading the intestinal parenchyma, etc. Interesting discussion focused on the mechanisms of pathogenic factors secretion leading to parasite-cells interactions and subsequent cell death (e.g. phagocytosis and trogocytosis). Finally, numerous conferences detailed the molecular mechanisms underlying the virulent phenotype of intestinal parasites. It was an extensive section devoted to parasite biology focusing on (i) genome organization and gene regulation; (ii) the processes of traffic and secretion of pathogenic factors allowing to feed on bacteria, kill human cells and to invade the tissue; (iii) the parasite organelles involved in the survival in anaerobic environments such as the human colon and (iv) the persistence of protozoan species due to both, encystation in the intestinal lumen and resistance to anti-parasitic drugs. Finally, forefront technologies as bioimaging and strategies for drug discovery were introduced and discussed.

Acknowledgements NG work was supported by European ERA-NET Infect-ERA program AMOEBAC (French National Agency for Research grant ANR-14-IFEC-0001-02).

References

Alfellani, M. A., Jacob, A. S., Perea, N. O., Krecek, R. C., Taner-Mulla, D., Verweij, J. J., et al. (2013). Diversity and distribution of *Blastocystis* sp. subtypes in non-human primates. *Parasitology, 140*(8), 966–971.

Alivisatos, A. P., Blaser, M. J., Brodie, E. L., Chun, M., Dangl, J. L., Donohue, T. J., et al. (2015). MICROBIOME. A unified initiative to harness Earth's microbiomes. *Science, 350*(6260), 507–508.

Atarashi, K., Tanoue, T., Shima, T., Imaoka, A., Kuwahara, T., Momose, Y., et al. (2011). Induction of colonic regulatory T cells by indigenous *Clostridium* species. *Science, 331*(6015), 337–341.

Audebert, C., Even, G., Cian, A., Loywick, A., Merlin, S., Viscogliosi, E., et al. (2016). Colonization with the enteric protozoa *Blastocystis* is associated with increased diversity of human gut bacterial microbiota. *Scientific Reports, 6,* 25255.

Barratt, J. L., Harkness, J., Marriott, D., Ellis, J. T., & Stark, D. (2011). A review of *Dientamoeba fragilis* carriage in humans: Several reasons why this organism should be considered in the diagnosis of gastrointestinal illness. *Gut Microbes, 2*(1), 3–12.

Bart, A., Wentink-Bonnema, E. M., Gilis, H., Verhaar, N., Wassenaar, C. J., van Vugt, M., et al. (2013). Diagnosis and subtype analysis of *Blastocystis* sp. in 442 patients in a hospital setting in the Netherlands. *BMC Infectious Diseases, 13,* 389.

Burgess, S. L., Buonomo, E., Carey, M., Cowardin, C., Naylor, C., Noor, Z., et al. (2014). Bone marrow dendritic cells from mice with an altered microbiota provide interleukin 17A-dependent protection against *Entamoeba histolytica* colitis. *MBio, 5*(6), e01817.

Burgess, S. L., Oka, A., Liu, B., Bolick, D. T., Oakland, D. N., Guerrant, R. L., et al. (2019). Intestinal parasitic infection alters bone marrow derived dendritic cell inflammatory cytokine production in response to bacterial endotoxin in a diet-dependent manner. *PLoS Neglected Tropical Diseases, 13*(7), e0007515.

Burrows, K., Ngai, L., Wong, F., Won, D., & Mortha, A. (2019). ILC2 activation by protozoan commensal microbes. *International Journal of Molecular Science, 20*(19).

Byers, J., & Eichinger, D. (2008). Acetylation of the *Entamoeba* histone H4 N-terminal domain is influenced by short-chain fatty acids that enter trophozoites in a pH-dependent manner. *International Journal for Parasitology, 38*(1), 57–64.

Byers, J., Faigle, W., & Eichinger, D. (2005). Colonic short-chain fatty acids inhibit encystation of *Entamoeba invadens*. *Cellular Microbiology, 7*(2), 269–279.

Cardenas, D., Bhalchandra, S., Lamisere, H., Chen, Y., Zeng, X. L., Ramani, S., et al. (2020). Two- and three-dimensional bioengineered human intestinal tissue models for *Cryptosporidium*. *Methods in Molecular Biology, 2052,* 373–402.

Chabe, M., Lokmer, A., & Segurel, L. (2017). Gut protozoa: Friends or foes of the human gut microbiota? *Trends in Parasitology, 33*(12), 925–934.

Chudnovskiy, A., Mortha, A., Kana, V., Kennard, A., Ramirez, J. D., Rahman, A., et al. (2016). Host-protozoan interactions protect from mucosal infections through activation of the inflammasome. *Cell, 167*(2), 444–456, e414.

Cobo, E. R., Kissoon-Singh, V., Moreau, F., Holani, R., & Chadee, K. (2017). MUC2 mucin and butyrate contribute to the synthesis of the antimicrobial peptide cathelicidin in response to *Entamoeba histolytica-* and dextran sodium sulfate-induced colitis. *Infection and Immunity, 85*(3).

Cooper, P., Walker, A. W., Reyes, J., Chico, M., Salter, S. J., Vaca, M., et al. (2013). Patent human infections with the whipworm, *Trichuris trichiura*, are not associated with alterations in the faecal microbiota. *PLoS ONE, 8*(10), e76573.

Cornick, S., Moreau, F., Gaisano, H. Y., & Chadee, K. (2017). Entamoeba histolytica-induced mucin exocytosis is mediated by VAMP8 and is critical in mucosal innate host defense. *MBio, 8*(5).

Costello, C. M., Sorna, R. M., Goh, Y. L., Cengic, I., Jain, N. K., & March, J. C. (2014). 3-D intestinal scaffolds for evaluating the therapeutic potential of probiotics. *Molecular Pharmaceutics, 11*(7), 2030–2039.

DeCicco RePass, M. A., Chen, Y., Lin, Y., Zhou, W., Kaplan, D. L., & Ward, H. D. (2017). Novel bio-engineered three-dimensional human intestinal model for long-term infection of *Cryptosporidium parvum*. *Infection and Immunity, 85*(3).

Desai, M. S., Seekatz, A. M., Koropatkin, N. M., Kamada, N., Hickey, C. A., Wolter, M., et al. (2016). A dietary fiber-deprived gut microbiota degrades the colonic mucus barrier and enhances pathogen susceptibility. *Cell, 167*(5), 1339–1353.e1321.

Dikongue, E., & Segurel, L. (2017). Latitude as a co-driver of human gut microbial diversity? *Bioessays, 39*(3).

Doll, J. P., & Franker, C. K. (1963). Experimental histomoniasis in gnotobiotic turkeys. I. Infection and histopathology of the bacteria-free host. *Journal of Parasitology, 49,* 411–414.

Du, Q., Schilde, C., Birgersson, E., Chen, Z. H., McElroy, S., & Schaap, P. (2014). The cyclic AMP phosphodiesterase RegA critically regulates encystation in social and pathogenic amoebas. *Cellular Signalling, 26*(2), 453–459.

Dutton, J. S., Hinman, S. S., Kim, R., Wang, Y., & Allbritton, N. L. (2019). Primary cell-derived intestinal models: Recapitulating physiology. *Trends in Biotechnology, 37*(7), 744–760.

El Safadi, D., Cian, A., Nourrisson, C., Pereira, B., Morelle, C., Bastien, P., et al. (2016). Prevalence, risk factors for infection and subtype distribution of the intestinal parasite *Blastocystis* sp. from a large-scale multi-center study in France. *BMC Infectious Diseases, 16*(1), 451.

Embley, T. M., & Martin, W. (2006). Eukaryotic evolution, changes and challenges. *Nature, 440*(7084), 623–630.

Escalante, N. K., Lemire, P., Cruz Tleugabulova, M., Prescott, D., Mortha, A., Streutker, C. J., et al. (2016). The common mouse protozoa *Tritrichomonas muris* alters mucosal T cell homeostasis and colitis susceptibility. *Journal of Experimental Medicine, 213*(13), 2841–2850.

Frisbee, A. L., Saleh, M. M., Young, M. K., Leslie, J. L., Simpson, M. E., Abhyankar, M. M., et al. (2019). IL-33 drives group 2 innate lymphoid cell-mediated protection during *Clostridium difficile* infection. *Nature Communications, 10*(1), 2712.

Galvan-Moroyoqui, J. M., Del Carmen Dominguez-Robles, M., Franco, E., & Meza, I. (2008). The interplay between *Entamoeba* and enteropathogenic bacteria modulates epithelial cell damage. *PLoS Neglected Tropical Diseases, 2*(7), e266.

Gerbe, F., Sidot, E., Smyth, D. J., Ohmoto, M., Matsumoto, I., Dardalhon, V., et al. (2016). Intestinal epithelial tuft cells initiate type 2 mucosal immunity to helminth parasites. *Nature, 529*(7585), 226–230.

Gilchrist, C. A., Petri, S. E., Schneider, B. N., Reichman, D. J., Jiang, N., Begum, S., et al. (2016). Role of the gut microbiota of children in Diarrhea due to the protozoan parasite *Entamoeba histolytica*. *Journal of Infectious Diseases, 213*(10), 1579–1585.

Glover, M., Colombo, S. A. P., Thornton, D. J., & Grencis, R. K. (2019). Trickle infection and immunity to *Trichuris muris*. *PLoS Pathogens, 15*(11), e1007926.

Goto, Y. (2019). Epithelial cells as a transmitter of signals from commensal bacteria and host immune cells. *Frontiers in Immunology, 10,* 2057.

Gouba, N., & Drancourt, M. (2015). Digestive tract mycobiota: A source of infection. *Médecine et Maladies Infectieuses, 45*(1–2), 9–16.

Hamad, I., Raoult, D., & Bittar, F. (2016). Repertory of eukaryotes (eukaryome) in the human gastrointestinal tract: Taxonomy and detection methods. *Parasite Immunology, 38*(1), 12–36.

Heo, I., Dutta, D., Schaefer, D. A., Iakobachvili, N., Artegiani, B., Sachs, N., et al. (2018). Modelling *Cryptosporidium* infection in human small intestinal and lung organoids. *Nature Microbiology, 3*(7), 814–823.

Hess, M., Liebhart, D., Bilic, I., & Ganas, P. (2015). *Histomonas meleagridis*—New insights into an old pathogen. *Veterinary Parasitology, 208*(1–2), 67–76.

Hooper, L. V., Littman, D. R., & Macpherson, A. J. (2012). Interactions between the microbiota and the immune system. *Science, 336*(6086), 1268–1273.

Hotez, P. J., Brindley, P. J., Bethony, J. M., King, C. H., Pearce, E. J., & Jacobson, J. (2008). Helminth infections: The great neglected tropical diseases. *The Journal of Clinical Investigation, 118*(4), 1311–1321.

Howitt, M. R., Lavoie, S., Michaud, M., Blum, A. M., Tran, S. V., Weinstock, J. V., et al. (2016). Tuft cells, taste-chemosensory cells, orchestrate parasite type 2 immunity in the gut. *Science, 351*(6279), 1329–1333.

Huang, Y., Mao, K., Chen, X., Sun, M. A., Kawabe, T., Li, W., et al. (2018). S1P-dependent interorgan trafficking of group 2 innate lymphoid cells supports host defense. *Science, 359*(6371), 114–119.

Iebba, V., Santangelo, F., Totino, V., Pantanella, F., Monsia, A., Di Cristanziano, V., et al. (2016). Gut microbiota related to *Giardia duodenalis*, *Entamoeba* spp. and *Blastocystis hominis* infections in humans from Cote d'Ivoire. *The Journal of Infection in Developing Countries, 10*(9), 1035–1041.

Ivanov, K., II, Atarashi, N., Manel, E. L., Brodie, T., Shima, U., Karaoz, D., et al. (2009). Induction of intestinal Th17 cells by segmented filamentous bacteria. *Cell, 139*(3), 485–498.

Iyer, L. R., Verma, A. K., Paul, J., & Bhattacharya, A. (2019). Phagocytosis of gut bacteria by *Entamoeba histolytica*. *Frontiers in Cellular Infection Microbiology, 9,* 34.

Jalili-Firoozinezhad, S., Gazzaniga, F. S., Calamari, E. L., Camacho, D. M., Fadel, C. W., Bein, A., et al. (2019). A complex human gut microbiome cultured in an anaerobic intestine-on-a-chip. *Nature Biomedical Engineering, 3*(7), 520–531.

Johansson, M. E., Ambort, D., Pelaseyed, T., Schutte, A., Gustafsson, J. K., Ermund, A., et al. (2011). Composition and functional role of the mucus layers in the intestine. *Cellular and Molecular Life Sciences, 68*(22), 3635–3641.

Kawabe, Y., Schilde, C., Du, Q., & Schaap, P. (2015). A conserved signalling pathway for amoebozoan encystation that was co-opted for multicellular development. *Scientific Reports, 5,* 9644.

Kim, H. J., Huh, D., Hamilton, G., & Ingber, D. E. (2012). Human gut-on-a-chip inhabited by microbial flora that experiences intestinal peristalsis-like motions and flow. *Lab on a Chip, 12*(12), 2165–2174.

Kim, H. J., Li, H., Collins, J. J., & Ingber, D. E. (2016). Contributions of microbiome and mechanical deformation to intestinal bacterial overgrowth and inflammation in a human gut-on-a-chip. *National Academy of Sciences of the United States of America, 113*(1), E7–15.

Lee, S. C., Tang, M. S., Lim, Y. A., Choy, S. H., Kurtz, Z. D., Cox, L. M., et al. (2014). Helminth colonization is associated with increased diversity of the gut microbiota. *PLoS Neglected Tropical Diseases, 8*(5), e2880.

Leon-Coria, A., Kumar, M., Moreau, F., & Chadee, K. (2018). Defining cooperative roles for colonic microbiota and Muc2 mucin in mediating innate host defense against *Entamoeba histolytica*. *PLoS Pathogens, 14*(11), e1007466.

Liu, S., Roellig, D. M., Guo, Y., Li, N., Frace, M. A., Tang, K., et al. (2016). Evolution of mitosome metabolism and invasion-related proteins in *Cryptosporidium*. *BMC Genomics, 17*(1), 1006.

Lokmer, A., Cian, A., Froment, A., Gantois, N., Viscogliosi, E., Chabe, M., et al. (2019). Use of shotgun metagenomics for the identification of protozoa in the gut microbiota of healthy individuals from worldwide populations with various industrialization levels. *PLoS ONE, 14*(2), e0211139.

Mahapatro, M., Foersch, S., Hefele, M., He, G. W., Giner-Ventura, E., McHedlidze, T., et al. (2016). Programming of intestinal epithelial differentiation by IL-33 derived from pericryptal fibroblasts in response to systemic infection. *Cell Reports, 15*(8), 1743–1756.

Maizels, R. M., Smits, H. H., & McSorley, H. J. (2018). Modulation of host immunity by Helminths: The expanding repertoire of parasite effector molecules. *Immunity, 49*(5), 801–818.

Manna, D., Lentz, C. S., Ehrenkaufer, G. M., Suresh, S., Bhat, A., & Singh, U. (2018).An NAD(+)-dependent novel transcription factor controls stage conversion in *Entamoeba*. *Elife, 7*.

Maynard, C. L., Elson, C. O., Hatton, R. D., & Weaver, C. T. (2012). Reciprocal interactions of the intestinal microbiota and immune system. *Nature, 489*(7415), 231–241.

McDole, J. R., Wheeler, L. W., McDonald, K. G., Wang, B., Konjufca, V., Knoop, K. A., et al. (2012). Goblet cells deliver luminal antigen to CD103+ dendritic cells in the small intestine. *Nature, 483*(7389), 345–349.

Mi-ichi, F., Miyamoto, T., Takao, S., Jeelani, G., Hashimoto, T., Hara, H., et al. (2015a). *Entamoeba* mitosomes play an important role in encystation by association with cholesteryl sulfate synthesis. *National Academy of Sciences of the United States of America, 112*(22), E2884–2890.

Mi-ichi, F., Nozawa, A., Yoshida, H., Tozawa, Y., & Nozaki, T. (2015b). Evidence that the *Entamoeba histolytica* mitochondrial carrier family links mitosomal and cytosolic pathways through exchange of 3′-phosphoadenosine 5′-phosphosulfate and ATP. *Eukaryotic Cell, 14*(11), 1144–1150.

Mi-ichi, F., Miyamoto, T., & Yoshida, H. (2017). Uniqueness of *Entamoeba* sulfur metabolism: Sulfolipid metabolism that plays pleiotropic roles in the parasitic life cycle. *Molecular Microbiology, 106*(3), 479–491.

Miller, A. M. (2011). Role of IL-33 in inflammation and disease. *Journal of Inflammation (London), 8*(1), 22.

Mmbaga, B. T., & Houpt, E. R. (2017). *Cryptosporidium* and *Giardia* infections in children: A review. *Pediatric Clinics of North America, 64*(4), 837–850.

Morton, E. R., Lynch, J., Froment, A., Lafosse, S., Heyer, E., Przeworski, M., et al. (2015). Variation in rural African gut microbiota is strongly correlated with colonization by entamoeba and subsistence. *PLoS Genetics, 11*(11), e1005658.

Nieves-Ramirez, M. E., Partida-Rodriguez, O., Laforest-Lapointe, I., Reynolds, L. A., Brown, E. M., Valdez-Salazar, A., et al. (2018). Asymptomatic intestinal colonization with protist *Blastocystis* is strongly associated with distinct microbiome ecological patterns. *MSystems, 3*(3).

Ohshima, K., Kanto, K., Hatakeyama, K., Ide, T., Wakabayashi-Nakao, K., Watanabe, Y., et al. (2014). Exosome-mediated extracellular release of polyadenylate-binding protein 1 in human metastatic duodenal cancer cells. *Proteomics, 14*(20), 2297–2306.

Parfrey, L. W., Walters, W. A., & Knight, R. (2011). Microbial eukaryotes in the human microbiome: Ecology, evolution, and future directions. *Frontiers in Microbiology, 2,* 153.

Parfrey, L. W., Walters, W. A., Lauber, C. L., Clemente, J. C., Berg-Lyons, D., Teiling, C., et al. (2014). Communities of microbial eukaryotes in the mammalian gut within the context of environmental eukaryotic diversity. *Frontiers in Microbiology, 5,* 298.

Pelaseyed, T., Bergstrom, J. H., Gustafsson, J. K., Ermund, A., Birchenough, G. M., Schutte, A., et al. (2014). The mucus and mucins of the goblet cells and enterocytes provide the first defense line of the gastrointestinal tract and interact with the immune system. *Immunological Reviews, 260*(1), 8–20.

Peng, L., Li, Z. R., Green, R. S., Holzman, I. R., & Lin, J. (2009). Butyrate enhances the intestinal barrier by facilitating tight junction assembly via activation of AMP-activated protein kinase in Caco-2 cell monolayers. *Journal of Nutrition, 139*(9), 1619–1625.

Proctor, L. M., Creasy, H. H., Fettweis, J. M., Lloyd-Price, J., Mahurkar, A., Zhou, W., et al. (2019). The integrative human microbiome project. *Nature, 569*(7758), 641–648.

Pullan, R. L., Smith, J. L., Jasrasaria, R., & Brooker, S. J. (2014). Global numbers of infection and disease burden of soil transmitted helminth infections in 2010. *Parasites and Vectors, 7,* 37.

Ramirez, J. D., Sanchez, L. V., Bautista, D. C., Corredor, A. F., Florez, A. C., & Stensvold, C. R. (2014). Blastocystis subtypes detected in humans and animals from Colombia. *Infection, Genetics and Evolution, 22,* 223–228.

Reinoso Webb, C., Koboziev, I., Furr, K. L., & Grisham, M. B. (2016). Protective and pro-inflammatory roles of intestinal bacteria. *Pathophysiology, 23*(2), 67–80.

Robertson, L. J., Clark, C. G., Debenham, J. J., Dubey, J. P., Kvac, M., Li, J., et al. (2019). Are molecular tools clarifying or confusing our understanding of the public health threat from zoonotic enteric protozoa in wildlife? *The International Journal for Parasitology: Parasites and Wildlife, 9,* 323–341.

Rojas, A. A., Castro, S. C., Matondo, M., Gianetto, Q. G., Varet, H., Sismeiro, O., Legendre, R., Fernandes, J., Hardy, D., Coppée, J. Y., Olivo-Marin, J. C., & Guillen, N. (2020). Insights into amebiasis using a human-intestinal model. *Cellular Microbiology,* Mar 16:e13203. https://doi.org/10.1111/cmi.13203.

Schneider, C., O'Leary, C. E., von Moltke, J., Liang, H. E., Ang, Q. Y., Turnbaugh, P. J., et al. (2018). A metabolite-triggered tuft cell-ILC2 circuit drives small intestinal remodeling. *Cell, 174*(2), 271–284, e214.

Shaulov, Y., Shimokawa, C., Trebicz-Geffen, M., Nagaraja, S., Methling, K., Lalk, M., et al. (2018). Escherichia coli mediated resistance of *Entamoeba histolytica* to oxidative stress is triggered by oxaloacetate. *PLoS Pathogens, 14*(10), e1007295.

Shiflett, A. M., & Johnson, P. J. (2010). Mitochondrion-related organelles in eukaryotic protists. *Annual Review of Microbiology, 64,* 409–429.

Shin, W., & Kim, H. J. (2018). Pathomimetic modeling of human intestinal diseases and underlying host-gut microbiome interactions in a gut-on-a-chip. *Methods in Cell Biology, 146,* 135–148.

Steele, S. P., Melchor, S. J., & Petri, W. A., Jr. (2016). Tuft cells: New players in colitis. *Trends in Molecular Medicine, 22*(11), 921–924.

Thursby, E., & Juge, N. (2017). Introduction to the human gut microbiota. *Biochemical Journal, 474*(11), 1823–1836.

Troeger, C., Forouzanfar, M., Rao, P. C., Khalil, I., Brown, A., & Reiner, R. C. Jr. (2017). Estimates of global, regional, and national morbidity, mortality, and aetiologies of diarrhoeal diseases: A systematic analysis for the Global Burden of Disease Study 2015. *Lancet Infectious Diseases, 17*(9), 909–948.

Tovaglieri, A., Sontheimer-Phelps, A., Geirnaert, A., Prantil-Baun, R., Camacho, D. M., Chou, D. B., et al. (2019). Species-specific enhancement of enterohemorrhagic *E. coli* pathogenesis mediated by microbiome metabolites. *Microbiome, 7*(1), 43.

Varet, H., Shaulov, Y., Sismeiro, O., Trebicz-Geffen, M., Legendre, R., Coppee, J. Y., et al. (2018). Enteric bacteria boost defences against oxidative stress in *Entamoeba histolytica. Scientific Reports, 8*(1), 9042.

Verma, A. K., Verma, R., Ahuja, V., & Paul, J. (2012). Real-time analysis of gut flora in *Entamoeba histolytica* infected patients of Northern India. *BMC Microbiology, 12,* 183.

von Moltke, J., Ji, M., Liang, H. E., & Locksley, R. M. (2016). Tuft-cell-derived IL-25 regulates an intestinal ILC2-epithelial response circuit. *Nature, 529*(7585), 221–5.

Wang, Y., Ahmad, A. A., Sims, C. E., Magness, S. T., & Allbritton, N. L. (2014). In vitro generation of colonic epithelium from primary cells guided by microstructures. *Lab on a Chip, 14*(9), 1622–1631.

Wang, Y., Kim, R., Sims, C. E., & Allbritton, N. L. (2019). Building a thick mucus hydrogel layer to improve the physiological relevance of in vitro primary colonic Epithelial models. *Cellular and Molecular Gastroenterology and Hepatology, 8*(4), 653–655.e655.

Watanabe, K., Gilchrist, C. A., Uddin, M. J., Burgess, S. L., Abhyankar, M. M., Moonah, S. N., et al. (2017). Microbiome-mediated neutrophil recruitment via CXCR2 and protection from amebic colitis. *PLoS Pathogens, 13*(8), e1006513.

Yatsunenko, T., Rey, F. E., Manary, M. J., Trehan, I., Dominguez-Bello, M. G., Contreras, M., et al. (2012). Human gut microbiome viewed across age and geography. *Nature, 486*(7402), 222–227.

Zindl, C. L., Lai, J. F., Lee, Y. K., Maynard, C. L., Harbour, S. N., Ouyang, W., et al. (2013). IL-22-producing neutrophils contribute to antimicrobial defense and restitution of colonic epithelial integrity during colitis. *National Academy of Sciences of the United States of America, 110*(31), 12768–12773.

Chapter 2
Resolving Amoebozoan Encystation from *Dictyostelium* Evo-Devo and Amoebozoan Comparative Genomics

Qingyou Du, Yoshinori Kawabe, and Pauline Schaap

Abstract Amoebozoa, such as *Entamoeba* and *Acanthamoeba,* survive environmental stress by encystment, but also comprise the Dictyostelia which form spores in multicellular fruiting bodies to survive starvation stress. Sporulation is triggered by cAMP activation of cAMP-dependent protein kinase (PKA), with cAMP levels being controlled by the adenylate cyclases ACG and ACR and the phosphodiesterase RegA. Many Dictyostelia can also alternatively encyst and we showed that environmental stress acts on ACG and ACR to increase cAMP and thereby activate PKA to trigger encystation, with RegA preventing precocious encystation and inducing cyst germination. RegA activity requires phosphorylation of its response regulator domain, which is controlled by Sensor Histidine Kinases/Phosphatases (SHKPs), which in *Dictyostelium* respond to developmental signals. Comparative genomics showed that RegA, AcrA and PKA and a wealth of SHKPs are deeply conserved in Amoebozoa, where SHKs may sense food and SHPs environmental stress.

Keywords Encystation · cAMP dependent protein kinase · RegA · *Acanthamoeba* · *Entamoeba* · *Giardia*

Encystation as a Universal Protist Survival Strategy

Unicellular protists populate all nine major eukaryote divisions, where several have given rise to multicellular forms. Protists can be free-living in a wide range of ecosystems or spend at least part of their life cycle as a commensal or parasite of plants or animals. In most of these habitats they are periodically subjected to stress in the form of starvation, drought and extreme temperatures, and lack of light for photosynthetic protists. To survive these life-threatening conditions the cells most commonly differentiate into a dormant encapsulated stage, called a cyst (Schaap and Schilde 2018). The divisions of Amoebozoa and Excavates contain several obligatory and opportunistic pathogens, such as *Entamoeba histolytica*, *Giardia lamblia* and *Acanthamoeba castellani* for which the cyst either propagates the disease outside the host

Q. Du · Y. Kawabe · P. Schaap (✉)
School of Life Sciences, University of Dundee, Dundee DD15EH, UK
e-mail: p.schaap@dundee.ac.uk

© Springer Nature Switzerland AG 2020

N. Guillen (ed.), *Eukaryome Impact on Human Intestine Homeostasis and Mucosal Immunology*, https://doi.org/10.1007/978-3-030-44826-4_2

body (*Entamoeba* and *Giardia*) (Mi-Ichi et al. 2016; Lauwaet et al. 2007) or, due to its high resilience, prevents the disease to be eradicated by either the immune system or antibiotics (*Acanthamoeba*) (Anwar et al. 2018; Lloyd 2014). For this reason, there are increasing efforts to understand the mechanisms and crucial proteins involved in the encystation process, with the ultimate goal to develop therapeutic agents to prevent encystation.

During encystment cells commonly loose motility and assume a spherical shape. Vesicles with cell wall precursors are exocytosed and a rigid cell wall is constructed. In *Dictyostelium* and *Acanthamoeba* cellulose is the main structural component of the cyst wall, while for *Entamoeba* this is chitin (β(1-4) linked *N*-acetyl glucosamine) (Arroyo-Begovich et al. 1980) and for *Giardia* a polymer of β(1-3)-*N*-acetyl-D-galactosamine (Gerwig et al. 2002). After wall synthesis is complete, the cells enter a long period of metabolic dormancy, which is broken in free-living protists by renewed food and water availability or removal of other stress factors, and in *Giardia* and *Entamoeba* by oral uptake.

Information on the mechanisms that control encystation in *Entamoeba* and *Acanthamoeba* is limited. In *Entamoeba invadens* encystation is induced by glucose depletion and hypo-osmolarity and is also stimulated by cholesteryl sulfate and by catecholamines (Vazquezdelara-Cisneros and Arroyo-Begovich 1984; Coppi et al. 2002; Mi-ichi et al. 2015). Cholesteryl sulfate synthesis is inhibited by chlorate, which also inhibits encystation. The catecholamine effects were specific for β1-adrenergic receptor agonists. Antagonists of the β-adrenergic receptors prevented catecholamine, but not di-butyryl-cAMP induced encystation, suggesting that, similar to mammalian β1-adrenergic receptors, the *Entamoeba* receptors activate an adenylate cyclase (Coppi et al. 2002). However, the *Entamoeba* genome contains neither adenylate cyclases nor mammalian-type β1-adrenergic receptors (Loftus et al. 2005; Wang et al. 2003), indicating that *Entamoeba* processes the catecholamine signal differently. Studies with inhibitors for specific heat-shock proteins and cysteine proteases indicated that HSP-90 prevents (Singh et al. 2015) and cysteine proteases promote encystation, respectively, although the latter were also required for growth (Sharma et al. 1996; de Meester et al. 1990).

Acanthamoeba encystation is problematic for the treatment of *Acanthamoeba* keratitis. Candidate proteins involved in regulation of encystation have in recent years emerged from transcriptome studies that identified genes upregulated in encystation. This is the case for autophagy genes, and a requirement for autophagy in encystation appeared evident from observations that encystation is inhibited by the autophagy inhibitors chloroquine and 3-methyladenine (Jha et al. 2014; Moon et al. 2015b) and by RNAi mediated silencing of the autophagy proteins Atg8 (Moon et al. 2013), Atg12 (Kim et al. 2015) and Atg16 (Song et al. 2012). PRMT5, an enzyme that methylates histones and other proteins, is also upregulated in encystation, and encystation is reduced upon PRMT5 gene silencing (Moon et al. 2016). Protein kinase C (PKC) was implicated in *Acanthamoeba* encystation because 21 out of its 27 PKC genes are upregulated in encystation (Moon et al. 2011) and silencing of one of those reduced encystation (Moon et al. 2012).

Insights into Encystation from *Dictyostelium* Sporulation

To survive environmental stress and to disperse to new feeding grounds, multicellular organisms such as fungi and mycetozoa form dormant encapsulated spores in fruiting bodies. The mycetozoan *Dictyostelium discoideum* is a genetically tractable amoebozoan model system for cell- and developmental research (Annesley and Fisher 2009). Upon starvation, its amoebas aggregate to form a fruiting body consisting of spores and stalk cells, a process in which the well-known second messenger cyclic AMP (cAMP) plays a dominant role. cAMP is firstly used as a secreted signal and chemoattractant that brings the cells together in aggregates (Konijn et al. 1967). Once aggregated, cAMP acting on cell surface cAMP receptors (cARs) also induces a proportion of cells to enter the spore differentiation pathway (Schaap and Van Driel 1985). cAMP additionally acts in a classical intracellular messenger role, activating cAMP-dependent protein kinase (PKA). In this role cAMP triggers the maturation of stalk cells and spores (Harwood et al. 1992; Hopper et al. 1993) and prevents spore germination under conditions that are not favourable for growth (Van Es et al. 1996). Intracellular cAMP levels are controlled by the adenylate cyclases AcaA (Pitt et al. 1992), AcrA (Soderbom et al. 1999) and AcgA (Pitt et al. 1992) and by the cAMP phosphodiesterase RegA (Shaulsky et al. 1998). AcaA is activated by cAMP in an autocatalytic feedback loop that produces the cAMP pulses that control aggregation (Pitt et al. 1992) and by the stalk-inducing signal c-di-GMP in prestalk cells (Chen et al. 2017). AcgA is activated by high ambient osmolarity in the spore head, which renders spores dormant (Van Es et al. 1996). RegA is activated by phosphorylation of its response regulator domain, which is the end-point of a forward or reverse phospho-relay system initiated by sensor histidine kinases/phosphatases (SHKPs) (Thomason et al. 1999). The SHKPs are in turn activated by secreted signals that are exchanged between the prestalk and prespore cells to regulate the timely maturation of spores and stalk cells (Wang et al. 1999; Anjard and Loomis 2005; Singleton et al. 1998) (Fig. 2.1).

D. discoideum is one of ~150 *Dictyostelium* species, which are subdivided into four major groups (Schaap et al. 2006). Group 4, to which *D. discoideum* belongs, has lost encystation of individual amoebas, but this strategy is still used by many species in the other three groups, when conditions for aggregation are not favourable (Romeralo et al. 2013). Comparative genomics of group-representative species (Heidel et al. 2011; Gloeckner et al. 2016; Urushihara et al. 2015) and of solitary Amoebozoa like *A. castellani* (Clarke et al. 2013), *Physarum polycephalum* (Schaap et al. 2015) and *Protostelium aurantium* (Hillmann et al. 2018a) allowed us to retrace the extent to which the genes that control sporulation are conserved throughout Dictyostelia and Amoebozoa.

Orthologs of the PKA catalytic (*pkaC*) and regulatory subunits (*pkaR*), *acaA*, *acrA*, *acgA* and *regA* and most of the 16 *D. discoideum* SHKP's were conserved throughout Dictyostelia (Gloeckner et al. 2016). *PkaC*, *pkaR*, *acrA* and *regA* were also conserved in the solitary amoebozoa mentioned above, with each having a broad range of adenylate cyclases and SHKPs, of which some were orthologs of the

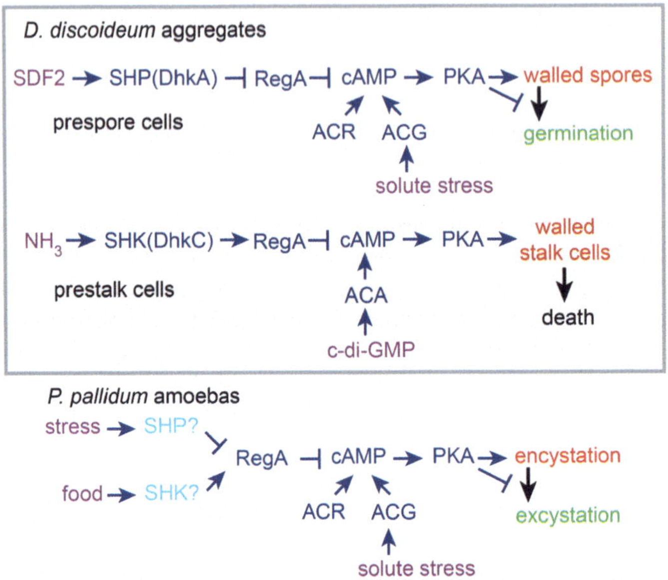

Fig. 2.1 A conserved cAMP signalling pathway for encystation and sporulation in Dictyostelia. The *D. discoideum* multicellular life cycle terminates in the formation of fruiting bodies with viable spores and dead stalk cells, both encapsulated in cellulose-rich walls. The maturation of spores and stalk cells is controlled by secreted stimuli (in violet). The peptide SDF2, acting on the sensor histidine phosphatase DhkA (Wang et al. 1999) inhibits RegA and thereby increases cAMP and PKA activity to induce spore maturation. High osmolarity acting on the adenylate cyclase ACG inhibits spore germination (Van Es et al. 1996). In prestalk cells, c-di-GMP acting on the adenylate cyclase ACA induces stalk maturation (Chen et al. 2017). In slugs stalk maturation is inhibited by ammonia, which acts on the sensor histidine kinase DhkC to activate RegA, thereby lowering intracellular cAMP (Singleton et al. 1998). A similar cAMP, RegA and PKA mediated pathway, operates to activate encystation and prevent excystation in response to stress in *P. pallidum* (Kawabe et al. 2015; Du et al. 2014), a dictyostelid which has retained the ancestral encystation pathway. RegA also negatively regulates encystation in the distantly related *A. castellani* (Du et al. 2014). Involvement of pathway components in dark blue was shown by gene knock-out. Those in light blue are inferred from their abundance in Amoebozoan genomes (see Table 2.1)

Dictyostelium enzymes (Fig. 2.2 and Table 2.1). However, we did not detect pkaC, pkaR, SHKP's, adenylate cyclases or cAMP phosphodiesterases in genome of the obligatory parasite *E. histolytica* (Loftus et al. 2005), indicating that this organism has lost cAMP mediated signalling altogether. Remarkably, the excavate amoeboid pathogen, *Naegleria gruberi* has regA, PKA and many SHKPs and adenylate cyclases (Fritz-Laylin et al. 2010) and on the whole cAMP signalling, PKA and/or SHKP mediated signalling are conserved across free-living eukaryote protists. The excavate obligate parasite *Giardia* appears to have only single PkaR, adenylate cyclase, and cAMP phosphodiesterase genes and similar to *Entamoeba*, no SHKPs. This suggests that obligate parasites lost most of their environmental cognition and possibly rely more on internal metabolic cues to regulate cell differentiation.

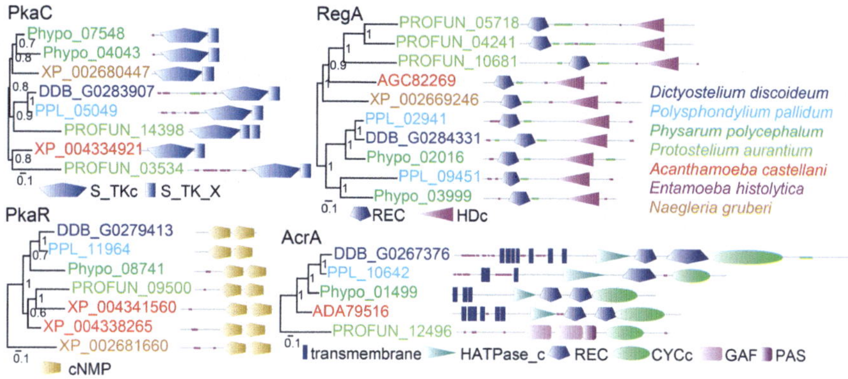

Fig. 2.2 Conservation of Dictyostelid encystation genes across Amoebozoa and Excavates. The genomes of the indicated amoebozoan and excavate species were queried with PkaC, PkaR, RegA and AcrA protein sequences. Phylogenies were constructed from the aligned sequences, which were annotated with the functional domain architectures of the proteins

Gene conservation is by no means conclusive evidence that the conserved genes perform the same biological role. Such evidence has to come from experiments that disrupt the function of the gene in question. Initially from an interest in understanding how developmental signalling evolved in the Dictyostelia, we developed gene modification procedures for *Polysphondylium pallidum*, a species belonging to taxon group 2, which in addition to sporulation in fruiting bodies, has retained the ancestral mechanism of encystation of individual amoebas in response to starvation and drought stress. In *P. pallidum*, inhibition of PKA function by overexpression of *pkaRm*, a dominant-negative pkaC inhibitor, or deletion of *pkaC* prevented both sporulation as well as encystation (Funamoto et al. 2003; Ritchie et al. 2008; Kawabe et al. 2015). Encystation was also blocked in a double null mutant in *acaA* and *acrA* (Kawabe et al. 2015). Deletion of RegA, which normally acts to reduce cAMP levels, had the opposite effect. In *regA* null mutants, the amoebas encysted precociously while still feeding (Du et al. 2014). In wild-type *P. pallidum* amoebas, cAMP levels increase upon starvation and upon exposure to increased osmolarity (a signal for approaching drought) (Ritchie et al. 2008), which, combined with the genetic evidence, strongly suggests that cAMP is a second messenger for perception of stress that acts on PKA to induce encystation. Evidently, the mechanisms that regulate sporulation in fruiting bodies are evolutionary derived from the mechanisms that cause encystation of single-celled amoebas.

Lack of gene modification procedures in solitary amoebas limited our ability to extrapolate these findings to other Amoebozoa. However, we expressed the *A. castellani regA* gene in *Escherichia coli*, purified the RegA protein and tested a range of PDE inhibitors. Two inhibitors, dipyridamole and trequinsin, were particularly effective. The inhibitors strongly increased solute-induced cAMP accumulation in intact *A. castellani* amoebas and caused precocious and excessive encystation (Du et al.

Table 2.1 Cyclic nucleotide and SHKP mediated signalling in Amoebozoa and Excavata

Category	*Dictyostelium discoideum*	*Dictyostelium fasciculatum*	*Acanthamoeba castellani*	*Entamoeba histolytica*	*Physarum polycephalum*	*Protostelium aurangium*	*Giardia lamblia*	*Naegleria gruberi*
Histidine kinases/phosphatases	16	14	48	0	51	71	0	27
Cyclic nucleotide signaling								
Adenylate/guanylate cyclases	5	5	67	0	64	52	1	108
cNMP binding (e.g. PkaR)	5	5	7	0	28	27	1	7
cNMP phosphodiesterases	7	6	10	1	11	16	1	7

Enumeration of sensor histidine kinases and proteins that synthesize, detect and hydrolyze cAMP or cGMP in six Amoebozoan genomes and two Excavate genomes. Data for *Acanthamoeba*, *D. discoideum* (group 4). *D. fasciculatum* (group 1), *A. castellani*, *E. histolytica*, *Phy. polycephalum*, *Pro. aurangium*, *G. lamblia* and *N. gruberi* were retrieved from (Eichinger et al. 2005; Heidel et al. 2011; Clarke et al. 2013; Loftus et al. 2005; Schaap et al. 2015; Hillmann et al. 2018b; Adam et al. 2013; Fritz-Laylin et al. 2010). *Entamoeba* and *Giardia* genomes were also queried by BLAST search using multiple representative genes in the different categories as bait

2014), similar to the effect of deleting *regA* from *P. pallidum*. This experiment demonstrated that cAMP also mediates stress-induced encystation in *A. castellani.* Within Amoebozoa, *A. castellani*, a member of the major branch of Discosea is relatively distant from Dictyostelia, which resides in Tevosea (Kang et al. 2017). This suggests that the use of cAMP as intermediate for stress-induced encystation is wide-spread in Amoebozoa. This appears to be confirmed by early studies that show enhancement of encystation by cAMP and dibutyryl-cAMP in the Amoebozoan *Hartmanella* (Raizada and Murti 1972). However, conclusions that cAMP acts downstream of catecholamine-induced encystation in *Entamoebidae* (Coppi and Eichinger 1999; Coppi et al. 2002) are unlikely to be valid in view of the lack of adenylate cyclase and *PkaR* genes in *Entamoeba*. A PKA inhibitor, amide 14–22, prevented excystation of *Giardia lamblia* (Abel et al. 2001), suggesting that in Excavates the effect of cAMP on encystment is opposite to that in Amoebozoa.

Further gene disruption studies in *P. pallidum* indicated a crucial role for one of its two cellulose synthases, DcsA, in encystation and sporulation, with the other, DcsB, being required for stalk cell differentiation (Du and Schaap 2014). Cellulose is the major cell wall component of the encapsulated cell types of Dictyostelia. The *Acanthamoeba* cell wall is also rich in cellulose (Moon and Kong 2012) and down-regulation of cellulose synthase expression by small interfering RNAs was shown to inhibit encystation (Moon et al. 2014; Aqeel et al. 2013). The cellulose synthesis inhibitors, 2,6-dichlorobenzonitrile and isoxaben that are widely used as weed-killers, also prevented *Acanthamoeba* encystation without showing cytotoxic effects on human cornea cells at the concentrations effective for cyst inhibition (Moon et al. 2015a). This offers possibilities for the treatment of *Acanthamoeba* keratites. While not all Amoebozoa use cellulose as major wall polymer, targeting cell wall synthesis is evidently an effective means to prevent viable cyst formation.

Glycogen synthase kinase 3 (GSK3), originally known to phosphorylate glycogen synthase and thereby reduce its activity, is also an intermediate in the wingless/wnt signalling pathway that determines cell fate in metazoa (Woodgett 1994). In *D. discoideum* GSK3 prevents dedifferentiation of prespore cells into basal disc cells (a subpopulation of stalk cells) (Harwood et al. 1995). This function of GSK3 was not conserved in *P. pallidum* (Kawabe et al. 2018), which like all non-group 4 species does not have a basal disc (Romeralo et al. 2013) and it probably only emerged in group 4. However, *P. pallidum gsk3* knockouts much more readily entered into encystation than into aggregation and sporulation (Kawabe et al. 2018). This suggest that GSK3 activity in starving amoebas normally determines the decision between unicellular encystation and multicellular sporulation.

By using a combination of forward and reverse genetic strategies and phosphoproteomics, we are pursuing our efforts to identify downstream targets for PKA and additional signalling intermediates that control and effectors that execute the encystation programme.

Acknowledgements Our research is funded by Welcome grant 100293/Z/12/Z, ERC Advanced grant 742288 and Leverhulme grant RPG-2016-220.

References

Abel, E. S., Davids, B. J., Robles, L. D., Loflin, C. E., Gillin, F. D., & Chakrabarti, R. (2001). Possible roles of protein kinase A in cell motility and excystation of the early diverging eukaryote *Giardia lamblia*. *Journal of Biological Chemistry, 276*(13), 10320–10329.

Adam, R. D., Dahlstrom, E. W., Martens, C. A., Bruno, D. P., Barbian, K. D., Ricklefs, S. M., et al. (2013). Genome sequencing of *Giardia lamblia* genotypes A2 and B isolates (DH and GS) and comparative analysis with the genomes of genotypes A1 and E (WB and Pig). *Genome Biology and Evolution, 5*(12), 2498–2511. https://doi.org/10.1093/gbe/evt197.

Anjard, C., & Loomis, W. F. (2005). Peptide signaling during terminal differentiation of dictyostelium. *Proceedings of the National Academy of Sciences of the United States of America, 102*(21), 7607–7611.

Annesley, S. J., & Fisher, P. R. (2009). Dictyostelium discoideum-a model for many reasons. *Molecular and Cellular Biochemistry, 329*(1–2), 73–91. https://doi.org/10.1007/s11010-009-0111-8.

Anwar, A., Khan, N. A., & Siddiqui, R. (2018). Combating *Acanthamoeba* spp. cysts: What are the options? *Parasit Vectors, 11*(1), 26. https://doi.org/10.1186/s13071-017-2572-z.

Aqeel, Y., Siddiqui, R., & Khan, N. A. (2013). Silencing of xylose isomerase and cellulose synthase by siRNA inhibits encystation in *Acanthamoeba castellani*. *Parasitology Research, 112*(3), 1221–1227. https://doi.org/10.1007/s00436-012-3254-6.

Arroyo-Begovich, A., Carabez-Trejo, A., & Ruiz-Herrera, J. (1980). Identification of the structural component in the cyst wall of *Entamoeba invadens*. *The Journal of Parasitology, 66*(5), 735–741.

Chen, Z. H., Singh, R., Cole, C., Lawal, H. M., Schilde, C., Febrer, M., et al. (2017). Adenylate cyclase A acting on PKA mediates induction of stalk formation by cyclic diguanylate at the Dictyostelium organizer. *Proceedings of the National Academy of Sciences of the United States of America, 114*(3), 516–521. https://doi.org/10.1073/pnas.1608393114.

Clarke, M., Lohan, A. J., Liu, B., Lagkouvardos, I., Roy, S., Zafar, N., et al. (2013). Genome of *Acanthamoeba castellani* highlights extensive lateral gene transfer and early evolution of tyrosine kinase signaling. *Genome Biology, 14*(2), R11. https://doi.org/10.1186/gb-2013-14-2-r11.

Coppi, A., & Eichinger, D. (1999). Regulation of *Entamoeba invadens* encystation and gene expression with galactose and N-acetylglucosamine. *Molecular and Biochemical Parasitology, 102*(1), 67–77.

Coppi, A., Merali, S., & Eichinger, D. (2002). The enteric parasite Entamoeba uses an autocrine catecholamine system during differentiation into the infectious cyst stage. *The Journal of Biological Chemistry, 277*(10), 8083–8090. https://doi.org/10.1074/jbc.M111895200.

de Meester, F., Shaw, E., Scholze, H., Stolarsky, T., & Mirelman, D. (1990). Specific labeling of cysteine proteinases in pathogenic and nonpathogenic *Entamoeba histolytica*. *Infection and Immunity, 58*(5), 1396–1401.

Du, Q., & Schaap, P. (2014). The social amoeba *Polysphondylium pallidum* loses encystation and sporulation, but can still erect fruiting bodies in the absence of cellulose. *Protist 165*(5), 569–579. https://doi.org/10.1371/journal.pcbi.1003817, https://doi.org/10.1016/j.protis.2014.07.003.

Du, Q., Schilde, C., Birgersson, E., Chen, Z. H., McElroy, S., & Schaap, P. (2014). The cyclic AMP phosphodiesterase RegA critically regulates encystation in social and pathogenic amoebas. *Cellular Signalling, 26*(2), 453–459. https://doi.org/10.1016/j.cellsig.2013.10.008.

Eichinger, L., Pachebat, J. A., Glockner, G., Rajandream, M. A., Sucgang, R., Berriman, M., et al. (2005). The genome of the social amoeba *Dictyostelium discoideum*. *Nature, 435*(7038), 43–57.

Fritz-Laylin, L. K., Prochnik, S. E., Ginger, M. L., Dacks, J. B., Carpenter, M. L., Field, M. C., et al. (2010). The genome of *Naegleria gruberi* illuminates early eukaryotic versatility. *Cell, 140*(5), 631–642. https://doi.org/10.1016/j.cell.2010.01.032.

Funamoto, S., Anjard, C., Nellen, W., & Ochiai, H. (2003). cAMP-dependent protein kinase regulates *Polysphondylium pallidum* development. *Differentiation, 71*(1), 51–61.

Gerwig, G. J., van Kuik, J. A., Leeflang, B. R., Kamerling, J. P., Vliegenthart, J. F., Karr, C. D., et al. (2002). The Giardia intestinalis filamentous cyst wall contains a novel beta(1-3)-N-acetyl-D-galactosamine polymer: A structural and conformational study. *Glycobiology, 12*(8), 499–505. https://doi.org/10.1093/glycob/cwf059.

Gloeckner, G., Lawal, H. M., Felder, M., Singh, R., Singer, G., Weijer, C. J., et al. (2016). The multicellularity genes of dictyostelid social amoebas. *Nature communications, 7,* 12085. https://doi.org/10.1038/ncomms12085.

Harwood, A. J., Hopper, N. A., Simon, M.-N., Driscoll, D. M., Veron, M., & Williams, J. G. (1992). Culmination in *Dictyostelium* is regulated by the cAMP-dependent protein kinase. *Cell, 69,* 615–624.

Harwood, A. J., Plyte, S. E., Woodgett, J., Strutt, H., & Kay, R. R. (1995). Glycogen synthase kinase 3 regulates cell fate in *Dictyostelium*. *Cell, 80,* 139–148.

Heidel, A., Lawal, H., Felder, M., Schilde, C., Helps, N., Tunggal, B., et al. (2011). Phylogeny-wide analysis of social amoeba genomes highlights ancient origins for complex intercellular communication. *Genome Research,* 1882–1891. https://doi.org/10.1101/gr.121137.111.

Hillmann, F., Forbes, G., Novohradska, S., Ferling, I., Riege, K., Groth, M., et al. (2018a). Multiple roots of fruiting body formation in amoebozoa. *Genome Biology and Evolution, 10*(2), 591–606. https://doi.org/10.1093/gbe/evy011.

Hillmann, F., Forbes, G., Novohradska, S., Ferling, I., Riege, K., Groth, M., et al. (2018b). Multiple roots of fruiting body formation in Amoebozoa. *Genome Biology and Evolution*. https://doi.org/10.1093/gbe/evy011.

Hopper, N. A., Harwood, A. J., Bouzid, S., Véron, M., & Williams, J. G. (1993). Activation of the prespore and spore cell pathway of *Dictyostelium* differentiation by cAMP-dependent protein kinase and evidence for its upstream regulation by ammonia. *EMBO Journal, 12,* 2459–2466.

Jha, B. K., Jung, H. J., Seo, I., Kim, H. A., Suh, S. I., Suh, M. H., et al. (2014). Chloroquine has cytotoxic effect in encysting Acanthamoeba through modulation of autophagy. *Antimicrobial Agents and Chemotherapy*. https://doi.org/10.1128/aac.03164-14.

Kang, S., Tice, A. K., Spiegel, F. W., Silberman, J. D., Panek, T., Cepicka, I., et al. (2017). Between a pod and a hard test: The deep evolution of amoebae. *Molecular Biology and Evolution, 34*(9), 2258–2270. https://doi.org/10.1093/molbev/msx162.

Kawabe, Y., Morio, T., Tanaka, Y., & Schaap, P. (2018). Glycogen synthase kinase 3 promotes multicellular development over unicellular encystation in encysting Dictyostelia. *Evodevo, 9,* 12. https://doi.org/10.1186/s13227-018-0101-6.

Kawabe, Y., Schilde, C., Du, Q., & Schaap, P. (2015). A conserved signalling pathway for amoebozoan encystation that was co-opted for multicellular development. *Scientific Reports, 5,* 9644. https://doi.org/10.1038/srep09644.

Kim, S. H., Moon, E. K., Hong, Y., Chung, D. I., & Kong, H. H. (2015). Autophagy protein 12 plays an essential role in Acanthamoeba encystation. *Experimental Parasitology, 159,* 46–52. https://doi.org/10.1016/j.exppara.2015.08.013.

Konijn, T. M., Van De Meene, J. G., Bonner, J. T., & Barkley, D. S. (1967). The acrasin activity of adenosine-3',5'-cyclic phosphate. *Proceedings of the National Academy of Sciences of the United States of America, 58*(3), 1152–1154.

Lauwaet, T., Davids, B. J., Reiner, D. S., & Gillin, F. D. (2007). Encystation of Giardia lamblia: A model for other parasites. *Current Opinion in Microbiology, 10*(6), 554–559. https://doi.org/10.1016/j.mib.2007.09.011.

Lloyd, D. (2014). Encystment in *Acanthamoeba castellani*: A review. *Experimental Parasitology, 145*(Suppl), S20–S27. https://doi.org/10.1016/j.exppara.2014.03.026.

Loftus, B., Anderson, I., Davies, R., Alsmark, U. C., Samuelson, J., Amedeo, P., et al. (2005). The genome of the protist parasite *Entamoeba histolytica*. *Nature, 433*(7028), 865–868.

Mi-ichi, F., Miyamoto, T., Takao, S., Jeelani, G., Hashimoto, T., Hara, H., et al. (2015). Entamoeba mitosomes play an important role in encystation by association with cholesteryl sulfate synthesis. *Proceedings of the National Academy of Sciences of the United States of America, 112*(22), E2884–E2890. https://doi.org/10.1073/pnas.1423718112.

Mi-Ichi, F., Yoshida, H., & Hamano, S. (2016). Entamoeba encystation: New targets to prevent the transmission of amebiasis. *PLoS Pathogens, 12*(10), e1005845. https://doi.org/10.1371/journal.ppat.1005845.
Moon, E. K., Chung, D. I., Hong, Y., & Kong, H. H. (2011). Expression levels of encystation mediating factors in fresh strain of *Acanthamoeba castellani* cyst ESTs. *Experimental Parasitology, 127*(4), 811–816. https://doi.org/10.1016/j.exppara.2011.01.003.
Moon, E. K., Chung, D. I., Hong, Y., & Kong, H. H. (2012). Protein kinase C signaling molecules regulate encystation of Acanthamoeba. *Experimental Parasitology, 132*(4), 524–529. https://doi.org/10.1016/j.exppara.2012.07.008.
Moon, E. K., Hong, Y., Chung, D. I., Goo, Y. K., & Kong, H. H. (2014). Down-regulation of cellulose synthase inhibits the formation of endocysts in Acanthamoeba. *Korean Journal of Parasitology, 52*(2), 131–135. https://doi.org/10.3347/kjp.2014.52.2.131.
Moon, E. K., Hong, Y., Chung, D. I., Goo, Y. K., & Kong, H. H. (2015a). Potential value of cellulose synthesis inhibitors combined with PHMB in the treatment of *Acanthamoeba keratitis*. *Cornea, 34*(12), 1593–1598. https://doi.org/10.1097/ico.0000000000000642.
Moon, E. K., Hong, Y., Chung, D. I., Goo, Y. K., & Kong, H. H. (2016). Identification of protein arginine methyltransferase 5 as a regulator for encystation of acanthamoeba. *Korean Journal of Parasitology, 54*(2), 133–138. https://doi.org/10.3347/kjp.2016.54.2.133.
Moon, E. K., Hong, Y., Chung, D. I., & Kong, H. H. (2013). Identification of atg8 isoform in encysting acanthamoeba. *Korean Journal of Parasitology, 51*(5), 497–502. https://doi.org/10.3347/kjp.2013.51.5.497.
Moon, E. K., Kim, S. H., Hong, Y., Chung, D. I., Goo, Y. K., & Kong, H. H. (2015b). Autophagy inhibitors as a potential antiamoebic treatment for *Acanthamoeba keratitis*. *Antimicrobial Agents and Chemotherapy, 59*(7), 4020–4025. https://doi.org/10.1128/aac.05165-14.
Moon, E. K., & Kong, H. H. (2012). Short-cut pathway to synthesize cellulose of encysting Acanthamoeba. *Korean Journal of Parasitology, 50*(4), 361–364. https://doi.org/10.3347/kjp.2012.50.4.361.
Pitt, G. S., Milona, N., Borleis, J., Lin, K. C., Reed, R. R., & Devreotes, P. N. (1992). Structurally distinct and stage-specific adenylyl cyclase genes play different roles in *Dictyostelium* development. *Cell, 69,* 305–315.
Raizada, M. K., & Murti, C. R. K. (1972). Transformation of trophic *Hartmannella culbertsoni* into viable cysts by cyclic 3',5'-adenosine monophosphate. *Journal of Cell Biology, 52*(3), 743–748.
Ritchie, A. V., van Es, S., Fouquet, C., & Schaap, P. (2008). From drought sensing to developmental control: Evolution of cyclic AMP signaling in social amoebas. *Molecular Biology and Evolution, 25*(10), 2109–2118. msn156 [pii], https://doi.org/10.1016/molbev/msn156.
Romeralo, M., Skiba, A., Gonzalez-Voyer, A., Schilde, C., Lawal, H., Kedziora, S., et al. (2013). Analysis of phenotypic evolution in Dictyostelia highlights developmental plasticity as a likely consequence of colonial multicellularity. *Proceedings of the Biological Sciences, 280*(1764), 20130976. https://doi.org/10.1098/rspb.2013.0976.
Schaap, P., Barrantes, I., Minx, P., Sasaki, N., Anderson, R. W., Benard, M., et al. (2015). The *Physarum polycephalum* genome reveals extensive use of prokaryotic two-component and metazoan-type tyrosine kinase signaling. *Genome Biology and Evolution, 8*(1), 109–125. https://doi.org/10.1093/gbe/evv237.
Schaap, P., & Schilde, C. (2018). Encystation: The most prevalent and under investigated differentiation pathway of eukaryotes. *Microbiology, 164*(5), 727–739. https://doi.org/10.1099/mic.0.000653.
Schaap, P., & Van Driel, R. (1985). Induction of post-aggregative differentiation in *Dictyostelium discoideum* by cAMP. Evidence of involvement of the cell surface cAMP receptor. *Experimental Cell Research, 159,* 388–398.
Schaap, P., Winckler, T., Nelson, M., Alvarez-Curto, E., Elgie, B., Hagiwara, H., et al. (2006). Molecular phylogeny and evolution of morphology in the social amoebas. *Science, 314*(5799), 661–663.

Sharma, M., Hirata, K., Herdman, S., & Reed, S. (1996). Entamoeba invadens: Characterization of cysteine proteinases. *Experimental Parasitology, 84*(1), 84–91. https://doi.org/10.1006/expr.1996.0092.

Shaulsky, G., Fuller, D., & Loomis, W. F. (1998). A cAMP-phosphodiesterase controls PKA-dependent differentiation. *Development, 125,* 691–699.

Singh, M., Sharma, S., Bhattacharya, A., & Tatu, U. (2015). Heat shock protein 90 regulates encystation in Entamoeba. *Frontiers in Microbiology, 6,* 1125. https://doi.org/10.3389/fmicb.2015.01125.

Singleton, C. K., Zinda, M. J., Mykytka, B., & Yang, P. (1998). The histidine kinase dhkC regulates the choice between migrating slugs and terminal differentiation in *Dictyostelium discoideum. Development Biology, 203,* 345–357.

Soderbom, F., Anjard, C., Iranfar, N., Fuller, D., & Loomis, W. F. (1999). An adenylyl cyclase that functions during late development of *Dictyostelium. Development, 126,* 5463–5471.

Song, S.-M., Han, B.-I., Moon, E.-K., Lee, Y.-R., Yu, H. S., Jha, B. K., et al. (2012). Autophagy protein 16-mediated autophagy is required for the encystation of *Acanthamoeba castellani. Molecular and Biochemical Parasitology, 183*(2), 158–165. https://doi.org/10.1016/j.molbiopara.2012.02.013.

Thomason, P. A., Traynor, D., Stock, J. B., & Kay, R. R. (1999). The RdeA-RegA system, a eukaryotic phospho-relay controlling cAMP breakdown. *Journal of Biological Chemistry, 274*(39), 27379–27384.

Urushihara, H., Kuwayama, H., Fukuhara, K., Itoh, T., Kagoshima, H., Shin, I. T., et al. (2015). Comparative genome and transcriptome analyses of the social amoeba *Acytostelium subglobosum* that accomplishes multicellular development without germ-soma differentiation. *BMC Genomics, 16,* 80. https://doi.org/10.1186/s12864-015-1278-x.

Van Es, S., Virdy, K. J., Pitt, G. S., Meima, M., Sands, T. W., Devreotes, P. N., et al. (1996). Adenylyl cyclase G, an osmosensor controlling germination of *Dictyostelium* spores. *Journal of Biological Chemistry, 271,* 23623–23625.

Vazquezdelara-Cisneros, L. G., & Arroyo Begovich, A. (1984). Induction of encystation of *Entamoeba invadens* by removal of glucose from the culture medium. *The Journal of Parasitology, 70*(5), 629–633.

Wang, Z., Samuelson, J., Clark, C. G., Eichinger, D., Paul, J., Van Dellen, K., et al. (2003). Gene discovery in the *Entamoeba invadens* genome. *Molecular and Biochemical Parasitology, 129*(1), 23–31.

Wang, N., Soderbom, F., Anjard, C., Shaulsky, G., & Loomis, W. F. (1999). SDF-2 induction of terminal differentiation in *Dictyostelium discoideum* is mediated by the membrane-spanning sensor kinase DhkA. *Molecular and Cellular Biology, 19*(7), 4750–4756.

Woodgett, J. R. (1994). Regulation and functions of the glycogen synthase kinase-3 subfamily. *Seminars in Cancer Biology, 5*(4), 269–275.

Chapter 3
Integrative Omics Analysis of the Effect of Bacteria on the Resistance of *Entamoeba histolytica* to Oxidative Stress

Yana Shaulov and Serge Ankri

Abstract *Entamoeba histolytica*, the etiological agent of amebiasis interacts with and feeds on the gut microbiota. Several studies have emphasized the close relationship between *E. histolytica* and the bacteria and its impact on the parasite's pathogenesis. In this work we present an integrated omics analysis of the effect of *E. coli* and other bacteria on the resistance of *E. histolytica* to oxidative stress.

Keywords Entamoeba · Enterobacterium · Stress · Intestinal flora · Malate dehydrogenase · Oxaloacetate · Leucine rich repeat proteins

Abbreviations

GAPDH	Glyceraldehyde-3-phosphate dehydrogenase putative
HKB	Heat killed bacteria
HKBOS	Heat killed bacteria exposed to oxidative stress
LB	Live bacteria
LRR	Leucine rich repeat proteins
MDH	Malate dehydrogenase
OS	Oxidative stress
OX	Oxidized proteins
OX-RAC	Resin assisted capture of oxidized proteins
NS	Nitrosative stress

Y. Shaulov · S. Ankri (✉)
Department of Molecular Microbiology, Ruth and Bruce Rappaport Faculty of Medicine, Technion, Haifa, Israel
e-mail: sankri@technion.ac.il

© Springer Nature Switzerland AG 2020

N. Guillen (ed.), *Eukaryome Impact on Human Intestine Homeostasis and Mucosal Immunology*, https://doi.org/10.1007/978-3-030-44826-4_3

Introduction

Entamoeba histolytica is an invasive, pathogenic protozoan which inhabits the human intestine and is the causative agent of amoebiasis. It is the second most lethal disease caused by a protozoan parasite, mainly in developing countries with poor socioeconomic and sanitary conditions (Marie and Petri 2014). It is estimated that amoebiasis accounted for 55,500 deaths and 2.237 million disability-adjusted life years in 2010 (Turkeltaub et al. 2015). Only 10% of the infected individuals will develop acute intestinal and extra-intestinal diseases. *E. histolytica's* life cycle consists of two stages; the trophozoite, an invasive form which can be found in the human large intestine and the cyst, the infective form, which is found in the external environment. Infection by *E. histolytica* occurs by ingestion of these cysts present in contaminated food or water. Upon ingestion of food contaminated with cysts, these cysts pass through the stomach and excyst in the small intestine where they produce trophozoites, which later can colonize the large intestine. In the colon, these trophozoites can either asymptomatically colonize the gut, re-encyst and be expelled in the feces or cause an invasive disease (Stanley and Reed 2001). As an anaerobic parasite, *E. histolytica* must withstand harsh environmental conditions once inside the host. During invasion of the intestine and the liver, the amoeba is confronted with reactive oxygen and nitrogen species (ROS and RNS) as a result of the host's immune system activation [for a recent review see (Nagaraja and Ankri 2018)]. *E. histolytica* uses different protective mechanisms to deal with ROS and RNS. These include thioredoxins, superoxide dismutase, peroxiredoxin, rubrerythrin, flavodiiron proteins, hybrid cluster proteins and NADPH: flavin oxidoreductase. On the other hand, *E. histolytica* lacks both catalase and glutathione, the major thiol in most eukaryotes. L-cysteine is the major thiol in *E. histolytica* and it is required by the parasite for its proliferation, survival, attachment and defense against oxidative stress (OS) (Jeelani and Nozaki 2016).

In the gut, the parasite resides, feeds, multiplies and interacts with the surrounding bacteria, whose contribution to the manifestation of the disease begins to be understood. Although the mechanisms underlying the molecular switch to a virulent phenotype have not been fully characterized yet, more and more evidences are supporting the fact that *E. histolytica* has a direct effect on the bacterial population and diversity of the gut microbiota. Patients suffering of intestinal amoebiasis in India have their predominant flora (*Bacteroides*, *Clostridium* sp. and *Lactobacillus*) altered by the presence of the parasite (Verma et al. 2012). In areas where amoebiasis is endemic, mixed intestinal infections of *E. histolytica* and enteropathogenic *Escherichia coli* are common (Paniagua et al. 2007). *Prevotella copri* is a part of the gut microbiota associated with inflammation (Scher et al. 2013). An elevated levels of *P. copri* were observed in children suffering from diarrheagenic *E. histolytica* infections in Bangladesh (Gilchrist et al. 2016). In contrast, the levels of *P. copri* were downregulated in individuals with intestinal amoebiasis in south west Cameroon. Recent metagenomic studies using the bacterial population present in human faecal samples (Iyer et al. 2019) found that *E. histolytica* prefers to phagocytose bacteria that are required for the maintenance of a healthy gut. One of these bacteria is *Lactobacillus ruminus*, a potential probiotic.

All these combined emphasize the importance of understanding the different mechanisms and outcomes of the interaction between *E. histolytica* and the gut microbiota. To shed light on this interplay we are presenting in this report an integration of two recent works of transcriptomic and redoxomic analysis aimed at studying the effect of *E. coli* on the resistance of *E. histolytica* to OS (Shaulov et al. 2018; Varet et al. 2018).

E. coli *Confers Resistance to OS to* E. histolytica *and Has an Effect on the Redox Proteome of the Parasite*

The effect of *E. coli* on the virulence of *E. histolytica* was first investigated in the 50s (Yanai and Okamoto 1952). Later in the 70s (Wittner and Rosenbaum 1970) it was found that bacteria have a boosting effect which is contact dependent on the virulence of *E. histolytica*. In the 80s Mirleman's group showed that *E. histolytica* trophozoites attach to and ingest bacteria by using their surface lectins or by using their mannose containing cell surface components as receptors for bacterial adhesins (Bracha et al. 1982). They confirmed that the boosting effect that bacteria have on the virulence of the parasite is contact dependent (Bracha and Mirelman 1984). They also reported that when *E. coli* O55 or *Shigella dysenteriae* are incubated with the parasite prior to its exposure to OS, the virulence of *E. histolytica* is preserved. Interestingly, this boosting effect is lost when the parasite is incubated with a catalase deficient *S. dysenteriae* strain (Bracha and Mirelman 1984). This work suggests that a crosstalk between *E. histolytica* and bacteria takes place during OS. These findings encouraged us to investigate more in details the response of *E. histolytica* to OS in the presence of bacteria. We found that live *E. coli* (LB) but not heat-killed *E. coli* (HKB) protect the parasite against OS induced by H_2O_2 (2.5 mM; 1 h) (Varet et al. 2018). This protective effect was specific to OS as no protection was observed against nitrosative stress (NS). To learn about the cellular mechanisms underlying this protective effect, we analyzed *E. histolytica*' s redox proteome using resin assisted capture (RAC) of oxidized proteins (OX) (OX-RAC) coupled to mass spectrometry. We determined the oxidation status of cysteine residues in proteins present in amoebas that were incubated with LB or with HKB. The purification of OX proteins using OX-RAC is based on the capture by reactive 2-thiopyridyl disulphide groups attached to Sepharose of oxidized cysteine residues [for more details see: (Shahi et al. 2016)]. Redox homeostasis, lipid metabolism and small molecule metabolism are some of the biological processes that were found to be enriched in *E. histolytica* trophozoites that were preincubated with HKB and exposed to OS. On the other hand, in trophozoites that were pre incubated with LB, the only enriched OX proteins found in this group belong to biological processes involved in redox homeostasis, represented by thioredoxins. Overall, these results indicate that preincubation of *E. histolytica* trophozoites with LB protects the parasite's proteins against oxidation. The fact that OX thioredoxins were present in trophozoites preincubated with LB or with HKT and then exposed to OS can be explained by the tendency of this family of proteins to be oxidized and reduced constantly in the cell as part of their antioxidant activity (Jung et al. 2014; Lu and Holmgren 2014).

The most surprising result of this work was the finding of 70 *E. coli* proteins that were co-purified with *E. histolytica*'s OX-proteins. 5 out of these 70 bacterial proteins were selected as potential candidates participating in the parasite's protective mechanism against OS. Malate dehydrogenase (MDH) which converts malate to oxaloacetate protects *E. histolytica* against OS through the action of oxaloacetate. Oxaloacetate and other alpha keto acids reduce H_2O_2 to water while undergoing nonenzymatic decarboxylation at the one-carbon position (Kim et al. 2016). The exact mechanism of this protection is not fully understood as it is not clear whether H_2O_2 is detoxified inside the bacteria or whether the detoxification occurs through secretion of MDH into the culture media and subsequently through its conversion to oxaloacetate. The latter is probably the most probable mechanism by which this protective effect takes place as MDH is part of *E. coli's* secretome (Nirujogi et al. 2017) and we have shown that addition of recombinant *E. coli* MDH in presence of malate confers resistance to *E. histolytica* trophozoites during OS. Secretion of oxaloacetate into the media is another possibility, but there is no information on such ability of *E. coli* to secrete oxaloacetate.

E. coli *Has an Effect on the Transcriptome of* **E. histolytica** *Exposed to OS*

Knowledge about the effect of bacteria on the transcriptome of *E. histolytica* was limited. Some *E. histolytica* genes that have their expression modulated in presence of *E. coli* 055 had been previously identified (Mendoza-Macias et al. 2009). Interestingly, their expression depends on the time of interaction between the parasite and the bacteria. For example, the gene coding for ubiquitin-conjugating enzyme was down regulated in short time interaction (1 h) and short term monoxenic culture (3 months) but had a normal level of expression following long term interaction (1 year). In contrast, genes that encode 40S ribosomal proteins S3a and S6 were found to be down regulated specifically within the long term monoxenic culture.

To gain more insights into the changes which occur in the gene expression of *E. histolytica* trophozoites exposed to *E. coli* O55, we performed a transcriptomics analysis. This analysis shows that short term interaction of *E. coli* 055 with *E. histolytica* (30 min) has a minor effect on the transcriptome of the parasite with only 43 modulated genes (Varet et al. 2018). This result suggests that the many effects that bacteria have on the physiology of *E. histolytica* (Bhattacharya et al. 1992; Ehrenkaufer et al. 2007) take place at the posttranslational level. Our work on the redoxome of *E. histolytica* exposed to OS and to *E. coli* did provide clues that this is indeed the case (Shaulov et al. 2018). However, we still need to gain insights into the changes occurring in the parasite at the protein level by performing quantitative proteomics.

Our recent work showed that *E. coli* has a significant impact on the transcriptome of the parasite when the parasite is exposed to OS (Varet et al. 2018). Nearly 50% of the parasite genome coding regions are modulated when comparing *E. histolytica* trophozoites that were incubated without *E. coli* versus *E. histolytica* trophozoites incubated with LB during OS. The transcriptomic response of the parasite exposed to LB and OS can be divided into two patterns: The first pattern (P1) includes genes that have their expression under OS (upregulated or downregulated) restored to a basal level when the trophozoites were pre-incubated with LB before OS. This pattern includes genes encoding antioxidant enzymes, such as peroxidases and thioredoxins, genes related to protein synthesis and homeostasis represented by ribosomal proteins and factors related to translation. The second pattern (P2) includes genes that have their expression under OS (upregulated or downregulated) reversed by pre-incubation with LB. Genes encoding enzymes related to glycolytic pathways and proteins belonging to the Leucine rich repeat protein (LRR) family belong to this pattern.

The majority of LRR in *E. histolytica* belong to the BspA family defined by their homology to bacterial LRR factors (Sharma 2010; Sharma et al. 1998). Not much is known about the role of LRRs in *E. histolytica.* A member of the BspA-like family named “cell surface protein” has a key role in *E. histolytica*’s ability to migrate towards a tumor necrosis factor gradient and to invade human colon explants (Silvestre et al. 2015). The facts that a common BspA expression profile was observed between the Enterobacteriaceae—*E. coli* and *S. enterica*, which protected the amoeba against OS and that a different expression profile was observed with *L. acidophilus*, which did not confer protection to the amoeba during OS suggest that the BspAs may serve for the recognition of the bacteria by the parasite under OS (Varet et al. 2018). The most significantly modulated amoebic LRRs genes correspond to LRR proteins that share structural homology with Toll like receptors (TLRs), specifically at the ectodomain needed for pathogen-associated molecular pattern recognition, exemplified by EHI_123820, EHI_119470, EHI_017710, EHI_139980 and EHI_087810. The amoebic LRRs may be cytoplasmic proteins since they lack transmembrane domain or they may use a co-receptor at the cell surface as it has been noticed for TLR-4 in B lymphocytes (Schweighoffer et al. 2017). Amoebic LRRs are transcriptionally regulated in the presence of bacteria and OS similarly to the transcription of TLR-4 gene in lipopolysaccharides activated leukocytes (Vaure and Liu 2014). The factors identified here suggest that despite its old age in evolutionary terms, the protozoan *E. histolytica* displays characteristics of higher eukaryotes’ innate immune system and that its BspAs are involved in the recognition of bacteria under OS. Some structural characteristics and properties of these BspAs seem to have been kept during evolution and transmitted to higher eukaryotes. All together these mechanisms of gene transcription regulation are of great interest to understand how bacteria (commensals and/or pathogens) influence the onset of amebiasis.

Integration of Redox Omics and Transcriptomics Data on the Response of *E. histolytica* to OS in Presence of *E. coli*

To gain additional information about the role of bacteria in the resistance of *E. histolytica* to OS, we decided to integrate our redox-omics and transcriptomics data. No significant overlapping or enrichment was found when comparing P1 or P2 with *E. histolytica* trophozoites incubated with LB during OS (data not shown). On the other hand we found that 115 genes are shared between P1 and the group of OXs in *E. histolytica* trophozoites incubated with heat killed bacteria and exposed to OS (HKBOS) (Table 3.1). PANTHER analysis (Thomas et al. 2003) of these 115 proteins indicates that they are enriched (Fold enrichment—7.52, *P* value—7.48E−04) for translation initiation family and ribosomal proteins, exemplified by translation initiation factor eIF-5A putative (EHI_151810), elongation factor 1 alpha 1 (EHI_011210) and 60S ribosomal protein L9 putative (EHI_193080). Oxidation of translation initiation factors has been reported to inhibit their activity in the yeast *Saccharomyces cerevisiae* (Shenton et al. 2006). Therefore, if the activity of translation initiation factors is also inhibited by oxidation in *E. histolytica*, then it may be possible that the parasite compensates their inhibition by overexpressing them. The same compensatory effect has been reported following the inhibition by nitrosylation of pyruvate: ferredoxin oxidoreductase (Santi-Rocca et al. 2012).

Another family of enriched protein is the tubulin family (Fold enrichment—41.53, *P* value—1.50E−04) exemplified by tubulin alpha chain (EHI_005950, EHI_010530) and tubulin beta chain (EHI_049920). It is known that cysteine oxidation of tubulin induces microtubule depolymerization, and increases the pool of soluble tubulin (Hinshaw et al. 1993; Valen et al. 1999). It is tempting to speculate that the increase of soluble tubulin is sensed by the cell as an excess of tubulin leading to the repression of its transcription. This hypothesis needs to be examined in *E. histolytica*.

We also found that the product of 64 genes are shared between P2 and HKBOS. One of them is the glyceraldehyde-3-phosphate dehydrogenase (GAPDH) (EHI_008200). This gene was found among genes that were up regulated without bacteria during OS and whose expression was reversed (downregulated) when incubated with LB. GAPDH expression is up regulated in *E. histolytica* trophozoites exposed to NS (Santi-Rocca et al. 2012) and its overexpression makes *E. histolytica* trophozoites more sensitive to NS (Shahi et al. 2016). The fact that GAPDH induces apoptosis in mammalian cells (Sawa et al. 1997) may imply that the induction of apoptosis as a result of the exposure to OS through GAPDH extends to *E. histolytica* trophozoites and is reversed when the trophozoites are incubated with LB.

Table 3.1 Shared genes between P1 and the group of OXs in *E. histolytica* trophozoites incubated with heat killed bacteria and exposed to OS

Id
EHI_000610
EHI_000670
EHI_004750
EHI_004810
EHI_005950
EHI_009410
EHI_010340
EHI_010530
EHI_011210
EHI_013220
EHI_013240
EHI_017630
EHI_022960
EHI_023110
EHI_023630
EHI_025380
EHI_026470
EHI_026480
EHI_028550
EHI_029540
EHI_035430
EHI_035730
EHI_039200
EHI_040400
EHI 041950
EHI_042920
EHI_044640
EHI_044740
EHI_048670
EHI_049920
EHI_050800
EHI_065710
EHI_068510
EHI_068520
EHI_068620
EHI_068650
EHI_069460

(continued)

Table 3.1 (continued)

Id
EHI_072060
EHI_073470
EHI_074150
EHI_077270
EHI_082520
EHI_083250
EHI_086080
EHI_086180
EHI_086530
EHI_093790
EHI_093850
EHI_096570
EHI_098190
EHI_100480
EHI_103450
EHI_104390
EHI_106690
EHI_110010
EHI_118880
EHI_120590
EHI_123280
EHI_130880
EHI_133960
EHI_133970
EHI_138520
EHI_138750
EHI_138770
EHI_143820
EHI_148040
EHI_148770
EHI_150130
EHI_151600
EHI_151800
EHI_151810
EHI_152220
EHI_152310
EHI_152340

(continued)

Table 3.1 (continued)

Id
EHI_152850
EHI_152960
EHI_153160
EHI_155420
EHI_155590
EHI_155710
EHI_156670
EHI_161030
EHI_161070
EHI_161650
EHI_162230
EHI_163540
EHI_164460
EHI_166900
EHI_169590
EHI_173460
EHI_173480
EHI_177400
EHI_177440
EHI_177650
EHI_178890
EHI_179330
EHI_179980
EHI_183540
EHI 186480
EHI_186770
EHI_186820
EHI_188120
EHI_190440
EHI_192590
EHI_193080
EHI_194040
EHI_194330
EHI_194450
EHI_194850
EHI_196980
EHI_198760

(continued)

Table 3.1 (continued)

Id
EHI_199050
EHI_199110
EHI_199120
EHI_201200

Perspectives

Despite the major recent findings that were presented in this work about the crosstalk between *E. histolytica* trophozoites and the gut bacterial flora and the recognition of this crosstalk as an important factor in the virulence of the amoeba, it is important to improve our understanding of this interaction and its exact mechanisms. The works presented here (summarized in Fig. 3.1) strongly suggest that enteric bacteria influence the course of an amoebic infection and boost the infectious process by supplying factors like oxaloacetate that lead to the parasite's survival in the host. In contrast, *L. acidophilus*, a probiotic bacteria, did not protect the amoeba against OS, probably due to its ability to produce endogenously hydrogen peroxide (Hertzberger et al. 2014). Nowadays, metronidazole is the most common standard therapy for amebiasis owing to its high efficacy and affordability (Lofmark et al. 2010). However, this drug has side effects (Hernandez Ceruelos et al. 2019) and there is a growing concern of

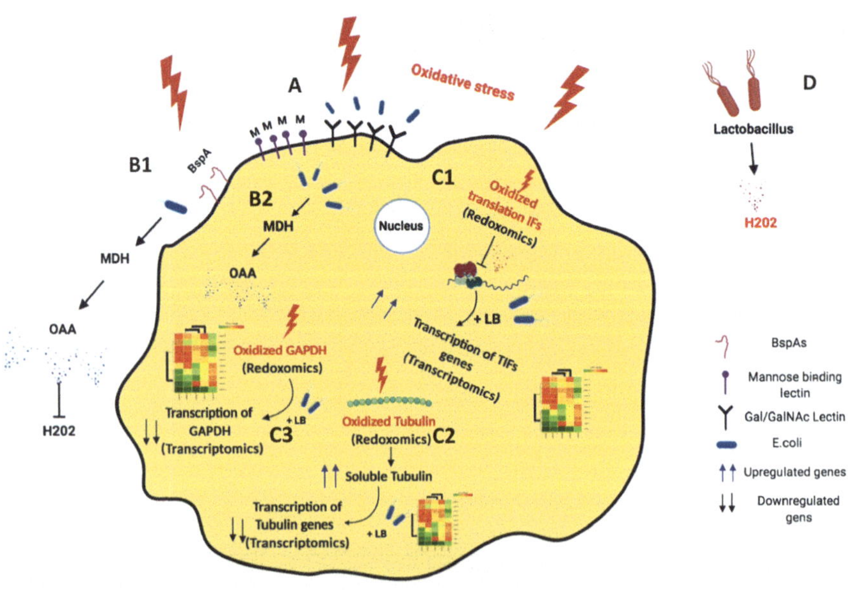

Fig. 3.1 Summary of the integration between transcriptomic and redoxomic studies on the role of *E. coli* and other bacteria on the resistance of *E. histolytica* to OS

the emergence of a metronidazole resistant *E. histolytica* strains (Bansal et al. 2004). Impairing the cross talk between the parasite and the bacteria and supporting the development of amebicide probiotics represent new strategies to fight amebiasis and the putative selection of metronidazole resistant *E. histolytica* strains.

The parasite contains different receptors on its surface that enables the bacteria to attach or to be ingested by the parasite (A) M—Mannose-containing receptors and Y—carbohydrate-binding (lectin) activity on the surface of the parasite. (B) (1) *E. coli* secretes MDH into the media. In presence of malate, MDH converts this substrate to oxaloacetate. Oxaloacetate detoxifies H_2O_2. (2) Alternatively, *E. coli* are ingested by the parasite and oxaloacetate produced by the bacteria is used by the parasite to detoxify H_2O_2. (C) changes occurring in the parasite at the proteome and transcriptional level during OS. C1—the translation initiation factor family was found to be oxidized during OS and its transcription is upregulated when the parasite is incubated with LB during OS. C2—the tubulin protein family is oxidized during OS and its gene transcription was found to upregulated. C3—GAPDH is oxidized during OS, in the transcriptomics analysis its transcription is downregulated upon incubation with LB during OS. (D) *Lactobacillus acidophilus* is not protecting the parasite against OS probably due to its ability to secrete endogenous H_2O_2.

Acknowledgements and Funding We would like to acknowledge the assistance of Ms. Shruti Nagaraja in the preparation of Fig. 3.1.

This study was supported by grants from the Israel Science Foundation (260/16), the US-Israel Binational Science Foundation (2015211), the Niedersachsen-Israel ZN3454 program and the European ERA-NET Infect-ERA program AMOEBAC (Israel Ministry of Health Grant 031L000).

References

Bansal, D., Sehgal, R., Chawla, Y., Mahajan, R. C., & Malla, N. (2004). In vitro activity of anti-amoebic drugs against clinical isolates of *Entamoeba histolytica* and *Entamoeba dispar*. *Annals of Clinical Microbiology and Antimicrobials, 3*, 27. https://doi.org/10.1186/1476-0711-3-27.

Bhattacharya, A., Ghildyal, R., Prasad, J., Bhattacharya, S., & Diamond, L. S. (1992). Modulation of a surface antigen of *Entamoeba histolytica* in response to bacteria. *Infection and Immunity, 60*(4), 1711–1713.

Bracha, R., Kobiler, D., & Mirelman, D. (1982). Attachment and ingestion of bacteria by trophozoites of *Entamoeba histolytica*. *Infection and Immunity, 36*(1), 396–406.

Bracha, R., & Mirelman, D. (1984). Virulence of *Entamoeba histolytica* trophozoites. Effects of bacteria, microaerobic conditions, and metronidazole. *Journal of Experimental Medicine, 160*(2), 353–368.

Ehrenkaufer, G. M., Haque, R., Hackney, J. A., Eichinger, D. J., & Singh, U. (2007). Identification of developmentally regulated genes in *Entamoeba histolytica*: Insights into mechanisms of stage conversion in a protozoan parasite. *Cellular Microbiology, 9*(6), 1426–1444. https://doi.org/10.1111/j.1462-5822.2006.00882.x.

Gilchrist, C. A., Petri, S. E., Schneider, B. N., Reichman, D. J., Jiang, N., Begum, S., … Petri, W. A., Jr. (2016). Role of the gut microbiota of children in diarrhea due to the protozoan parasite *Entamoeba histolytica*. *The Journal of Infectious Diseases, 213*(10), 1579–1585. https://doi.org/10.1093/infdis/jiv772 (jiv772 [pii]).

Hernandez Ceruelos, A., Romero-Quezada, L. C., Ruvalcaba Ledezma, J. C., & Lopez Contreras, L. (2019). Therapeutic uses of metronidazole and its side effects: An update. *European Review for Medical and Pharmacological Sciences, 23*(1), 397–401. https://doi.org/10.26355/eurrev_201901_16788.

Hertzberger, R., Arents, J., Dekker, H. L., Pridmore, R. D., Gysler, C., Kleerebezem, M., et al. (2014). H(2)O(2) production in species of the *Lactobacillus acidophilus* group: A central role for a novel NADH-dependent flavin reductase. *Applied and Environment Microbiology, 80*(7), 2229–2239. https://doi.org/10.1128/AEM.04272-13.

Hinshaw, D. B., Miller, M. T., Omann, G. M., Beals, T. F., & Hyslop, P. A. (1993). A cellular model of oxidant-mediated neuronal injury. *Brain Research, 615*(1), 13–26. https://doi.org/10.1016/0006-8993(93)91110-e.

Iyer, L. R., Verma, A. K., Paul, J., & Bhattacharya, A. (2019). Phagocytosis of gut bacteria by *Entamoeba histolytica. Frontiers in Cellular and Infection Microbiology, 9,* 34. https://doi.org/10.3389/fcimb.2019.00034.

Jeelani, G., & Nozaki, T. (2016). Entamoeba thiol-based redox metabolism: A potential target for drug development. *Molecular and Biochemical Parasitology, 206*(1–2), 39–45. https://doi.org/10.1016/j.molbiopara.2016.01.004.

Jung, T., Hohn, A., & Grune, T. (2014). The proteasome and the degradation of oxidized proteins: Part II—Protein oxidation and proteasomal degradation. *Redox Biology, 2,* 99–104. https://doi.org/10.1016/j.redox.2013.12.008.

Kim, J. G., Park, S. J., Sinninghe Damste, J. S., Schouten, S., Rijpstra, W. I., Jung, M. Y., et al. (2016). Hydrogen peroxide detoxification is a key mechanism for growth of ammonia-oxidizing archaea. *Proceedings of the National Academy of Sciences of the United States of America, 113*(28), 7888–7893. https://doi.org/10.1073/pnas.1605501113.

Lofmark, S., Edlund, C., & Nord, C. E. (2010). Metronidazole is still the drug of choice for treatment of anaerobic infections. *Clinical Infectious Diseases, 50*(Suppl 1), S16–S23. https://doi.org/10.1086/647939.

Lu, J., & Holmgren, A. (2014). The thioredoxin antioxidant system. *Free Radical Biology and Medicine, 66,* 75–87. https://doi.org/10.1016/j.freeradbiomed.2013.07.036.

Marie, C., & Petri, W. A., Jr. (2014). Regulation of virulence of *Entamoeba histolytica. Annual Review of Microbiology, 68,* 493–520. https://doi.org/10.1146/annurev-micro-091313-103550.

Mendoza-Macias, C. L., Barrios-Ceballos, M. P., de la Pena, L. P., Rangel-Serrano, A., Anaya-Velazquez, F., Mirelman, D., et al. (2009). *Entamoeba histolytica*: Effect on virulence, growth and gene expression in response to monoxenic culture with *Escherichia coli* 055. *Experimental Parasitology, 121*(2), 167–174. https://doi.org/10.1016/j.exppara.2008.10.011.

Nagaraja, S., & Ankri, S. (2018). Utilization of different omic approaches to unravel stress response mechanisms in the parasite *Entamoeba histolytica. Frontiers in Cellular and Infection Microbiology, 8,* 19. https://doi.org/10.3389/fcimb.2018.00019.

Nirujogi, R. S., Muthusamy, B., Kim, M. S., Sathe, G. J., Lakshmi, P. T., Kovbasnjuk, O. N., … Jabbour, R. E. (2017). Secretome analysis of diarrhea-inducing strains of *Escherichia coli. Proteomics, 17*(6). https://doi.org/10.1002/pmic.201600299.

Paniagua, G. L., Monroy, E., Garcia-Gonzalez, O., Alonso, J., Negrete, E., & Vaca, S. (2007). Two or more enteropathogens are associated with diarrhoea in Mexican children. *Annals of Clinical Microbiology and Antimicrobials, 6,* 17. https://doi.org/10.1186/1476-0711-6-17.

Santi-Rocca, J., Smith, S., Weber, C., Pineda, E., Hon, C. C., Saavedra, E., et al. (2012). Endoplasmic reticulum stress-sensing mechanism is activated in *Entamoeba histolytica* upon treatment with nitric oxide. *PLoS ONE, 7*(2), e31777. https://doi.org/10.1371/journal.pone.0031777.

Sawa, A., Khan, A. A., Hester, L. D., & Snyder, S. H. (1997). Glyceraldehyde-3-phosphate dehydrogenase: Nuclear translocation participates in neuronal and nonneuronal cell death. *Proceedings of the National Academy of Sciences of the United States of America, 94*(21), 11669–11674. https://doi.org/10.1073/pnas.94.21.11669.

Scher, J. U., Sczesnak, A., Longman, R. S., Segata, N., Ubeda, C., Bielski, C., … Littman, D. R. (2013). Expansion of intestinal *Prevotella copri* correlates with enhanced susceptibility to arthritis. *Elife, 2*, e01202. https://doi.org/10.7554/eLife.01202.

Schweighoffer, E., Nys, J., Vanes, L., Smithers, N., & Tybulewicz, V. L. J. (2017). TLR4 signals in B lymphocytes are transduced via the B cell antigen receptor and SYK. *Journal of Experimental Medicine, 214*(5), 1269–1280. https://doi.org/10.1084/jem.20161117.

Shahi, P., Trebicz-Geffen, M., Nagaraja, S., Alterzon-Baumel, S., Hertz, R., Methling, K., … Ankri, S. (2016). Proteomic identification of oxidized proteins in *Entamoeba histolytica* by resin-assisted capture: Insights into the role of arginase in resistance to oxidative stress. *PLoS Neglected Tropical Diseases, 10*(1), e0004340. https://doi.org/10.1371/journal.pntd.0004340.

Sharma, A. (2010). Virulence mechanisms of Tannerella forsythia. *Periodontology 2000, 54*, 106–116.

Sharma, A., Sojar, H. T., Glurich, I., Honma, K., Kuramitsu, H. K., & Genco, R. J. (1998). Cloning, expression, and sequencing of a cell surface antigen containing a leucine-rich repeat motif from Bacteroides forsythus ATCC 43037. *Infection and immunity, 66*, 5703–5710.

Shaulov, Y., Shimokawa, C., Trebicz-Geffen, M., Nagaraja, S., Methling, K., Lalk, M., … Ankri, S. (2018). *Escherichia coli* mediated resistance of *Entamoeba histolytica* to oxidative stress is triggered by oxaloacetate. *PLoS Pathogens, 14*(10), e1007295. https://doi.org/10.1371/journal.ppat.1007295.

Shenton, D., Smirnova, J. B., Selley, J. N., Carroll, K., Hubbard, S. J., Pavitt, G. D., … Grant, C. M. (2006). Global translational responses to oxidative stress impact upon multiple levels of protein synthesis. *Journal of Biological Chemistry, 281*(39), 29011–29021. https://doi.org/10.1074/jbc.M601545200.

Silvestre, A., Plaze, A., Berthon, P., Thibeaux, R., Guillen, N., & Labruyere, E. (2015). In *Entamoeba histolytica*, a BspA family protein is required for chemotaxis toward tumour necrosis factor. *Microbial Cell, 2*(7), 235–246. https://doi.org/10.15698/mic2015.07.214.

Stanley, S. L., Jr., and Reed, S. L. (2001). Microbes and microbial toxins: Paradigms for microbial-mucosal interactions. VI. *Entamoeba histolytica*: Parasite-host interactions. *The American Journal of Physiology: Gastrointestinal and Liver Physiology, 280*(6), G1049–G1054. https://doi.org/10.1152/ajpgi.2001.280.6.g1049.

Thomas, P. D., Campbell, M. J., Kejariwal, A., Mi, H., Karlak, B., Daverman, R., et al. (2003). PANTHER: A library of protein families and subfamilies indexed by function. *Genome Research, 13*(9), 2129–2141. https://doi.org/10.1101/gr.772403.

Turkeltaub, J. A., McCarty, T. R., & Hotez, P. J. (2015). The intestinal protozoa: Emerging impact on global health and development. *Current Opinion in Gastroenterology, 31*(1), 38–44. https://doi.org/10.1097/MOG.0000000000000135.

Valen, G., Sonden, A., Vaage, J., Malm, E., & Kjellstrom, B. T. (1999). Hydrogen peroxide induces endothelial cell atypia and cytoskeleton depolymerization. *Free Radical Biology and Medicine, 26*(11–12), 1480–1488. https://doi.org/10.1016/s0891-5849(99)00009-x.

Varet, H., Shaulov, Y., Sismeiro, O., Trebicz-Geffen, M., Legendre, R., Coppee, J. Y., et al. (2018). Enteric bacteria boost defences against oxidative stress in *Entamoeba histolytica*. *Scientific Reports, 8*(1), 9042. https://doi.org/10.1038/s41598-018-27086-w.

Vaure, C., & Liu, Y. (2014). A comparative review of toll-like receptor 4 expression and functionality in different animal species. *Frontiers in immunology, 5*, 316.

Verma, A. K., Verma, R., Ahuja, V., and Paul, J. (2012). Real-time analysis of gut flora in *Entamoeba histolytica* infected patients of Northern India. *BMC Microbiology, 12*, 183. https://doi.org/10.1186/1471-2180-12-183 (-2180-12-183 [pii]).

Wittner, M., and Rosenbaum, R. M. (1970). Role of bacteria in modifying virulence of *Entamoeba histolytica*. Studies of amebae from axenic cultures. *The American Journal of Tropical Medicine and Hygiene, 19*(5), 755–761. https://doi.org/10.4269/ajtmh.1970.19.755.

Yanai, T., & Okamoto, J. (1952). Relationship between the number of *Escherichia coli* in stools and the infectivity of rats to *Entamoeba histolytica* in the experimental infection of the amoebae. *Kitasato Archives of Experimental Medicine, 25*(1), 67–68.

Chapter 4
Histomonas meleagridis Molecular Traits—From Past to Future

Ivana Bilic and Michael Hess

Abstract The protozoan parasite *Histomonas meleagridis* is the causative agent of histomonosis, a poultry disease whose significance is underlined by the absence of any licenced prophylaxis or treatment. Considering the parasite-bacteria interplay, histomonosis represents an intriguing interaction in medicine. The present review targets the molecular research on *H. meleagridis*, covering the period after re-emergence of histomonosis at the beginning of the twenty-first century until today, from early phylogenetic studies to recent investigations applying omics techniques.

Keywords *Histomonas meleagridis* · Histomonosis · Blackhead disease · Bacteria · *Escherichia coli* · Poultry · Virulence · Proteome · Transcriptome · Exoproteome

Abbreviations

CP	cysteine peptidase
FBA	fructose bisphosphate aldolase
GAPDH	glyceraldehyde-3-phospate dehydrogenase
MW	molecular weight
pI	isoelectric point
TPI	triosephosphate isomerase

I. Bilic (✉) · M. Hess
Clinic for Poultry and Fish Medicine, University of Veterinary Medicine, Vienna, Austria
e-mail: ivana.bilic@vetmeduni.ac.at

© Springer Nature Switzerland AG 2020

N. Guillen (ed.), *Eukaryome Impact on Human Intestine Homeostasis and Mucosal Immunology*, https://doi.org/10.1007/978-3-030-44826-4_4

Introduction—*Histomonas meleagridis* and Histomonosis

Histomonas meleagridis is a microaerophilic flagellated extracellular protozoan parasite of the order Tritrichomonadida (Cepicka et al. 2010). It causes histomonosis (syn. Blackhead disease or histomoniasis), an important disease of gallinaceous birds (Tyzzer 1920). The destructive outcome of histomonosis is evident in turkeys, in which it can lead to mortalities of up to 100% (McDougald 2005). In chickens, the disease is less severe although it might result in substantial production losses, characterized by drop in egg production and increased mortalities (Hess et al. 2015). *Histomonas meleagridis* initially infects the caecum, but the crucial point destining the disastrous outcome for the host is the translocation of the parasite to the liver. With this feature histomonosis is similar to amoebiasis in humans, caused by *Entamoeba histolytica*, but this is where the similarity recedes. Both protozoan parasites differ in various aspects of their morphology and are phylogenetically quite distant with one belonging to the phylum Parabasalia and the other to Amoebozoa (Cepicka et al. 2010; Kang et al. 2017). The transmission of *H. meleagridis* can be i) indirectly via an intermediate host, the caecal worm *Heterakis gallinarum*, whose eggs harbour the protozoan and ii) directly from bird to bird as shown in turkeys and chickens (Hess et al. 2006a). Remarkably, the sole presence of *H. meleagridis* is insufficient to induce the disease, as the infection of turkeys and chickens is dependent on the presence of bacteria in the gut. These findings are based upon studies performed in gnotobiotic birds infected with bacteria-free *H. gallinarum* eggs harbouring the protozoan and it was shown that both *H. meleagridis* and bacteria are needed for histomonosis to occur (Doll and Franker 1963). Histomonosis is reported worldwide and was efficiently controlled for decades by the use of chemotherapeutics as preventive and curative drugs (Liebhart et al. 2017). In the last 15 years, changes in drug legislation in the European Union and the USA, together with the increasing popularity of free-range housing, have led to the reoccurrence of the disease (Hess et al. 2015). Up till now, only a prototype live vaccine, based on an in vitro attenuated strain, has been shown to be effective against *H. meleagridis* (Hess et al. 2008). However, the vaccine is not yet commercially available and it might still take some time until it will be registered. Licensing of new drugs is also not a realistic scenario considering the demands on safety and efficacy (Regmi et al. 2016). Currently, in attempt to prevent mortalities veterinarians can only rely on the implementation of proper flock management and a very early administration of a licenced aminoglycoside antibiotic Paramomycin (Clark and Kimminau 2017; Liebhart et al. 2017). Following decades without any research, the ban of effective drugs against *H. meleagridis* and the subsequent reappearance of histomonosis revived the research on the parasite and the disease itself. In the present review we focus on the molecular research on *H. meleagridis*, which started only recently as an aftermath to the re-emergence of histomonosis.

Defined in Vitro Cultures—An Essential Fundament for Molecular Research

Histomonas meleagridis can be propagated in vitro only in the presence of bacteria (Hess et al. 2015). The most common *H. meleagridis* in vitro cultures are so called xenic cultures, in which the parasite is propagated together with turkey or chicken caecal flora. Such cultures are usually set up by inoculating the intestinal content of a bird suffering from histomonosis into appropriate media. Since *H. meleagridis* is often not the sole parasite in chicken or turkey caecum, cultures frequently contain *Tetratrichomonas gallinarum* or *Blastocystis* sp.. In an attempt to create more defined in vitro cultures, a mono-eukaryotic or "clonal" culture of *H. meleagridis* was developed, in which a single parasitic cell was transferred via micromanipulation into appropriate medium containing caecal bacteria (Hess et al. 2006b). Another approach showed the successful establishment of a mono-eukaryotic, but not necessarily a "clonal", culture by inoculating liver lesions of *H. meleagridis*-infected birds into culture media pre-incubated with caecal bacteria (Pham et al. 2013). The advantage of mono-eukaryotic cultures is evident in both animal experiments and studies investigating the biology of the parasite (Hess et al. 2015). However, dynamic and unclear composition of bacterial strains co-cultivated with *H. meleagridis* impeded detailed molecular investigations. A breakthrough was the establishment of a monoxenic mono-eukaryotic (clonal) *H. meleagridis* culture, by replacing the caecal flora with a single bacterial strain (Ganas et al. 2012). In this way, the complexity and obscurity of the initial xenic culture was strongly minimized, resulting in a composite culture in which all micro-organisms are well-defined.

Since any *H. meleagridis* in vitro culture generally originates from an outbreak of histomonosis, initial isolates are virulent strains. A long-term cultivation results in the attenuation of *H. meleagridis* (Hess et al. 2008). Considering this feature, a system consisting of a virulent and an attenuated *H. meleagridis* monoxenic mono-eukaryotic (clonal) culture, both originating from the same parasite cell, was developed (Ganas et al. 2012).

Histomonas meleagridis—*Phylogenetic Positioning*

The re-emergence of histomonosis at the beginning of this century incited molecular research on *H. meleagridis*. The majority of the early studies focused on the phylogenetic position of *H. meleagridis* and the development of molecular methods to detect the parasite (Bleyen et al. 2007; Gerbod et al. 2001; Grabensteiner and Hess 2006; Hafez et al. 2005; Hauck et al. 2006; Hauck and Hafez 2009, 2010; Liebhart et al. 2006; Mantini et al. 2009; van der Heijden et al. 2006). Based on the 18S rRNA and beta tubulin sequences, *H. meleagridis* was shown to be closely related to the disputed human parasite *Dientamoeba fragilis*. Despite their amoeboid-like morphology under the light microscope both taxa are considered trichomonads and are placed in the class Tritrichomonadea (Cepicka et al. 2010; Gerbod et al. 2001; Malik et al. 2011).

The first molecular study aside to phylogenetic investigations, confirmed the presence of hydrogenosomes in *H. meleagridis*, mitochondrion-related organelles involved in anaerobic energy metabolism (Mazet et al. 2008). The study reported three genes encoding proteins involved in the hydrogenosomal carbon metabolism and showed the localization of this organelle by immunohistochemistry. Another study characterized in detail the α-actinins of *H. meleagridis* (Leberl et al. 2010). For the first time three α-actinin proteins, bearing different length of a central rod domain, were described for a protozoan. Furthermore, the immunogenicity of α-actinins for the most important hosts of *H. meleagridis*, turkeys and chickens, could be demonstrated. Further progress in molecular characterization of *H. meleagridis* was achieved in a study reporting a broad spectrum of partial protein-coding sequences with homology to both intracellular and surface proteins (Bilic et al. 2009). By this, the panel of available sequences could be extended considerably. Since the majority of identified sequences showed the highest homology to *Trichomonas vaginalis*, the study indicated that in many cellular processes *H. meleagridis* might be very similar to this human parasite.

Applying the extended spectrum of available sequences, multi-locus analysis of over 250 diagnostic samples positive for *H. meleagridis* demonstrated the presence of two genotypes (Bilic et al. 2014). The majority of investigated diagnostic samples from histomonosis outbreaks grouped into genotype 1, a tendency confirmed throughout different later case reports (Sulejmanovic et al. 2019a, b) and generally in the field (personal observation). Genotype 2 has been detected in only a small subset of samples largely from France. A case report indicated that the clinical appearance of histomonosis caused by strains belonging to the genotype 2 seems to be aberrant, with the majority of lesions restricted to the caeca and a lower overall mortality (Grafl et al. 2015).

Histomonas meleagridis—*Omics Studies*

The first study on "big data" of *H. meleagridis* revealed a cDNA library obtained from a xenic culture material (Klodnicki et al. 2013). It reported 3425 *H. meleagridis* sequences of which the majority remained not annotated with the exception of 81 genes encoding putative hydrogenosomal proteins. A high AT content of *H. meleagridis* sequences was noticed and authors determined a codon usage frequency for the parasite, a valuable tool for molecular studies. A more recent investigation on de novo transcriptome of *H. meleagridis* sequenced normalized cDNA libraries of a virulent and an attenuated strain, both propagated as clonal monoxenic culture (Mazumdar et al. 2017). Considering the background of such *H. meleagridis* cultures, data analysis was straightforward and an assignment of *H. meleagridis*-specific contigs was facilitated. Even though the study sequenced two phenotypically different strains of the parasite, due to the normalization of sequencing libraries no specific differences between two strains were described and a single transcriptome reference database containing 3356 contigs was disclosed. Gene ontology analysis of obtained sequences combined with data mining provided novel biological insights into various cellular processes of *H. meleagridis*, such as proteostasis, cytoskeleton network, metabolism, environmental adaptation and potential pathogenic mechanisms. Additionally, transcriptome data were used to perform an in silico drug screen to identify potential anti-histomonal compounds. A list of promising effective compounds was noticed, however as licencing of drugs for food producing animals extends far beyond the in vivo efficacy, results were suggested to be taken as a starting point for identification of possible treatment options.

The first study on the proteome used *H. meleagridis* cells enriched by flow cytometry, an elegant approach to separate the parasite from co-cultivating caecal bacteria (Pham et al. 2016). However, it seems that acquired *H. meleagridis* material was below the amount optimal for two dimensional gel electrophoresis, as only 44 protein spots were detected. Mass spectrometric analysis of spots identified the majority of spots as putative actin, with the exception of two spots which were mapped to putative α-actinin and superoxide-dismutase. The dominance of different actin spots in *H. meleagridis* proteome was brought in connection to potential existence of different actin genes in the parasite's genome and an effect of post-translational modification or protein degradation. Indeed, transcriptome analysis reported a complex transcript composition regarding the cytoskeleton composition, especially actin, and processes concerning reorganization of actin microfilaments (Mazumdar et al. 2017). The notion of active cytoskeletal reorganization is also corroborated by changes in *H. meleagridis* morphology observed both, in vitro and in vivo (Gruber et al. 2017; Tyzzer 1919). The abundance of different actin spots was recently confirmed by comparative proteome studies (Monoyios et al. 2017, 2018). However, in contrast to the first proteome study, actin spots were detected (i) as up-regulated in a virulent strain and (ii) as an addition to the major actin spot which, contrary to numerous actin spots, displayed a predicted gel position in respect to the calculated molecular weight (MW) and the isoelectric point (pI). It was noticed that all actin spots mapped to the same contig, indicating that different forms of actin probably derived from post-translational modification and/or degradation of actin protein encoded by the same gene, rather than different genes (Table 4.1). Furthermore, the up-regulation of actin in the virulent strain was detected only in gel-based approaches, whereas a label-free gel-free method failed to detect this phenomenon (Monoyios et al. 2018). This indicated rather subtle changes that involved certain post-translational modifications and/or degradation of actin, visible only when proteins were separated on the gel, but were masked in the gel-free approach by a bulk of unmodified actin. Recently, N-terminal acetylation of actin was reported as the most abundant post-translational modification of this protein in *E. histolytica*, with an implication in cell morphology and division, actin filament organisation and parasite invasion in an in vitro model of amoebic human infection (Hernandez-Cuevas et al. 2019). The nature and the role of actin modifications observed in the virulent *H. meleagridis* are unknown, however considering that certain biological relationship between two parasites exists, one can hypothesize that a comparable process is likely to occur.

Table 4.1 Summary of *H. meleagridis* proteins identified in proteomic studies

Protein	Conventional 2D experiment	2D-DIGE α experiment	2D-DIGE β experiment	SWATH proteome	Exoproteome
Cytoskeleton					
Actin (HAGI01002078) PTM*	↑Virulent	↑Virulent	↑Virulent		
Actin (HAGI01002078)					Identified
α-actinin 2 (HAGI01000004) PTM*	↑Virulent				
Actin family protein (HAGI01000438)			↑Virulent		
α-actinin (HAGI01000323)				↑Virulent	
Actin-binding protein (HAGI01000491)				↑Virulent	Identified
Coronin (HAGI01002072)	↑Attenuated	↑Attenuated	↑Attenuated		Identified
Coronin (HAGI01001177)			↑Attenuated		Identified
Coronin (HAGI01001043)				↑Attenuated	Identified
α-actinin (HAGI01002778)	↑Attenuated				
Adenylyl cyclase-associated protein (HAGI01001260) PTM*	↑Attenuated				
Adenylyl cyclase-associated protein (HAGI01001260)					Identified
Peptidases					
Clan CD, family 13, asparaginyl endopeptidase-like c\ysteine peptidase (HAGI01000042)	↑Virulent	↑Virulent			
Clan CD, family 13, cparaginyl endopeptidase-like cysteine peptidase (HAGI01002270)				↑Virulent	
Clan MG, family M24, aminopeptidase P-like metallopeptidase (HAGI01000029) PTM*			↑Virulent		
Clan MG, family M24, aminopeptidase P-like metallopeptidase (HAGI01000029)					Identified
Clan CA, family C1, cathepsin L-like cysteine peptidase (HAGI01002930)				↑Virulent	
Clan CA, family C1, cathepsin L-like cysteine peptidase (HAGI01002235)				↑Virulent	Identified

(continued)

Table 4.1 (continued)

Protein	Conventional 2D experiment	2D-DIGE α experiment	2D-DIGE β experiment	SWATH proteome	Exoproteome
Clan CA, family C1, cathepsin L-like cysteine peptidase (HAGI01001652)				↑Virulent	
NlpC/P60 superfamily cysteine peptidase domain-containing protein (HAGI01001764)				↑Virulent	Identified
NlpC/P60 superfamily cysteine peptidase domain-containing protein (HAGI01002646)				↑Virulent	Identified
LysM peptidoglycan domain containing peptidase (HAGI01001637)				↑Virulent	Identified
Clan MG, family M24, aminopeptidase P-like metallopeptidase (HAGI01001269)	↑Attenuated				Identified
Clan MH, family M20, peptidase T-like metallopeptidase (HAGI01001053)		↑Attenuated			Identified
Adaptation to stress					
Endoplasmic reticulum HSP70 (HAGI01001009)	↑Virulent				Identified
Cytosolic HSP70 (HAGI01000755)	↑Virulent	↑Virulent			Identified
14-3-3 (HAGI01001863)	↑Virulent	↑Virulent			Identified
Chaperonin containing TCP-1 (CCT) subunit γ (HAGI01002092)	↑Virulent				
Cell division					
Cell division cycle protein 48 –like (Cdc48-like) (HAGI01000996) PTM*	↑Attenuated	↑Attenuated			
α-Tubulin 1 (HAGI01001402)	↑Attenuated				
β-Tubulin (HAGI01000230)	↑Attenuated				
Chromosome partitioning protein ATPase Mrp-like (HAGI01001677)				↑Attenuated	
Metabolic processes					
Glyceraldehyde-3-phosphate dehydrogenase-GAPDH (HAGI01000443) PTM*		↑Virulent	↑Virulent		

(continued)

Table 4.1 (continued)

Protein	Conventional 2D experiment	2D-DIGE α experiment	2D-DIGE β experiment	SWATH proteome	Exoproteome
Glyceraldehyde-3-phosphate dehydrogenase –GAPDH (HAGI01000443)		↑Attenuated	↑Attenuated		Identified
Phosphoribulokinase uiridine kinase family protein (HAGI01000159)	↑Attenuated				
Enolase (HAGI01001135) PTM*	↑Virulent				
Enolase (HAGI01001135)	↑Attenuated	↑Attenuated			Identified
Fructose biphosphate aldolase (FBAL) (HAGI01000595)			↑Virulent		Identified
Pyruvate phosphate dikinase (PPDK) (HAGI01001086)		↑Virulent			Identified
Iron-containing alcohol dehydrogenase (HAGI01000356)			↑Virulent		Identified
Phosphoenolpyruvate carboxykinase (PEPCK) PTM* (HAGI01001872)		↑Virulent	↑Virulent		
Phosphoenolpyruvate carboxykinase (PEPCK) (HAGI01001872)					Identified
Alcohol dehydrogenase iron-containing family protein (HAGI01000245)				↑Attenuated	
Phosphomannomutase (HAGI01001842)		↑Virulent			Identified
Phosphoglycerate mutase (HAGI01001716)			↑Virulent		Identified
Adenylate kinase family protein (HAGI01002611)				↑Virulent	
Dihydroorotate dehydrogenase family protein (HAGI01000099)	↑Attenuated				
Dihydroorotate dehydrogenase family protein (HAGI01000288)				↑Attenuated	Identified
Amidohydrolase family protein (HAGI01000242)	↑Attenuated			↑Attenuated	
Phosphofructokinase family protein (HAGI01001514)				↑Attenuated	
Phosphofructokinase family protein (HAGI01000564)				↑Attenuated	
Pyridoxal-phosphate dependent enzyme family protein (HAGI01000329)	↑Attenuated				
Fumarate hydratase (HAGI01000271)	↑Attenuated				Identified

(continued)

Table 4.1 (continued)

Protein	Conventional 2D experiment	2D-DIGE α experiment	2D-DIGE β experiment	SWATH proteome	Exoproteome
Iron hydrogenase (HAGI01001174)				↑Attenuated	Identified
Iron hydrogenase (HAGI01002119)				↑Attenuated	
Iron hydrogenase 64kDa (HAGI01000826)				↑Attenuated	
Iron hydrogenase 64kDa (HAGI01001036)				↑Attenuated	
NADH dehydrogenase 51kDa (HAGI01000290)				↑Attenuated	
AMP-binding enzyme family protein (HAGI01000098)	↑Attenuated				
Transketolase family protein (HAGI01001062)			↑Attenuated		Identified
Xantine dehydrogenase (HAGI01000700)				↑Attenuated	Identified
Xantine dehydrogenase (HAGI01000019)				↑Attenuated	Identified
Amidohydrolase family protein (HAGI01000242)				↑Attenuated	
Pyridoxal-phosphate dependent enzyme family protein (HAGI01000329)				↑Attenuated	
Aminotransferase classes I and II family protein (HAGI01002389)				↑Attenuated	
Malic enzyme AP-65/ adhesin AP-65 (HAGI01002133)				↑Attenuated	Identified
Transcription					
Polyadenylate-binding protein (HAGI01002937)	↑Attenuated				
Translation/protein synthesis					
Seryl-tRNA synthetase family protein (HAGI01000166)	↑Attenuated				
Glycyl-tRNA isoform A (HAGI01000007)	↑Attenuated				
Elongation factor 2 (HAGI01001476) PTM*	↑Attenuated				
Aspartyl-tRNA synthetase (HAGI01001075)	↑Attenuated				Identified

(continued)

Table 4.1 (continued)

Protein	Conventional 2D experiment	2D-DIGE α experiment	2D-DIGE β experiment	SWATH proteome	Exoproteome
60S acidic ribosomal protein P1 (HAGI01002520)				↑Virulent	
Protein transport/membrane trafficking					
Sec23/Sec24 trunk domain containing protein (HAGI01001023)	↑Attenuated				
C2 domain containing protein (HAGI01000999) PTM*	↑Attenuated				
C2 domain containing protein (HAGI01000999)					Identified
C2 domain containing protein (HAGI01001617)				↑Virulent	
SNARE domain-containing protein (HAGI01002630)				↑Virulent	
Pore-forming protein					
Surfactant protein B-like (HAGI01003015)				↑Virulent	
Protein degradation					
Ubiquitin-activating enzyme (E1) (HAGI01001756)	↑Attenuated				Identified
Signal transduction					
Rab family GTPase (Rab11clike) (HAGI01001951)		↑Virulent			
EF hand family protein (HAGI01001994)				↑Virulent	
G protein α subunit (HAGI01001440)					
Unknown function					
Hypothetical protein (HAGI01001468)		↑Virulent			
Hypothetical protein (HAGI01003301)				↑Virulent	
Hypothetical protein (HAGI01002296)				↑Virulent	
Hypothetical protein (HAGI01000587)				↑Virulent	
Hypothetical protein (HAGI01002399)				↑Virulent	
Hypothetical protein (HAGI01000628)				↑Attenuated	
Hypothetical protein (HAGI01002115)				↑Attenuated	

* PTM post-translational modification

Similarly to actin, the glyceraldehyde-3-phospate dehydrogenase (GAPDH) was detected as up-regulated in the virulent strain on multiple gel positions, which did not correspond to the predicted position according to calculated MW and pI (Table 4.1) (Monoyios et al. 2018). The possible contribution of modified or truncated actin and GAPDH to processes not related to their original function, such as binding of host plasminogen was speculated. An analogy was found in some other parasitic pathogens in which the localization on the parasites' surface and plasminogen-binding for these two ubiquitous cellular proteins was shown (Gonzalez-Miguel et al. 2015; Lama et al. 2009). The binding of plasminogen to the surface receptor molecules is an important step in its activation to plasmin, a potent serine peptidase that degrades extracellular matrix (Miles and Parmer 2013). By executing this function, plasmin assists the pathogens in dissemination to other organs and establishing a systemic infection (Gonzalez-Miguel et al. 2016). Considering that in the final stages of an infection, *H. meleagridis* spreads to numerous organs in the turkey (Singh et al. 2007; Tyzzer 1934), the ability to bind host plasminogen and its subsequent activation to plasmin might be part of the underlying mechanism.

In addition to actin and GAPDH, comparative proteome studies reported a number of other proteins up-regulated in either the virulent or attenuated *H. meleagridis* strain (Monoyios et al. 2017, 2018). The expression pattern generally mirrored the adaptation process of the parasite to the in vitro conditions, which was not surprising as in vitro propagated *H. meleagridis* was used for analyses. The majority of proteins up-regulated in the virulent strain were related to cellular stress management, whereas proteins up-regulated in the attenuated strain reflected increased metabolism, amoeboid morphology and the adaptation to in vitro propagation (Table 4.1). In addition, possible functional contribution of some proteins, especially legumain cysteine peptidase, towards virulence was hypothesized. In contrast to gel-based analyses, which often re-identified up-regulated proteins, gel-free shotgun proteomic analysis identified considerably different set of proteins as up-regulated in each strain (Table 4.1) (Monoyios et al. 2018). Overall, the tendency towards cellular stress management in the virulent strain versus metabolic proteins in the attenuated parasite remained, but many more proteins with potential role in virulence, such as cytolytic factors and proteins involved in cytoadherence mechanisms were identified as up-regulated in the virulent strain. Particularly, the detection of three clan CA, family C1 cathepsin L-like cysteine peptidases (CPs), as significantly up regulated in the virulent strain, is noteworthy (Table 4.1). Cathepsin L-like cysteine peptidases were also identified as the most represented group of peptidases in the *H. meleagridis* transcriptome (Mazumdar et al. 2017), and are brought in connection to the virulence mechanisms for several trichomonad parasites including *Trichomonas gallinae*, an aetiological agent of avian trichomonosis (Amin et al. 2012; Hirt et al. 2011). Whether *H. meleagridis* CPs are involved in virulence mechanisms at all and what are the underlying mechanisms still needs to be shown. Regardless of this, it seems that the virulent *H. meleagridis* strain has stronger intrinsic proteolytic activity than the attenuated parasite. This is particularly apparent in gel-based experiments where a higher abundance of low molecular weight proteins in the virulent as compared to the attenuated strain can be perceived.

A first study is also available focusing on the parasite's exoproteome, whose shotgun analysis identified 176 proteins in the extracellular milieu of *H. meleagridis* (Mazumdar et al. 2019). Only a minor part of identified proteins displayed a secretion signal motif, which indicated the dominance of non-classical secretion pathways in *H. meleagridis*. It was already hypothesized for *Giardia intestinalis*, a protozoan parasite mainly present in mammals; that alternate secretory pathways might dominate, as the majority of secreted proteins do not possess a premeditated export signal peptide motif (Ma'ayeh et al. 2017). Furthermore, proteins not guided by the secretion signal motif might undertake two unrelated functions in different cellular compartments and/or extracellularly, displaying their moonlighting feature. This, in turn, could aid to *H. meleagridis* survival and/or virulence by expanding the function of its' proteome. In that respect, the identification of actin and several glycolytic enzymes, such as enolase, fructose bisphosphate aldolase (FBA), GAPDH, and triosephosphate isomerase (TPI) is particularly interesting (Table 4.1). Prevailing traits of all these proteins are their alternate localization on parasite's surface and the ability to bind host plasminogen. Comparative proteome studies already suggested that post-translationally modified forms of these enzymes, which were found as up-regulated in virulent *H. meleagridis*, might be part of an alternative mechanism (moonlighting) unrelated to their original functions (Monoyios et al. 2017, 2018).

Aside from *H. meleagridis* specific proteins, proteome and exoproteome analyses identified a number of bacterial proteins with differential up-regulation depending on the phenotype of the co-cultivating parasite (Mazumdar et al. 2019; Monoyios et al. 2017, 2018). Remarkably, in the exoproteomic study the majority of differences were of *E. coli* origin, whereas variations in up-regulation of *H. meleagridis* proteins were limited to two proteins in the virulent strain. The identification of some *E. coli* proteins was expected due to monoxenic cultivation of *H. meleagridis*, but a considerable diversity of their abundance which could be linked to the phenotype of the co-cultivating parasite was surprising and intriguing. Despite that a fresh subculture of *E. coli* was added at every passage and inoculation of an individual experiment, results demonstrated that even a relatively short period of co-cultivation with the parasite was enough to induce a differential gene expression. All this indicates that the relation between *H. meleagridis* and bacteria is complex and of more mutualistic nature than previously recognized (Bilic and Hess 2020).

Conclusions and Future Aspects

Over the past decade molecular data on *H. meleagridis* have been accumulating, especially recent omics studies contributed extensively to this field. With this, proteins (or their coding sequences) involved in metabolism (carbohydrate, amino acid, nucleotide), energy production (hydrogenosomal proteins), ribosomal biogenesis, oxidative stress, degradome, cell signalling and cytoskeleton formation and function have been identified. Based on the information from homologous protozoa, proteins potentially involved in the virulence of *H. meleagridis*, such as adhesins, cysteine peptidases and pore-forming toxins were described. Comparative proteome analyses primarily reflected the level of adaption to the in vitro growth, although for some proteins a possible role in the virulence of the parasite was suggested. The exoproteome analysis supplied the information on the *H. meleagridis* extracellular milieu indicating its' active microenvironment. Aside to information on *H. meleagridis* proteins, quantitative proteomics supplied the first molecular support for the interdependence and complex interactions between *H. meleagridis* and bacteria. It is undeniable that acquired data opened new avenues in the research on this important poultry parasite, and these should be pursued. However, despite its high significance for future research, a fundamental part for molecular research, the complete genomic sequence of the parasite, is not yet publicly available. Apart from being a superior database to the existing one, it will facilitate an array of experiments involving genome manipulation of the parasite, an essential approach to tackle functional dissection of genes involved in virulence. Determining the genomic sequences from the clonal monoxenic virulent and attenuated strain used in the studies mentioned above should be advantageous to elucidate the genetic basis of attenuation. Given the fact that the virulent strain originated from a single *H. meleagridis* cell and that its' prolonged in vitro cultivation resulted in the attenuation; both strains are genetically almost identical and represent an ideal material to pursue aforementioned investigations.

Acknowledgements The authors thank all colleagues from the Clinic for Poultry and Fish Medicine and Christian Doppler Laboratory for Innovative Poultry Vaccines (IPOV) at the Vetmeduni Vienna for their unceasing support. Furthermore, the authors want to thank the Austrian Science Fund (FWF) for the substantial funding of the research on *H. meleagridis* and histomonosis.

References

Amin, A., Nobauer, K., Patzl, M., Berger, E., Hess, M., & Bilic, I. (2012). Cysteine peptidases, secreted by *Trichomonas gallinae*, are involved in the cytopathogenic effects on a permanent chicken liver cell culture. *PLoS ONE, 7,* e37417.

Bilic, I. & Hess, M. (2020). Interplay between *Histomonas meleagridis* and bacteria: mutualistic or predator-prey?. *Trends Parasitol, 36*(3), 232–235. https://doi.org/10.1016/j.pt.2019.12.015. Epub 2020 Jan 22.

Bilic, I., Leberl, M., & Hess, M. (2009). Identification and molecular characterization of numerous *Histomonas meleagridis* proteins using a cDNA library. *Parasitology, 136,* 379–391.

Bilic, I., Jaskulska, B., Souillard, R., Liebhart, D., & Hess, M. (2014). Multi-locus typing of *Histomonas meleagridis* isolates demonstrates the existence of two different genotypes. *PLoS ONE, 9*(3), e92438.

Bleyen, N., De, G. K., De, G. J., & Goddeeris, B. M. (2007). Specific detection of *Histomonas meleagridis* in turkeys by a PCR assay with an internal amplification control. *Veterinary Parasitology, 143,* 206–213.

Cepicka, I., Hampl, V., & Kulda, J. (2010). Critical taxonomic revision of Parabasalids with description of one new genus and three new species. *Protist, 161,* 400–433.

Clark, S., & Kimminau, E. (2017). Critical review: Future control of blackhead disease (Histomoniasis) in Poultry. *Avian Diseases, 61,* 281–288.

Doll, J. P., & Franker, C. K. (1963). Experimental histomoniasis in gnotobiotic turkeys. I. Infection and histopathology of the bacteria-free host. *Journal of Parasitology, 49,* 411–414.

Ganas, P., Liebhart, D., Glosmann, M., Hess, C., & Hess, M. (2012). *Escherichia coli* strongly supports the growth of *Histomonas meleagridis*, in a monoxenic culture, without influence on its pathogenicity. *International Journal for Parasitology, 42,* 893–901.

Gerbod, D., Edgcomb, V. P., Noel, C., Zenner, L., Wintjens, R., Delgado-Viscogliosi, P., et al. (2001). Phylogenetic position of the trichomonad parasite of turkeys, *Histomonas meleagridis* (Smith) Tyzzer, inferred from small subunit rRNA sequence. *Journal of Eukaryotic Microbiology, 48,* 498–504.

Gonzalez-Miguel, J., Morchon, R., Siles-Lucas, M., & Simon, F. (2015). Fibrinolysis and proliferative endarteritis: two related processes in chronic infections? The model of the blood-borne pathogen Dirofilaria immitis. *PLoS ONE, 10,* e0124445.

Gonzalez-Miguel, J., Siles-Lucas, M., Kartashev, V., Morchon, R., & Simon, F. (2016). Plasmin in parasitic chronic infections: Friend or foe? *Trends in Parasitology, 32,* 325–335.

Grabensteiner, E., & Hess, M. (2006). PCR for the identification and differentiation of *Histomonas meleagridis*, *Tetratrichomonas gallinarum* and *Blastocystis* spp. *Veterinary Parasitology, 142,* 223–230.

Grafl, B., Weise, H., Le, B. J., Liebhart, D., Bilic, I., & Hess, M. (2015). Aberrant clinical appearance and pathomorphology noticed during an outbreak of Histomonosis indicates a different pathogenesis of *Histomonas meleagridis* genotype 2. *Avian Diseases, 59,* 452–458.

Gruber, J., Ganas, P., & Hess, M. (2017). Long-term in vitro cultivation of *Histomonas meleagridis* coincides with the dominance of a very distinct phenotype of the parasite exhibiting increased tenacity and improved cell yields. *Parasitology, 144,* 1253–1263.

Hafez, H. M., Hauck, R., Luschow, D., & McDougald, L. (2005). Comparison of the specificity and sensitivity of PCR, nested PCR, and real-time PCR for the diagnosis of histomoniasis. *Avian Diseases, 49,* 366–370.

Hauck, R., & Hafez, H. M. (2009). Partial sequence of the beta-tubulin of *Histomonas meleagridis* and the activity of benzimidazoles against *H. meleagridis* in vitro. *Parasitology Research, 104,* 1183–1189.

Hauck, R., & Hafez, H. M. (2010). Systematic position of *Histomonas meleagridis* based on four protein genes. *Journal of Parasitology, 96,* 396–400.

Hauck, R., Luschow, D., & Hafez, H. M. (2006). Detection of Histomonas meleagridis DNA in different organs after natural and experimental infections of meat turkeys. *Avian Diseases, 50,* 35–38.

Hernandez-Cuevas, N. A., Jhingan, G. D., Petropolis, D., Vargas, M., & Guillen, N. (2019). Acetylation is the most abundant actin modification in *Entamoeba histolytica* and modifications of actin's amino-terminal domain change cytoskeleton activities. *Cellular Microbiology, 21,* e12983.

Hess, M., Grabensteiner, E., & Liebhart, D. (2006a). Rapid transmission of the protozoan parasite *Histomonas meleagridis* in turkeys and specific pathogen free chickens following cloacal infection with a mono-eukaryotic culture. *Avian Pathol, 35,* 280–285.

Hess, M., Kolbe, T., Grabensteiner, E., & Prosl, H. (2006b). Clonal cultures of *Histomonas meleagridis*, *Tetratrichomonas gallinarum* and a *Blastocystis* sp. established through micromanipulation. *Parasitology, 133,* 547–554.

Hess, M., Liebhart, D., Grabensteiner, E., & Singh, A. (2008). Cloned *Histomonas meleagridis* passaged in vitro resulted in reduced pathogenicity and is capable of protecting turkeys from histomonosis. *Vaccine, 26,* 4187–4193.

Hess, M., Liebhart, D., Bilic, I., & Ganas, P. (2015). *Histomonas meleagridis*—New insights into an old pathogen. *Veterinary Parasitology, 208,* 67–76.

Hirt, R. P., de Miguel, N., Nakjang, S., Dessi, D., Liu, Y. C., Diaz, N., et al. (2011). *Trichomonas vaginalis* pathobiology new insights from the genome sequence. *Advances in Parasitology, 77,* 87–140.

Kang, S., Tice, A. K., Spiegel, F. W., Silberman, J. D., Panek, T., Cepicka, I., et al. (2017). Between a pod and a hard test: The deep evolution of amoebae. *Molecular Biology and Evolution, 34,* 2258–2270.

Klodnicki, M. E., McDougald, L. R., & Beckstead, R. B. (2013). A genomic analysis of *Histomonas meleagridis* through sequencing of a cDNA library. *J Parasitol, 99,* 264–269.

Lama, A., Kucknoor, A., Mundodi, V., & Alderete, J. F. (2009). Glyceraldehyde-3-phosphate dehydrogenase is a surface-associated, fibronectin-binding protein of *Trichomonas vaginalis*. *Infection and Immunity, 77,* 2703–2711.

Leberl, M., Hess, M., & Bilic, I. (2010). *Histomonas meleagridis* possesses three alpha-actinins immunogenic to its hosts. *Molecular and Biochemical Parasitology, 169,* 101–107.

Liebhart, D., Weissenbock, H., & Hess, M. (2006). In-situ hybridization for the detection and identification of *Histomonas meleagridis* in tissues. *Journal of comparative pathology, 135,* 237–242.

Liebhart, D., Ganas, P., Sulejmanovic, T., & Hess, M. (2017). Histomonosis in poultry: previous and current strategies for prevention and therapy. *Avian Pathology, 46,* 1–18.

Ma'ayeh, S. Y., Liu, J., Peirasmaki, D., Hornaeus, K., Bergstrom, L. S., Grabherr, M., et al. (2017). Characterization of the *Giardia intestinalis* secretome during interaction with human intestinal epithelial cells: The impact on host cells. *PLoS Neglected Tropical Diseases, 11,* e0006120.

Malik, S. B., Brochu, C. D., Bilic, I., Yuan, J., Hess, M., Logsdon, J. M., et al. (2011). Phylogeny of parasitic parabasalia and free-living relatives inferred from conventional markers versus *Rpb1*, a single-copy gene. *PLoS One, 6,* e20774.

Mantini, C., Dalia-Cornette, J., Noda, S., van der Heijden, H. M., Capron, M., Dei-Cas, E., et al. (2009). Molecular identification and phylogenetic relationships of trichomonad isolates of galliform birds inferred from nuclear small subunit rRNA gene sequences. *Parasitology Research, 106,* 163–170.

Mazet, M., Diogon, M., Alderete, J. F., Vivares, C. P., & Delbac, F. (2008). First molecular characterisation of hydrogenosomes in the protozoan parasite *Histomonas meleagridis*. *International Journal for Parasitology, 38,* 177–190.

Mazumdar, R., Endler, L., Monoyios, A., Hess, M., & Bilic, I. (2017). Establishment of a de novo reference transcriptome of *Histomonas meleagridis* reveals basic insights about biological functions and potential pathogenic mechanisms of the parasite. *Protist, 168,* 663–685.

Mazumdar, R., Nobauer, K., Hummel, K., Hess, M., & Bilic, I. (2019). Molecular characterization of *Histomonas meleagridis* exoproteome with emphasis on protease secretion and parasite-bacteria interaction. *PLoS ONE, 14,* e0212429.

McDougald, L. R. (2005). Blackhead disease (histomoniasis) in poultry: A critical review. *Avian Diseases, 49,* 462–476.

Miles, L. A., & Parmer, R. J. (2013). Plasminogen receptors: The first quarter century. *Seminars in Thrombosis and Hemostasis, 39,* 329–337.

Monoyios, A., Patzl, M., Schlosser, S., Hess, M., & Bilic, I. (2017). Unravelling the differences: comparative proteomic analysis of a clonal virulent and an attenuated *Histomonas meleagridis* strain. *International Journal for Parasitology*. https://doi.org/10.1016/j.ijpara.2017.08.017.

Monoyios, A., Hummel, K., Nobauer, K., Patzl, M., Schlosser, S., Hess, M., et al. (2018). An alliance of gel-based and gel-free proteomic techniques displays substantial insight into the proteome of a virulent and an attenuated *Histomonas meleagridis* Strain. *Front Cell Infect Microbiol, 8,* 407.

Pham, A. D., Mast, J., De Gussem, J. K., Mcdougald, L. R., & Goddeeris, B. M. (2013). Establishing mono-eukaryotic *Histomonas meleagridis* cultures from in vivo infection contaminated with *Tetratrichomonas gallinarum* and *Blastocystis* spp. *Parasitology, 140,* 1266–1274.

Pham, A. D., Mast, J., Magez, S., Goddeeris, B. M., & Carpentier, S. C. (2016). The enrichment of *Histomonas meleagridis* and its pathogen-specific protein analysis: A first step to shed light on its virulence. *Avian Diseases, 60,* 628–636.

Regmi, P. R., Shaw, A. L., Hungerford, L. L., Messenheimer, J. R., Zhou, T., Pillai, P., et al. (2016). Regulatory considerations for the approval of drugs against histomoniasis (blackhead disease) in Turkeys, chickens, and game birds in the United States. *Avian Diseases, 60,* 725–730.

Singh, A., Weissenbock, H. & Hess, M. (2007). *Histomonas meleagridis*: Immunohistochemical localization of parasitic cells in formalin-fixed, paraffin-embedded tissue sections of experimentally infected turkeys demonstrates the wide spread of the parasite in its host. *Exp Parasitol.*

Sulejmanovic, T., Grafl, B., Bilic, I., Jaskulska, B., & Hess, M. (2019a). PCR and serology confirm the infection of turkey hens and their resilience to histomonosis in mixed flocks following high mortalities in toms. *Parasit Vectors, 12,* 228.

Sulejmanovic, T., Turblin, V., Bilic, I., Jaskulska, B., & Hess, M. (2019b). Detection of *Histomonas meleagridis* DNA in dust samples obtained from apparently healthy meat turkey flocks without effect on performance. *Avian Pathology, 48,* 329–333.

Tyzzer, E. E. (1919). Developmental phases of protozoon of "blackhead" in turkeys. *The Journal of Medical Research, 40,* 1–30.

Tyzzer, E. E. (1920). The flagellate character and reclassification of the parasite producing "Blackhead" in turkeys- *Histomonas* (gen. nov.) *meleagridis* (Smith). *Journal of Parasitology, 6,* 124–131.

Tyzzer, E. E. (1934). Studies in histomoniasis, or blackhead infection, in chicken and the turkey. *Proceedings of the American Academy of Arts and Science, 69,* 189–264.

van der Heijden, H. M., Landman, W. J., Greve, S., & Peek, R. (2006). Genotyping of *Histomonas meleagridis* isolates based on Internal Transcribed Spacer-1 sequences. *Avian Pathol, 35,* 330–334.

Chapter 5
Exploring the Biology and Evolution of *Blastocystis* and Its Role in the Microbiome

Anastasios D. Tsaousis, Emma L. Betts, Abby McCain, Jamie M. Newton, Vasana Jinatham, and Eleni Gentekaki

Abstract *Blastocystis* is a microbial eukaryote, considered to be the most prevalent microbe in the human gut, colonizing approximately one billion individuals worldwide. *Blastocystis* is extremely genetically diverse with 17 distinct genetic subtypes found in birds and mammals. Although *Blastocystis* presence has been linked to intestinal disorders, its pathogenicity still remains controversial due to its high prevalence in asymptomatic carriers. *Blastocystis* can withstand fluctuations of oxygen in the gut and as a result harbors peculiar mitochondrion-related organelles (MROs). These are considered to be an intermediate form between a typical aerobic mitochondrion and an obligate anaerobic hydrogenosome. Genomic analysis has shown that 2.5% of *Blastocystis* genes have been laterally acquired from eukaryotes and prokaryotes. These acquired genes are associated with carbohydrate scavenging and metabolism, anaerobic amino acid and nitrogen metabolism, oxygen-stress resistance, and pH homeostasis. In addition, *Blastocystis* has genes associated with secretion that are potentially involved in infection, escaping host defense and even affect composition of the prokaryotic microbiome and inflammation of the gut. In this chapter, we provide an overview of the state-of-the-art *Blastocystis* knowledge, and present published data that can be used to understand the genomic adaptations of this microbial organism and its role within the microbiome of the hosts.

Keywords *Blastocystis* · Eukaryome · Genetic diversity · Microbiome · Mitochondrion-related organelles (MROs) · Pathogenicity · Prevalencel subtyping

---

A. D. Tsaousis (✉) · E. L. Betts · J. M. Newton
Laboratory of Molecular and Evolutionary Parasitology, RAPID Group, School of Biosciences, University of Kent, Canterbury, Kent, UK
e-mail: A.Tsaousis@kent.ac.uk; tsaousis.anastasios@gmail.com

A. McCain · V. Jinatham · E. Gentekaki (✉)
School of Science and Gut Microbiome Research Group, Mae Fah Luang University, Chiang Rai, Thailand
e-mail: gentekaki.ele@mfu.ac.th

© Springer Nature Switzerland AG 2020

N. Guillen (ed.), *Eukaryome Impact on Human Intestine Homeostasis and Mucosal Immunology*, https://doi.org/10.1007/978-3-030-44826-4_5

General Introduction

Blastocystis is a microbial eukaryote colonizing the gastrointestinal tract of a variety of hosts, including humans, artiodactyls, marsupials, perissodactyls, proboscideans and rodents, avian species, reptiles, fish and some insects (Boreham and Stenzel 1993; Stenzel and Boreham 1996). Alexeieff was the first to provide a detailed description of *Blastocystis* from a number of hosts including rats, chickens and reptiles (Alexeieff 1911). At that time, the organism was classified as yeast and named *Blastocystis enterocola*. A number of possible earlier accounts of *Blastocystis* dating as far back as the 1840 London cholera epidemic also exist (Zierdt 1991). Following its discovery in humans, Brumpt (1912) adopted the name *Blastocystis hominis*. The organism is now classified within the stramenopiles, a diverse group of eukaryotes that includes diatoms and oomycetes (Silberman et al. 1996). Specifically, in phylogenetic trees *Blastocystis* groups with opalinids, *Karotomorpha* and *Proteromonas*, all of which reside in the gastrointestinal tracts of metazoans. Though stramenopiles have at some stage of their life cycles a flagellum *Blastocystis* has lost its flagellar apparatus along with any related protein coding genes (Fig. 5.1) (Gentekaki et al. 2017). Four morphologically distinct stages of *Blastocystis* have been identified so far, including the vacuolar, granular, cyst and amoeboid forms (Tan and Suresh 2006; Tan et al. 2010; Tan 2008; Clark and Stensvold 2016). Due to its very small size (being as small as 5 μm and averaging 8–12 μm), immobility and lack of descriptive morphological

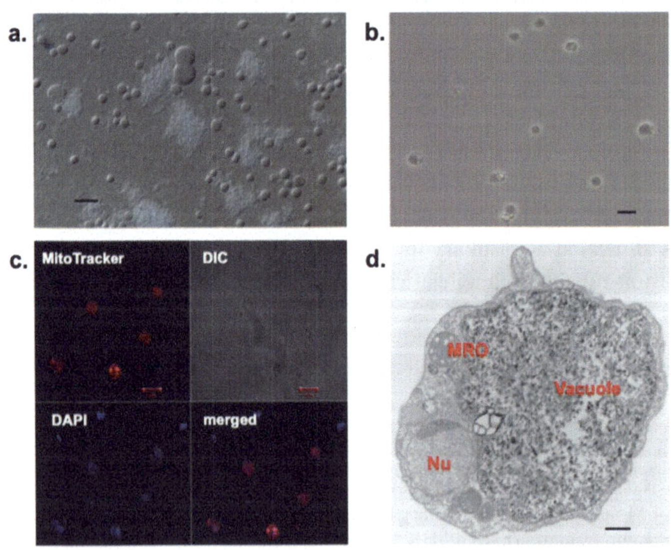

Fig. 5.1 *Blastocystis* cells. **a** Light microscopy of a *Blastocystis* xenic culture in Jones media. Scale bar: 25 μm. **b** Light microscopy of a neutral-red stained *Blastocystis* axenic culture. Scale bar: 20 μm. **c** Staining of *Blastocystis* mitochondrion-related organelles with MitoTracker red, DAPI staining of the nucleus and MROs and a differential interference contrast (DIC) image of *Blastocystis* cells. Scale bar: 10 μm. **d** Transmission electron microscopy picture shown a *Blastocystis* cell with its nucleus (Nu), various MROs and a large vacuole. Scale bar: 500 nm

characters, *Blastocystis* has been often overlooked or mistaken for cell debris when observed using microscopy.

The *Blastocystis* life cycle still requires some elucidation. However, it is widely understood that *Blastocystis* enters the host in the metabolically inactive dormant cyst form via the faecal oral route (Tan 2004). The faecal oral route involves a food or waterborne source being the primary cause of the spread of *Blastocystis*. There is, however, evidence for zoonotic transmission as well. Regardless of the source, once transmission is complete excystation occurs in the large intestine into the vacuolar form and can then further morph into the granular or ameboid form. At this stage *Blastocystis* can replicate by binary fission and can start to proliferate. The organism encysts in the large intestine. The cyst further develops in the faeces losing the fibrillar layer it initially possesses. Once it is released into the faeces, the cyst is free to enter another host.

Blastocystis is one of the most commonly encountered protists in the human gut with an estimated prevalence of one billion (Stensvold and Clark 2016; Clark et al. 2013). Its colonization rate ranges from 20% in Europe (Bart et al. 2013) to over 30% in some developing countries (Alfellani et al. 2013a; Ramirez et al. 2014), with one study showing incidence of 100% in a group of children in Senegal (El Safadi et al. 2014). Numerous animal studies suggest that prevalence of *Blastocystis* is higher in animals (Alfellani et al. 2013c; Betts et al. 2018; Cian et al. 2017). Because asymptomatic carriage is frequent and since presence of other intestinal parasites is not excluded definitively in patients with gastrointestinal symptoms, questions have been raised regarding *Blastocystis* biology, pathogenicity, transmission and possible impacts on the host and its gut microbiota.

Prevalence, Diversity and Biogeography

The advent of molecular methods has revealed unexpected genetic diversity that does not correspond to the morphological stasis of *Blastocystis*; based on SSU rRNA sequences, 17 known subtypes (STs) that colonize mammals and birds (Fig. 5.2) along with various isolates from ectothermic hosts have been identified (Noel et al. 2005; Stensvold and Clark 2016; Yoshikawa et al. 2016). These are all considered separate species. It is highly probable that the genetic diversity of *Blastocystis* is greater, but as yet uncovered due to sampling bias towards specific hosts. Specifically, sampling efforts have centered on humans and animals of importance to us (pets, zoo animals and livestock), while insects, arthropods and other ectotherms remain only sparsely sampled (Alfellani et al. 2013c; Cian et al. 2017; Masuda et al. 2018; Betts et al. 2018; Paulos et al. 2018). Regardless, differences observed in the SSU rRNA gene are also reflected in the genomes of the various *Blastocystis* subtypes. Specifically, not only is the genetic distance between subtypes high, but those also differ in their GC percent content and gene complement (Gentekaki et al. 2017).

At first look, *Blastocystis* does not appear to be host specific (Fig. 5.2). Subtypes 1–9 colonize humans, but these have also been found in several other hosts

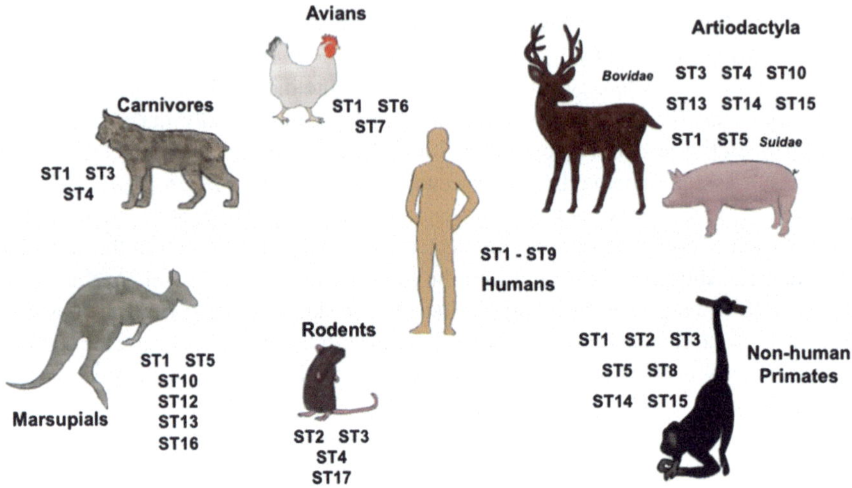

Fig. 5.2 *Blastocystis* subtypes with various host specificities

(Stensvold and Clark 2016). The exception seems to be ST9, which has only been found in humans. ST10 to ST17 have only been found in animals with the exception of ST12, which has been found in humans from South America (Ramirez et al. 2016). Nonetheless, when considering hosts at higher taxonomic levels, then some degree of specificity exists (Alfellani et al. 2013b). For instance, ST10 and ST14 are hallmark subtypes of artiodactyls (Cian et al. 2017; Betts et al. 2018). Notably, sequences from insects and some other ectothermic hosts have yet to be isolated from endotherms. Thus temperature might pose a constraint on host specificity of this organism as it has been speculated for other protists (Jinatham et al. 2019).

The various subtypes colonizing humans are distributed globally, however some patterns have emerged. Subtype 3 is the most abundant and also the most widespread worldwide, while ST4 is mostly restricted in Europe potentially suggesting its recent origin (Stensvold and Clark 2016) (Fig. 5.3). Nonetheless, data suggest that STs present in fowl (ST6, ST7) and porcine hosts (ST5) are more often present in inhabitants from rural regions. Frequent contact between animals and humans in these regions likely leads to transmission of these subtypes between hosts. In contrast, such contact is limited in urban centers.

While many subtyping studies from developed countries exist, those lag behind in low- and middle-income countries (LMICs) (Noradilah et al. 2017; Ramirez et al. 2016; Thathaisong et al. 2013; Yowang et al. 2018). Typically, diagnosis of *Blastocystis* in LMICs is carried out microscopically and is part of large-scale routine parasitology surveys. To our knowledge, currently there are less than 40 reports investigating prevalence and subtyping of *Blastocystis* in LMICs, many of which are focusing on immunocompromised patients or patients with gastrointestinal symptoms (Fig. 5.3). Based on all these studies, the overall prevalence of *Blastocystis* in LMICs of Africa, South America, South East Asia, rest of Asia, and the Middle

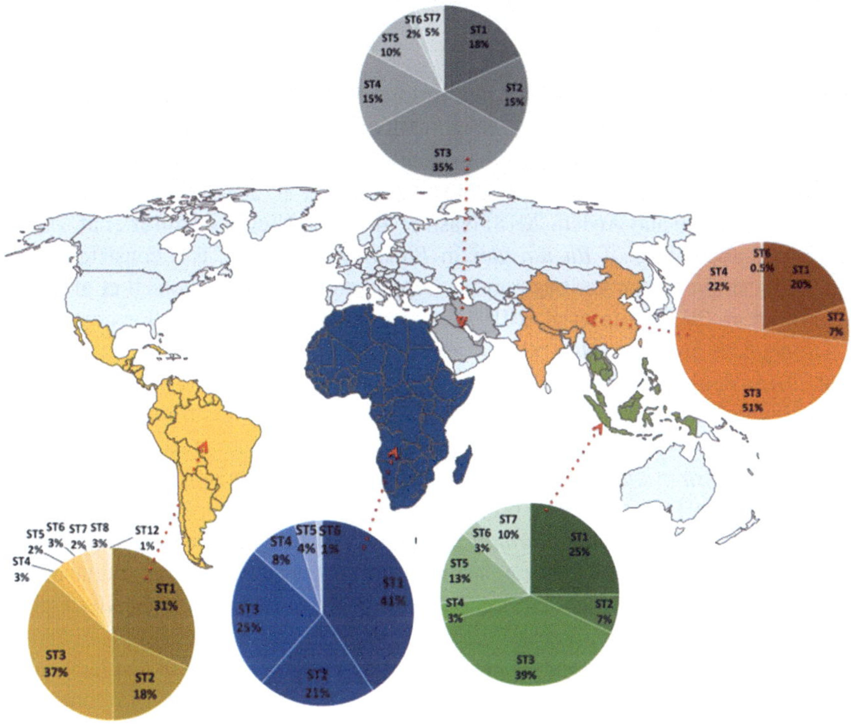

Fig. 5.3 Prevalence and biogeographical distribution of human *Blastocystis* subtypes in low and middle income countries

East averages at around 30% (Fig. 5.3). Matching the global trend, ST3 is the predominant subtype distributed in LMICs, excluding countries in Africa, where ST1 predominates (Di Cristanziano et al. 2019; Abdulsalam et al. 2013). In South America, prevalence rates of ST3 and ST1 are very similar with ST3 being marginally more dominant (Villamizar et al. 2019; Forsell et al. 2017; Oliveira-Arbex et al. 2018). Subtypes ST5 to ST7 are commonly found in all LMICs often in association with fowl and pigs highlighting their zoonotic potential.

Typically, humans seem to be colonized by a single subtype with some exceptions (Meloni et al. 2011; Whipps et al. 2010). Whether this is a methodological issue (direct sequencing versus cloning) or if indeed a single subtype dominates, remains unclear. Though similar information on animals is just emerging, it seems that mixed colonization is more common than in humans (AbuOdeh et al. 2019; Cian et al. 2017; Betts et al. 2018). This information along with ST-specific functional repertoires implies distinct interactions in the human gut ecosystem, an area of study that remains little explored.

Blastocystis and Microbiome

Several recent studies have found that presence of *Blastocystis* is strongly correlated to specific microbial profiles. Specifically, *Blastocystis* carriage in individuals without gastrointestinal symptoms has been associated with higher bacterial richness and diversity (Andersen et al. 2015; Audebert et al. 2016; Forsell et al. 2017; Laforest-Lapointe and Arrieta 2018; Nash et al. 2017; Nieves-Ramirez et al. 2018). The lower abundance of *Bacteroides* in *Blastocystis* carriers is a consistent finding across studies from various regions (Andersen et al. 2015; Forsell et al. 2017; Beghini et al. 2017; Tito et al. 2019). In contrast, positive associations exist with *Ruminococcus* and other clostridia, *Prevotella*, and *Methanobrevibacter* (Andersen et al. 2015; Beghini et al. 2017; Nash et al. 2017). The negative association of *Blastocystis* with *Bacteroides* has been attributed to the latter not contributing enough to a "*Blastocystis*-favorable" environment (Stensvold and van der Giezen 2018). When subtype of *Blastocystis* was taken into account in microbiome studies differential associations of STs with specific prokaryotic taxa were noted. Specifically, ST3 was negatively correlated with *Akkermansia*, while ST4 had the opposite relationship (Tito et al. 2019). *Akkermansia* is a mucin utilising bacterium that is considered beneficial. Abundant evidence suggests that this bacterium is indicative of good intestinal health and has protective action against metabolic disorders (Cani and de Vos 2017; Dao et al. 2019; Hanninen et al. 2018). Collectively these results suggest that subtype characterization is essential for accurately determining the relationship between *Blastocystis*, microbiota profiles, and host health.

An important caveat that needs to be taken into account is causality of the observed microbiota changes. Whether it is *Blastocystis* that alters the host prokaryotic microbiota, or another reason (e.g. low-grade inflammation) is as yet unknown (Nieves-Ramirez et al. 2018). Differentiating between the two will greatly advance our understanding of gut ecology and roles of individual component taxa. A good starting point is to undertake studies similar to those of Yason et al. (2019). The authors performed in vitro competitive assays of a laboratory grown strain of ST7 and individual bacteria and extended experiments to cell lines and mice (Yason et al. 2019). The authors consistently showed decrease of *Bifidobacterium longum*, which was attributed to oxidative stress caused by either the host immune system or metabolic activity of *Blastocystis* (Yason et al. 2019). Thus the role of *Blastocystis* in the gut remains unclear hinting at complex interactions that have yet to be defined. A major challenge in accurately defining the role of *Blastocystis* in the gut lies in finding ways to use native microbial flora, which better represents the gut ecosystem rather than lab grown strains.

In that vein, in-depth understanding of *Blastocystis* biology is essential to disentangle its role in the gut and interactions with microbiota.

Blastocystis *Biology*

Currently, the genomes of nine subtypes are available in public databases, but only a few have been investigated in detail. Comparative genomics of as many subtypes as possible would provide considerable insights into their biology and pathogenicity. A recent study showed that up to ~2.5% of *Blastocystis* genes have been laterally acquired from prokaryotic and eukaryotic organisms (Eme et al. 2017). These genes relate to infection and evasion of host defenses, anaerobic amino acid and nitrogen metabolism, oxygen-stress resistance and carbohydrate scavenging, and have likely played key roles in *Blastocystis* adapting to the gut ecosystem (Eme et al. 2017). Of those laterally acquired genes, only one has been cellularly localized and biochemically characterized. The *SufCB* gene encodes for a protein that is involved in the assembly of Fe/S clusters in *Blastocystis.* Similarly to the archaeal taxon Methanomicrobiales, the SUF system of *Blastocystis* has a *SufCB* fusion gene (Tsaousis et al. 2012). In phylogenetic trees, the *Blastocystis* and archaeal homologues cluster together into a strongly supported clade, suggesting a lateral gene transfer event from the Methanomicrobiales. SufCB is also found in the genomes of all *Blastocystis* subtypes, as well as *Proteromonas lacertae*, a close relative of *Blastocystis.* The *Blastocystis* SufCB protein is cytosolic, binds [4Fe-4S] clusters, has ATPase activity, and is overexpressed under conditions of oxygen stress (Tsaousis et al. 2012). This mirrors findings in various bacteria, where the SUF machinery is also overexpressed under oxygen stress and iron depletion (Tsaousis et al. 2014; Mettert et al. 2008; Rangachari et al. 2002).

The alternative oxidase (AOX) is another protein that has played a role to *Blastocystis* dealing with variable oxygen concentrations in the gut. The AOX was shown to localize in the mitochondrion-related organelles (MROs) of *Blastocystis* (Tsaousis et al. 2018). The localization and functional characterization of this protein in *Blastocystis* suggest that the cells themselves do respire oxygen (Tsaousis et al. 2018), questioning previous hypotheses about the "obligate" anaerobic nature of this organism. By having both SufCB and AOX proteins, *Blastocystis* might be able to quickly buffer transient fluctuations of oxygen in the gut. This could also explain the absence of the organism in patients with established inflammatory bowel syndrome (IBS), whereby the gut environment is highly permissive to oxygen (Ramirez et al. 2014; Tito et al. 2019). The peculiarities of *Blastocystis* MROs likely contribute to the organism's ability to survive in the extreme environment of the gut.

Blastocystis *MROs*

In addition to the AOX, *Blastocystis* MROs contain many peculiar and potentially recurrent functions that reflect its unique lifestyle (Fig. 5.4). It is worth mentioning that the MRO is currently the only organelle of *Blastocystis* that has been highly characterized (Makiuchi and Nozaki 2014). These MROs combine metabolic properties

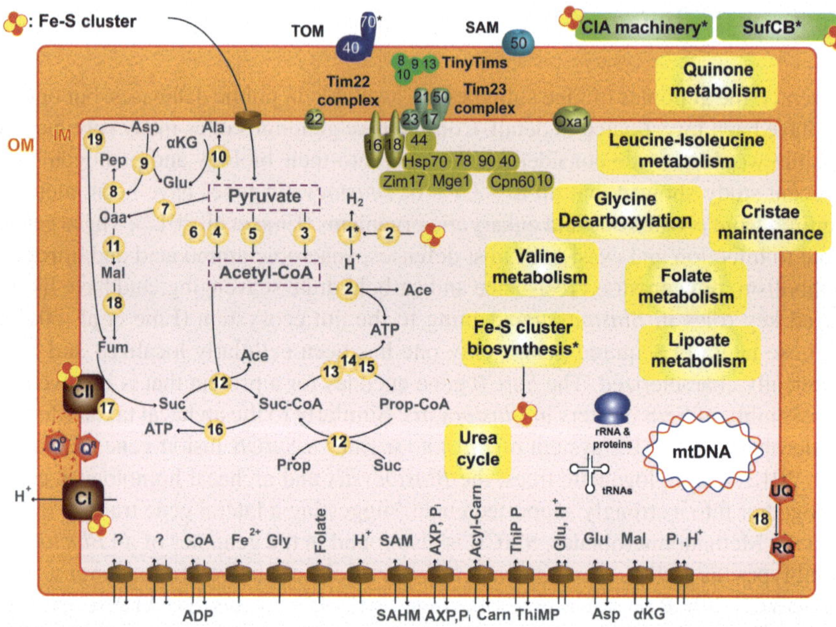

Fig. 5.4 Proposed metabolic map of *Blastocystis* mitochondrion-related organelles (MROs) based on the genome predictions [figure modified from (Gentekaki et al. 2017; Tsaousis et al. 2019)]. Various metabolic features of the *Blastocystis* MRO's roles in energy generation, Fe/S cluster assembly, amino acid and lipid metabolism. Numbers associated with protein descriptions are outlined below: (1) FeFe-hydrogenase, Flavodoxin; (2) iron-only hydrogenase maturation rSAM protein HydE; (3) por, nifJ; pyruvate-ferredoxin/flavodoxin oxidoreductase; (4) dihydrolipoamide succinyltransferase; (5) PDK2_3_4; pyruvate dehydrogenase kinase 2/3/4; (6) 2-oxoglutarate dehydrogenase; (7) Pyruvate carboxylase, alpha subunit; (8) phosphoenolpyruvate carboxykinase (ATP; (9) Aspartate aminotransferase; (10) Alanine aminotransferase; (11) Malate dehydrogenase; (12) Acetate: Succinate CoA transferase; (13) Methylmalonyl-CoA mutase; (14) Methylmalonyl-CoA epimerase; (15) Propionyl-CoA carboxylase alpha subunit; (16) Succinyl-CoA Synthetase; (17) Succinate dehydrogenase subunit 5; (18) Rhodoquinone Biosynthesis enzyme RquA; (19) Aspartate ammonia lyase. Proteins/pathways labelled with an asterix (*) were shown to be localised in *Blastocystis* MROs using immunofluorescence microscopy. Standard amino-acid abbreviations are used: Ace, acetate; ACP, acyl carrier protein; aKG, alpha-ketoglutarate; BCD, branched chain amino acid degradation; CI, Complex I; CII, Complex II; Carn, Carnitine; CDP-DAG, cytidine diphosphate diacylglycerol; CIA machinery: Cytosolic Iron/Sulphur cluster Assembly machinery; CL, cardiolipin; DHAP, dihydroxyacetone phosphate; DHOro, dihydroorotate; Fd, Ferredoxin; Fum, fumarate; Gly3P, glycerol-3-phosphate; Mal, malate; MMC, methyl-malonyl-CoA; Nd(p), NAD(P); mtDNA, mitochondrial DNA; Oaa, oxaloacetate; Oro, orotate; PA, phosphatidic acid; PE phosphatidylethanolamine; Pep, phosphoenol pyruvate; PI, phosphatidylinositol; Prop, propionate; PS, phosphatidylserine; QO/R, quinone/quinol, oxidized or reduced; RQ, rhodoquinone; SAHC, S-adenosylhomocysteine; SAM, S-adenosylmethionine; Suc, succinate; SUF; Sulphur mobilization; THF, tetrahydrafolate; ThiMP, thiamine monophosphate; ThiPP, thiamine pyrophosphate; UQ, ubiquinone

of aerobic mitochondria, anaerobic mitochondria and hydrogenosomes (Stechmann et al. 2008) thus blurring the boundaries between all these organelles, as previously defined (Embley and Martin 2006). In silico predictions have demonstrated that *Blastocystis* organelles do harbor a mitochondrial genome, have elements of both aerobic and anaerobic metabolism having an incomplete tricarboxylic acid (TCA) cycle, pathways of amino acid metabolism, Fe/S cluster biosynthesis machinery, mitochondrial protein import, urea cycle, transporters for exchange of metabolites and quinone metabolism (Fig. 5.4) (Stechmann et al. 2008; Gentekaki et al. 2017). So far, candidates from the following pathways have been localized in *Blastocystis* MROs: the mitochondrial protein import (Tom70) (Tsaousis et al. 2011), components of the Fe/S cluster biosynthesis machinery (Tsaousis et al. 2012; Long et al. 2011; Tsaousis et al. 2014), the anaerobic metabolism (FeFe-Hydrogenase) (Stechmann et al. 2008), the TCA cycle [succinyl-CoA synthesase (SCS)] (Hamblin et al. 2008) and parts of the glycolytic pathway (Río Bártulos et al. 2018) (Fig. 5.4).

Biochemically, a large amount of the current knowledge on *Blastocystis* metabolism has been resolved by functional characterization of its MROs (Lantsman et al. 2008). An unusual feature of its metabolism (Fig. 5.4) is a TCA cycle that runs in reverse to the canonical mitochondrial cycle and with only half of the pathway being present. This pathway terminates with the reduction of fumarate to succinate with fumarate acquiring electrons from fumarate reductase, which is ligated to the membrane bound electron transporter complex II (Tsaousis et al. 2019). The electron transport chain (ETC) also works in reverse to the canonical mitochondrial chain with complex III, complex IV and ATP synthase being absent. Complex I acquires its electrons from NADH (Stechmann et al. 2008), which are then transported to complex II by the reduction of ubiquinone (Q) to ubiquinol (QH_2). QH_2 is oxidised back to Q at complex II and the cycle keeps repeating (Stechmann et al. 2008; Gentekaki et al. 2017). There is no aerobic synthesis of ATP, since ATP synthase is absent, thus the ETC could only be responsible for the production of the proton gradient in the organelle (Denoeud et al. 2011).

Since no chemiosmotic ATP synthesis takes place in *Blastocystis,* ATP is synthesized anaerobically (Denoeud et al. 2011; Stechmann et al. 2008). Though the ATP synthesizing mechanisms of *Blastocystis* are common in anaerobic protists (Muller et al. 2012), its pyruvate metabolism is quite unique (see Fig. 5.4) in that there are three enzymes, which convert pyruvate to acetyl-CoA: pyruvate: ferrodoxin oxidoreductase (PFO) and pyruvate: $NADP^+$ oxidoreductase (PNO), both of which are common in anaerobic protists and pyruvate dehydrogenase (PDH), which is almost universally present in aerobic organisms and canonical mitochondria (Gentekaki et al. 2017; Tsaousis et al. 2019). Although PFO is present in *Blastocystis*, activity of this enzyme has yet to be detected (Eme et al. 2017).

Nonetheless, most biochemical pathways in *Blastocystis* mitochondria remain uncharacterized. An ‘omics-based approach, such as metabolomics, could be applied to map the organism's metabolic pathways allowing identification of the metabolites produced by the organism. A metabolomics NMR study has already been conducted to analyze the metabolism of the protozoan parasite *Giardia lamblia* (Vermathen et al. 2018), which would be a practical tool to explore *Blastocystis* metabolism as well. Previous studies have suggested that *Blastocystis* has a significant impact on the gut microbiome, thus it is not unlikely that its metabolites might underpin specific interactions with the microbiota (Andersen et al. 2015; Hanninen et al. 2018; Yason et al. 2019). Therefore, metabolomic-based studies of *Blastocystis* positive fecal samples, as well as, in vitro metabolomics will contribute significant information.

Conclusion

Blastocystis is a microbial eukaryote that has attracted considerable research interest in the last decade. Despite this, we have only scratched the surface on the role of *Blastocystis* in the complex and extreme habitat of the gut. Significantly, the question of its pathogenicity, which was raised decades ago, remains unanswered. While hypotheses have been brought forth about differential pathogenicity of subtypes, this has not been shown unequivocally. Future studies should include all subtypes of *Blastocystis* and combine in silico analysis, in vitro culturomics and classical cell biology and biochemistry approaches. New tools that will characterize *Blastocystis* and interactions within its native environment (the gut) are urgently needed. Only then can valid conclusions be drawn about the pathogenicity of the different subtypes.

Acknowledgements This research was supported by BBSRC research grant (BB/M009971/1) to Dr. Anastasios D. Tsaousis. Dr. Eleni Gentekaki is supported by The Thailand Research Fund (RSA6080048). We would like to thank the University of Kent for sponsoring ADT's trip to Thailand under a GCRF grant. We are grateful for the support of the Gut Microbiome Research Group to VJ and AM. ELB was supported by a Ph.D. studentship from the School of Biosciences at the University of Kent and JMN is supported by a Kent Health Ph.D. scholarship.

References

Abdulsalam, A. M., Ithoi, I., Al-Mekhlafi, H., et al. (2013). Subtype distribution of *Blastocystis* isolates in Sebha, Libya. *PLOS ONE, 8*(12), e84372.

AbuOdeh, R., Ezzedine, S., Madkour, M., et al. (2019). Molecular subtyping of *Blastocystis* from diverse animals in the United Arab Emirates. *Protist, 170*(5), 125679. https://doi.org/10.1016/j.protis.2019.125679.

Alexeieff, A. (1911). Sur la nature des formations dites "kystes de *Trichomonas intestinalis'*. Cr Soc Biol.

Alfellani, M. A., Jacob, A. S., Perea, N. O., et al. (2013a). Diversity and distribution of *Blastocystis* sp. subtypes in non-human primates. *Parasitology, 140*(8), 966–971. https://doi.org/10.1017/s0031182013000255.

Alfellani, M. A., Stensvold, C. R., Vidal-Lapiedra, A., et al. (2013b). Variable geographic distribution of *Blastocystis* subtypes and its potential implications. *Acta Trop, 126*(1), 11–18. https://doi.org/10.1016/j.actatropica.2012.12.011.

Alfellani, M. A., Taner-Mulla, D., Jacob, A. S., et al. (2013c). Genetic diversity of *Blastocystis* in livestock and zoo animals. *Protist, 164*(4), 497–509. https://doi.org/10.1016/j.protis.2013.05.003.

Andersen, L. O., Bonde, I., Nielsen, H. B. et al. (2015). A retrospective metagenomics approach to studying *Blastocystis*. *FEMS Microbiology Ecology 91*(7). Epub Jun 29, 2015. https://doi.org/10.1093/femsec/fiv072.

Audebert, C., Even, G., Cian, A., et al. (2016). Colonization with the enteric protozoa *Blastocystis* is associated with increased diversity of human gut bacterial microbiota. *Sci Rep, 6,* 25255. https://doi.org/10.1038/srep25255.

Bart, A., Wentink-Bonnema, E., Gilis, H., et al. (2013). Diagnosis and subtype analysis of *Blastocystis* sp. in 442 patients in a hospital setting in the Netherlands. *BMC Infectious Diseases, 13*(1), 389. https://doi.org/10.1186/1471-2334-13-389.

Beghini, F., Pasolli, E., Truong, T. D., et al. (2017). Large-scale comparative metagenomics of *Blastocystis*, a common member of the human gut microbiome. *ISME J, 11*(12), 2848–2863. https://doi.org/10.1038/ismej.2017.139.

Betts, E. L., Gentekaki, E., Thomasz, A., et al. (2018). Genetic diversity of *Blastocystis* in non-primate animals. *Parasitology, 145*(9), 1228–1234. https://doi.org/10.1017/s0031182017002347.

Boreham, P. F., & Stenzel, D. J. (1993). *Blastocystis* in humans and animals: Morphology, biology, and epizootiology. *Advances in Parasitology, 32,* 1–70.

Cani, P. D., & de Vos, W. M. (2017). Next-generation beneficial microbes: The case of *Akkermansia muciniphila*. *Frontiers in Microbiology, 8,* 1765. https://doi.org/10.3389/fmicb.2017.01765.

Cian, A., El Safadi, D., Osman, M., et al. (2017). Molecular epidemiology of *Blastocystis* sp. in various animal groups from two French zoos and evaluation of potential zoonotic risk. *PLoS One, 12*(1), e0169659. https://doi.org/10.1371/journal.pone.0169659.

Clark, C. G., & Stensvold, C. R. (2016). *Blastocystis*: Isolation, xenic cultivation, and cryopreservation. *Curr Protoc Microbiol, 43,* 20A.1.1–20A.1.8. https://doi.org/10.1002/cpmc.18.

Clark, C. G., van der Giezen, M., Alfellani, M. A., et al. (2013). Recent developments in *Blastocystis* research. *Advances in Parasitology, 82,* 1–32. https://doi.org/10.1016/b978-0-12-407706-5.00001-0.

Dao, M. C., Belda, E., Prifti, E., et al. (2019). *Akkermansia muciniphila* abundance is lower in severe obesity, but its increased level after bariatric surgery is not associated with metabolic health improvement. *Am J Physiol Endocrinol Metab, 317*(3), E446–E459. https://doi.org/10.1152/ajpendo.00140.2019.

Denoeud, F., Roussel, M., Noel, B., et al. (2011). Genome sequence of the stramenopile *Blastocystis*, a human anaerobic parasite. *Genome Biol, 12*(3), R29-2011-12-3-r29. https://doi.org/10.1186/gb-2011-12-3-r29.

Di Cristanziano, V., Alfonso, D. Ä., Berrilli, F., et al. (2019). Correction: Lower prevalence of *Blastocystis* sp. infections in HIV positive compared to HIV negative adults in Ghana. *PLOS ONE, 14*(10), e0224000.

El Safadi, D., Gaayeb, L., Meloni, D., et al. (2014). Children of Senegal River Basin show the highest prevalence of *Blastocystis* sp. ever observed worldwide. *BMC Infect Dis, 14,* 164-2334-14-164. https://doi.org/10.1186/1471-2334-14-164.

Embley, T. M., Martin, W. (2006). Eukaryotic evolution, changes and challenges. *Nature 440*(7084):623–630. nature04546 [pii].

Eme, L., Gentekaki, E., Curtis, B. et al. (2017). Lateral gene transfer in the adaptation of the anaerobic parasite *Blastocystis* to the gut. *Current Biology* 27(6):807–820. S0960-9822(17)30138-0 [pii].

Forsell, J., Bengtsson-Palme, J., Angelin, M., et al. (2017). The relation between *Blastocystis* and the intestinal microbiota in Swedish travellers. *BMC in Microbiology, 17*(1), 231-017-1139-7. https://doi.org/10.1186/s12866-017-1139-7.

Gentekaki, E., Curtis, B. A., Stairs, C. W., et al. (2017). Extreme genome diversity in the hyper-prevalent parasitic eukaryote *Blastocystis*. *PLoS Biol, 15*(9), e2003769. https://doi.org/10.1371/journal.pbio.2003769.

Hamblin, K., Standley, D. M., Rogers, M. B., et al. (2008). Localization and nucleotide specificity of *Blastocystis* succinyl-CoA synthetase. *Mol Microbiol, 68*(6), 1395–1405. https://doi.org/10.1111/j.1365-2958.2008.06228.x.

Hanninen, A., Toivonen, R., Poysti, S., et al. (2018). *Akkermansia muciniphila* induces gut microbiota remodelling and controls islet autoimmunity in NOD mice. *Gut, 67*(8), 1445–1453. https://doi.org/10.1136/gutjnl-2017-314508.

Jinatham, V., Popluechai, S., Clark, C. G., et al. (2019). *Entamoeba chiangraiensis* n. sp. (Amoebozoa: Entamoebidae) isolated from the gut of Asian swamp eel (Monopterus albus) in northern Thailand. *Parasitology, 146,* 1–23. https://doi.org/10.1017/s0031182019000775.

Laforest-Lapointe, I., Arrieta, M. C. (2018). Microbial eukaryotes: A missing link in gut microbiome studies. *mSystems 3*(2):e00201. https://doi.org/10.1128/msystems.00201-17. eCollection 2018 Mar-Apr.

Lantsman, Y., Tan, K. S., Morada, M., et al. (2008). Biochemical characterization of a mitochondrial-like organelle from *Blastocystis* sp. subtype 7. *Microbiology, 154*(Pt 9), 2757–2766. https://doi.org/10.1099/mic.0.2008/017897-0.

Long, S., Changmai, P., Tsaousis, A. D., et al. (2011). Stage-specific requirement for Isa1 and Isa2 proteins in the mitochondrion of *Trypanosoma brucei* and heterologous rescue by human and *Blastocystis* orthologues. *Molecular Microbiology, 81*(6), 1403–1418. https://doi.org/10.1111/j.1365-2958.2011.07769.x.

Makiuchi, T., & Nozaki, T. (2014). Highly divergent mitochondrion-related organelles in anaerobic parasitic protozoa. *Biochimie, 100,* 3–17. https://doi.org/10.1016/j.biochi.2013.11.018.

Masuda, A., Sumiyoshi, T., Ohtaki, T. et al. (2018). Prevalence and molecular subtyping of *Blastocystis* from dairy cattle in Kanagawa, Japan. *Parasitology International 67(*6):702–705. S1383-5769(18)30098-9 [pii].

Meloni, D., Sanciu, G., Poirier, P., et al. (2011). Molecular subtyping of *Blastocystis* sp. isolates from symptomatic patients in Italy. *Parasitology Research, 109*(3), 613–619. https://doi.org/10.1007/s00436-011-2294-7.

Mettert, E. L., Outten, F. W., Wanta, B., et al. (2008). The impact of O(2) on the Fe–S cluster biogenesis requirements of *Escherichia coli* FNR. *Journal of Molecular Biology, 384*(4), 798–811. https://doi.org/10.1016/j.jmb.2008.09.080.

Muller, M., Mentel, M., van Hellemond, J. J., et al. (2012). Biochemistry and evolution of anaerobic energy metabolism in eukaryotes. *Microbiology Molecular Biology Reviews, 76*(2), 444–495. https://doi.org/10.1128/mmbr.05024-11.

Nash, A. K., Auchtung, T. A., Wong, M. C., et al. (2017). The gut mycobiome of the human microbiome project healthy cohort. *Microbiome, 5*(1), 153. https://doi.org/10.1186/s40168-017-0373-4.

Nieves-Ramirez, M. E., Partida-Rodriguez, O., Laforest-Lapointe, I., et al. (2018). Asymptomatic intestinal colonization with protist *Blastocystis* Is strongly associated with distinct microbiome ecological patterns. *MSystems, 3*(3), e00007. https://doi.org/10.1128/msystems.00007-18.

Noel, C., Dufernez, F., Gerbod, D. et al. (2005). Molecular phylogenies of *Blastocystis* isolates from different hosts: Implications for genetic diversity, identification of species, and zoonosis. *Journal of Clinical Microbiology 43*(1):348–355. 43/1/348 [pii].

Noradilah, S. A., Moktar, N., Anuar, T. S., et al. (2017). Molecular epidemiology of blastocystosis in Malaysia: Does seasonal variation play an important role in determining the distribution and risk factors of *Blastocystis* subtype infections in theÂ Aboriginal community? *Parasites & Vectors, 10*(1), 360. https://doi.org/10.1186/s13071-017-2294-2.

Oliveira-Arbex, A. P., David, É. B., & Guimarães, S. (2018). *Blastocystis* genetic diversity among children of low-income daycare center in Southeastern Brazil. *Infection, Genetics and Evolution, 57,* 59–63.

Paulos, S., Köster, P. C., de Lucio, A., et al. (2018). Occurrence and subtype distribution of *Blastocystis* sp. in humans, dogs and cats sharing household in northern Spain and assessment of zoonotic transmission risk. *Zoonoses Public Health, 65*(8), 993–1002. https://doi.org/10.1111/zph.12522.

Ramirez, J. D., Sanchez, A., Hernandez, C., et al. (2016). Geographic distribution of human *Blastocystis* subtypes in South America. *Infection Genetics and Evolution, 41,* 32–35. https://doi.org/10.1016/j.meegid.2016.03.017.

Ramirez, J. D., Sanchez, L. V., Bautista, D. C., et al. (2014). *Blastocystis* subtypes detected in humans and animals from Colombia. *Infection Genetics Evolution, 22,* 223–228. https://doi.org/10.1016/j.meegid.2013.07.020.

Rangachari, K., Davis, C. T., Eccleston, J. F. et al. (2002). SufC hydrolyzes ATP and interacts with SufB from *Thermotoga maritima. FEBS Letters 514*(2–3):225–228. S0014579302023694 [pii].

Río Bártulos, C., Rogers, M. B., Williams, T. A., et al. (2018). Mitochondrial targeting of glycolysis in a major lineage of eukaryotes. *Genome Biology and Evolution, 10*(9), 2310–2325. https://doi.org/10.1093/gbe/evy164.

Silberman, J. D., Sogin, M. L., Leipe, D. D., et al. (1996). Human parasite finds taxonomic home. *Nature, 380*(6573), 398. https://doi.org/10.1038/380398a0.

Stechmann, A., Hamblin, K., Perez-Brocal, V., et al. (2008). Organelles in *Blastocystis* that blur the distinction between mitochondria and hydrogenosomes. *Curr Biol, 18*(8), 580–585. https://doi.org/10.1016/j.cub.2008.03.037.

Stensvold, C. R., & van der Giezen, M. (2018). Associations between gut microbiota and common luminal intestinal parasites. *Trends in Parasitology, 34*(5), 369–377. https://doi.org/10.1016/j.pt.2018.02.004.

Stensvold, C. R., Clark, C. G. (2016). Current status of *Blastocystis*: A personal view. *Parasitology Internatioanl 65*(6 Pt B):763–771. S1383-5769(16)30154-4 [pii].

Stenzel, D. J., & Boreham, P. F. (1996). *Blastocystis hominis* revisited. *Clinical Microbiology Reviews, 9*(4), 563–584.

Tan, K. S. (2008). New insights on classification, identification, and clinical relevance of *Blastocystis* spp. *Clinical Microbiology Reviews, 21*(4), 639–665. https://doi.org/10.1128/cmr.00022-08.

Tan, K. S. W. (2004). *Blastocystis* in humans and animals: New insights using modern methodologies. *Veterinary Parasitology, 126*(1–2), 121–144. https://doi.org/10.1016/j.vetpar.2004.09.017.

Tan, K. S., Mirza, H., Teo, J. D., et al. (2010). Current views on the clinical relevance of *Blastocystis* spp. *Current Infectious Disease Reports, 12*(1), 28–35. https://doi.org/10.1007/s11908-009-0073-8.

Tan, T. C., & Suresh, K. G. (2006). Amoeboid form of *Blastocystis hominis*—A detailed ultrastructural insight. *Parasitology Research, 99*(6), 737–742. https://doi.org/10.1007/s00436-006-0214-z.

Thathaisong, U., Siripattanapipong, S., Mungthin, M., et al. (2013). Identification of *Blastocystis* subtype 1 variants in the home for girls, Bangkok, Thailand. *The American Journal of Tropical Medicine and Hygiene, 88*(2), 352–358. https://doi.org/10.4269/ajtmh.2012.12-0237.

Tito, R. Y., Chaffron, S., Caenepeel, C., et al. (2019). Population-level analysis of *Blastocystis* subtype prevalence and variation in the human gut microbiota. *Gut, 68*(7), 1180–1189. https://doi.org/10.1136/gutjnl-2018-316106.

Tsaousis, A. D., Gaston, D., Stechmann, A., et al. (2011). A functional Tom70 in the human parasite *Blastocystis* sp.: Implications for the evolution of the mitochondrial import apparatus. *Molecular Biology and Evolution, 28*(1), 781–791. https://doi.org/10.1093/molbev/msq252.

Tsaousis, A. D., Gentekaki, E., Eme, L., et al. (2014). Evolution of the cytosolic iron-sulfur cluster assembly machinery in *Blastocystis* species and other microbial eukaryotes. *Eukaryotic Cell, 13*(1), 143–153. https://doi.org/10.1128/ec.00158-13.

Tsaousis, A. D., Hamblin, K. A., Elliott, C. R., et al. (2018). The human gut colonizer *Blastocystis* respires using complex II and alternative oxidase to buffer transient oxygen fluctuations in the gut. *Frontiers in Cellular Infectious Microbiology, 8,* 371. https://doi.org/10.3389/fcimb.2018.00371.

Tsaousis, A. D., Ollagnier de Choudens, S., Gentekaki, E., et al. (2012). Evolution of Fe/S cluster biogenesis in the anaerobic parasite *Blastocystis*. *Proceedings of Natlioanl Academic Sciences U S A, 109*(26), 10426–10431. https://doi.org/10.1073/pnas.1116067109.

Tsaousis, A. D., Yarlett, N. R., Tan, K. S. (2019). The mitochondrion-related organelles of *Blastocystis*. In: J. Tachezy (Ed.), *Hydrogenosomes and mitosomes: Mitochondria of anaerobic eukaryotes* (2nd edn., vol MICROMONO, vol 9), pp 267–286. Cham, Switzerland: Springer.

Vermathen, M., Muller, J., Furrer, J., et al. (2018). ^{1}H HR-MAS NMR spectroscopy to study the metabolome of the protozoan parasite *Giardia lamblia*. *Talanta, 188*, 429–441. https://doi.org/10.1016/j.talanta.2018.06.006.

Villamizar, X., Higuera, A., Herrera, G., et al. (2019). Molecular and descriptive epidemiology of intestinal protozoan parasites of children and their pets in Cauca, Colombia: A cross-sectional study. *BMC Infectious Diseases, 19*(1), 190. https://doi.org/10.1186/s12879-019-3810-0.

Whipps, C. M., Boorom, K., Bermudez, L. E., et al. (2010). Molecular characterization of *Blastocystis* species in Oregon identifies multiple subtypes. *Parasitology Research, 106*(4), 827–832. https://doi.org/10.1007/s00436-010-1739-8.

Yason, J. A., Liang, Y. R., Png, C. W., et al. (2019). Interactions between a pathogenic *Blastocystis* subtype and gut microbiota: in vitro and in vivo studies. *Microbiome, 7*(1), 30. https://doi.org/10.1186/s40168-019-0644-3.

Yoshikawa, H., Koyama, Y., Tsuchiya, E. et al. (2016). *Blastocystis* phylogeny among various isolates from humans to insects. *Parasitology International 65*(6 Pt B):750–759. S1383-5769(16)30057-5 [pii].

Yowang, A., Tsaousis, A. D., Chumphonsuk, T. et al. (2018). High diversity of *Blastocystis* subtypes isolated from asymptomatic adults living in Chiang Rai, Thailand. *Infectious Genetics Evolution 65*, 270–275. S1567-1348(18)30616-6 [pii].

Zierdt, C. H. (1991). *Blastocystis hominis*-past and future. *Clinical Microbiology Reviews, 4*(1), 61–79. https://doi.org/10.1128/cmr.4.1.61.

Chapter 6
Advances in the Study of *Blastocystis* spp. in Mexico: Prevalence, Genetic Diversity, Clinical Association and Their Possible Role in the Human Intestine

P. Morán-Silva, M. E. Nieves-Ramírez, L. Rojas-Velázquez, A. Serrano-Vázquez, E. González-Rivas, E. Hernández-Hernández, H. Pérez-Juárez, O. Partida-Rodríguez, M. C. Arieta, U. Magaña, M. Santín, B. B. Finlay, and C. Ximénez-García

Abstract *Blastocystis* spp. has a worldwide distribution, and it has been associated with gastrointestinal symptomatology; however, its role in health or disease remains unclear. Subtype 3 is the most frequently reported subtype in different populations, with a high haplotype diversity. The recent diversity of this protist may be related to the migration of the human population. The ST3 haplotype network shows that haplotype 1 is ancestor from which the other haplotypes are derived. In the studied community (Xoxocotla, Morelos), a direct association exists between the presence of *Blastocystis* spp. and the changes in the bacterial and eukaryotic intestinal microbiota

P. Morán-Silva · M. E. Nieves-Ramírez · L. Rojas-Velázquez · A. Serrano-Vázquez · E. González-Rivas · E. Hernández-Hernández · H. Pérez-Juárez · O. Partida-Rodríguez · U. Magaña · C. Ximénez-García (✉)
Laboratorio de Inmunología, Unidad de Investigación en Medicina Experimental, Facultad de Medicina, Universidad Nacional Autónoma de México, Mexico City, Mexico
e-mail: cximenez@unam.com

L. Rojas-Velázquez · U. Magaña
Unidad de Posgrado, Universidad Nacional Autónoma de México (UNAM), Ciudad de México, Mexico

M. C. Arieta · B. B. Finlay
Michael Smith Laboratories, Department of Microbiology and Immunology, University of British Columbia, Vancouver, British Columbia, Canada

M. C. Arieta
Department of Physiology and Pharmacology, University of Calgary, Calgary, Alberta, Canada

Department of Pediatrics, University of Calgary, Calgary, Alberta, Canada

M. Santín
Environmental Microbial and Food Safety Laboratory, Department of Agriculture, United States, Agricultural Research Service, Beltsville, MD, USA

B. B. Finlay
Department of Biochemistry and Molecular Biology, University of British Columbia, Vancouver, British Columbia, Canada

© Springer Nature Switzerland AG 2020

N. Guillen (ed.), *Eukaryome Impact on Human Intestine Homeostasis and Mucosal Immunology*, https://doi.org/10.1007/978-3-030-44826-4_6

in the absence of gastrointestinal or inflammatory diseases, indicating that *Blastocystis* ST3 favors the diversity and richness of bacterial populations and decrease the inflammatory processes. Thus, there is important evidence that suggests that *Blastocystis* spp. plays an important role as a mutualist in the regulation of the inflammatory response in the studied healthy individuals.

Keywords Blastocystis spp infection · Subtypes · Subtype genetic diversity · Bacterial and eukaryotic microbiota changes

Background

Blastocystis spp. (Heterokonta, Stramenopiles) is and enteric protist with a worldwide distribution that can be inhabiting the digestive tract of several metazoans, such as fishes, amphibians, birds, reptiles, rodents, and humans (Yoshikawa et al. 2007; Silberman et al. 1996; Stechmann et al. 2008; Stenzel and Boreham 1996; Tan 2004). *Blastocystis* spp. has been proposed as an emerging pathogen for humans, replacing traditionally endemic parasites in developing countries, such as *Entamoeba histolytica* and *Giardia duodenalis* (Rodríguez et al. 2008). In the last two decades, there has been an increase in the reported data relative to the frequency of *Blastocystis* spp. worldwide (Tan 2008). However, the microscopic examination of fresh fecal samples is a routine diagnostic method for *Blastocystis* spp. detection; that can lead to erroneous diagnosis and the underestimation or subestimation of the prevalence of this protest (Baldo et al. 2004).

The prevalence of *Blastocystis* spp. varies between 0.5 and 62% (Clark et al. 2013). The real frequency of *Blastocystis* spp. cannot be established based on this range because of two main factors, the predisposition of the individual to be colonized and the diagnostic tool used microscopic analysis or molecular biology through polymerase chain reaction (PCR); studies with both methods were not comparable in sensitivity or specificity, resulting in inconsistent parasite prevalence data (Clark et al. 2013).

This increase in frequency of *Blastocystis* spp. infections has been the most reported for an intestinal protist, and *Blastocystis* spp. infections are more predominant in underdeveloped countries where the hygiene conditions, exposure to domestic and peridomestic animals, as well as the consumption of food and/or contaminated water, represent an ideal environment for Blastocystosis (Clark et al. 2013). However, *Blastocystis* spp. is not exclusive to underdeveloped countries; it is also prevalent in developed countries (Beyhan et al. 2015). The presence of *Blastocystis* spp. has been associated with a variety of symptoms, mainly gastrointestinal symptoms, such as diarrhea and abdominal pain, which in most cases are self-limiting. In addition, irritable bowel syndrome (IBS) has been associated with *Blastocystis* infection (Yakoob et al. 2004).

Studies on the taxonomic classification of *Blastocystis* spp. have helped to better understand this organism; however, knowledge on its genetic diversity, its geography distribution, the specificity of the host and its role in the human intestine is still lacking. Molecular evidence based on the small subunit ribosomal RNA (SSU rDNA) gene suggests that at least 17 genetic subtypes can be recognized within *Blastocystis* spp. species and nine of these subtypes are found in humans (Silberman et al. 1996; Stenzel and Boreham 1996). Therefore, molecular epidemiology studies of infection by this protist would help us to better understand the role of *Blastocystis* spp. in health and disease, clarifying whether *Blastocystis* spp. is truly a parasite or pathogen or rather a commensal or even a mutualist in the human intestine.

Advances in the Study of *Blastocystis* spp. in Mexico

Five years ago, clinical studies reported in Mexican individuals with gastrointestinal symptoms or irritable bowel síndrome is associated with *Blastocystis* spp. infection authors suggest that this protist could be the etiological factor (Villalobos et al. 2014; Vargas-Sanchez et al. 2015). To date, there are studies in Mexican populations that consider *Blastocystis* spp. infection as an important factor for the presence of specific symptoms such as chronic abdominal pain (Toro Monjaraz et al. 2018) or consequences of poliparasitism (Galvan-Ramírez et al. 2019; García-Flores et al. 2019).

The massive treatment of the scholar to eradicate helminths infection in Mexico seems to creative an empty ecological niche that has been occupied by other intestinal protozoa including *Blastocystis* spp. (Rodríguez et al. 2008; Rojas et al. 2016). In the last three years, has been reported an important interaction of the intestinal protozoa (Eukaryome) and the intestinal microbiota (Partida-Rodríguez et al. 2017). Eventhough, the mechanisms of interaction between the intestinal microbiota and *Blastocystis* spp. are not well understood, efforts have been made to identify the STs of *Blastocystis* spp. and their potential pathogenic proteins that could be used as genetic biomarkers (Villalobos et al. 2014; Vargas-Sanchez et al. 2015; Rojas-Velázquez et al. 2018). Furthermore, there are reports that mention the posssible role of *Blastocystis* spp. in the modulation of some intestinal bacterial populations and the Immune response (Nieves-Ramírez et al. 2018).

Genetic Diversity and Geographical Distribution of *Blastocystis* Subtypes

Blastocystis spp. is globally distributed, showing a high rate of human infection from underdeveloped to developed countries (Clark et al. 2013; Alfellani et al. 2013). In countries such as Mexico, the phenomenon of epidemiological change of *Blastocystis* spp. increasing the prevalence of infection in the general population (Villalobos et al.

2014; Vargas-Sanchez et al. 2015; Villegas-Gómez et al. 2016; Rojas-Velázquez et al. 2018).

Due to the improvements in molecular biology and the use of polymerase chain reaction (PCR) and sequencing, *Blastocystis* spp. has been studied for its genetic diversity, and currently 17 subtypes have been identified using the small subunit ribosomal RNA gene (SSUrRNA) as a target. Within the 17 subtypes described, with a couple exceptions, only the first 9 subtypes are capable of infecting humans. The molecular analysis of *Blastocystis* spp. showed the genetic variation among *Blastocystis* spp. isolates with 3–5% divergence between subtypes, which increases the possibility that more than one species can infect a host (Silberman et al. 1996; Kukoschke and Múller 1991; Boreham et al. 1992). Despite the great genetic diversity, the distribution of *Blastocystis* spp. subtypes is not homogeneous worldwide. There is a predominant distribution of subtypes; subtypes 1, 2, 3 and 4 have a higher frequency and have been associated with gastrointestinal symptoms (Alfellani et al. 2013; Engsbro et al. 2014; Forsell et al. 2012). In the literature, there are reports that subtypes 1, 2 and 3 have a worldwide distribution, unlike subtype 4, which seems to be more frequent in the European and Asian continent (Alfellani et al. 2013; Ramírez et al. 2016).

Recently, *Blastocystis* spp. subtypes 1, 2 and 3 were identified in five states from Mexico (Ciudad de México, Michoacán, Sonora, Estado de Mexico and Morelos) (Villalobos et al. 2014; Vargas-Sanchez et al. 2015; Villegas-Gómez et al. 2016; Rojas-Velázquez et al. 2018; Alarcon-Valdes et al. 2018). In the last state is where our research group has investigated the genetic diversity, geographic distribution and the possible implications that this protist may have on the human intestine, taking as a study model a rural population of the state of Morelos in Mexico. We collected fecal samples from 182 volunteers (86 men and 96 women) from Xoxocotla, State of Morelos (Mexico), aged between from 2 to 51 years. We found a high frequency of *Blastocystis* spp. in 39.56% of the samples through PCR analysis and sequencing, which also showed the presence of three different subtypes of *Blastocystis*. These three subtypes of *Blastocystis* (ST) were recorded according to the following frequencies: *Blastocystis* ST1, 9.7% ($n = 7$ samples); ST2, 15.3% ($n = 11$ samples); and ST3, 75% ($n = 54$ samples) (Rojas-Velázquez et al. 2018) (Fig. 6.1).

Additionally, we provided new data on the geographic distribution and genetic diversity of *Blastocystis* ST3 of the rural human population in Xoxocotla, Morelos, Mexico and we compare its diversity and structure genetics with what was previously observed in the populations of *Blastocystis* ST3 from other regions of the planet (Rojas-Velázquez et al. 2018) and through the construction of a network of haplotypes of *Blastocystis* subtype 3 isolates from humans, we discovered that this subtype has a high diversity of haplotypes and a high genetic structure in many countries of the world, in contrast to rural populations in Mexico such as the state of Morelos, which seems to have a low diversity of haplotypes. This study also revealed evidence that suggests a recent increase in the diversity of *Blastocystis* ST3 worldwide and may be related to the migration of human populations (Fig. 6.2). One haplotype (Haplotype 1) was the most frequently detected haplotype, and it is perhaps the ancestral type

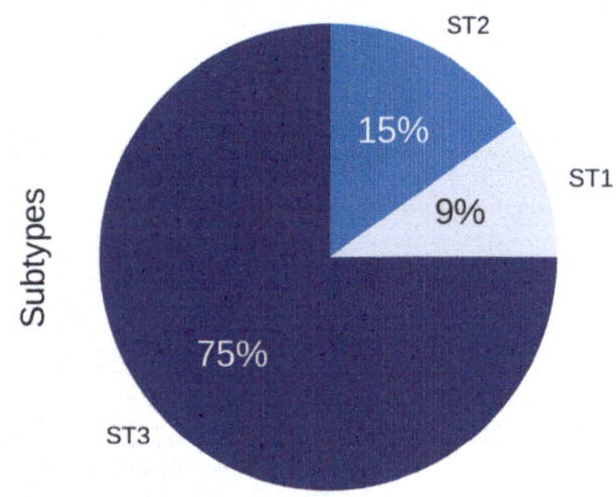

Fig. 6.1 Frequency of *Blastocystis*subtypes in the Xoxocotla, Morelos population. Targeting the SSU-rDNA according to DNA-barcoding. Three *Blastocystis* subtypes (ST) were recorded according to the following frequencies: *Blastocystis* ST1, 9.7% (n = 7 samples); ST2, 15.3% (n = 11 samples); and ST3, 75% (n = 54 samples) (Rojas-Velázquez et al. 2018)

from which all the other haplotypes have been generated recently (Rojas-Velázquez et al. 2018).

Clinical Association or Commensal Behavior of *Blastocystis* spp.

This parasite is often transmitted via the oral-fecal route to people who work directly or indirectly with animals, such as those involved in animal farming or industrial livestock production, and it is estimated that one billion humans are colonized by *Blastocystis* spp. (Tan 2008).

Previously, *Blastocystis* spp. was considered an opportunistic parasite only reported in immunocompromised and immunosuppressed patients, such as HIV-infected patients, who, due to immunodeficiency, could be more susceptible to infection by parasites than general population (Zali et al. 2004; Hailemariam et al. 2004; Gassama et al. 2001). In clinical practice, *Blastocystis* spp. has been considered as non pathogenic infectious agent without clinical importance. However, a number of publications have referred the association between *Blastocystis* spp. and irritable bowel syndrome (Yakoob et al. 2004; Poirier et al. 2012; Beatty et al. 2014; Jimenez-Gonzalez et al. 2012; Ramirez-Miranda et al. 2010). Although *Blastocystis* spp. has been detected in patients with gastrointestinal symptoms such as diarrhea, abdominal pain and distention the prevalence of this infection in asymptomatic individuals is considerably high (Rodríguez et al. 2008; Alfellani et al. 2013; Pandey et al. 2015; Scanlan et al. 2014) in our results, we reported a high frequency of infection in asymptomatic healthy individuals (Rojas-Velázquez et al. 2018).

The existence of genetically diverse infection sources has been recently reported and the human host can be co-colonized with multiple different subtypes or haplotypes. Previous studies have opened the possibility of competition or cooperation

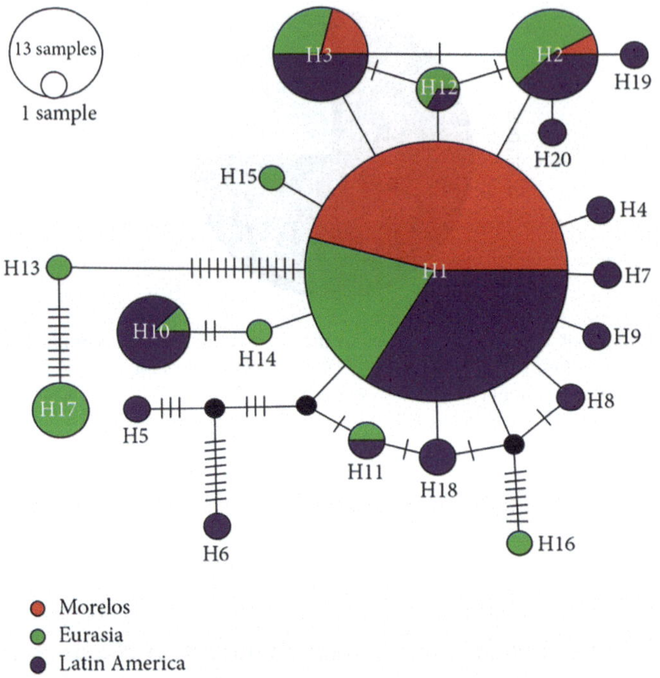

Populations	N	S	h	Hd	π	±SD	Tajima's D
Morelos	54	1	3	0.14186	0.00107	0.068	-1.68258^{ns}
Latin America	69	25	15	0.67775	0.01334	0.058	-2.36261^{**}
Eurasia	46	35	11	0.74010	0.04473	0.052	-1.30307^{ns}
All populations	169	44	20	0.56276	0.01886	0.044	-2.20000^{**}

Fig. 6.2 Haplotype network of *Blastocystis* ST3 of human populations and the statistics data of genetic diversity observed within different geographical populations around the world. N: number of sequences; S: number of segregating sites; h: number of haplotypes; Hd: haplotype diversity; π: nucleotide diversity; ns: not significant. $^{**}p < 0.01$. Latin America: *Blastocystis* populations of North and South America (i.e., Mexico, Colombia, Brazil, Ecuador, Bolivia, Peru, and Argentina), except that of Morelos. Eurasia: *Blastocystis* populations of Europa and Asia (i.e., Nepal, Switzerland, Iraq, Italy, and France). Haplotype network of *Blastocystis* ST3 of human populations at different regions from Latin America, Europe, and Asia. Each circle represents a haplotype, and each color represents the place where it was obtained. The size of each circle is proportional to the frequency of the haplotype in each population, where it was found.The circles in black stand for missing haplotypes and the short lines show the mutational steps (Rojas-Velázquez et al. 2018)

between *Blastocystis* spp. subtypes for a successful gut colonization (Scanlan et al. 2015). *Blastocystis* mixed infections has been estimated to be 6% worldwide (Alfellani et al. 2013). Scanlan et al. (2015) developed and applied *Blastocystis* ST-specific PCRs for the investigation of the most common subtypes of *Blastocystis* (ST1 to ST4) to a healthy human cohort (n = 50). They reported mixed infections in 22% of the cases, all of which had been identified as single-ST infections in a previous study using state-of-the-art methods (Scanlan et al. 2015).

Recently, had been reported that next generation amplicon sequencing (NGS) was a powerful tool to investigate mixed infections and detect low abundance subtypes of *Blastocystis* (Maloney et al. 2019). Using this technology, we identified mixed *Blastocystis* infections in the aforementioned population (Xoxocotla, Morelos). The study identified 17 mixed infections representing 13.7% of all *Blastocystis* infections. Mixed subtype infections are underrepresented compared to expectations from subtype prevalence (Rojas-Velázquez et al. 2019). However, mixed infections in this study were within the range of observations from other human studies (Alfellani et al. 2013; Scanlan et al. 2015). In our work, we reported co-infections of ST1/ST3, ST2/ST3, and ST1/ST2/ST3. This study provides important information about the epidemiology of *Blastocystis* and represents the first application of a *Blastocystis*-specific NGS protocol to study *Blastocystis* in humans. More studies are needed to characterize mixed subtype infections and intrasubtype variation to understand the transmission dynamics, epidemiology, and potential pathogenicity of *Blastocystis* in humans and animals hostess, and NGS provides a valuable tool for achieving this goal (Rojas-Velázquez et al. 2019).

Interaction of *Blastocystis* spp. and the Gut Microbiota of Asymptomatic Individuals

The relationship between human-associated gut protists and the resident gut bacterial community has only recently begun to be explored (Barash et al. 2017; Burgess and Petri 2016). A number of human-associated microbiome studies have been focused on the role of the microbiome in health and in disease and or the factors that might influence its diversity and composition (Brown et al. 2013). The intestinal microbiota is highly variable among human host, and its diversity is affected by factors such as diet, sociogeographic setting, antibiotic use, disease, age, and to a lesser degree, the human genetics (Yatsunenko et al. 2012; Goodrich et al. 2014; Cho and Blaser 2012). A direct association between intestinal parasites, and human intestinal microbiota both in composition and diversity has been previously reported (Morton et al. 2015; Zaiss et al. 2015). We found that colonization with *Blastocystis* spp. is strongly associated with broad shifts in the gut-resident bacterial community and an increase in bacterial alpha and beta diversity. A correlation analysis between the abundances of the 100 most abundant taxa of the 16S versus 18S genes showed that *Blastocystis* sp. ST3 was positively correlated with members of the *Ruminococcaceae* family and negatively correlated with *Prevotella copri* (Fig. 6.3). However we observed a more discrete correlation between the colonization by *Blastocystis* spp. and eukaryotic microbiota diversity, detecting statistically significant differences in diversity (Shanon) but no in the abundance (Chao). Therefore, the rural community of asymptomatic individuals from the state of Morelos, Mexico, provides an ideal study cohort to better understand how the most commonly found protist in

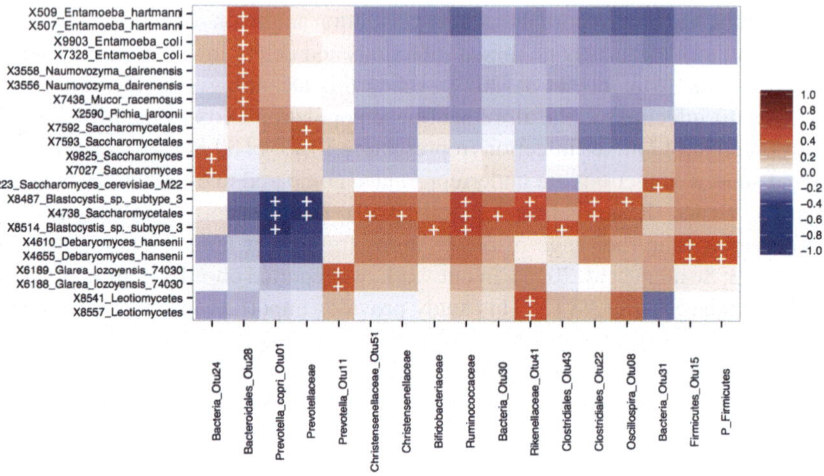

Fig. 6.3 Heatmap of biweight correlations (bicor method) between top 100 bacterial (x axis) and top 100 taxon (y axis) OTUs in fecal simples from study participants. Colors denote positive (red) and negative (blue) correlation values. Significant correlations are denoted with a plus sing ($p < 0.05$, FDR) (Nieves-Ramírez et al. 2018)

the gastrointestinal tract alters the ecology of the bacterial and eukaryotic microbiome in the absence of symptoms of inflammation, which correlates with the most abundant phylotypes in the microbiome of the human intestine. Additionally, this suggests that *Blastocystis* spp. might exert a predatory grazing effect on the more abundant bacteria, as observed with *P. copri* (Nieves-Ramírez et al. 2018). This is supported by food web theory and is an example of macro- and microecology where the increase in community diversity through grazing or predation occurs through a top-down control on the strongest competitors, which consequently allows for the colonization and persistence of weaker competitors in the community (Kato et al. 2018; McDonald-Madden et al. 2016; Paine et al. 2016).

Blastocystis *spp. and Their Effect on the Immune Response*

The role of *Blastocystis* spp. in the intestine is very important since there is a direct interaction between the intestinal epithelium and the underlying immune system (Belkaid and Hand 2015). We found that the immunological effects in individuals without gastrointestinal symptoms colonized by *Blastocystis* spp. displayed lower levels of fecal calprotectin, which is a marker of intestinal inflammation and is derived from the secretion of cytosolic proteins from neutrophils (Walsham and Sherwood 2016), compared to non-colonized individuals. Additionally, individuals colonized with *Blastocystis* spp. show low levels of IgA (the most abundant mucosal antibody) and have a fundamental role in maintaining homeostasis with the microbiome

(Nieves-Ramírez et al. 2018; Gutzeit et al. 2014) by binding and neutralizing invading pathogens and eliminating microbes near the mucus layer (Johansen et al. 1999).

In summary, there are aspects of this parasitosis in the human host that are highly controversial and that, will be resolved as the populations studied increase, both in groups of asymptomatic carriers and in patients with various gastrointestinal pathologies. On the other hand, our results in the population studied, suggest that, colonization by this interesting protist could have a regulatory role of the intestinal bacterial structure, metabolic processes and the immune response that would seem to have a more protective role of intestinal homeostasis than being conditioning of a pathological state.

Acknowledgements We are grateful to Marco Gudiño for his excellent technical assistance. The authors are grateful to the Health Ministry of the state of Morelos, Mexico for supporting this study in the community of Xoxocotla. Liliana Rojas-Velázquez is a doctoral student from Programa de Doctorado en Ciencias Biomédicas, Universidad Nacional Autónoma de México (UNAM) and was supported by the fellowship 348424/239901 from CONACYT. This paper constitutes a partial fulfilment of the Graduate Program.

Compliance with Ethical Standards

Funding This study was funded by Grants IN215018, from PAPIIT (DGAPA), Universidad Nacional Autónoma de México (UNAM) and 272601 from the National Council for Science and Technology in Mexico (CONACYT).

Disclosure of Interests All authors declare they have no conflict of interest.

Ethical Approval All procedures performed in studies involving human participants were in accordance with the ethical standards of the institutional and/or national research committee and with the 1964 Helsinki declaration and its later amendments or comparable ethical standards. Approval was granted by the Mexican Commission on Ethics and Research of the Health Ministry of the state of Morelos (Comisiones de Ética y de Investigación del Ministerio de Salud del Estado de Morelos); and the Commission on Ethics in Research of the Facultad de Medicina of the Universidad Nacional Autónoma de México (UNAM) (Comité de Ética de Investigación de la Facultad de Medicina de la Universidad Nacional Autónoma de México).

Informed Consent Informed consent for participation and publication was obtained from all individual participants included in the study.

References

Alarcon-Valdes, P., Villalobos, G., Martinez-Flores, W. A., et al. (2018). Can the pyruvate: Ferredoxin oxidoreductase (PFOR) gene be used as an additional marker to discriminate among *Blastocystis* strains or subtypes? *Parasities and Vectors, 11,* 564. https://doi.org/10.1186/s13071-018-3141-9.

Alfellani, M. A., Stensvold, C. R., Vidal-Lapiedra, A., et al. (2013). Variable geographic distribution of *Blastocystis* subtypes and its potential implications. *Acta Tropica, 126,* 11–18. https://doi.org/10.1016/j.actatropica.2012.12.011.

Baldo, E. T., Belizario, V. Y., De Leon, W. U., et al. (2004). Infection status of intestinal parasites in children living in residential institutions in Metro Manila the Philippines. *Korean Journal of Parasitology, 42,* 67–70.

Barash, N. R., Maloney, J. G., Singer, S. M., et al. (2017). Giardia alters commensal microbial diversity throughout the murine gut. *Infection and Immunity*. https://doi.org/10.1128/IAI.00948-16.

Beatty, J. K., Bhargava, A., & Buret, A. G. (2014). Post-infecctious irritable bowel syndrome: mechanistic insights into chronic disturbances following enteric infection. *World Journal of Gastroenterology, 20,* 3976–3985.

Belkaid, Y., & Hand, T. (2015). Role of the microbiota in immunity and inflammation. *Cell, 157,* 121–141.

Beyhan, Y. E., Yilmaz, H., Cengiz, Z. T., et al. (2015). Clinical significance and prevalence of *Blastocystis hominis* in Van, Turkey. *Saudi Medical Journal, 36,* 1118–1121.

Boreham, P. F., Upcroft, J. A., & Dunn, L. A. (1992). Protein and DNA evidence for two demes of *Blastocystis hominis* from humans. *International Journal for Parasitology, 22,* 49–53.

Brown, E. M., Sadarangani, M., & Finlay, B. B. (2013). The role of the immune system in governing host-microbe interactions in the intestine. *Nature Immunology, 7,* 660–667.

Burgess, S. L., & Petri, W. A. (2016). The intestinal bacterial microbiome and E. *histolytica* infection. *Current Tropical Medicine Reports, 3,* 71–74.

Cho, I., & Blaser, M. J. (2012). The human microbiome: at the interface of health and disease. *Nature Reviews Genetics, 13,* 260–270.

Clark, C. G., van der Giezen, M., Alfellani, M. A., et al. (2013). Recent developments in *Blastocystis* research. *Advances in Parasitology, 82,* 1–32.

Engsbro, A. L., Stensvold, C. R., Vedel, N., et al. (2014). Prevalence, incidence and risk factors of intestinal parasites in Danish primary care patients with irritable bowel syndrome. *Scandinavian Journal of Infectious Diseases, 46,* 204–209. https://doi.org/10.3109/00365548.2013.861609.

Forsell, J., Granlund, M., Stenvols, C. R., et al. (2012). Subtype analysis of *Blastocystis* isolates in Swedish patients. *European Journal of Clinical Microbiology and Infectious Diseases, 31,* 689–1696.

Galvan-Ramírez M., Madriz Elisondo, A. L., Ramirez, C. G. T., et al. (2019). Enteroparasitism and risk factors associated with clinical manifestation in children and adults of Jalisco state in Werstern Mexico. *Osong Public Health Research Perspective, 10,* 39–48. https://doi.org/10.24171/j.phrp.2019.10.1.08.

García-Flores, L., Santillán-Benítez, J. G., Cuevas-Yañéz, E., et al. (2019). Evaluation of the effect of 1,3-bis-(4-phenyl-[1,2,3] triazole-1-il)2-propanol in comparison with metronidaole in an in vitro culture of *Blastocystis* in simples of patients with irritable bowerl syndrome. *Journal of Parasitic Diseases, 43,* 506–512. https://doi.org/10.1007/s12639-019-01118-2.

Gassama, A., Sow, P. S., Fall, F., et al. (2001). Ordinary and opportunistic enteropathogens associated with diarrhea in Senegalese adults in relation to human immunodeficiency virus serostatus. *International Journal Infectious Diseases, 5,* 192–198.

Goodrich, J. K., Waters, J. L., Poole, A. C., et al. (2014). Human genetics shape the gut microbiome. *Cell, 159,* 789–799. https://doi.org/10.1016/j.cell.2014.09.053.

Gutzeit, C., Magri, G., & Cerutti, A. (2014). Intestinal IgA production and its role in host-microbe interaction. *Immunological Reviews, 260,* 76–85.

Hailemariam, G., Kasuu, A., Abebe, G., et al. (2004). Intestinal parasitic infections in HIV/AIDS and HIV seronegative individuals in a teaching hospital, Ethiopia. *Japanese Journal of Infectious Diseases, 57,* 41–43.

Jimenez-Gonzalez, D. E., Martinez-Flores, W. A., Reyes-Gordillo, J., et al. (2012). *Blastocystis* infection is associated with irritable bowel syndrome in a Mexican patient population. *Parasitology Research, 110,* 1269–1275. https://doi.org/10.1007/s00436-011-2626-7.

Johansen, F. E., Pekna, M., Norderhaug, I. N., et al. (1999). Absence of epitelial immunoglobulin A transport, with increased mucosal leakiness, in peolymeric immunoglobulin receptor/secretory component-deficient mice. *Journal of Experimental Medicine, 190,* 915–922.

Kato, Y., Kondoh, M., Ishikawa, N. F., et al. (2018). Using food network unfolding to evaluate foddwe complexity in terms of biodiversity: theory and applications. *Ecology Letters, 7,* 1065–1074.

Kukoschke, K. G., & Müller, H. E. (1991). SDS-PAGE and immunological analysis of different axenic *Blastocystis hominis* strains. *Journal of Medical Microbiology, 35,* 35–39.

Maloney, J. G., Lombard, J. E., Shivley, C. B., et al. (2019). Zoonotic and genetically diverse subtypes of *Blastocystis* in US pre-weaned dairy heifer calves. *Parasitology Research, 118,* 575–582.

McDonald-Madden, E., Sabbadin, R., Game, E., et al. (2016). Using food-web theory to conserve ecosystems. *National Communication*. https://doi.org/10.1038/ncomms10245.

Morton, E. R., Lynch, J., Froment, A., et al. (2015). Variation in rural African gut microbiota is strongly correlated with colonization by Entamoeba and subsistence. *PLoS Genetics*. https://doi.org/10.1371/journal.pgen.100565811.

Nieves-Ramírez, M. E., Partida-Rodriguez, O., Laforest-Lapointe, I. et al. (2018). Asymptomatic intestinal colonization with protist *Blastocystis* is strongly associated with distinct microbiome ecological patterns. mSystems. https://doi.org/10.1128/msystems.00007-18.

Paine, C. E., Beck, H., & Terborgh, J. (2016). How mammalian predation contributes to tropical tree community structure. *Ecology, 97,* 3326–3336.

Pandey, P. K., Verma, P., Marathe, N., et al. (2015). Prevalence and subtype analysis of *Blastocystis* in healthy Indian individuals. *Infectious Genetics Evolution, 31,* 296–299.

Partida-Rodríguez, O., Serrano-Vázquez, A., Nieves-Ramírez, M. E., et al. (2017). Human intestinal microbiota: Interaction between parasites and the host immune response. *Archives of Medical Research, 48,* 690–700. https://doi.org/10.1016/j.arcmed.2017.11.015.

Poirier, P., Wawryzyniak, C. P., Vivares, F., et al. (2012). New insights into *Blastocystis* spp.: A potential link with irritable bowel syndrome. *PLoS Pathogens, 8,* 1–4.

Ramírez, J. D., Sánchez, A., Hernández, C., et al. (2016). Geographic distribution of human *Blastocystis* subtypes in South America. *Infectious Genetics Evolution, 41,* 32–35. https://doi.org/10.1016/j.meegid.2016.03.017.

Ramirez-Miranda, M. E., Hernandez-Castellanos, R., Lopez-Escamilla, E., et al. (2010). Parasites in Mexican patients with irritable bowel syndrome: A case-control study. *Parasites and Vectors*. https://doi.org/10.1186/1756 3305 3-96.

Rodríguez, E., Mateos, B., González, J. C., et al. (2008). Transición parasitaria a *Blastocystis hominis* en niños de la zona centro del estado de Guerrero, México. *Parasitologia Latinoamericana, 63,* 20–28.

Rojas, L., Moran, P., Valadez, A., et al. (2016). *Entamoeba histolytica* and *Entamoeba dispar* infection in Mexican school children: genotyping and phylogenetic relationship. *BMC Infectious Diseases, 16,* 485. https://doi.org/10.1186/s12879-016-1812-8.

Rojas-Velázquez, L., Maloney, J. G., Molokin, A., et al. (2019). Use of next-generation amplicon sequencing to study *Blastocystis* genetic diversity in a rural human population from Mexico. *Parasites and Vectors*. https://doi.org/10.1186/s13071-019-3814-z.

Rojas-Velázquez, L., Morán, P., Serrano-Vázquez, A., et al. (2018). Genetic diversity and distribution of *Blastocystis* subtype 3 in human populations, with special reference to a rural population in central Mexico. *BioMedical Research International*. https://doi.org/10.1155/2018/3916263.

Scanlan, P. D., Stensvold, C. R., & Cotter, P. D. (2015). Development and application of a *Blastocystis* subtype-specific PCR assay reveals that mixed-subtype infections are common in a healthy human population. *Applied and Environment Microbiology, 81,* 4071–4076.

Scanlan, P. D., Stensvold, C. R., Rajilic-Stojanovic, M., et al. (2014). The microbial eukaryote *Blastocystis* is a prevalent and diverse member of the healthy human gut microbiota. *FEMS Microbiology Ecology, 90,* 326–330.

Silberman, J. D., Sogin, M. L., Leipe, D. D., et al. (1996). Human parasite finds taxonomic home. *Nature, 380,* 398.

Stechmann, A., Hamblin, K., Pérez-Brocal, V., et al. (2008). Organelles in *Blastocystis* that blur the distinction between mitochondria and hydrogenosomes. *Current Biology, 18,* 580–585.

Stenzel, D. J., & Boreham, P. F. (1996). *Blastocystis hominis* revisited. *Clinical Microbiology Reviews, 9,* 563–584.

Tan, K. S. (2004). *Blastocystis* in humans and animals: New insights using modern methodologies. *Veterinary Parasitology, 126,* 121–143.

Tan, K. S. W. (2008). New insights on classification, identification, and clinical relevance of *Blastocystis* spp. *Clinical Microbiology Reviews, 21,* 639–665.

Toro Monjaraz, E. M., Vichido Luna, M. A., Montijo Barrios, E., et al. (2018). *Blastocystis hominis* and chronic abdominal pain in children: Is there an association between them? *Journal of Tropical Pediatrics, 64,* 279–283. https://doi.org/10.1093/tropej/fmx060.

Vargas-Sanchez, G. B., Romero-Valdovinos, M., Ramirez-Guerrero, C., et al. (2015). *Blastocystis* isolates from patients with irritable bowel syndrome and from asymptomatic carriers exhibit similar parasitological loads, but significantly different generation times and genetic variability across multiple subtypes. *PLoS ONE.* https://doi.org/10.1371/journal.pone.0124006.

Villalobos, G., Orozco-Mosqueda, G. E., Lopez-Perez, M., et al. (2014). Suitability of internal trascribed spacers (ITS) as markers for the genetic structure of *Blastocystis* spp. *Parasites Vectors, 3*(7), 461. https://doi.org/10.1186/s13071-014-0461-2.7.

Villegas-Gómez, I., Martínez-Hernández, F., Urrea-Quezada, A., et al. (2016). Comparison of the genetic variability of *Blastocystis* subtypes between human carriers from two contrasting climatic regions of México. *Infectious Genetics Evolution, 44,* 334–340. https://doi.org/10.1016/j.meegid.2016.07.036.

Walsham, N. E., & Sherwood, R. A. (2016). Fecal calprotectin in inflammatory bowel disease. *Clinical and Experimental Gastroenterology, 9,* 21–29.

Yakoob, J., Jafri, W., Jafri, N., et al. (2004). Irritable bowel syndrome: in search of an etiology: role of *Blastocystis hominis*. *American Journal of Tropical Medicine and Hygeine, 70,* 383–385.

Yatsunenko, T., Rey, F. E., Manary, M. J., et al. (2012). Human gut microbiome viewed across age and geography. *Nature, 486,* 222–227. https://doi.org/10.1038/nature11053.

Yoshikawa, H., Wu, Z., Howe, J., et al. (2007). Ultrastructural and phylogenetic studies on Blastocystis isolates from cockroaches. *Journal of Eukaryotic Microbiology, 54,* 33–37.

Zaiss, M. M., Rapin, A., Lebon, L., et al. (2015). The intestinal microbiota contributes to the ability of helminths to modulate allergic inflammation. *Immunity, 43,* 998–1010. https://doi.org/10.1016/j.immuni.2015.09.012.

Zali, M. R., Mehr, A. J., Rezaian, M., et al. (2004). Prevalence of intestinal parasitic pathogens among HIV positive individuals in Iran. *Japanese Journal of Infectious Diseases, 57,* 268–270.

Chapter 7
Cryptosporidium Infection in Bangladesh Children

Carol A. Gilchrist

Abstract In humans in addition to diarrheal disease and sub-clinical infections the *Cryptosporidium* parasite can cause developmental delays and growth failure in malnourished infants. We hypothesized that genetic variants may be responsible for the differences in pathogen phenotypes. Resequencing thirty-two Bangladesh *C. hominis* isolates identified both polymorphic regions and evidence of frequent genetic recombination. This increased the genetic diversity of the parasites. Additional studies are now needed to identify the genetic changes predisposing the parasite to the different phenotypes.

Keywords *Cryptosporidium* · Genome · gp60 · Whole genome sequencing

Introduction

The eukaryotic protozoan *Cryptosporidium* is a member of the apicomplexan phylum of parasitic alveolates. Unlike some of the other parasitic members of this phyla (*Plasmodium* and *Toxoplasma*) *Cryptosporidium* can complete its lifecycle in a single host (Kosek et al. 2001). The lifecycle of this parasite features a sexual reproductive stage which produces environmentally resistant and infective sporulated oocysts which are shed in the host faeces. While several different *Cryptosporidium* species can infect humans three species *Cryptosporidium hominis*, *C. meleagridis* and *C. parvum* are the main pathogens (Pumipuntu and Piratae 2018). The source of new *Cryptosporidium hominis* and *Cryptosporidium parvum anthroponosum* subtype infections is expected to be contamination of drinking-water or food by human fecal material containing the infective environmentally resistant oocysts (Rossle and Latif 2013; King et al. 2019; Nader et al. 2019). Similarly, the source of new *C. meleagridis* and *C. parvum* infections is thought to involve either human or zoonotic fecal contamination (*C. meleagridis*: avian; *C. parvum*: a wide range of other mammals) (Pumipuntu and Piratae 2018).

C. A. Gilchrist (✉)
Departments of Medicine, University of Virginia, Charlottesville, VA, USA
e-mail: cg2p@virginia.edu

© Springer Nature Switzerland AG 2020

N. Guillen (ed.), *Eukaryome Impact on Human Intestine Homeostasis and Mucosal Immunology*, https://doi.org/10.1007/978-3-030-44826-4_7

Human infections with members of the *Cryptosporidium* species was previously viewed as a cause of self-limited mild diarrhea, and a concern only in poorly controlled HIV-positive patients (Cama et al. 2007). One of the surprises of the recent multicenter enteric disease surveys (Global Enteric Multicenter Study (GEMS) and Malnutrition and Enteric Disease Study (MAL-ED)) was that cryptosporidiosis was one of the top causes of diarrhea in malnourished infants (Kotloff et al. 2013; Acosta et al. 2014; Platts-Mills et al. 2014; Checkley et al. 2015; Korpe et al. 2016; Sow et al. 2016). In low and middle income countries where the death rate of children under 5 years of age is high it was estimated that at least 5% of the deaths of children under 5 were due to water borne diarrheal disease such as cryptosporidiosis (Prüss-Ustün et al. 2019).

Genomic Diversity in Bangladesh *Cryptosporidium hominis*

In our Bangladesh longitudinal birth cohort 77% of children experience at least one infection with *Cryptosporidium* spp. in their first two years of life (Korpe et al. 2016). We recently investigated the species causing human cryptosporidiosis in Bangladeshi infants (Steiner et al. 2018). The children involved in this study were born into an urban slum of Dhaka, Bangladesh (Section 11 of Mirpur Thana). This neighbourhood was densely populated, and the annual median household income of participants was 12,950 Taka (approximately €138 EUR). Surveillance samples were collected monthly and from every diarrheal infection until the infants were 2 years of age (Steiner et al. 2018). In this work we defined a new infection as occurring when we identified a discordant genotype or when more than 2 months had elapsed from the prior positive stool sample. The species responsible for cryptosporidiosis at this urban site was mainly *C. hominis* and the frequency was dependent on the age of the child and the time of year (Fig. 7.1) (Gilchrist et al. 2018). In our cohort children less than 6 months of age were exclusively breastfed (no supplementary water) and therefore were presumably either simply not exposed to the waterborne *Cryptosporidium* parasite or protected by maternal antibodies (Korpe et al. 2013; Steiner et al. 2018). The risk of exposure to contaminated drinking water thereafter increased (Pal et al. 1989; Sirajul Islam et al. 2007; Gilchrist et al. 2018).

The majority of the households of the children enrolled into our study were connected to the municipal water supply (99.6%) but, as the municipal water supply is frequently interrupted in Mirpur, community water tanks are used to store water prior to household use (Steiner et al. 2018). The sampling of this municipal water has shown that it is commonly contaminated with fecal material possibly due to the frequencies of water leaks and the low water pressure which allows surface water to enter the water pipes (Islam et al. 2010; Pickering et al. 2019). Dhaka surface water is contaminated with human faecal material especially during the rainy season (June–August) when the Mirpur area of Dhaka is prone to flooding (Alam 2009). It is therefore not surprising that *Cryptosporidium* infections were more frequent during

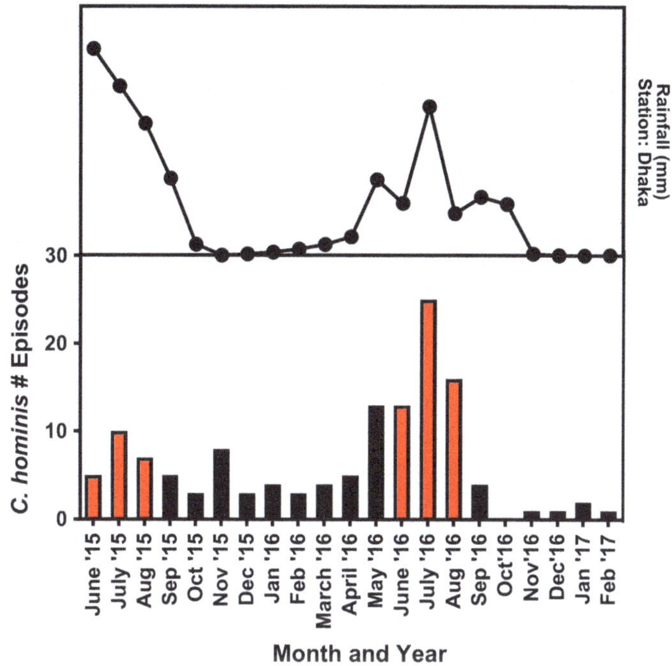

Fig. 7.1 In infants >6 months of age *Cryptosporidium hominis* infections occur more frequently during the rainy season. The common x axis indicates the month and year the y axis in the upper panel indicates total monthly rainfall (mm) in the Dhaka area. The lower panel indicates the number of typed new *C. hominis* infections (subclinical or diarrheal) occurring within in that month. Monsoon months are indicated by the red bars. Figure modified and reproduced from Gilchrist et al. (2018) with permission

the monsoon (early June–late September of 2015 and 2016) (Fig. 7.1: red bars) (Pal et al. 1989; Gilchrist et al. 2018).

Boiling or filtering water kills or removes *Cryptosporidium* oocysts and we observed less cryptosporidiosis in the 56% of our study households which did this prior to consumption of municipal water ($p = 0.044$) (1995; Steiner et al. 2018). While chlorine decontamination of the community water tanks or at point of use has been proposed the *Cryptosporidium* parasite is resistant to chlorine (Sirajul Islam et al. 2007; Pickering et al. 2019). Chlorination tablets were in fact used by one of our study households and the child living in this household was one of the ones infected with *Cryptosporidium* (Steiner et al. 2018). It is however worth noting that the Dhaka municipal water supply is not the only source of cryptosporidiosis in our study population (Korpe et al. 2018). In our study 14% of the children from households where drinking water was treated were still infected with the *Cryptosporidium* parasite.

Recent advances in genomic sequencing coupled with the development of new protocols for *Cryptosporidium* research are allowing us to study one of the fundamental sources of information on these organisms, the genomic DNA sequence (Hadfield et al. 2015; Gilchrist et al. 2018). The dry season population bottleneck may result in random fixation of alleles in the parasite population irrespective of their effect on parasite virulence (Lequime et al. 2016). This bottleneck could also however result in the selection for of mutations facilitating transmission during the dry season and increasing the virulence of the parasite over time (Rafaluk et al. 2015; Van den Bergh et al. 2018). *Cryptosporidium* parasites are variable with symptomatic disease occurring in only a subset of cases, all children therefore do not experience cryptosporidial diarrhea upon infection (Korpe et al. 2016). However the consequences of an infection with the *Cryptosporidium* parasite in the early years of life in low income households (subclinical or symptomatic) can also exacerbate the effects of malnutrition, and result in long term effects on child health (Korpe et al. 2016; Steiner et al. 2018). The differences in disease severity could be the result of genetically encoded changes in parasite virulence (Bouzid et al. 2013; Li et al. 2013).

Bangladesh *Cryptosporidium* oocysts were isolated and DNA purified from the fecal material of children infected during 2016–2017 (Gilchrist et al. 2018). Illumina sequence libraries were prepared and the resulting Bangladesh sequence reads were mapped against the *Cryptosporidium* reference genome assemblies (Abrahamsen et al. 2004; Xu et al. 2004; Hadfield et al. 2015; Ifeonu et al. 2016). This work provided us with information on the genetic variation present in 32 Bangladesh *C. hominis* isolates (Gilchrist et al. 2018). The allele ratio in the heterozygote SNPs were similar to those expected for a single diploid genotype indicating that the infection arose from a single oocyst which contained sporozoites with different alleles.

Polymorphic DNA

The heterozygosity was calculated across contiguous bp genomic intervals and several DNA regions where the nucleotide diversity was particularly high were identified in the Bangladesh parasite population. The genes within these DNA regions encoded several membrane proteins one of which was the highly polymorphic *gp60* gene (Cama et al. 2007). The polymorphic sequences could not however be associated with differences in parasite virulence but this may be due to the fact that currently this genetic information on all these areas is only available from a relatively small number of *C. hominis* isolates (Abrahamsen et al. 2004; Xu et al. 2004; Isaza et al. 2015; Guo et al. 2015a, b; Hadfield et al. 2015; Ifeonu et al. 2016; Sikora et al. 2017; Gilchrist et al. 2018). Historically the polymorphic region within the *gp60* gene has been used to genotype *Cryptosporidium* (Cama et al. 2007). While the *gp60* genotype has been correlated with clinical phenotype in some studies in our Bangladesh population we have not found any link between the *gp60* genotype and parasite virulence (Cama et al. 2008; Jex and Gasser 2010; Widmer and Lee 2010; Insulander et al. 2013; Korpe et al. 2016).

Recombination

Interestingly the phylogenetic trees derived from the polymorphic DNA were not congruent with the phylogenetic trees derived from the entire genome of the isolate. This suggested that sexual recombination might occur frequently in these parasites. To investigate this possibility the decay in linkage between common SNPs (SNPs with a frequency ≥0.2) was measured and it was found that on average SNPs separated by greater than 300 bp were inherited independently from each other (Fig. 7.2). The size of the DNA in linkage disequilibrium (LD) is dependent on the amount interbreeding between populations (admixture), SNP density and the amount of genetic recombination. In humans the functional mapping of SNPs requires that the analysis is done in a large population (Consortium et al. 2015) as there often large regions of DNA in which the single nucleotide polymorphisms (SNPs) are all in strong LD with each other in the human genome "haplotype blocks" (Wang et al. 2002). *Cryptosporidium* seems to have a rapid break-down in linkage disequilibrium similar to

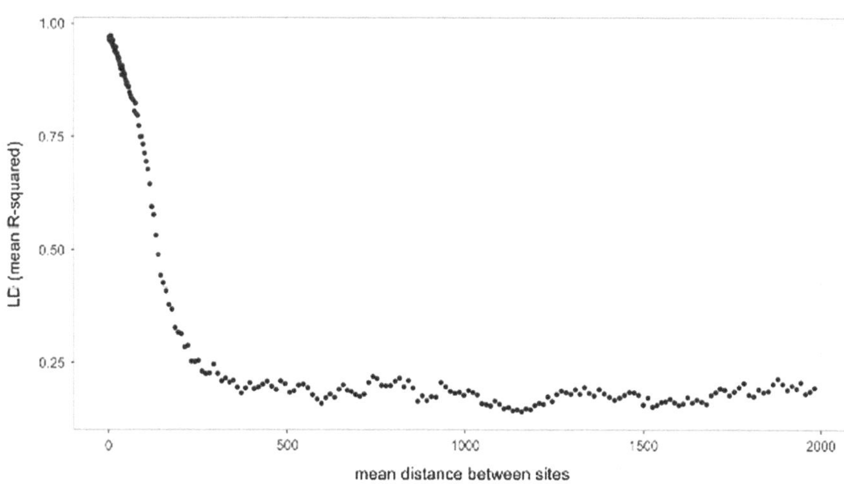

Fig. 7.2 High rates of recombination are apparent in this *C. hominis* population. Average linkage disequilibrium (LD) (y-axis) between neighboring (SNPs) as a function of the distance (bp) separating them in the genome (x-axis). The graph was generated using the using data from 32 Bangladesh *C. hominis* genomes which had >80% genome coverage of at least 10x. The pairwise values were calculated using the SNPs with a minor allele frequency >0.20. Y axis indicates the average R^2 value (the square of the correlation coefficient of two SNPs) and X axis the physical separation 0–2000 bp. Figure reproduced from Gilchrist et al. (2018) with permission

that occurring in the well-studied *Plasmodium falciparum,* another apicomplexan (Mu et al. 2005).

In *Plasmodium*, this high rate of recombination has facilitated identification of mutations responsible for drug resistance (Haldar et al. 2018). Currently, the only approved drug for the treatment of diarrhea caused by *Cryptosporidium* infection is nitazoxanide (NTZ) (U.S. Department of Health and Human Services 2003). While no resistance to this drug has been reported it is not in broad use as it has been shown to have an low efficacy in the malnourished infants and the AIDS patients whose need for treatment is greatest (Amadi et al. 2002; Rossignol et al. 2006; Manjunatha et al. 2016). Additional drugs are however undergoing development and strategies that block adaption of the parasite to these anti-cryptosporidial drugs may be essential to protect this investment (Antony and Parija 2016; Manjunatha et al. 2017).

Conclusions

Cryptosporidium intra-host changes in the infecting parasite population may occur either via selection of parasite sporozoites or merozoites or by genetic recombination. The relatively short DNA in linkage disequilibrium indicates that high rates of recombination take place during parasite sexual reproduction. This is evident even when the parasites are collected within a small community and over a relatively short period of time. This finding is not only biologically interesting in its own right it might greatly facilitate the identification of virulence associated mutations.

Acknowledgements We are grateful for the willing participation of the parents and children at the icddr,b study sites and wish to thank the field workers, nurses, laboratory staff of the Parasitology Laboratory of icddr,b without whom we could not have completed this research.

Conflict of Interest Statement The author has no conflict of interest to disclose.

Funding The work mentioned in this chapter was supported by the National Institute of Allergy and Infectious Diseases (NIAID) R21 AI142656 and R01 AI-043596, the Bill and Melinda Gates Foundation Grant (OPP1100514). Sequencing was at the Wellcome Trust Sanger Institute which is supported by the Wellcome Trust. Work at icddr,b is supported by the core donors (Government of the People's Republic of Bangladesh, GAC, Sida, UKAid). The funders had no participation in the writing of this chapter or the decision to publish.

Ethical Considerations The work reported here as reviewed by the Ethical and Research Review Committees of the International Centre for Diarrhoeal Disease Research, Bangladesh and by the Institutional Review Board of the University of Virginia. Informed written consent was obtained from the parents or guardians for the participation of their child in the study.

References

Abrahamsen, M. S., Templeton, T. J., Enomoto, S., Abrahante, J. E., Zhu, G., Lancto, C. A., et al. (2004). Complete genome sequence of the apicomplexan, *Cryptosporidium parvum*. *Science, 304,* 441–445. https://doi.org/10.1126/science.1094786.

Acosta, A. M., Chavez, C. B., Flores, J. T., Olotegui, M. P., Pinedo, S. R., Trigoso, D. R., et al. (2014). The MAL-ED study: A multinational and multidisciplinary approach to understand the relationship between enteric pathogens, malnutrition, gut physiology, physical growth, cognitive development, and immune responses in infants and children up to 2 years of age in resource-poor environments. *Clinical Infectious Diseases, 59,* S193–S206. https://doi.org/10.1093/cid/ciu653.

Alam, G. M. J. (2009). Environmental pollution of Bangladesh—It's effect and control. In *International Conference on Mechanical Engineering 2009*, Dhaka, pp. 1–6.

Amadi, B., Mwiya, M., Musuku, J., Watuka, A., Sianongo, S., Ayoub, A., et al. (2002). Effect of nitazoxanide on morbidity and mortality in Zambian children with cryptosporidiosis: A randomised controlled trial. *Lancet, 360,* 1375–1380. https://doi.org/10.1016/S0140-6736(02)11401-2.

Antony, H., & Parija, S. (2016). Antimalarial drug resistance: An overview. *Tropical Parasitoloy, 6,* 30. https://doi.org/10.4103/2229-5070.175081.

Bouzid, M., Hunter, P. R., Chalmers, R. M., & Tyler, K. M. (2013). *Cryptosporidium* pathogenicity and virulence. *Clinical Microbiology Reviews, 26,* 115–134. https://doi.org/10.1128/CMR.00076-12.

Cama, V. A., Bern, C., Roberts, J., Cabrera, L., Sterling, C. R., Ortega, Y., et al. (2008). *Cryptosporidium* species and subtypes and clinical manifestations in children, Peru. *Emerging Infectious Diseases, 14,* 1567–1574. https://doi.org/10.3201/eid1410.071273.

Cama, V. A., Ross, J. M., Crawford, S., Kawai, V., Chavez-Valdez, R., Vargas, D., et al. (2007). Differences in clinical manifestations among *Cryptosporidium* species and subtypes in HIV-infected persons. *The Journal of Infectious Diseases, 196,* 684–691. https://doi.org/10.1086/519842.

Checkley, W., White, A. C., Jaganath, D., Arrowood, M. J., Chalmers, R. M., Chen, X.-M., et al. (2015). A review of the global burden, novel diagnostics, therapeutics, and vaccine targets for *Cryptosporidium*. *The Lancet Infectious Diseases, 15,* 85–94. https://doi.org/10.1016/S1473-3099(14)70772-8.

Consortium T 1000 GP, 1000 Genomes Project Consortium, Auton, A., Brooks, L. D., Durbin, R. M., Garrison, E. P., Kang, H. M., Korbel, J. O., et al. (2015). A global reference for human genetic variation. *Nature, 526,* 68–74. https://doi.org/10.1038/nature15393.

Gilchrist, C. A., Cotton, J. A., Burkey, C., Arju, T., Gilmartin, A., Lin, Y., et al. (2018). Genetic diversity of *Cryptosporidium hominis* in a Bangladeshi community as revealed by whole-genome sequencing. *Journal of Infectious Diseases, 218,* 259–264. https://doi.org/10.1093/infdis/jiy121.

Guo, Y., Li, N., Lysén, C., Frace, M., Tang, K., Sammons, S., et al. (2015a). Isolation and enrichment of *Cryptosporidium* DNA and verification of DNA purity for whole-genome sequencing. *Journal of Clinical Microbiology, 53,* 641–647. https://doi.org/10.1128/JCM.02962-14.

Guo, Y., Tang, K., Rowe, L. A., Li, N., Roellig, D. M., Knipe, K., et al. (2015b). Comparative genomic analysis reveals occurrence of genetic recombination in virulent *Cryptosporidium hominis* subtypes and telomeric gene duplications in *Cryptosporidium parvum*. *BMC Genomics, 16,* 320. https://doi.org/10.1186/s12864-015-1517-1.

Hadfield, S. J., Pachebat, J. A., Swain, M. T., Robinson, G., Cameron, S. J., Alexander, J., et al. (2015). Generation of whole genome sequences of new *Cryptosporidium hominis* and *Cryptosporidium parvum* isolates directly from stool samples. *BMC Genomics, 16,* 650. https://doi.org/10.1186/s12864-015-1805-9.

Haldar, K., Bhattacharjee, S., & Safeukui, I. (2018). Drug resistance in *Plasmodium*. *Nature Reviews Microbiology, 16,* 156–170. https://doi.org/10.1038/nrmicro.2017.161.

Ifeonu, O. O., Chibucos, M. C., Orvis, J., Su, Q., Elwin, K., Guo, F., et al. (2016). Annotated draft genome sequences of three species of *Cryptosporidium*: *Cryptosporidium meleagridis* isolate

UKMEL1, *C. baileyi* isolate TAMU-09Q1 and *C. hominis* isolates TU502_2012 and UKH1. *Pathogens and Disease, 74*, ftw080. https://doi.org/10.1093/femspd/ftw080.

Insulander, M., Silverlås, C., Lebbad, M., Karlsson, L., Mattsson, J. G., & Svenungsson, B. (2013). Molecular epidemiology and clinical manifestations of human cryptosporidiosis in Sweden. *Epidemiology and Infection, 141,* 1009–1020. https://doi.org/10.1017/S0950268812001665.

Isaza, J. P., Galván, A. L., Polanco, V., Huang, B., Matveyev, A. V., Serrano, M. G., et al. (2015). Revisiting the reference genomes of human pathogenic *Cryptosporidium* species: Reannotation of *C. parvum* Iowa and a new *C. hominis* reference. *Scientific Reports, 5,* 16324. https://doi.org/10.1038/srep16324.

Islam, S., Begum, H. A., & Nili, N. Y. (2010). Bacteriological safety assessment of municipal tap water and quality of bottle water in Dhaka City: Health hazard analysis. *Bangladesh Journal of Medical Microbiology, 4,* 9–13. https://doi.org/10.3329/bjmm.v4i1.8462.

Jex, A. R., & Gasser, R. B. (2010). Genetic richness and diversity in *Cryptosporidium hominis* and *C. parvum* reveals major knowledge gaps and a need for the application of "next generation" technologies—Research review. *Biotechnology Advances, 28,* 17–26. https://doi.org/10.1016/j.biotechadv.2009.08.003.

King, P., Tyler, K. M., & Hunter, P. R. (2019). Anthroponotic transmission of *Cryptosporidium parvum* predominates in countries with poorer sanitation: A systematic review and meta-analysis. *Parasites & Vectors, 12,* 16. https://doi.org/10.1186/s13071-018-3263-0.

Korpe, P. S., Gilchrist, C. A., Burkey, C., Taniuchi, M., Ahmed, E., Madan, V., et al. (2018). Case-control study of *Cryptosporidium* transmission in Bangladeshi households. *Clinical Infectious Diseases*. https://doi.org/10.1093/cid/ciy593.

Korpe, P. S., Haque, R., Gilchrist, C. A., Valencia, C., Niu, F., Lu, M. M., et al. (2016). Natural history of cryptosporidiosis in a longitudinal study of slum-dwelling Bangladeshi children: Association with severe malnutrition. *PLoS Neglected Tropical Diseases, 10,* e0004564. https://doi.org/10.1371/journal.pntd.0004564.

Korpe, P. S., Liu, Y., Siddique, A., Kabir, M., Ralston, K., Ma, J. Z., et al. (2013). Breast milk parasite-specific antibodies and protection from amebiasis and cryptosporidiosis in Bangladeshi infants: A prospective cohort study. *Clinical Infectious Diseases, 56,* 988–992. https://doi.org/10.1093/cid/cis1044.

Kosek, M., Alcantara, C., Lima, A. A., & Guerrant, R. L. (2001). Cryptosporidiosis: An update. *The Lancet Infectious Diseases, 1,* 262–269. https://doi.org/10.1016/S1473-3099(01)00121-9.

Kotloff, K. L., Nataro, J. P., Blackwelder, W. C., Nasrin, D., Farag, T. H., Panchalingam, S., et al. (2013). Burden and aetiology of diarrhoeal disease in infants and young children in developing countries (the Global Enteric Multicenter Study, GEMS): A prospective, case-control study. *Lancet, 382,* 209–222. https://doi.org/10.1016/S0140-6736(13)60844-2.

Lequime, S., Fontaine, A., Ar Gouilh, M., Moltini-Conclois, I., & Lambrechts, L. (2016). Genetic drift, purifying selection and vector genotype shape dengue virus intra-host genetic diversity in mosquitoes. *PLoS Genetics, 12,* e1006111. https://doi.org/10.1371/journal.pgen.1006111.

Li, N., Xiao, L., Cama, V. A., Ortega, Y., Gilman, R. H., Guo, M., et al. (2013). Genetic recombination and *Cryptosporidium hominis* virulent subtype IbA10G2. *Emerging Infectious Diseases, 19,* 1573–1582. https://doi.org/10.3201/eid1910.121361.

Manjunatha, U. H., Chao, A. T., Leong, F. J., & Diagana, T. T. (2016). Cryptosporidiosis drug discovery: Opportunities and challenges. *ACS Infectious Diseases, 2,* 530–537. https://doi.org/10.1021/acsinfecdis.6b00094.

Manjunatha, U. H., Vinayak, S., Zambriski, J. A., Chao, A. T., Sy, T., Noble, C. G., et al. (2017). A *Cryptosporidium* PI(4)K inhibitor is a drug candidate for cryptosporidiosis. *Nature, 546,* 376–380. https://doi.org/10.1038/nature22337.

Mu, J., Awadalla, P., Duan, J., McGee, K. M., Joy, D. A., McVean, G. A. T., et al. (2005). Recombination hotspots and population structure in *Plasmodium falciparum*. *PLoS Biology, 3,* e335. https://doi.org/10.1371/journal.pbio.0030335.

Nader, J. L., Mathers, T. C., Ward, B. J., Pachebat, J. A., Swain, M. T., Robinson, G., et al. (2019). Evolutionary genomics of anthroponosis in *Cryptosporidium*. *Nature Microbiology*. https://doi.org/10.1038/s41564-019-0377-x.

Pal, S., Bhattacharya, S. K., Das, P., Chaudhuri, P., Dutta, P., De, S. P., et al. (1989). Occurrence and significance of *Cryptosporidium* infection in Calcutta. *Transactions of the Royal Society of Tropical Medicine and Hygiene, 83,* 520–521.

Pickering, A. J., Crider, Y., Sultana, S., Swarthout, J., Goddard, F. G., Anjerul Islam, S., et al. (2019). Effect of in-line drinking water chlorination at the point of collection on child diarrhoea in urban Bangladesh: A double-blind, cluster-randomised controlled trial. *The Lancet Global Health, 7,* e1247–e1256. https://doi.org/10.1016/S2214-109X(19)30315-8.

Platts-Mills, J. A., McCormick, B. J. J., Kosek, M., Pan, W. K., Checkley, W., Houpt, E. R., et al. (2014). Methods of analysis of enteropathogen infection in the MAL-ED cohort study. *Clinical Infectious Diseases, 59,* S233–S238. https://doi.org/10.1093/cid/ciu408.

Prüss-Ustün, A., Wolf, J., Bartram, J., Clasen, T., Cumming, O., Freeman, M. C., et al. (2019). Burden of disease from inadequate water, sanitation and hygiene for selected adverse health outcomes: An updated analysis with a focus on low- and middle-income countries. *International Journal of Hygiene and Environmental Health, 222,* 765–777. https://doi.org/10.1016/j.ijheh.2019.05.004.

Pumipuntu, N., & Piratae, S. (2018). Cryptosporidiosis: A zoonotic disease concern. *Veterinary World, 11*, 681–686. https://doi.org/10.14202/vetworld.2018.681-686.

Rafaluk, C., Gildenhard, M., Mitschke, A., Telschow, A., Schulenburg, H., & Joop, G. (2015). Rapid evolution of virulence leading to host extinction under host-parasite coevolution. *BMC Evolutionary Biology, 15,* 112. https://doi.org/10.1186/s12862-015-0407-0.

Rossignol, J., Kabil, S. M., El-Gohary, Y., & Younis, A. M. (2006). Effect of nitazoxanide in diarrhea and enteritis caused by *Cryptosporidium* species. *Clinical Gastroenterology and Hepatology, 4,* 320–324. https://doi.org/10.1016/j.cgh.2005.12.020.

Rossle, N. F., & Latif, B. (2013). Cryptosporidiosis as threatening health problem: A review. *Asian Pacific Journal of Tropical Biomedicine, 3,* 916–924. https://doi.org/10.1016/S2221-1691(13)60179-3.

Sikora, P., Andersson, S., Winiecka-Krusnell, J., Hallström, B., Alsmark, C., Troell, K., et al. (2017). Genomic variation in IbA10G2 and other patient-derived *Cryptosporidium hominis* subtypes. *Journal of Clinical Microbiology, 55,* 844–858. https://doi.org/10.1128/JCM.01798-16.

Sirajul Islam, M., Brooks, A., Kabir, M. S., Jahid, I. K., Shafiqul Islam, M., Goswami, D., et al. (2007). Faecal contamination of drinking water sources of Dhaka city during the 2004 flood in Bangladesh and use of disinfectants for water treatment. *Journal of Applied Microbiology, 103,* 80–87. https://doi.org/10.1111/j.1365-2672.2006.03234.x.

Sow, S. O., Muhsen, K., Nasrin, D., Blackwelder, W. C., Wu, Y., Farag, T. H., et al. (2016). The burden of *Cryptosporidium* diarrheal disease among children <24 months of age in moderate/high mortality regions of Sub-Saharan Africa and South Asia, utilizing data from the global enteric multicenter study (GEMS). *PLoS Neglected Tropical Diseases, 10,* e0004729. https://doi.org/10.1371/journal.pntd.0004729.

Steiner, K. L., Ahmed, S., Gilchrist, C. A., Burkey, C., Cook, H., Ma, J. Z., et al. (2018). Species of cryptosporidia causing subclinical infection associated with growth faltering in rural and urban Bangladesh: A birth cohort study. *Clinical Infectious Diseases, 67,* 1347–1355. https://doi.org/10.1093/cid/ciy310.

U.S. Department of Health and Human Services. (2003). New drug for parasitic infections in children. *FDA Consumer, 37*, 4.

Van den Bergh, B., Swings, T., Fauvart, M., & Michiels, J. (2018). Experimental design, population dynamics, and diversity in microbial experimental evolution. *Microbiology and Molecular Biology Reviews, 82*. https://doi.org/10.1128/MMBR.00008-18.

Wang, N., Akey, J. M., Zhang, K., Chakraborty, R., & Jin, L. (2002). Distribution of recombination crossovers and the origin of haplotype blocks: The interplay of population history, recombination, and mutation. *American Journal of Human Genetics, 71,* 1227–1234. https://doi.org/10.1086/344398.

Widmer, G., & Lee, Y. (2010). Comparison of single- and multilocus genetic diversity in the protozoan parasites *Cryptosporidium parvum* and *C. hominis*. *Applied and Environment Microbiology, 76,* 6639–6644. https://doi.org/10.1128/AEM.01268-10.

Xu, P., Widmer, G., Wang, Y., Ozaki, L. S., Alves, J. M., Serrano, M. G., et al. (2004). The genome of *Cryptosporidium hominis*. *Nature, 431,* 1107–1112. https://doi.org/10.1038/nature02977.

Working Group on Waterborne Cryptosporidiosis. (1995). *Cryptosporidium and Water: A Public Health Handbook*. Georgia: Atlanta.

Chapter 8
The Current Molecular Epidemiological Scenario of *Cryptosporidium*, *Giardia* and *Blastocystis* in Spain. Implication for Public Health

Pamela C. Köster and David Carmena

Abstract The enteric protozoan parasites *Cryptosporidium* spp. and *Giardia duodenalis* are major contributors to the burden of gastrointestinal diseases globally. Both pathogens primarily affect children living in resource-poor settings with limited or no access to clean water and sanitation facilities, but are also significant public health threats in developed countries. Additionally, *Cryptosporidium* spp. and *G. duodenalis* are common causes of waterborne and foodborne outbreaks of gastrointestinal disease globally. Besides, the Stramenopile *Blastocystis* sp. is the most common eukaryotic organism reported in the human gut. Although its pathogenicity is a topic of debate, there is increasing evidence demonstrating that this protist can be associated with gastrointestinal disorders (diarrhoea, irritable bowel syndrome) and extra-intestinal manifestations, including urticaria. Because *Cryptosporidium* spp., *G. duodenalis* and *Blastocystis* sp. share the same transmission (faecal-oral) route, are able to infect a wide range of animal species other than humans with variable host specificities, and their infective forms are environmentally resilient, the study of these pathogens should be ideally approached under the One Health umbrella. In this context, molecular-based methods including PCR and sequencing provide powerful tools to investigate the epidemiology and transmission of these parasites. In Spain, cryptosporidiosis and giardiosis, but not blastocystosis, are notifiable diseases. However, the true incidence of these infections remain largely unknown because underdiagnosing and underreporting. Symptomatic cryptosporidiosis and giardiosis disproportionally affect children under four years of age, but we know now that subclinical infections are also common in apparently healthy individuals of all age groups. However, molecular data regarding the frequency and diversity of these pathogens are limited and spatially and temporally discontinuous. This chapter aims to provide, from a public veterinary health perspective, an updated account on the epidemiology of *Cryptosporidium*, *G. duodenalis* and *Blastocystis* in Spain, with an emphasis on the description of the species/genotypes circulating in symptomatic and asymptomatic human populations. Current knowledge on the presence of these pathogens in production (livestock), companion (dogs and cats) and wildlife animal

P. C. Köster · D. Carmena (✉)
Parasitology Reference and Research Laboratory, National Centre for Microbiology, Ctra. Majadahonda-Pozuelo Km 2, 28220 Majadahonda, Madrid, Spain
e-mail: dacarmena@isciii.es

© Springer Nature Switzerland AG 2020
N. Guillen (ed.), *Eukaryome Impact on Human Intestine Homeostasis and Mucosal Immunology*, https://doi.org/10.1007/978-3-030-44826-4_8

species is also discussed, including their potential role as natural reservoirs of human infections, and the available evidence of zoonotic (and anthroponotic) transmission events.

Keywords *Blastocystis* · *Cryptosporidium* · *Giardia* · Diarrhoea · Epidemiology · Molecular characterization · Host specificity · Speciation · Experimental infection · Natural infection · Transmission

Introduction

The enteric protozoan parasites *Cryptosporidium* spp. (Apicomplexa: Cryptosporidiidae), and *Giardia duodenalis* (Metamonada: Hexamitidae) are among the most important causes of protozoan-diarrheal illness in humans and animals globally. Clinical manifestations vary from asymptomatic carriage to acute diarrhoea and chronic disease. Transmission of giardiosis and cryptosporidiosis is through the faecal-oral route, either indirectly via accidental ingestion of contaminated water or food, or directly via human-to-human or animal-to-human contact. Approximately 200 million people in Asia, Africa and Latin America present symptomatic giardiosis with some 500,000 new cases reported each year (WHO 1996). Cryptosporidiosis is a leading cause of diarrhoeal death in children younger than 5 years in sub-Saharan Africa and Southeast Asia, only second after rotaviral enteritis (GBD 2016). Both giardiosis and cryptosporidiosis primarily affect young children living in low-income countries and are strongly linked with malnutrition, growth faltering, and cognitive impairment (Guerrant et al. 1999; Halliez and Buret 2013). These very same disorders, although to a much lower extent, have been described in some Spanish paediatric populations (Azcona-Gutiérrez et al. 2017). The Stramenopile *Blastocystis* sp. is one of the most common enteric parasites found in human stool samples. It is estimated that up to 1 billion people would be colonized/infected with this protozoan species globally (Clark et al. 2013). The clinical and public health relevance of *Blastocystis* sp. remains controversial, as it is commonly found in both apparently healthy individuals and patients suffering from intestinal (diarrhoea, irritable bowel syndrome) and extra-intestinal (urticaria) manifestations (Scanlan et al. 2014). Hence, it is currently not possible to distinguish between colonization and infection.

At least 38 *Cryptosporidium* species have been named today, of which *C. hominis* and *C. parvum* account for ~90% of human cryptosporidiosis cases globally (Ryan et al. 2014). The only *Giardia* species that is pathogenic to humans, *G. duodenalis*, is indeed a species complex consisting of eight (A to H) assemblages; assemblages A and B, which commonly infect humans and other mammals, assemblages C and D in dogs and other canids, assemblage E in hoofed animals, assemblage F in cats, assemblage G in rodents and assemblage H in pinnipeds (Cacciò et al. 2018). Extensive genetic diversity has also been demonstrated within *Blastocystis* sp., allowing the differentiation of at least 17 subtypes (ST), of which STs 1–9 and 12 infect humans, among other mammal species (Stensvold et al. 2007; Ramírez et al. 2016).

Blastocystis ST1–4 are the subtypes more frequently identified in humans. Because *Cryptosporidium* spp., *G. duodenalis*, and *Blastocystis* sp. genetic variants have marked differences in host range and specificity, molecular epidemiological surveys are essential not only to determine the frequency and diversity of species/genotypes in a given population or geographical region, but also to characterize transmission pathways, to identify the range of host species able to harbour these pathogens, and to assess zoonotic (or anthroponotic) potential.

The Spanish Epidemiological Scenario

The epidemiology of *G. duodenalis* and *Cryptosporidium* spp. in Spain has been the subject of previous reviews (Navarro-i-Martinez et al. 2011; Carmena et al. 2012), so the interested reader is referred to them for in-depth information. Regarding human infections, most of the available data comes from observational, transversal studies conducted in schoolchildren populations by conventional (microscopy) diagnostic methods. In brief, reported prevalence rates of human giardiosis ranged from 3–7% in asymptomatic individuals to 13–25% in symptomatic individuals. *Giardia duodenalis* infections have also been identified in ovine (3–44%), bovine (5–30%), canine (5–20%), feline (14%), and wildlife (1–16%) animal populations (Table 8.1). Similarly, *Cryptosporidium* infections were more frequently found in individuals declaring gastrointestinal complaints (5–30%) than in asymptomatic subjects (1–5%). *Cryptosporidium* spp. infections were a common cause of diarrhoea in young calves (4–60%) and lambs (5–90%). In contrast, *Cryptosporidium* spp. infections in domestic dogs (7–15%) and cats (5–10%) were mostly reported in asymptomatic animals (Table 8.1).

Importantly, *Giardia* cysts and *Cryptosporidium* oocysts have been found in up to 100% of the surface waters intended for human or agricultural consumption (Carmena et al. 2007; Castro-Hermida et al. 2009), including recreational river areas (Castro-Hermida et al. 2010). When the viability of the (oo)cysts was assessed with vital dyes, infective forms of *Giardia* and *Cryptosporidium* were detected in 57–61% of the samples analysed (Castro-Hermida et al. 2010). (Oo)cysts of both pathogens were also a frequent finding (16–60%) in packed salads and other fresh produce (e.g. Amorós et al. 2010). The fact that infective *G. duodenalis* cysts and *Cryptosporidium* oocysts are regularly found in drinking waters and ready-to-eat, fresh meals explain why these pathogens are common causes of waterborne and foodborne outbreaks of diarrhoeal illness globally (Efstratiou et al. 2017; Ryan et al. 2018, 2019).

In addition, *Giardia* cysts and *Cryptosporidium* oocysts have been isolated from soil samples in Spanish public parks (Martínez-Moreno et al. 2007; Dado et al. 2012). Although these studies did not assess neither the viability/infectivity of the parasite forms nor their species/genotypes, environmental contamination with faecal material from companion animals or even humans can be a source of infection to people in close contact with contaminated soil, particularly children playing in sand pits on public grounds.

Table 8.1 Main surveillance, epidemiological, and research features of *Giardia duodenalis*, *Cryptosporidium* spp., and *Blastocystis* sp. in Spain

	Giardia duodenalis	*Cryptosporidium* spp.	*Blastocystis* sp.
Surveillance			
Notifiable disease?	Yes[c]	Yes[c]	No
Official number of cases[a]	1627	646	Unknown
Time series?	No	No	No
Outbreak investigation	Limited	Limited	No
Prevalence[b] *(%)*			
Asymptomatic humans	3–7	1–5	3–22
Symptomatic humans	13–25	5–30	8–23
Ovine	3–44	5–90	Unknown
Bovine	5–30	4–60	2
Dogs	5–20	7–15	0
Cats	14	5–10	0
Wildlife	1–16	1–27	2
Surface waters	25–100	15–90	Unknown
Fresh produce	16–50	20–60	Unknown
Research			
Genotyping studies?	Limited	Limited	Limited
Transmission studies?	Limited	Limited	Limited

[a]Confirmed cases reported to the European Centre for Disease Prevention and Control in 2015
[b]Composite data from references Carmena et al. (2012), Navarro-i-Martinez et al. (2011) and Paulos et al. (2018)
[c]Since 2015

Comparatively far lesser epidemiological information is currently available on *Blastocystis* carriage/infection in Spain (reviewed in Paulos et al. 2018). As in the case of *G. duodenalis* and *Cryptosporidium* infections, most of the research conducted on *Blastocystis* sp. has been generated in microscopy-based prospective (Del Aguila et al. 1997; Paulos et al. 2018) or clinical retrospective (González-Moreno et al. 2011; Salvador et al. 2016) studies (Table 8.1). *Blastocystis* sp. has also been identified in intensively reared pigs (Navarro et al. 2008) and farmed ostriches and rheas (Ponce Gordo et al. 2002).

Current Surveillance and Research Limitations

Cryptosporidiosis and giardiosis, but not blastocystosis, are compulsory notifiable diseases in Spain since March 2015 (Table 8.1). Based on official data, a total of 1627 and 646 confirmed cases of giardiosis and cryptosporidiosis were reported in Spain in

2015 (ECDC 2018a, b). Of note, these figures represent a fraction of the true number of cases. Variable health care seeking behaviour by patients, poor awareness among primary care physicians, underdiagnosing, and underreporting are all important contributing factors to this problem. When uncertainties associated to those factors are considered, the true number of cases are expected to be 2–3 order of magnitude higher than the declared number of cases (Cacciò and Chalmers 2016). Information regarding the situation of human blastocystosis is even scarcer. In practical terms this general lack of information means that the seasonal distribution of these enteric parasites along the year are largely unknown. Historical case series at national scale are also lacking. Seasonal patterns have been investigated at regional level during relatively short periods of time. These studies have evidenced that *Cryptosporidium* cases peak during late summer and early autumn (Abal-Fabeiro et al. 2015; Azcona-Gutiérrez et al. 2017). The very same temporal trend has been previously reported in other European countries including Germany, The Netherlands, and UK and has been associated to increased travel and exposure to recreational water (Chalmers et al. 2009; Fournet et al. 2013).

Importantly, molecular studies investigating the occurrence and genetic variability of *Cryptosporidium* spp., *G. duodenalis* and *Blastocystis* sp. infections in human and animal populations and in environmental (water, soil) samples in Spain are limited (reviewed in Navarro-i-Martinez et al. 2011; Carmena et al. 2012; Paulos et al. 2018). Active molecular-based epidemiological surveillance has been demonstrated as a very useful tool in outbreak investigations (e.g. Fuentes et al. 2015) and to identify novel or rare species/genotypes of enteric parasites in clinical samples (Martínez-Ruiz et al. 2016; Merino et al. 2019; Millán et al. 2019).

No routine surveillance or monitoring programs to detect the presence of the infective stages of *Cryptosporidium* spp., *G. duodenalis*, and *Blastocystis* sp. in drinking water and fresh produce are currently in place in Spain. Indeed, current legislation does not specify the maximum concentration levels of these pathogens in water and food matrices.

Molecular Epidemiological Research in Humans

The molecular diversity of *G. duodenalis*, *Cryptosporidium* spp., and *Blastocystis* sp. in human infections in Spain has been mainly assessed in paediatric populations with or without clinical manifestations. In some instances, these parasites were accidentally identified during routine testing at clinical laboratories. This is particularly true for *Blastocystis* sp. Available genotyping data indicate that *Giardia duodenalis* assemblage B is more prevalent than assemblage A in most of the surveys conducted (Sahagún et al. 2008; de Lucio et al. 2015; Gabín-García et al. 2017), although both genetic variants have been detected at similar frequencies in other studies (Azcona-Gutiérrez et al. 2017). At the sub-genotype level BIV is the most common sub-assemblage circulating in those clinical patient populations. Similarly, sub-assemblage BIV has been the predominant genotype found in children attending day care centres (Mateo et al. 2014), in schoolchildren (Cardona et al. 2011), and in community surveys (de Lucio et al. 2017). These data suggest that, besides parasite genotype, other determinants (e.g. co-infections, host immune status, microbiome) may play a role in the progression from asymptomatic carriage to symptomatic infection. A recent study investigating *G. duodenalis* infections in individuals of all ages with and without clinical symptoms revealed that children were more commonly infected by assemblage B, whereas asymptomatic infection was more common in patients with assemblage A than in those with assemblage B (Wang et al. 2019). Importantly, asymptomatic carriage of *G. duodenalis* has been demonstrated by PCR in a larger than expected proportion of apparently healthy schoolchildren attending primary and secondary schools in Madrid (Reh et al. 2019). This finding may have important consequences as unnoticed school and household transmission events could represent a public health issue for at-risk populations such as young children, the elderly and immunocompromised individuals.

Regarding cryptosporidiosis, human cases in Spain are primarily caused by *C. hominis* (80–90%) and *C. parvum* (10–18%) (de Lucio et al. 2015; Segura et al. 2015; Azcona-Gutiérrez et al. 2017), but these frequencies may vary depending on the region considered. For instance, in the Autonomous region of Galicia (northwest Spain), it was *C. parvum* and not *C. hominis* the main *Cryptosporidium* species detected in summer peaks during the period 2000–2008, particularly in rural areas (Abal-Fabeiro et al. 2015). Considering that this Spanish region breeds and feeds large populations of dairy and beef cattle, the finding mentioned above has been interpreted as the consequence of zoonotic *C. parvum* transmission. Infections by less common *Cryptosporidium* species including *C. meleagridis* (Abal-Fabeiro et al. 2014; Segura et al. 2015), *C. canis* (Abal-Fabeiro et al. 2014; de Lucio et al. 2016), and *C. cuniculus* (Martínez-Ruiz et al. 2016) have also been reported sporadically, some of them reflecting zoonotic events. These surveys revealed that IbA10G2 and IIdA15G2 were the sub-genotypes more prevalent within *C. hominis* and *C. parvum*, respectively. Taken together, available molecular data seem to indicate that human giardiosis and cryptosporidiosis in Spain are primarily of anthroponotic origin, with a comparatively smaller contribution from zoonotic reservoirs in production and companion animals (Abal-Fabeiro et al. 2014, 2015; Azcona-Gutiérrez et al. 2017).

In Spain, ST4 has been demonstrated as the most prevalent (94%) *Blastocystis* subtype circulating in symptomatic, mono-infected patients (Domínguez-Márquez et al. 2009). Interestingly, ST4 has been detected at considerable lower rates than ST1–3 in general (Paulos et al. 2018) and schoolchildren (Reh et al. 2019) populations in different Spanish geographical regions, suggesting that ST4 may be more pathogenic than other *Blastocystis* subtypes commonly seen in humans.

Molecular Epidemiological Research in Livestock

In Spain, molecular genotyping studies have been conducted in *G. duodenalis*-positive samples from cattle, sheep, and goats; these surveys revealed that assemblage E was the most prevalent genotype in all the species studied, ranging from 64 to 100% in cattle and sheep and 100% in goats (Castro-Hermida et al. 2006, 2007, 2011a; Cardona et al. 2015). Zoonotic assemblage B was identified in 8–35% of sheep, whereas assemblage A sub-assemblage AI was found in 1% of goats. In cattle, genetic analyses for *G. duodenalis* have been conducted in asymptomatic animals including neonatal calves, heifers, and cows (Castro-Hermida et al. 2007, 2011a). Besides assemblage E, sub-assemblage AI was also identified in 36% of cattle.

Cryptosporidium infections in cattle have a marked age-related pattern; thus, *C. parvum* is more prevalent in pre-weaned calves, whereas *C. bovis*, *C. andersoni*, and *C ryanae* are more frequently found in post-weaned calves, heifers, and adult animals (Santín et al. 2008). Molecular epidemiological data from Spanish herds confirm this trend (Castro-Hermida et al. 2007, 2011a; Quílez et al. 2008b; Cardona et al. 2015). Calves, lambs, and goats infected with *C. parvum* have been demonstrated to harbour zoonotic sub-genotypes (e.g. IIaA15G2) of the parasite and should, therefore, considered as natural reservoirs of human cryptosporidiosis (Quílez et al. 2008a; Díaz et al. 2010). Intriguingly, feline-specific *G. duodenalis* assemblage G and *C. felis* have been detected in cattle in northern Spain (Cardona et al. 2015). This finding reveals that inter-species transmission of apparently host-specific species/genotypes is possible when environmental conditions are favourable.

Overall, molecular epidemiological data from Spain were in line with those suggesting that the actual role of livestock as major reservoirs for *G. duodenalis* and *Cryptosporidium* spp. infections to humans might be much less relevant than initially anticipated (O'Handley 2007; Feng and Xiao 2011). The molecular diversity of *Blastocystis* sp. carriage/infections in livestock is unknown at present.

Molecular Epidemiological Research in Companion Animals

Molecular epidemiological studies investigating the genetic diversity of *G. duodenalis*, *Cryptosporidium* spp., and *Blastocystis* sp. in domestic dogs and cats in Spain are particularly scarce. These studies (conducted in stray, sheltered, and owned animals) showed that both *G. duodenalis* and *Cryptosporidium* infections were primarily caused by canine-specific (e.g. *G. duodenalis* assemblages C and D, *C. canis*) or feline-specific (e.g. *G. duodenalis* assemblage F, *C. felis*) genetic variants of these parasites (Ortuño et al. 2014; de Lucio et al. 2017). However, other surveys have demonstrated that dogs can harbour infections by potentially zoonotic *G. duodenalis* sub-assemblages AII, BIII, and BIV. However, most of these genetic variants contained single nucleotide polymorphisms that were not present in sequences of human origin belonging to the same sub-assemblages. Taken together, these facts seem to suggest that sub-assemblages AII, BIII, and BIV may be naturally circulating in canine populations (Gil et al. 2017; Adell-Aledón et al. 2018; Sanchez-Thevenet et al. 2019). Zoonotic transmission of giardiasis and cryptosporidiosis could not been demonstrated between pet dogs and cats and their owners in northern Spain, indicating that this kind of events should be relatively infrequent. Intriguingly , *C. hominis* of unknown sub-genotype was detected in a sheltered dog in this very same geographical region (Gil et al. 2017). Whether this finding represents a true infection or an accidental acquisition and mechanical carriage of *C. hominis* oocysts of anthroponotic origin via environmental contamination remains to be clarified in further molecular studies. Neither companion dogs nor cats seem suitable hosts for *Blastocystis* sp. (Paulos et al. 2018).

Molecular Epidemiological Research in Wildlife

Previous epidemiological studies have reported the occurrence of *Giardia duodenalis* and *Cryptosporidium* spp. infections in small rodents and insectivores (Torres et al. 2000), birds (Reboredo-Fernández et al. 2015; Cano et al. 2016), ungulates including roe deer and wild boars (Castro-Hermida et al. 2011a, b), and several carnivore species including badgers, red foxes, genets, beech martens, otters, and wolves (Méndez-Hermida et al. 2007; Mateo et al. 2017). Available molecular data revealed that zoonotic *G. duodenalis* assemblage B was present in buzzards, quails, and magpies (Reboredo-Fernández et al. 2015). The finding of BIV (the *G. duodenalis* sub-assemblage more prevalent in Spanish human populations) in waterfowl may represent a public health concern, as these birds commonly nest, breed, and feed in source waters intended for human consumption (Cano et al. 2016).

Since its naming in 2002 (Morgan-Ryan et al. 2002), *C. hominis* has been largely recognized as a human-specific species. However, this notion has been challenged by recent experimental and molecular epidemiological data demonstrating that *C. hominis* is able to successfully infect a wide range of mammal species other than human and non-human primates including cattle (e.g. Razakandrainibe et al. 2018), sheep (e.g. Connelly et al. 2013), donkeys (Jian et al. 2016), kangaroos (Zahedi et al. 2018), and field mice (Čondlová et al. 2018), among others. In Spain, no evidence of *C. hominis* infections circulating in livestock has been described yet, but *C. hominis* oocysts have been identified in the intestinal tract of a domestic dog (Gil et al. 2017), a badger (Mateo et al. 2017) and four red foxes (Montoya et al. 2019). These findings may be just the consequence of passive carriage of ingested oocysts but may also represent true infections as consequence of spill-over events of anthroponotic *C. hominis* from humans to companion and wildlife (Fig. 8.1).

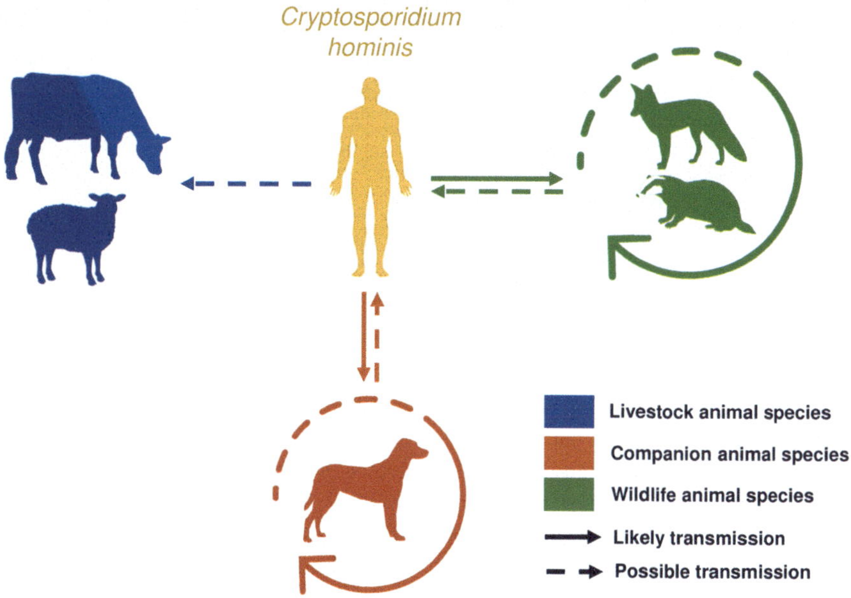

Fig. 8.1 Potential transmission pathways of *Cryptosporidium hominis* among human, livestock, companion, and wildlife animal species in Spain according to current molecular epidemiological data

Molecular Epidemiological Research in Surface Waters

Giardia cysts and *Cryptosporidium* oocysts have been reported in up to 100% of the surface waters intender for human consumption, and in 27% of tap water from municipalities in northern Spain with chlorination treatment only (Carmena et al. 2007). Both parasites have also been detected in the final effluent of drinking water treatment plants at concentrations of 0–4 (oo)cysts per litre in northwest Spain (Castro-Hermida et al. 2008, 2015), and at concentrations <1 (oo)cyst per litre in north-eastern Spain (Ramo et al. 2017). Vital staining dyes revealed that up to 95% of the (oo)cysts identified were viable (Castro-Hermida et al. 2015). Sequence analyses allowed the identification of a wide range of species/genotypes, including *C. hominis*, *C. parvum*, *C. andersoni*, *C. ubiquitum*, and *C. muris* within *Cryptosporidium*, and assemblages A (including sub-assemblages AI and AII) and E within *G. duodenalis* (Galván et al. 2014; Castro-Hermida et al. 2015). In addition, *Cryptosporidium* oocysts have also been detected in 19% of public swimming pools with a maximum concentration of 13 oocysts per litre (Gracenea et al. 2018). Considered together, these results demonstrate that both *Cryptosporidium* spp. and *G. duodenalis* are ubiquitous in environmental and recreative waters in Spain. This fact, together with the ineffectiveness of treatments in drinking water treatment plants in eliminating/inactivating the (oo)cysts of both protozoan species represent a perceptible risk for waterborne outbreaks of giardiosis and/or cryptosporidiosis, and a serious concern for the water industry and the public health authorities responsible for assuring the safety and quality of drinking water.

Concluding Remarks

Despite the unquestionable progress achieved in the last two decades, the epidemiology of the diarrhoea-causing enteric parasites *G. duodenalis*, *Cryptosporidium* spp., and *Blastocystis* sp. in Spain remain insufficiently understood. Current molecular data demonstrate that the transmission of these pathogens and the factors that determine their pathogenicity are more intricate than initially anticipated and involve interconnected human, animal, and environmental reservoirs. The species/genotypes involved in the infections are an important factor in determining the pathogenicity of the infection, but it is not the only determinant in the equation. Other variables including co-infections, host age and immune status, and microbiota abundance and diversity are also key actors in tipping the balance between health and disease.

Funding Research summarized in this chapter and conducted at the Spanish National Centre for Microbiology was funded by the Health Institute Carlos III (ISCIII), Ministry of Economy and Competitiveness (Spain), under projects CP12/03081 and PI16CIII/00024. The funder had no role in study design, data collection and analysis, decision to publish or preparation of the manuscript.

Disclosure of Interests All authors declare they have no conflict of interest.

References

Abal-Fabeiro, J. L., Maside, X., Llovo, J., & Bartolomé, C. (2015). Aetiology and epidemiology of human cryptosporidiosis cases in Galicia (NW Spain), 2000–2008. *Epidemiology and Infection, 143,* 3022–3035.

Abal-Fabeiro, J. L., Maside, X., Llovo, J., Bello, X., Bello, X., Torres, M., et al. (2014). High-throughput genotyping assay for the large-scale genetic characterization of *Cryptosporidium* parasites from human and bovine samples. *Parasitology, 141,* 491–500.

Adell-Aledón, M., Köster, P. C., de Lucio, A., Puente, P., Hernández-de-Mingo, M., Sánchez-Thevenet, P., et al. (2018). Occurrence and molecular epidemiology of *Giardia duodenalis* infection in dog populations in eastern Spain. *BMC Veterinary Research, 14,* 26.

Amorós, I., Alonso, J. L., & Cuesta, G. (2010). *Cryptosporidium* oocysts and *Giardia* cysts on salad products irrigated with contaminated water. *Journal of Food Protection, 73,* 1138–1140.

Azcona-Gutiérrez, J. M., de Lucio, A., Hernández-de-Mingo, M., García-García, C., Soria-Blanco, L. M., Morales, L., et al. (2017). Molecular diversity and frequency of the diarrheagenic enteric protozoan *Giardia duodenalis* and *Cryptosporidium* spp. in a hospital setting in Northern Spain. *PLoS One, 12,* e0178575.

Cacciò, S. M., & Chalmers, R. M. (2016). Human cryptosporidiosis in Europe. *Clinical Microbiology & Infection, 22,* 471–480.

Cacciò, S. M., Lalle, M., & Svärd, S. G. (2018). Host specificity in the *Giardia duodenalis* species complex. *Infection, Genetics and Evolution, 66,* 335–345.

Cano, L., de Lucio, A., Bailo, B., Cardona, G. A., Muadica, A. S., Lobo, L., et al. (2016). Identification and genotyping of *Giardia* spp. and *Cryptosporidium* spp. isolates in aquatic birds in the Salburua wetlands, Álava, Northern Spain. *Veterinary Parasitology, 221,* 144–148.

Cardona, G. A., Carabin, H., Goñi, P., Arriola, L., Robinson, G., Fernández-Crespo, J. C., et al. (2011). Identification and molecular characterization of *Cryptosporidium* and *Giardia* in children and cattle populations from the province of Álava, North of Spain. *Science of the Total Environment, 412–413,* 101–108.

Cardona, G. A., de Lucio, A., Bailo, B., Cano, L., de Fuentes, I., & Carmena, D. (2015). Unexpected finding of feline-specific *Giardia duodenalis* assemblage F and *Cryptosporidium felis* in asymptomatic adult cattle in Northern Spain. *Veterinary Parasitology, 209,* 258–263.

Carmena, D., Aguinagalde, X., Zigorraga, C., Fernández-Crespo, J. C., & Ocio, J. A. (2007). Presence of *Giardia* cysts and *Cryptosporidium* oocysts in drinking water supplies in northern Spain. *Journal of Applied Microbiology, 102,* 619–629.

Carmena, D., Cardona, G. A., & Sánchez-Serrano, L. P. (2012). Current situation of *Giardia* infection in Spain: Implications for public health. *World Journal of Clinical Infectious Diseases, 2,* 1–12.

Castro-Hermida, J. A., Almeida, A., González-Warleta, M., Correia da Costa, J. M., Rumbo-Lorenzo, C., & Mezo, M. (2007). Occurrence of *Cryptosporidium parvum* and *Giardia duodenalis* in healthy adult domestic ruminants. *Parasitology Research, 101,* 443–448.

Castro-Hermida, J. A., Almeida, A., González-Warleta, M., Da Costa, J. M., & Mezo, M. (2006). Prevalence and preliminary genetic analysis of *Giardia* isolated from adult sheep in Galicia (northwest Spain). *Journal of Eukaryotic Microbiology, 53,* S172–S173.

Castro-Hermida, J. A., García-Presedo, I., Almeida, A., González-Warleta, M., Correia Da Costa, J. M., & Mezo, M. (2008). Presence of *Cryptosporidium* spp. and *Giardia duodenalis* through drinking water. *Science of the Total Environment, 405,* 45–53.

Castro-Hermida, J. A., García-Presedo, I., Almeida, A., González-Warleta, M., Correia Da Costa, J. M., & Mezo, M. (2011a). *Cryptosporidium* spp. and *Giardia duodenalis* in two areas of Galicia (NW Spain). *Science of the Total Environment, 409,* 2451–2459.

Castro-Hermida, J. A., García-Presedo, I., Almeida, A., González-Warleta, M., Da Costa, J. M., & Mezo, M. (2009). Detection of *Cryptosporidium* spp. and *Giardia duodenalis* in surface water: A health risk for humans and animals. *Water Research, 43,* 4133–4142.

Castro-Hermida, J. A., García-Presedo, I., González-Warleta, M., & Mezo, M. (2010). *Cryptosporidium* and *Giardia* detection in water bodies of Galicia, Spain. *Water Research, 44,* 5887–5896.

Castro-Hermida, J. A., García-Presedo, I., González-Warleta, M., & Mezo, M. (2011b). Prevalence of *Cryptosporidium* and *Giardia* in roe deer (*Capreolus capreolus*) and wild boars (*Sus scrofa*) in Galicia (NW, Spain). *Veterinary Parasitology, 179,* 216–219.

Castro-Hermida, J. A., González-Warleta, M., & Mezo, M. (2015). *Cryptosporidium* spp. and *Giardia duodenalis* as pathogenic contaminants of water in Galicia, Spain: The need for safe drinking water. *International Journal of Hygiene and Environmental Health, 218,* 132–138.

Chalmers, R. M., Elwin, K., Thomas, A. L., Guy, E. C., & Mason, B. (2009). Long-term *Cryptosporidium* typing reveals the aetiology and species-specific epidemiology of human cryptosporidiosis in England and Wales, 2000 to 2003. *Eurosurveillance Weekly, 14,* 19086.

Clark, C. G., van der Giezen, M., Alfellani, M. A., & Stensvold, C. R. (2013). Recent developments in *Blastocystis* research. *Advances in Parasitology, 82,* 1–32.

Čondlová, Š., Horčičková, M., Sak, B., Květoňová, D., Hlásková, L., Konečný, R., et al. (2018). *Cryptosporidium apodemi* sp. n. and *Cryptosporidium ditrichi* sp. n. (Apicomplexa: Cryptosporidiidae) in *Apodemus* spp. *European Journal of Protistology, 63,* 1–12.

Connelly, L., Craig, B. H., Jones, B., & Alexander, C. L. (2013). Genetic diversity of *Cryptosporidium* spp. within a remote population of Soay Sheep on St. Kilda Islands, Scotland. *Applied and Environmental Microbiology, 79,* 2240–2246.

Dado, D., Izquierdo, F., Vera, O., & Montoya, A. (2012). Detection of zoonotic intestinal parasites in public parks of Spain. Potential epidemiological role of microsporidia. *Zoonoses and Public Health, 59,* 23–28.

de Lucio, A., Bailo, B., Aguilera, M., Cardona, G. A., Fernández-Crespo, J. C., & Carmena, D. (2017). No molecular epidemiological evidence supporting household transmission of zoonotic *Giardia duodenalis* and *Cryptosporidium* spp. from pet dogs and cats in the province of Álava, Northern Spain. *Acta Tropica, 170,* 48–56.

de Lucio, A., Martínez-Ruiz, R., Merino, F. J., Bailo, B., Aguilera, M., Fuentes, I., et al. (2015). Molecular genotyping of *Giardia duodenalis* isolates from symptomatic individuals attending two major public hospitals in Madrid, Spain. *PLoS One, 10,* e0143981.

de Lucio, A., Merino, F. J., Martínez-Ruiz, R., Bailo, B., Aguilera, M., Fuentes, I., et al. (2016). Molecular genotyping and sub-genotyping of *Cryptosporidium* spp. isolates from symptomatic individuals attending two major public hospitals in Madrid, Spain. *Infection, Genetics and Evolution, 37,* 49–56.

Del Aguila, C., Navajas, R., Gurbindo, D., Ramos, J. T., Mellado, M. J., Fenoy, S., et al. (1997). Microsporidiosis in HIV-positive children in Madrid (Spain). *Journal of Eukaryotic Microbiology, 44,* 84S–85S.

Díaz, P., Quílez, J., Chalmers, R. M., Panadero, R., López, C., Sánchez-Acedo, C., et al. (2010). Genotype and subtype analysis of *Cryptosporidium* isolates from calves and lambs in Galicia (NW Spain). *Parasitology, 137*(8), 1187–1193.

Domínguez-Márquez, M. V., Guna, R., Muñoz, C., & Borrás, R. (2009). High prevalence of subtype 4 among isolates of *Blastocystis hominis* from symptomatic patients of a health district of Valencia (Spain). *Parasitology Research, 105,* 949–955.

Efstratiou, A., Ongerth, J. E., & Karanis, P. (2017). Waterborne transmission of protozoan parasites: Review of worldwide outbreaks—An update 2011-2016. *Water Research, 114,* 14–22.

European Centre for Disease Prevention and Control. (2018a). Cryptosporidiosis—Annual Epidemiological Report for 2015. https://ecdc.europa.eu/en/publications-data/cryptosporidiosis-annual-epidemiological-report-2015. Accessed December 6, 2019.

European Centre for Disease Prevention and Control. (2018b). Giardiasis—Annual Epidemiological Report for 2015. https://ecdc.europa.eu/en/publications-data/giardiasis-annual-epidemiological-report-2015. Accessed December 6, 2019.

Feng, Y., & Xiao, L. (2011). Zoonotic potential and molecular epidemiology of *Giardia* species and giardiasis. *Clinical Microbiology Reviews, 24,* 110–140.

Fournet, N., Deege, M. P., Urbanus, A. T., Nichols, G., Rosner, B. M., Chalmers, R. M., et al. (2013). Simultaneous increase of *Cryptosporidium* infections in the Netherlands, the United Kingdom and Germany in late summer season, 2012. *Euro Surveillance, 18*(2).

Fuentes, I., Martín, C., Beristain, X., Mazón, A., Saugar, J. M., Blanco, A., et al. (2015). *Cryptosporidium hominis* genotypes involved in increased incidence and clusters of cases, Navarra, Spain, 2012. *Epidemiology and Infection, 143,* 1033–1036.

Gabín-García, L. B., Bartolomé, C., Abal-Fabeiro, J. L., Méndez, S., Llovo, J., & Maside, X. (2017). Strong genetic structure revealed by multilocus patterns of variation in *Giardia duodenalis* isolates of patients from Galicia (NW-Iberian Peninsula). *Infection, Genetics and Evolution, 48,* 131–141.

Galván, A. L., Magnet, A., Izquierdo, F., Fernández Vadillo, C., Peralta, R. H., Angulo, S., et al. (2014). A year-long study of *Cryptosporidium* species and subtypes in recreational, drinking and wastewater from the central area of Spain. *Science of the Total Environment, 468–469,* 368–375.

GBD 2015 Mortality and Causes of Death Collaborators. (2016). Global, regional, and national life expectancy, all-cause mortality, and cause-specific mortality for 249 causes of death, 1980–2015: A systematic analysis for the Global Burden of Disease Study 2015. *Lancet, 388,* 1459–1544.

Gil, H., Cano, L., de Lucio, A., Bailo, B., de Mingo, M. H., Cardona, G. A., et al. (2017). Detection and molecular diversity of *Giardia duodenalis* and *Cryptosporidium* spp. in sheltered dogs and cats in Northern Spain. *Infection, Genetics and Evolution, 50,* 62–69.

González-Moreno, O., Domingo, L., Teixidor, J., & Gracenea, M. (2011). Prevalence and associated factors of intestinal parasitisation: A cross-sectional study among outpatients with gastrointestinal symptoms in Catalonia, Spain. *Parasitology Research, 108,* 87–93.

Gracenea, M., Castaño, S., Méndez, J., & Lucena, F. (2018). Faecal contamination in public pools in Barcelona province: *Cryptosporidium* spp. and bacterial indicators. *Journal of Water and Health, 16,* 762–772.

Guerrant, D. I., Moore, S. R., Lima, A. A., Patrick, P. D., Schorling, J. B., & Guerrant, R. L. (1999). Association of early childhood diarrhea and cryptosporidiosis with impaired physical fitness and cognitive function four-seven years later in a poor urban community in northeast Brazil. *American Journal of Tropical Medicine and Hygeine, 61,* 707–713.

Halliez, M. C., & Buret, A. G. (2013). Extra-intestinal and long term consequences of *Giardia duodenalis* infections. *World Journal of Gastroenterology, 19,* 8974–8985.

Jian, F., Liu, A., Wang, R., Zhang, S., Qi, M., Zhao, W., et al. (2016). Common occurrence of *Cryptosporidium hominis* in horses and donkeys. *Infection, Genetics and Evolution, 43,* 261–266.

Martínez-Moreno, F. J., Hernández, S., López-Cobos, E., Becerra, C., Acosta, I., & Martínez-Moreno, A. (2007). Estimation of canine intestinal parasites in Córdoba (Spain) and their risk to public health. *Veterinary Parasitology, 143,* 7–13.

Martínez-Ruiz, R., de Lucio, A., Fuentes, I., & Carmena, D. (2016). Autochthonous *Cryptosporidium cuniculus* infection in Spain: First report in a symptomatic paediatric patient from Madrid. *Enfermedades Infecciosas y Microbiologia Clinica, 34,* 532–534.

Mateo, M., de Mingo, M. H., de Lucio, A., Morales, L., Balseiro, A., Espí, A., et al. (2017). Occurrence and molecular genotyping of *Giardia duodenalis* and *Cryptosporidium* spp. in wild mesocarnivores in Spain. *Veterinary Parasitology, 235,* 86–93.

Mateo, M., Mateo, M., Montoya, A., Bailo, B., Saugar, J. M., Aguilera, M., et al. (2014). Detection and molecular characterization of *Giardia duodenalis* in children attending day care centers in Majadahonda, Madrid, Central Spain. *Medicine (Baltimore), 93,* e75.

Méndez-Hermida, F., Gómez-Couso, H., Romero-Suances, R., & Ares-Mazás, E. (2007). *Cryptosporidium* and *Giardia* in wild otters (*Lutra lutra*). *Veterinary Parasitology, 144,* 153–156.

Merino, F. J., Köster, P. C., Fuentes, I., & Carmena, D. (2019). Imported cryptosporidiosis caused by *Cryptosporidium hominis* IbA13G3 in Spain. The relevance of molecular-based surveillance. *Enfermedades Infecciosas y Microbiologia Clinica, 37,* 552–554.

Millán, R., Köster, P. C., Fuentes, I., & Carmena, D. (2019). *Cryptosporidium hominis* IbA12G3: First report of a rare sub-genotype in Spain. *Enfermedades Infecciosas y Microbiologia Clinica, 37,* 279–281.

Montoya, A., Checa, R., Rodríguez, E., et al. (2019). *El zorro rojo (Vulpes vulpes), potencial reservorio natural de Cryptosporidium hominis en el noroeste de España*. Paper presented at the XXI Congress of the Spanish Society of Parasitology, Pontevedra, July 3–5, 2019.

Morgan-Ryan, U. M., Fall, A., Ward, L. A., Hijjawi, N., Sulaiman, I., Fayer, R., et al. (2002). *Cryptosporidium hominis* n. sp. (Apicomplexa: Cryptosporidiidae) from *Homo sapiens*. *Journal of Eukaryotic Microbiology, 49,* 433–440.

Navarro, C., Domínguez-Márquez, M. V., Garijo-Toledo, M. M., Vega-García, S., Fernández-Barredo, S., Pérez-Gracia, M. T., et al. (2008). High prevalence of *Blastocystis* sp. in pigs reared under intensive growing systems: Frequency of ribotypes and associated risk factors. *Veterinary Parasitology, 153,* 347–358.

Navarro-i-Martinez, L., del Águila, C., & Bornay-Llinares, F. (2011). *Cryptosporidium*: un género en revisión. Situación en España. *Enfermedades Infecciosas y Microbiologia Clinica, 29,* 135–143.

O'Handley, R. M. (2007). *Cryptosporidium parvum* infection in cattle: Are current perceptions accurate? *Trends in Parasitology, 23,* 477–480.

Ortuño, A., Scorza, V., & Castellà, J. (2014). Prevalence of intestinal parasites in shelter and hunting dogs in Catalonia, Northeastern Spain. *Veterinary Journal, 199,* 465–467.

Paulos, S., Köster, P. C., de Lucio, A., Hernández-de-Mingo, M., Cardona, G. A., Fernández-Crespo, J. C., et al. (2018). Occurrence and subtype distribution of *Blastocystis* sp. in humans, dogs and

cats sharing household in northern Spain and assessment of zoonotic transmission risk. *Zoonoses and Public Health, 65,* 993–1002.
Ponce Gordo, F., Herrera, S., & Castro, A. T. (2002). Parasites from farmed ostriches (*Struthio camelus*) and rheas (*Rhea americana*) in Europe. *Veterinary Parasitology, 107,* 137–160.
Quílez, J., Torres, E., Chalmers, R. M., Hadfield, S. J., del Cacho, E., & Sánchez-Acedo, C. (2008a). *Cryptosporidium* genotypes and subtypes in lambs and goat kids in Spain. *Applied and Environment Microbiology, 74,* 6026–6031.
Quílez, J., Torres, E., Chalmers, R. M., Robinson, G., Del Cacho, E., & Sanchez-Acedo, C. (2008b). *Cryptosporidium* species and subtype analysis from dairy calves in Spain. *Parasitology, 135,* 1613–1620.
Ramírez, J. D., Sánchez, A., Hernández, C., Flórez, C., Bernal, M. C., Giraldo, J. C., et al. (2016). Geographic distribution of human *Blastocystis* subtypes in South America. *Infection, Genetics and Evolution, 41,* 32–35.
Ramo, A., Del Cacho, E., Sánchez-Acedo, C., & Quílez, J. (2017). Occurrence of *Cryptosporidium* and *Giardia* in raw and finished drinking water in north-eastern Spain. *Science of the Total Environment, 580,* 1007–1013.
Razakandrainibe, R., Diawara, E. H. I., Costa, D., Goff, L. L., Lemeteil, D., Ballet, J. J., et al. (2018). Common occurrence of *Cryptosporidium hominis* in asymptomatic and symptomatic calves in France. *PLOS Neglected Tropical Diseases, 12,* e0006355.
Reboredo-Fernández, A., Ares-Mazás, E., Cacciò, S. M., & Gómez-Couso, H. (2015). Occurrence of *Giardia* and *Cryptosporidium* in wild birds in Galicia (Northwest Spain). *Parasitology, 142,* 917–925.
Reh, L., Muadica, A. S., Köster, P. C., Balasegaram, S., Verlander, N. Q., Chércoles, E. R., et al. (2019). Substantial prevalence of enteroparasites *Cryptosporidium* spp., *Giardia duodenalis* and *Blastocystis* sp. in asymptomatic schoolchildren in Madrid, Spain, November 2017 to June 2018. *Euro Surveillance, 24,* 1900241.
Ryan, U., Fayer, R., & Xiao, L. (2014). *Cryptosporidium* species in humans and animals: Current understanding and research needs. *Parasitology, 141,* 1667–1685.
Ryan, U., Hijjawi, N., Feng, Y., & Xiao, L. (2019). *Giardia*: An under-reported foodborne parasite. *International Journal for Parasitology, 49,* 1–11.
Ryan, U., Hijjawi, N., & Xiao, L. (2018). Foodborne cryptosporidiosis. *International Journal for Parasitology, 48,* 1–12.
Sahagún, J., Clavel, A., Goñi, P., Seral, C., Llorente, M. T., Castillo, F. J., et al. (2008). Correlation between the presence of symptoms and the *Giardia duodenalis* genotype. *European Journal of Clinical Microbiology and Infectious Diseases, 27,* 81–83.
Salvador, F., Sulleiro, E., Sánchez-Montalvá, A., Alonso, C., Santos, J., Fuentes, I., et al. (2016). Epidemiological and clinical profile of adult patients with *Blastocystis* sp. infection in Barcelona, Spain. *Parasites & Vectors, 9,* 548.
Sanchez-Thevenet, P., Carmena, D., Adell-Aledón, M., Dacal, E., Arias, E., Saugar, J. M., et al. (2019). High prevalence and diversity of zoonotic and other intestinal parasites in dogs from Eastern Spain. *Vector-Borne and Zoonotic Diseases, 19,* 915–922.
Santín, M., Trout, J. M., & Fayer, R. (2008). A longitudinal study of cryptosporidiosis in dairy cattle from birth to 2 years of age. *Veterinary Parasitology, 155,* 15–23.
Scanlan, P. D., Stensvold, C. R., Rajilić-Stojanović, M., Heilig, H. G., De Vos, W. M., O'Toole, P. W., et al. (2014). The microbial eukaryote *Blastocystis* is a prevalent and diverse member of the healthy human gut microbiota. *FEMS Microbiology Ecology, 90,* 326–330.
Segura, R., Prim, N., Montemayor, M., Valls, M. E., & Muñoz, C. (2015). Predominant virulent IbA10G2 subtype of *Cryptosporidium hominis* in human isolates in Barcelona: A five-year study. *PLoS ONE, 10,* e0121753.
Stensvold, C. R., Suresh, G. K., Tan, K. S., Thompson, R. C., Traub, R. J., Viscogliosi, E., et al. (2007). Terminology for *Blastocystis* subtypes—A consensus. *Trends in Parasitology, 23,* 93–96.

Torres, J., Gracenea, M., Gómez, M. S., Arrizabalaga, A., & González-Moreno, O. (2000). The occurrence of *Cryptosporidium parvum* and *C. muris* in wild rodents and insectivores in Spain. *Veterinary Parasitology, 92,* 253–260.

Wang, Y., González-Moreno, O., Roellig, D. M., Oliver, L., Huguet, J., Guo, Y., et al. (2019). Epidemiological distribution of genotypes of *Giardia duodenalis* in humans in Spain. *Parasites & Vectors, 12,* 432.

World Health Organization. (1996). The World Health Report. Fighting Disease Fostering Development. http://www.who.int/whr/1996/en/. Accessed December 6, 2019.

Zahedi, A., Monis, P., Gofton, A. W., Oskam, C. L., Ball, A., Bath, A., et al. (2018). *Cryptosporidium* species and subtypes in animals inhabiting drinking water catchments in three states across Australia. *Water Research, 134,* 327–340.

Chapter 9
The Protozoome of the Periodontal Sulcus: From Health to Disease

Julien Santi-Rocca

Abstract Periodontitis is an infectious disease that affects about half the adult population worldwide. While the periodontal bacterial microbiome is well described, its non-bacterial component in health and disease needs a better characterization. Two parasites have been identified in this environment: the amoeba *Entamoeba gingivalis* and the flagellate *Trichomonas tenax*. In this review, the literature about these two parasites is discussed. The analysis of epidemiological data revealed that there is a consistent link between each parasite and the health status of the patient, their prevalence increasing from health to periodontitis. The role of the protozoome in the oral ecology during the development of periodontal disease should be studied and its importance, reconsidered.

Keywords Protozoome · Microbiome · Periodontitis · Gingivitis · *Entamoeba gingivalis* · *Trichomonas tenax* · Parasite · Infection

Abbreviations

CAL	Clinical attachment loss
GI	Gingival index
HIV-1	Human immunodeficiency virus 1
PCR	Polymerase chain reaction
PD	Pocket depth
PI	Plaque index

J. Santi-Rocca (✉)
Science and Healthcare for Oral Welfare, Toulouse, France
e-mail: jsr@periodontitis.show

© Springer Nature Switzerland AG 2020

N. Guillen (ed.), *Eukaryome Impact on Human Intestine Homeostasis and Mucosal Immunology*, https://doi.org/10.1007/978-3-030-44826-4_9

Introduction: The Prokaryotic Microbiome

Microbiome and Dysbiosis

The human individual bears a large number of foreign organisms, altogether forming a biological unit interacting with the environment, known as "holobiont". The interactions between the multicellular, macroscopic host and the symbiotic unicellular microbes, as well as the interactions between the microbes themselves, are essential for the fitness of the holobiont. As expected, the symbiotic microbes are preferentially located at the surface of the epithelial barriers, which almost exclusively delimit the interface between the host and the environment. The mouth makes no exception and the oral microbiome has been extensively studied at the bacterial level, with more than 600 identified prokaryotic species (Dewhirst et al. 2010). Changes in the composition of the oral microbiome can cause a disequilibrium, called "dysbiosis", and lead to several diseases, like candidiasis or periodontitis.

The Bacteriome of Periodontitis

Periodontitis leads to the destruction of the alveolar bone surrounding and supporting the teeth, resulting in their loosening and eventually their loss. Periodontitis is not a disease resulting from the infection of a healthy host by one pathogenic organism. In contrast, it evolves from an existing gingivitis (a gum inflammation without tissue lysis), whose cause can be infectious (microbe colonization), or mechanical (irritation or wound by foreign bodies like tartar or prostheses). While the trigger to switch from a controlled gingivitis to a destructive and invasive periodontitis is still unclear, the sulcus is sequentially colonized by bacterial species while it becomes deeper (a periodontal pocket) and more anaerobic, until it becomes able to host the anaerobic bacteria of the "red complex": *Porphyromonas gingivalis*, *Treponema denticola*, and *Tannerella forsythia* (Socransky et al. 1998). Though the importance of these periodontopathogenic bacteria in some cases is undoubted, the abundance of other prokaryotic organisms can be correlated to periodontitis, like *Aggregatibacter actinomycetemcomitans* (Zambon et al. 1983), and depend on the age of the patient (Feres et al. 2016). Furthermore, the microbiomes associated to periodontitis also depend on the patient's genetic background (Offenbacher et al. 2016). Altogether, these conclusions remind us the need for a complete, unbiased description of the microbiome of the healthy and diseased sulcus.

Other Kingdoms of Life

The microorganisms are not limited to bacteria. The first microorganisms to be observed were fungi by Robert Hooke in 1665 (Hooke et al. 1665). Then, Antonie van Leeuwenhoek described the first protozoans in 1674 (van Leeuwenhoek 1674), and finally bacteria in 1676 (van Leeuwenhoek 1676). These discoveries were made possible by the technical improvements of microscopy, reaching a magnification of 250×. Thanks to new genetic and phylogenetic techniques, archaea were classified as a separate domain of prokaryotes in 1977 (Woese and Fox 1977) and as a kingdom in 1990 (Woese et al. 1990).

Recent evolutions in sequencing technologies allow to easily address the question of the presence of non-bacterial organisms in the pathobiome of periodontitis. Though parasites were detected in dental plaque very early in the modern scientific literature (Müller 1773; Gros 1849), they have raised poor interest in the analysis of metagenomics and metatranscriptomics data: only one metagenomic study sought only one parasite: *Entamoeba gingivalis* (Deng et al. 2017). Though based on a low number of subjects (4 patients with periodontitis, 10 healthy individuals), this study is a breakthrough, introducing quantitative evaluation of the parasite abundance in healthy sulci and periodontal pockets. Further research is needed to understand the meaning of these results.

In this review, we will focus on the two protozoans that have been identified in the mouth: the amoeba *Entamoeba gingivalis* and the flagellate *Trichomonas tenax*.

Entamoeba gingivalis

Discovery of the Species Entamoeba gingivalis

While free amoebae were described since the middle of the eighteenth century (Rösel von Rosenhof 1755), the association of microorganisms with humans and, in particular, diseases, remained mainly hypothetic. Antonie van Leeuwenhoek, in several instances in 1707–8, mentioned the presence of "animalcules" in the secretions from patients with fever: from the deposits on the tongue and in spitted phlegms (van Leeuwenhoek 1708). The causal link between infection and disease was accepted at the end of the nineteenth century thanks to the works of Louis Pasteur and Robert Koch. The observation of *Entamoeba gingivalis* (then "*Amoeba gengivalis*") in dental plaque samples by G. Gros (Fig. 9.1) did not imply a possible link between amoeba colonization and disease (Gros 1849):

> Au milieu des productions du tartre des dents, on voit des vibrions, une sorte de végétation qui est quelque fois très régulière; mais on n'avait pas encore mentionné les vésicules que nous avons représentées Pl. VI. C. Ces vésicules ont un mouvement si lent et si obscur qu'il faut en être averti pour remarquer qu'elles prennent toutes les formes, par une extension et contraction amoebéenne, laissant toujours voir à l'intérieur des globules qui semblent se

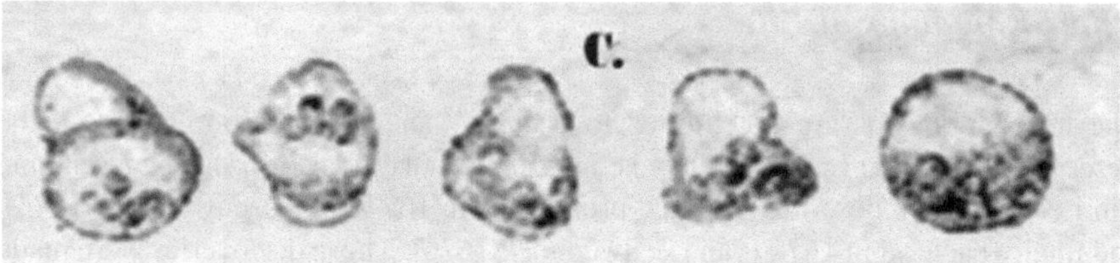

Fig. 9.1 *Amoeba gengivalis* (Gros 1849), first description of *Entamoeba gingivalis*. Figure from Table 6c in Gros (1849)

> déplacer un peu, et être l'analogue de ce que nous connaissons chez de certains infusoires soi-disant polygastriques. Leur origine, leur rôle et leur fin sont ignorés. Elles se trouvent surtout à la face interne des dents. Est-ce encore une génération spontanée?

Biology of Entamoeba gingivalis

The parasite *E. gingivalis* belongs to the genus *Entamoeba*, in which several species are amenable to culture: fastidious and polyxenic for *E. gingivalis* itself, easy and monoxenic for the non-pathogenic human-hosted *E. dispar* (cultured with the trypanosomatid *Crithidia fasciculata*), routine and axenic for the human pathogen *E. histolytica*, the reptile-borne *E. invadens*, and the free-living *E. moshkovskii*. Most of the biology of *E. gingivalis* is extrapolated from observations from other members of the genus *Entamoeba* and caution must be taken when concluding about its traits.

Only few conclusions about the physiology of *E. gingivalis* are certain, mainly due to the incompatibility of its culture conditions with most experimental setups. However, some observations were made by light microscopy in the dental plaque from patients and could testify that this organism had an amoeboid motility (Gros 1849). The phagocytosis of leukocytic nuclei by *E. gingivalis* has been initially controversial, since authors mainly extrapolated from observations they had poorly—or not—reported (Prowazek 1904). Though dying of the samples supported the nuclear nature of the material ingested by the amoeba (Goodey and Wellings 1917) and the phagocytosis process was precisely described (Child 1926), the possible mechanisms by which *E. gingivalis* would be able to phagocytose a specific part—the nucleus—of the target human cells are still to be deciphered.

Morphological depictions of the mouth amoebae have been countless at the beginning of the twentieth century, with several morphotypes. Some authors speculated that several species might be present in the mouth, like *Entamoeba gingivalis* and *Amoeba buccalis* (Fig. 9.3 from Drew and Griffin (1917)). Some descriptions that were made after culture on agar plates should be considered with caution, since they remind features of the *Naegleria* genus, like transformation to flagellated or cystic forms (Drew and Griffin 1917). Cysts from uncultured *E. gingivalis* were also

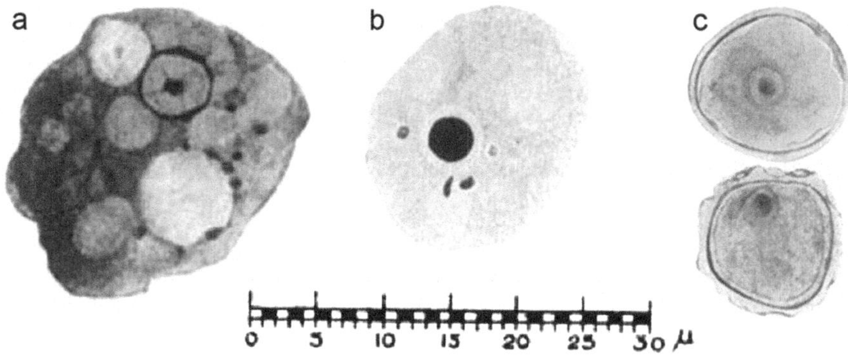

Fig. 9.2 Trophozoite of *Entamoeba gingivalis* (**a**), trophozoite of *Amoeba buccalis* (**b**), and cysts of *Amoeba buccalis* (**c**). Iron-haematoxylin stain. Modified from Drew and Griffin (1917)

reported 3 times (Chiavaro 1914; Smith and Barrett 1915b; Craig 1916), but were never documented since then.

A consensus was reached in the early 1920s, that there was only one morphologically-discernable species of *E. gingivalis* and that this parasite did not produce cysts (Howitt 1925) (Fig. 9.2).

Entamoeba gingivalis *and Disease*

The etiological link between *Entamoeba histolytica* and dysentery (Lösch 1875) paved the way for the studies of the involvement of *Entamoeba gingivalis* in periodontitis (Kartulis 1893). Though periodontal damage was correlated to the colonization by *E. gingivalis* (Hinshaw and Simonton 1928), periodontitis pathophysiogenesis genesis was not convincingly mimicked in animals, in spite of experimental attempts in various species, including in the canine model (Kofoid et al. 1929).

The reverse demonstration was attempted, by specifically killing the parasite to cure the disease. A treatment with emetine was efficient for amoebic dysentery (Rogers 1912). Emetine was then used against periodontitis and revealed itself to be equally effective in getting rid of *E. gingivalis* and in quickly curing inflammation and bleeding (Bass and Johns 1914; Barrett 1914a). This correlation was not a proof per se and raised a legitimate debate about the specificity of the treatment, as later related by Howitt (1925). Though some authors reported changes in the parasite load but not in the bacterial flora (Sturridge 1915), this debate remained open. Finally, this therapeutic strategy was abandoned after emetine toxicity, possibly leading to death by heart failure after overdosage, was documented (Dale 1915; Chopra and Ghosh 1922).

With no alternative strategies for efficient cure and no animal models, almost all the studies about the possible link between *E. gingivalis* and periodontitis became epidemiological studies, seeking a correlation between the presence of the parasite and the disease.

Prevalence of Entamoeba gingivalis

Since the end of the nineteenth century, reports about the prevalence of *E. gingivalis* were almost unanimous about its ubiquitous character in periodontal pockets, and its absence from healthy sulci. For instance, amoebae were perfectly correlated to the disease in all of the 46 cases of periodontitis and of the 7 healthy individuals (Barrett 1914b; Smith and Barrett 1915a). The amoeba *E. gingivalis* was also detected in 95.5% (192/201) of sites of pyorrhea alveolaris, while in none of the 17 healthy controls (Fisher 1927). In another publication, *E. gingivalis* was detected in 97.7% (85/87) of patients with Riggs' disease, but the authors also mentioned that amoebae could be detected in inflamed sites (gingivitis), but never in healthy ones (Bass and Johns 1914). Interestingly, another article mentioned that *E. gingivalis* was detected in 20.6% (14/68) of healthy sites, in parallel to 100% of diseased sites (Chiavaro 1914), questioning the absence of the parasite in health. Finally, the prevalence of amoebae in periodontitis patients could be as low as 78% (39/50) in a study (Sturridge 1915), revealing that, in some pyorrhea cases, *E. gingivalis* could not be detected.

The presence of a potential pathogen in healthy individuals and its absence in some patients directly invalidated the first of the Koch's postulates (1884–1890). The main criticism—and defense—was the inaccuracy of microscopy: not sensitive enough if the microscopist does not scan enough fields or misses some *E. gingivalis* forms, not specific enough if leukocytes are confused for trophozoites, for instance, and user-dependent. Before the development of phase contrast microscopy by Fritz Zernike (1933) and this of differential interference contrast microscopy by Georges Nomarski (1953), the *bona fide* discrimination of *E. gingivalis* was possible only after staining fixed samples and visualizing its typical nucleus. The will to avoid false positives—the main criticism—by microscopy encouraged the use of staining, a methodology that required more technicity and implied additional steps of fixation, staining, washing. This rendered technical biases and variability between operators more probable.

Except few reports concerning parasites (see Table 9.1), almost all the publications about periodontitis focused on its bacteriome. Interestingly, a histology study confirmed a link between the severity of the disease and the presence of amoebae, though with low numbers of samples (Gottlieb and Miller 1971). Microscopy confirmed low prevalence of amoebae in healthy sites (from 0 to 34% in healthy mouths, up to 81.3% in healthy sites within diseased mouths, see Table 9.1) and its high prevalence in periodontal pockets (from 12% to 100%) (Jaskoski 1963; Keyes and Rams 1983; Linke et al. 1989; el Azzouni and el Badry 1994; Lucht et al. 1998; Zdero et al. 1999; Athari et al. 2007; Ghabanchi et al. 2010; Bonner et al. 2014; Yazar et al. 2016; Hassan et al. 2019). Molecular methods like PCR (Kikuta et al. 1996; Trim et al. 2011; Bonner et al. 2014; Garcia et al. 2018a) or metagenomics

Table 9.1 Prevalence of *Entamoeba gingivalis* in periodontitis compared to health and/or intermediate stages

Publication		Detection	Prevalence			Definition			Comment
First author	Year		Health	Intermediate/gingivitis	Periodontitis	Health	Intermediate/gingivitis	Periodontitis	
Hassan	2019	Microscopy	22.5% (9/40)	40% (16/40)	15% (6/40)	"Clinically healthy gingiva", PI = 0, GI = 0, CAL = 0 mm, PD ≤ 3 mm	Gingivitis: "inflammation with no clinical attachment loss"	PD ≥ 3 mm and CAL ≥ 1 mm	Fixed, trichrome/hematoxylin-eosin
Garcia	2018	PCR: ST1	48.6% (51/105)	47.5% (38/80)	57.8% (59/102)	"Healthy individuals not undergoing any oral treatment"	Gingivitis: "Orthodontic treatment without periodontal disease"	PD ≥ 6 mm	Coinfection is frequent: 24%, 40%, 34%, in health, gingivitis, and periodontitis, respectively
		PCR: ST2	29.5% (31/105)	73.8% (59/80)	50% (51/102)				
		PCR: ST1 or ST2	54.3% (57/105)	81.3% (65/80)	73.5% (75/102)				
Deng	2017	Metagenomics	60% (6/10)		100% (4/4)	PD = 2–3 mm		PD = 4-10 mm	Lower abundance in the healthy
Yazar	2016	Microscopy		33.8% (23/68)	37.4% (40/107)		Gingivitis: inflammation and bleeding	PD = 5-7 mm, horizontal bone loss affecting at least 2 teeth	Live, saline-mounted
Bonner	2014	PCR	33.3% (11/33)		80.6% (58/72)	PD < 3mm and maximum 1 of following: oedema, bleeding, gum recession ≥ 1 mm, mobility [0–3] ≥ 2		PD ≥ 3 mm and at least 2 of following: oedema, bleeding, gum recession ≥ 1 mm, mobility [0-3] ≥ 2	PCR specific of ST1 (Zaffino et al., 2019)
		Microscopy	9.1% (3/33)		86.1% (62/72)				Live, saliva-mounted
Ibrahim	2012	Microscopy	16% (4/25)	46.7% (14/30)	70% (21/30)	"Healthy"	"Gingivitis"	"Periodontitis"	Fixed, Giemsa
Al-Hamiary	2011	Microscopy		42.1% (98/233)	45.5% (35/77)	"Healthy"	"Gingivitis"	"Periodontitis"	Live (Ringer's solution)
Trim	2011	PCR	0% (0/12)		69.2% (18/26)	PD ≤ 3 mm, no inflammation		PD > 4 mm	Quantitative PCR, detection of ST1 and ST2 (Zaffino et al., 2019)
			0% (0/12)		26.9% (7/26)	PD ≤ 3 mm, no inflammation		PD > 4 mm	Conventional PCR, detection of ST1 and ST2 (Zaffino et al., 2019)
Ghabanchi	2010	Microscopy	2% (1/50)		12% (6/50)	"No periodontal disease and healthy gingiva"		"Periodontal pocket, calculus", gum inflammation	Fixed, Giemsa
Athari	2007	Microscopy	6.9% (11/160)	43.1% (69/160)		"Healthy control"	"Gingivitis or periodontitis"		Live (Ringer's solution)
Zdero	1999	Microscopy	48% (24/50)	76% (38/50)		"Healthy mouth"	"Gingivitis or periodontitis"		Fixed, trichrome
Lucht	1998	Microscopy	7.7% (1/13)		0% (0/2)	Not defined	"Gingivitis without necrosis"	"Marginal periodontitis, necrotic periodontitis, necrotic gingivitis"	Fixed, trichrome, HIV-1 negative patients
			0% (0/32)		76.9% (10/13)				Fixed, trichrome, HIV-1 positive patients
Kikuta	1996	PCR	0% (0/20)	6.3% (2/32)		"Healthy"	"Marginal periodontitis or gingivitis"		PCR specific of ST1 - long amplicon
El Azzouni	1994	Microscopy	34% (17/50)		64% (32/50)	"Good oral condition"		"Advanced periodontitis"	Fixed, trichrome
Linke	1989	Microscopy	81.3% (13/16)		59.3% (32/54)	PD ≤ 3 mm in periodontitis patients		PD ≥ 4 mm	Live, saline-mounted
Keyes	1983	Microscopy	0% (0/20)	0% (0/16)	100% (65/65)	GI = 0; PD ≤ 3 mm	GI = 2-3; PD ≤ 4 mm	GI = 2-3; PD ≥ 7 mm	Live, water-mounted
Gottlieb	1971	Histology		16.7% (1/6)	90% (9/10)		"Moderate periodontitis"	"Advanced periodontitis"	Iron hematoxylin
Jaskoski	1963	Microscopy	49.6% (61/123)	73.5% (86/117)		"Healthy oral condition"	"Poor oral condition"		MIF method; "Dental group" was used

Studies about *Entamoeba gingivalis* prevalence after 1950 were selected if they allowed to compare periodontitis to health and/or intermediate stages. Intermediate stages referred either to gingivitis or to low-grade periodontitis, as indicated in the column "Definition: Intermediate/gingivitis". Precision about the method are put in the "Comment" column. Data are presented as the prevalence in the group followed by (number of positive samples/total number of samples)
In the publication from Jaskoski, the numbers for the "Student group" cited in the text could not allow to calculate the numbers in the table, so only the "Dental group" was used

Table 9.2 Prevalence of *Entamoeba gingivalis* in health, gingivitis, and periodontitis: comparison of microscopy and PCR

Groups	Methods	Health	Gingivitis	Periodontitis
Per individual	Microscopy	13.3% (35/263)	39% (151/387)	54.7% (277/506)
	PCR	45.3% (68/150)	81.3% (65/80)	75.5% (151/200)
	All	24.9% (103/413)	46.3% (216/467)	60.6% (428/706)
Per article	Microscopy	12.3% (±11.7%)	32.5% (±16.8%)	55.2% (±28.5%)
	PCR	29.2% (±22.4%)	81.3% (±0%)	74.4% (±4.7%)
	All	17.3% (±17.5%)	40.6% (±23.8%)	60% (±26.1%)

Data for Table 9.1 were curated as explained below. Data were treated *per* individual, independently of their publication, presenting the average prevalence and (number of positive samples/total number of samples). Then, individuals were grouped *per* article, meaning that the average prevalence presented in the lower part of the table is the mean of the averages *per* group in each article (±standard deviation), allowing to evaluate the reproducibility between articles

Garcia: Data specific for only ST1 and only ST2 were redundant with the total "ST1 or ST2" and removed

Deng: Only metagenomics study. Removed

Trim: The conventional PCR was redundant, less accurate, and thus removed

Gottlieb: Only histology study. Removed

Athari: Mixed categories for the diseased. Removed

Zdero: Mixed categories for the diseased. Removed

Lucht: Data between HIV-1 positive and HIV-1 negative patients were merged

Kikuta: The PCR efficiency may have been impacted by the length of the amplicon (1.4 kb). This is not comparable to other methods presented here and was removed

Linke: Data from the healthy samples were excluded because they were taken in diseased mouths

Jaskoski: Mixed categories for the diseased. Removed

(Deng et al. 2017) also showed the same pattern: low prevalence in health (from 0 to 54.3%) and high prevalence in periodontitis (from 6.3 to 100%).

Means from curated data (see Table 9.2) show a lower prevalence of *E. gingivalis* in health (24.9%) than in periodontitis (60.6%). Taking into consideration the three health stages and all the methods for all the subjects ("*per* individual", "all methods"), there is a consistent link between the infection by *E. gingivalis* and health status (Chi-square test, $p < 10^{-3}$). Dispersion measures in data treated "per article" confirm a high variability between the studies, revealed by high standard deviations. Though prevalence seems higher with PCR than with microscopy, the difference is not significant, and these results cannot be properly compared. Indeed, the methods are very variable: microscopy can be with live or fixed material, the number of scanned fields is not indicated, the density of the sample is not homogeneous between users. For the PCR, the three studies included in the Table 9.2 used different primers: the genetically diverse *E. gingivalis* (Cembranelli et al. 2013) has been classified in two subtypes (Garcia et al. 2018b) and the sets of primers have different specificities for ST1 and ST2 (Zaffino et al. 2019). Microscopy should overcome this issue, since these subtypes seem to be morphologically identical, notwithstanding the risk of false positives with leukocytes already mentioned. Interestingly, the metagenomics study identified *E. gingivalis* in 60% of healthy individuals, but in lower abundance as compared to periodontitis (Deng et al. 2017). This may explain why, according to the PCR sensitivity, the results vary so much, in particular in healthy individuals.

Concluding Remarks About E. gingivalis

The parasite *E. gingivalis* has been at the interface of clinical and scientific research for 170 years and is undoubtedly linked to periodontitis. However, almost nothing is known about the biology of the parasite, and its place in the oral microbiome is still questioned by clinicians and scientists. Interestingly, a part of the literature has been long ignored and experimental studies, in particular epidemiological ones, are redundant and always lead to the same conclusion: the parasite is correlated to the disease. Still, most results cannot be compared, and there is an urgent need to standardize methodologies and to adopt the same definition for the disease, like the universal classification of staging and grading of periodontitis (Tonetti et al. 2018). Finally, the role of *E. gingivalis* in periodontitis will be better understood when functional studies about the parasite in culture and in the infectious context will be possible.

Trichomonas tenax

Discovery of the Species Trichomonas tenax

The first written description of *Trichomonas tenax* was done by Otto Friedrich Müller, from dental plaque, who named it "*Cercaria tenax*" (Müller 1773):

> Membranous *Cercaria*, well-defined and somewhat more solid anterior part, rear-end three times as short. Substantially bigger than *Monas lens*. Transparent, inverted egg-shaped membrane with a neat and strong front; it ends as a sharp at its posterior extremity or in a very short tail. No internal apparatus. Moves in all the directions with quick turns [...] In infusion of tooth dirt within four days. (translation from Latin by the author)

This description was completed in 1786 by drawings of the parasite (Fig. 9.3 from Müller (1786)). The current genus name was given by Steinberg in 1862 (reported in Dobell (1939)), sharing similarities with the pathogen of the feminine sexual tract *Trichomonas vaginalis* (Donné 1836). However, it was then believed that several species were present in the mouth. Finally, the species was unified under the current binomial name proposed by Dobell in 1939 (Dobell 1939).

For complete information about the history of research about *T. tenax*, please consult the following reviews (Kofoid et al. 1929; Dobell 1939; Pardi et al. 2002).

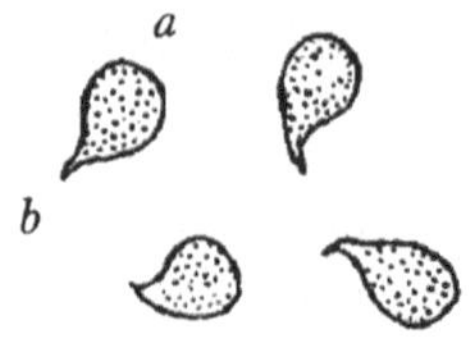

Fig. 9.3 *Cercaria tenax* (Müller 1786), reproduction by Dobell (1939). Can be appreciated the "stronger front" (**a**) and the "small tail" (**b**)

Biology and Pathogenicity of Trichomonas tenax

Since its axenic cultivation in 1962 (Diamond 1962), the parasite *T. tenax* was as amenable for studies as the model *T. vaginalis*. However, only few studies were directly made with the oral parasites, and, as for *E. gingivalis*, most of what we think we know about this parasite is just an extrapolation from its cousin from the feminine sexual tract.

The parasite of the mouth is present in the dental sulcus, surrounded by other microorganisms and in possible contact with the human epithelium. Interestingly, it harbours a fibronectin-like molecule at its surface that may promote adherence to the microbiome and to the host cells (Ribaux et al. 1983). Upon interaction with targets, *T. tenax* has possible weapons to kill them: proteases (Bozner and Demes 1991a, b; Yamamoto et al. 2000; El Sibaei et al. 2012) and hemolysins (Nagao et al. 2000). However, pathogenicity has been long debated and the challenge of *T. tenax* with some types of target cells (cultures HeLa or Hep-2 cells) resulted, in the corresponding experimental conditions, to undetectable to very low adherence and cytotoxicity (Alderete and Pearlman 1984; Alderete and Garza 1985). However, *T. tenax* adhered and was cytotoxic towards HeLa and other types of cells (Ribeiro et al. 2015). Differences could be explained by differences in experimental setups and loss of virulence according to culture conditions for some parasite isolates (Santi-Rocca et al. 2008).

Interestingly, no cyst forms were identified except in one report (Bensen 1910), with important consequences for the transmission mode of the parasite.

Prevalence of Trichomonas tenax

The prevalence of *T. tenax* in health in disease was recently reviewed in a systematic fashion (Marty et al. 2017). To compare results from the studies in the last 20 years, we chose publications reporting *T. tenax* prevalence in adult populations, with a control group (Table 9.3). Detection was made either by culture (Pardi et al. 2002; Kurnatowska et al. 2004; Bernaola Paredes et al. 2012), by microscopy (Mahdi and al-Saeed 1993; Zdero et al. 1999; Athari et al. 2007; Ghabanchi et al. 2010; Al-hamiary et al. 2011; Ibrahim and Abbas 2012; Al-Khayat 2016; Bisson et al. 2018), or by PCR (Athari et al. 2007; Mehr et al. 2015; Bracamonte-Wolf et al. 2019). The prevalence ranged from 0 to 20% in health, from 0 to 56.9% in gingivitis, and from 6

Table 9.3 Prevalence of *Trichomonas tenax* in periodontitis compared to health and/or gingivitis

Publication		Detection	Prevalence			Definition			Comment
First author	Year		Health	Gingivitis	Periodontitis	Health	Gingivitis	Periodontitis	
Bracamonte	2019	PCR		35% (7/20)	70% (21/30)		"Dental plaque-induced gingivitis", AAP/EFP	"Periodontitis", AAP/EFP	Conventional PCR
Bisson	2018	Microscopy	0% (0/50)		11% (16/145)	PD ≤ 3 mm in periodontitis patient		PD ≥ 4 mm	Live, saline-mounted
Al-Khayat	2016	Microscopy	6.9% (4/58)	56.9% (33/58)		"Healthy"	"Gingivitis"		Live , saline-mounted
Mehr	2015	PCR	9.6% (5/52)		26.9% (14/52)	"Healthy"		"Periodontitis", Down syndrome	Conventional PCR
Bernaola-Paredes	2012	Culture	17% (9/53)		24.5% (41/110)	Other oral diseases		"Chronic periodontitis"	Checked by microscopy
Ibrahim	2012	Microscopy	20% (5/25)	30% (9/30)	33.3% (10/30)	"Healthy"	"Gingivitis"	"Periodontitis"	Fixed, Giemsa
Al-Hamiary	2011	Microscopy	1.9% (6/310)	14.2% (33/233)	40.3% (31/77)	"Healthy"	"Gingivitis"	"Periodontitis"	Live (Ringer's solution)
Ghabanchi	2010	Microscopy	0% (0/50)		6% (3/50)	"no periodontal disease and healthy gingiva"		"Periodontal pocket, calculus", gum inflammation	Fixed, Giemsa
Athari	2007	Microscopy	1.9% (3/160)	17.5% (28/160)		"Healthy"	"Gingivitis or periodontitis"		Live (Ringer's solution) or Giemsa
		PCR	1.9% (3/160)	14.2% (17/120)	40% (16/40)	"Healthy"	"Gingivitis"	"Periodontitis"	Conventional PCR
Kurnatowska	2004	Culture		0% (0/10)	41% (32/78)		"Gingivitis catharralis"	"Adult periodontitis"	Checked by microscopy
Pardi	2002	Culture	3.3% (1/30)		30% (9/30)	"Healthy"		"Chronic marginal periodontitis"	Checked by microscopy
Zdero	1999	Microscopy	10% (5/50)	20% (10/50)		"Healthy mouth"	"Gingivitis or periodontitis"		Fixed, trichrome
Mahdi	1993	Microscopy	4.1% (11/271)		8.4% (12/143)	"Healthy mouth"		"Diseased mouth" with caries or pyorrhea	Live , saline-mounted

Studies about *Trichomonas tenax* prevalence since 1999 were selected if they allowed to compare periodontitis to health and/or gingivitis
Data are presented as the prevalence in the group followed by (number of positive samples/total number of samples)

Table 9.4 Prevalence of *Trichomonas tenax* in health, gingivitis, and periodontitis: comparison of microscopy and PCR

Groups	Methods	Health	Gingivitis	Periodontitis
Per individual	Culture	12% (10/83)	0% (0/10)	37.6% (82/218)
	Microscopy	3.4% (26/764)	23.4% (75/321)	16.2% (72/445)
	PCR	3.8% (8/212)	17.1% (24/140)	41.8% (51/122)
	All	4.2% (44/1059)	21% (99/471)	26.1% (205/785)
Per article	Culture	10.2% (±6.8%)	0% (±0%)	36.1% (±4.6%)
	Microscopy	5.5% (±6.9%)	33.7% (±17.6%)	19.8% (±14.1%)
	PCR	5.7% (±3.9%)	24.6% (±10.4%)	45.6% (±18%)
	All	6.5% (±6.7%)	25% (±18.3%)	31.3% (±17.5%)

Data for Table 9.3 were curated as explained below. Data were treated like for Table 9.2
The study by microscopy from Athari and colleagues, and this of Zdero and colleagues were removed because the data did not allow to differentiate gingivitis from periodontitis patients

to 70% in periodontitis. The sum-up Table 9.4 with curated data shows that the same trend is for all the methods: the prevalence of *T. tenax* is higher in periodontitis than in health. Taking into consideration the three health stages and all the methods for all the subjects ("*per* individual", "all methods"), there is a consistent link between the infection by *T. tenax* and health status (Chi-square test, $p < 10^{-3}$). Finally, the analysis "*per* article" is only indicative for it is based on very low number of papers for PCR and culture.

Concluding Remarks About Trichomonas tenax

While the etiological link between *T. tenax* and periodontitis is still to be proven, a correlation exists between the disease and the presence of the parasite. There is also an association between some clones of *T. tenax* and the severity of the periodontitis in the same geographical area (Benabdelkader et al. 2019). Considering that, in diseased sites, tooth mobility, accumulation of plaque, and clinical attachment loss are correlated to the presence of *T. tenax* (Bisson et al. 2018), the hypotheses about a possible saprophytic way of life of the parasite in the sulcus or in the periodontal pocket are poorly supported by the current literature.

Discussion

Other Non-bacterial Entities

The possible link between archaea and periodontitis has already been discussed recently (Nguyen-Hieu et al. 2013; Perez-Chaparro et al. 2014) and very few data are available since then. In a study, archaea accounted for 0.22% of the total microbiome (Deng et al. 2017), questioning their relevance in the etiology of periodontitis, even if they are correlated. However, organisms in low abundance could modulate or even orchestrate changes in the oral microbiome, and such "keystone pathogen" could be of major importance for the evolution towards disease (Hajishengallis et al. 2011).

Fungi are frequently carried in the healthy and the diseased: they make part of the mouth microbiome and have an impact on oral health (a comprehensive review (Sardi et al. 2010)). Even though their role in periodontitis may not be direct, they should be included in microbiome studies.

Viruses can be key actors of the pathobiome: they can modulate host responses, permeability of epithelia, and infect bacteria. For instance, viruses from the *Redondoviridae* study were associated to periodontitis and their abundance is decreased with treatment (Pinto et al. 2016). Herpes viruses, potent immune-modulator and triggering tissue damage, are also correlated to the disease (Binshabaib et al. 2018). Finally, bacteriophages are of special interest for periodontitis (Pinto et al. 2016): while they can enhance the virulence of the bacteria they infect (Ly et al. 2014), they also bring hope for future therapeutic developments.

The Importance of Dark Matter

Among the most accurate methods for diagnosis, the ones based on next-generation sequencing, like metagenomics, allow the fine quantification of sequences that are characteristic of taxonomic entities. The success of the strategy depends on the databases that are used to identify sequences, sensitivity, on the depth of sequencing, and specificity, on the informatic treatment. Unfortunately, it is often impossible to assign a sequence to an organism (and to a gene), despite colossal efforts to update databases: it can account for around half the sequences (Deng et al. 2017). This "dark matter" can bear essential information (virtually half the information of the experiment) and its treatment in the future with appropriate methods will open new perspectives.

Part of the Same Pathobiome?

The parasites *E. gingivalis* and *T. tenax* can be found in the same periodontal pockets, which are in general sampled with a probe, a curette, or a paper point. They may not live in a very close vicinity, but samples are mixed. Nevertheless, both protozoans may be anaerobic or microaerophilic and thus share the same ecological niches. Data about their biology will shed new light on their possible interactions and competition or cooperation in the periodontal pocket.

Strikingly, the prevalence of these parasites is not the same in health and disease, respectively: 4.2% and 26.1% for *T. tenax*, 17.3% and 60% for *E. gingivalis*. Biases can be responsible for these differences, or it may be of biological relevance. Indeed, these parasites may not be part of the same pathobiomes: they may be present in different places (niches), at different times (stage of the disease) or in a different context (interacting entities, microbiome, host characteristics). Future studies will address these questions by conserving the original 3D structure of the subgingival biofilm (Mark Welch et al. 2016) and will thus offer a new view of the periodontal pocket environment.

To conclude, *E. gingivalis* and *T. tenax* are suspected of being important organisms of the periodontal pathobiome, active in the evolution towards periodontal disease, though in different ways. The scientific reports about the correlation of their presence or abundance in the microbiome and periodontal diseases are accumulating since more than a century. However, they have benefited until now from a certain "presumption of innocence", resulting from a wrong interpretation: since the formal proof of their etiological link with the pathogenesis was not provided, they were considered saprophytic and not a priority. In contrast, there is no scientific evidence proving they are commensal. Conclusions should be supported by scientific proofs and further research is the only way to get out of this status quo.

Acknowledgements Thanks are due to Dr. David Martin for the revision of the manuscript.

References

Al-hamiary, A. K., Kezar, M. Y., & Al-Khafaji, Y. A. (2011). Prevalence of oral protozoa in periodontitis and gingivitis patients whose attended to clinics periodontics, Dentistry College\Babylon Univ. *Magazin of Al-Kufa University for Bio, 1*(3).

Al-Khayat, F. A. A.-M. (2016). The prevalence of oral protozoa *Trichomonas tenax* in some patients with gingivitis. *Journal of Baghdad College of Dentistry, 325*(3154), 1–4.

Alderete, J. F., & Garza, G. E. (1985). Specific nature of *Trichomonas vaginalis* parasitism of host cell surfaces. *Infection and Immunity, 50*(3), 701–708.

Alderete, J. F., & Pearlman, E. (1984). Pathogenic *Trichomonas vaginalis* cytotoxicity to cell culture monolayers. *The British Journal of Venereal Diseases, 60*(2), 99–105. https://doi.org/10.1136/sti.60.2.99.

Athari, A., Soghandi, L., Haghighi, A., & Kazemi, B. (2007). Prevalence of oral trichomoniasis in patients with periodontitis and gingivitis using PCR and direct smear. *Iranian Journal of Public Health*, 33–37.

Barrett, M. (1914a). Clinical report upon amoebic pyorrhea. *Dental Cosmos, 56*(12), 1345–1350.

Barrett, M. (1914b). The protozoa of the mouth in relation to pyorrhea alveolaris. *Dental Cosmos, 56,* 948–953.

Bass, C. C., & Johns, F. M. (1914). The specific cause and the prompt specific cure of pyorrhea alveolaris or Rigg's disease. *New Orleans Medical and Surgical Journal, 67*(5), 456–469.

Benabdelkader, S., Andreani, J., Gillet, A., Terrer, E., Pignoly, M., Chaudet, H., et al. (2019). Specific clones of *Trichomonas tenax* are associated with periodontitis. *PLoS ONE, 14*(3), e0213338. https://doi.org/10.1371/journal.pone.0213338.

Bensen, W. (1910). Trichomonas intestinalis und vaginalis des Menschen. *Archiv für Protistenkunde, 18,* 115–127.

Bernaola Paredes, E., Gutiérrez Romero, F., & Guillén Oneeglio, A. (2012). Aislamiento de *Trichomonas tenax* en pacientes con periodontitis crónica al medio de cultivo de tioglicolato modificado.

Binshabaib, M., AL Harthi, S. S., Salehpoor, D., Michelogiannakis, D., & Javed, F. (2018). Contribution of herpesviruses in the progression of periodontal and peri-implant diseases in systemically healthy individuals. *Reviews in Medical Virology, 28*(5), e1996. https://doi.org/10.1002/rmv.1996.

Bisson, C., Lec, P. H., Blique, M., Thilly, N., & Machouart, M. (2018). Presence of trichomonads in subgingival biofilm of patients with periodontitis: Preliminary results. *Parasitology Research, 117*(12), 3767–3774. https://doi.org/10.1007/s00436-018-6077-2.

Bonner, M., Amard, V., Bar-Pinatel, C., Charpentier, F., Chatard, J. M., Desmuyck, Y., et al. (2014). Detection of the amoeba *Entamoeba gingivalis* in periodontal pockets. *Parasite, 21,* 30. https://doi.org/10.1051/parasite/2014029.

Bozner, P., & Demes, P. (1991a). Cell-associated and extracellular proteolytic activity of an oral flagellate, *Trichomonas tenax*. *Archives of Oral Biology, 36*(1), 77–83. https://doi.org/10.1016/0003-9969(91)90056-z.

Bozner, P., & Demes, P. (1991b). Degradation of collagen types I, III, IV and V by extracellular proteinases of an oral flagellate *Trichomonas tenax*. *Archives of Oral Biology, 36*(10), 765–770. https://doi.org/10.1016/0003-9969(91)90043-t.

Bracamonte-Wolf, C., Orrego, P. R., Munoz, C., Herrera, D., Bravo, J., Gonzalez, J., et al. (2019). Observational cross-sectional study of *Trichomonas tenax* in patients with periodontal disease attending a Chilean university dental clinic. *BMC Oral Health, 19*(1), 207. https://doi.org/10.1186/s12903-019-0885-3.

Cembranelli, S. B., Souto, F. O., Ferreira-Paim, K., Richinho, T. T., Nunes, P. L., Nascentes, G. A., et al. (2013). First evidence of genetic intraspecific variability and occurrence of *Entamoeba gingivalis* in HIV(+)/AIDS. *PLoS ONE, 8*(12), e82864. https://doi.org/10.1371/journal.pone.0082864.

Chiavaro, A. (1914). Researches upon *Entamoeba buccalis*. *Dental Review, 28,* 1122–1135.

Child, H. J. (1926). Studies on the ingestion of leucocytes and on mitosis in *Endamoeba gingivalis*. *University of California Publications in Zoology, 28,* 251–284.

Chopra, R. N., & Ghosh, B. N. (1922). The therapeutics of emetine. *Indian Medical Gazette, 57*(7), 248–253.

Craig, C. F. (1916). Observations upon the *Endamebae* of the mouth. I. *Endamoeba gingivalis* (*buccalis*). *The Journal of Infectious Diseases, 18*(2), 220–238. https://doi.org/10.1093/infdis/18.2.220.

Dale, H. H. (1915). A preliminary note on chronic poisoning by emetine. *British Medical Journal,* 2(2868), 895. https://doi.org/10.1136/bmj.2.2868.895.

Deng, Z. L., Szafranski, S. P., Jarek, M., Bhuju, S., & Wagner-Dobler, I. (2017). Dysbiosis in chronic periodontitis: Key microbial players and interactions with the human host. *Scientific Reports, 7*(1), 3703. https://doi.org/10.1038/s41598-017-03804-8.

Dewhirst, F. E., Chen, T., Izard, J., Paster, B. J., Tanner, A. C. R., Yu, W.-H., et al. (2010). The human oral microbiome. *Journal of Bacteriology, 192*(19), 5002–5017. https://doi.org/10.1128/jb.00542-10.

Diamond, L. S. (1962). Axenic cultivation of *Trichomonas tenax*, the oral flagellate of man. I. Establishment of cultures. *Journal of Protozoology, 9*, 442–444. https://doi.org/10.1111/j.1550-7408.1962.tb02650.x.

Dobell, C. (1939). The common flagellate of the human mouth, *Trichomonas tenax* (O.F.M.): Its discovery and its nomenclature. *Parasitology, 31*(1):138–146. https://doi.org/10.1017/s0031182000012671.

Donné, A. (1836). Animalcules observés dans les matières purulentes et le produit des sécrétions des organes génitaux de l'homme et de la femme. Comptes Rendus Hebdomadaires des Séances de l'Académie des Sciences (Vol. 3), Paris.

Drew, A. H., & Griffin, U. D. (1917). Transactions of the society: II. The parasitology of pyorrhœa alveolaris. *Journal of the Royal Microscopical Society, 37*(2), 185–197. https://doi.org/10.1111/j.1365-2818.1917.tb00682.x.

el Azzouni, M. Z., & el Badry, A. M. (1994). Frequency of *Entamoeba gingivalis* among periodontal and patients under chemotherapy. *Journal of the Egyptian Society of Parasitology, 24*(3), 649–655.

El Sibaei, M. M., Abdel-Fattah, N. S., Ahmed, S. A., & Abou-Seri, H. M. (2012). Growth kinetics, antigen profiling, and proteinase activity of Egyptian *Trichomonas tenax* isolates derived from patients having oral infections. *Experimental Parasitology, 130*(4), 416–422. https://doi.org/10.1016/j.exppara.2012.01.018.

Feres, M., Teles, F., Teles, R., Figueiredo, L. C., & Faveri, M. (2016). The subgingival periodontal microbiota of the aging mouth. *Periodontology 2000, 72*(1), 30–53. https://doi.org/10.1111/prd.12136.

Fisher, J. H. (1927). Pyorrhea alveolaris: The rôle of certain microörganisms found in the lesions. *The American Journal of Pathology, 3*(2), 169–178.161.

Garcia, G., Ramos, F., Maldonado, J., Fernandez, A., Yanez, J., Hernandez, L., et al. (2018a). Prevalence of two *Entamoeba gingivalis* ST1 and ST2-kamaktli subtypes in the human oral cavity under various conditions. *Parasitology Research, 117*(9), 2941–2948. https://doi.org/10.1007/s00436-018-5990-8.

Garcia, G., Ramos, F., Martinez-Hernandez, F., Hernandez, L., Yanez, J., & Gaytan, P. (2018b). A new subtype of *Entamoeba gingivalis*: "*E. gingivalis* ST2, kamaktli variant". *Parasitology Research, 117*(4), 1277–1284. https://doi.org/10.1007/s00436-018-5798-6.

Ghabanchi, J., Zibaei, M., Afkar, M. D., & Sarbazie, A. H. (2010). Prevalence of oral *Entamoeba gingivalis* and *Trichomonas tenax* in patients with periodontal disease and healthy population in Shiraz, southern Iran. *Indian Journal of Dental Research, 21*(1), 89–91. https://doi.org/10.4103/0970-9290.62821.

Goodey, T., & Wellings, A. W. (1917). Observations on *Entamoeba gingivalis* from the human mouth, with a note on the trichomonad flagellate *Tetratrichomonas buccalis* n. sp. *Parasitology, 9*(4), 537–559. https://doi.org/10.1017/s0031182000006193.

Gottlieb, D. S., & Miller, L. H. (1971). *Entamoeba gingivalis* in periodontal disease. *Journal of Periodontology, 42*(7), 412–415. https://doi.org/10.1902/jop.1971.42.7.412.

Gros, G. (1849). Fragments d'helminthologie et de physiologie microscopique. *Bulletin de la Société Impériale des Naturalistes de Moscou, 22*, 549–573.

Hajishengallis, G., Liang, S., Payne, M. A., Hashim, A., Jotwani, R., Eskan, M. A., et al. (2011). Low-abundance biofilm species orchestrates inflammatory periodontal disease through the commensal microbiota and complement. *Cell Host & Microbe, 10*(5), 497–506. https://doi.org/10.1016/j.chom.2011.10.006.

Hassan, S., Madkour, G., Henin, R., Gad, S., & Abd El-Aal, A. (2019). Is *Entamoeba gingivalis* a risk factor for periodontal diseases? A case-control study. *Journal of Periodontology, 3*, 18–28. https://doi.org/10.26810/perioj.2019.a3.

Hinshaw, H. C., & Simonton, F. (1928). The amoebas of the mouth in relation to incidence of pyorrhea, clinical symptoms, extent of lesions, and age of patients. *Journal of Dental Research, 8*(4), 507–528.

Hooke, R., Allestry, J., & Martyn, J. (1665). Micrographia, or, some physiological descriptions of minute bodies made by magnifying glasses: With observations and inquiries thereupon. Printed by Jo. Martyn and Ja. Allestry, printers to the Royal Society, London.

Howitt, B. F. (1925). The cultivation of *Endamoeba gingivalis* (Gros). *University of California Publications in Zoology, 28*(4), 65–126.

Ibrahim, S., & Abbas, R. (2012). Evaluation of *Entamoeba gingivalis* and *Trichomonas tenax* in patients with periodontitis and gingivitis and its correlation with some risk factors. *Journal of Baghdad College of Dentistry, 24*(3), 158–162.

Jaskoski, B. J. (1963). Incidence of oral protozoa. *Transactions of the American Microscopical Society, 82*(4), 418–420. https://doi.org/10.2307/3224205.

Kartulis, S. (1893). Muskelparasiten beim Menschen. *Zeitschrift für Hygiene und Infektionskrankheiten, 13,* 2.

Keyes, P. H., & Rams, T. E. (1983). A rationale for management of periodontal diseases: Rapid identification of microbial 'therapeutic targets' with phase-contrast microscopy. *The Journal of the American Dental Association, 106*(6), 803–812.

Kikuta, N., Yamamoto, A., & Goto, N. (1996). Detection and identification of *Entamoeba gingivalis* by specific amplification of rRNA gene. *Canadian Journal of Microbiology, 42*(12), 1248–1251. https://doi.org/10.1139/m96-161.

Kofoid, C. A., Hinshaw, H. C., & Johnstone, H. G. (1929). Animal parasites of the mouth and their relation to dental disease*. *Journal of the American Dental Association, 16*(8), 1436–1455. https://doi.org/10.14219/jada.archive.1929.0207.

Kurnatowska, A. J., Dudko, A., & Kurnatowski, P. (2004). Invasion of *Trichomonas tenax* in patients with periodontal diseases. *Wiadomosci Parazytologiczne, 50*(3), 397–403.

Linke, H. A., Gannon, J. T., & Obin, J. N. (1989). Clinical survey of *Entamoeba gingivalis* by multiple sampling in patients with advanced periodontal disease. *International Journal for Parasitology, 19*(7), 803–808.

Lösch, F. (1875). Massenhafte Entwicklung von Amöben im Dickdarm. *Virchows Archiv, 65,* 196–211.

Lucht, E., Evengard, B., Skott, J., Pehrson, P., & Nord, C. E. (1998). *Entamoeba gingivalis* in human immunodeficiency virus type 1-infected patients with periodontal disease. *Clinical Infectious Diseases, 27*(3), 471–473. https://doi.org/10.1086/514709.

Ly, M., Abeles, S. R., Boehm, T. K., Robles-Sikisaka, R., Naidu, M., Santiago-Rodriguez, T., et al. (2014). Altered oral viral ecology in association with periodontal disease. *mBio, 5*(3), e01133–01114. https://doi.org/10.1128/mbio.01133-14.

Mahdi, N. K., & al-Saeed, A. T. (1993). *Trichomonas tenax* in Basrah, Iraq. *Journal of Pakistan Medical Association, 43*(12), 261–262.

Mark Welch, J. L., Rossetti, B. J., Rieken, C. W., Dewhirst, F. E., & Borisy, G. G. (2016). Biogeography of a human oral microbiome at the micron scale. *Proceedings of the National Academy of Sciences of the United States of America, 113*(6), E791–E800. https://doi.org/10.1073/pnas.1522149113.

Marty, M., Lemaitre, M., Kemoun, P., Morrier, J. J., & Monsarrat, P. (2017). *Trichomonas tenax* and periodontal diseases: A concise review. *Parasitology, 144*(11), 1417–1425. https://doi.org/10.1017/S0031182017000701.

Mehr, A. K., Zarandi, A., & Anush, K. (2015). Prevalence of oral *Trichomonas tenax* in periodontal lesions of down syndrome in Tabriz, Iran. *Journal of Clinical and Diagnostic Research, 9*(7), Zc88–90. https://doi.org/10.7860/jcdr/2015/14725.6238.

Müller, O. F. (1773). Vermium terrestrium et fluviatilium, seu animalium infusoriorum, helminthicorum et testaceorum, non marinorum, succincta historia. Heineck and Faber, Copenhagen and Leipzig.

Müller, O. F. (1786). Animalcula infusoria fluviatilia et marina. Copenhagen.

Nagao, E., Yamamoto, A., Igarashi, T., Goto, N., & Sasa, R. (2000). Two distinct hemolysins in *Trichomonas tenax* ATCC 30207. *Oral Microbiology and Immunology, 15*(6), 355–359. https://doi.org/10.1034/j.1399-302x.2000.150603.x.

Nguyen-Hieu, T., Khelaifia, S., Aboudharam, G., & Drancourt, M. (2013). Methanogenic archaea in subgingival sites: A review. *APMIS, 121*(6), 467–477. https://doi.org/10.1111/apm.12015.

Offenbacher, S., Divaris, K., Barros, S. P., Moss, K. L., Marchesan, J. T., Morelli, T., et al. (2016). Genome-wide association study of biologically informed periodontal complex traits offers novel insights into the genetic basis of periodontal disease. *Human Molecular Genetics, 25*(10), 2113–2129. https://doi.org/10.1093/hmg/ddw069.

Pardi, G., Perrone, M., & Mazzali de Ilja, R. (2002). Incidencia de *Trichomonas tenax* en pacientes con periodontitis marginal crónica. *Acta Odontológica Venezolana, 40,* 152–159.

Perez-Chaparro, P. J., Goncalves, C., Figueiredo, L. C., Faveri, M., Lobao, E., Tamashiro, N., et al. (2014). Newly identified pathogens associated with periodontitis: A systematic review. *Journal of Dental Research, 93*(9), 846–858. https://doi.org/10.1177/0022034514542468.

Pinto, G., Silva, M. D., Peddey, M., Sillankorva, S., & Azeredo, J. (2016). The role of bacteriophages in periodontal health and disease. *Future Microbiology, 11,* 1359–1369. https://doi.org/10.2217/fmb-2016-0081.

Prowazek, S. V. (1904). *Entamoeba buccalis* n. sp. *Arb ad kaiserl Gesundheitsamte, 21,* 42–44.

Ribaux, C. L., Magloire, H., Joffre, A., & Morrier, J. J. (1983). Immunohistochemical localization of fibronectin-like protein on the cell surface of the oral flagelatte *Trichomonas tenax*. *Journal de Biologie Buccale, 11*(1), 41–51.

Ribeiro, L. C., Santos, C., & Benchimol, M. (2015). Is *Trichomonas tenax* a parasite or a commensal? *Protist, 166*(2), 196–210. https://doi.org/10.1016/j.protis.2015.02.002.

Rogers, L. (1912). The rapid cure of amoebic dysentery and hepatitis by hypodermic injections of soluble salts of emetine. *British Medical Journal, 1*(2686), 1424–1425. https://doi.org/10.1136/bmj.1.2686.1424.

Rösel von Rosenhof, A. J. (1755). *Der monatlich-herausgegebenen Insecten-Belustigung* (Vol. 3). Fleischmann, J.J., Nürnberg.

Santi-Rocca, J., Weber, C., Guigon, G., Sismeiro, O., Coppee, J. Y., & Guillen, N. (2008). The lysine- and glutamic acid-rich protein KERP1 plays a role in *Entamoeba histolytica* liver abscess pathogenesis. *Cellular Microbiology, 10*(1), 202–217. https://doi.org/10.1111/j.1462-5822.2007.01030.x.

Sardi, J. C., Duque, C., Mariano, F. S., Peixoto, I. T., Hofling, J. F., & Goncalves, R. B. (2010). *Candida* spp. in periodontal disease: A brief review. *Journal of Oral Science, 52*(2), 177–185. https://doi.org/10.2334/josnusd.52.177.

Smith, A., & Barrett, M. (1915a). Emetin in the treatment of peridental suppurations. *Dental Cosmos, 57,* 1201.

Smith, A. J., & Barrett, M. (1915b). The parasite of oral endamebiasis, *Endameba gingivalis* (Gros). *The Journal of Parasitology, 1*(4), 159–174.

Socransky, S. S., Haffajee, A. D., Cugini, M. A., Smith, C., & Kent, R. L., Jr. (1998). Microbial complexes in subgingival plaque. *Journal of Clinical Periodontology, 25*(2), 134–144.

Sturridge, E. (1915). The discovery of *Entamoeba buccalis* in pyorrhoea pus, and treatment by emetin hydrochloride. *Proceedings of the Royal Society of Medicine, 8*(Odontol Sect), 80–83.

Tonetti, M. S., Greenwell, H., & Kornman, K. S. (2018). Staging and grading of periodontitis: Framework and proposal of a new classification and case definition. *Journal of Periodontology, 89*(Suppl 1), S159–S172. https://doi.org/10.1002/JPER.18-0006.

Trim, R. D., Skinner, M. A., Farone, M. B., Dubois, J. D., & Newsome, A. L. (2011). Use of PCR to detect *Entamoeba gingivalis* in diseased gingival pockets and demonstrate its absence in healthy gingival sites. *Parasitology Research, 109*(3), 857–864. https://doi.org/10.1007/s00436-011-2312-9.

van Leeuwenhoek, A. (1674). Letter 6.

van Leeuwenhoek, A. (1676). Letter 18.

van Leeuwenhoek, A. (1708).

Woese, C. R., & Fox, G. E. (1977). Phylogenetic structure of the prokaryotic domain: The primary kingdoms. *Proceedings of the National Academy of Sciences of the United States of America, 74*(11), 5088–5090. https://doi.org/10.1073/pnas.74.11.5088.

Woese, C. R., Kandler, O., & Wheelis, M. L. (1990). Towards a natural system of organisms: Proposal for the domains Archaea, Bacteria, and Eucarya. *Proceedings of the National Academy of Sciences of the United States of America, 87*(12), 4576–4579. https://doi.org/10.1073/pnas.87.12.4576.

Yamamoto, A., Asaga, E., Nagao, E., Igarashi, T., & Goto, N. (2000). Characterization of the cathepsin B-like proteinases of *Trichomonas tenax* ATCC 30207. *Oral Microbiology and Immunology, 15*(6), 360–364. https://doi.org/10.1034/j.1399-302x.2000.150604.x.

Yazar, S., Cetinkaya, U., Hamamci, B., Alkan, A., Sisman, Y., Esen, C., et al. (2016). Investigation of *Entamoeba gingivalis* and *Trichomonas tenax* in periodontitis or gingivitis patients in Kayseri. *Türkiye Parazitoloji Dergisi, 40*(1), 17–21. https://doi.org/10.5152/tpd.2016.4351.

Zaffino, M., Dubar, M., Debourgogne, A., Bisson, C., & Machouart, M. (2019). Development of a new TaqMan PCR assay for the detection of both *Entamoeba gingivalis* genotypes. *Diagnostic Microbiology and Infectious Disease, 95*(4), 114886. https://doi.org/10.1016/j.diagmicrobio.2019.114886.

Zambon, J. J., Christersson, L. A., & Slots, J. (1983). *Actinobacillus actinomycetemcomitans* in human periodontal disease. Prevalence in patient groups and distribution of biotypes and serotypes within families. *Journal of Periodontology, 54*(12), 707–711. https://doi.org/10.1902/jop.1983.54.12.707.

Zdero, M., Ponce de León, P., Vasconi, M. D., Nocito, I., & Lucca, A. (1999). Entamoeba gingivalis y *Trichomonas tenax*: Hallazgo en poblaciones humanas con y sin patología bucal. *Acta Bioquímica Clínica Latinoamericana, 33*(3), 359–365.

Chapter 10
Tritrichomonas spp. and Their Impact on Gut Immune Homeostasis

Eric Yixiao Cao and Arthur Mortha

Abstract The intestinal microbiota, the collection of all microorganisms residing in our gastrointestinal tract, provides a plethora of microbial diversity. Bacteria, viruses, fungi, worms and protozoa are the primary microorganisms comprising our microbiota. These microbes include representative species capable of both causing severe harm or symbiotic benefits to their host. In relation, the host immune system has evolved complex detection systems to identify members of each microbial faction and interpret their harmful or peaceful capabilities. As a consequence, our immune system mounts appropriate responses to either eliminate or tolerate members of our microbiota. Protozoa are an underappreciated kingdom within our microbiota and the interactions of these microbes with our immune system remain understudied. Several recent reports have demonstrated that the presence of *Tritrichomonas* spp. in the intestinal tract of mice and men facilitates novel interactions with our host immune system. Within this chapter, we are summarizing the most recent knowledge on how *Tritrichomonads*, as a newly emerging group of intestinal protozoan commensals, shape and communicate with our intestinal immune system.

Keywords Eukaryome · Microbiome · ILC2 · *Tritrichomonas* · *T. mu* · *T. muris* · Tuft cells

The Gut Microbiome, A Multi-faceted Ecosystem

Accumulating reports demonstrate that fungi, worms, viruses and protozoa are additional phylogenetic taxa of our intestinal microbiota (Belkaid and Hand 2014; Filyk and Osborne 2016; Li et al. 2019). Studies in animal models and patients have highlighted the impact of these microbes on the body's physiology (Gilbert et al. 2018). The advancements in microbial research revealed unexpected involvements of these microbes reaching beyond the intestinal tract, inducing systemic effects on organ development, immunity, metabolism, physical fitness, mental health, intelligence and behaviour (Hernandez et al. 2016; Dinan and Cryan 2017; Maruvada et al. 2017;

E. Y. Cao · A. Mortha (✉)
Department of Immunology, University of Toronto, M5S 1A8 Toronto, Canada
e-mail: arthur.mortha@utoronto.ca

© Springer Nature Switzerland AG 2020

N. Guillen (ed.), *Eukaryome Impact on Human Intestine Homeostasis and Mucosal Immunology*, https://doi.org/10.1007/978-3-030-44826-4_10

Gould et al. 2018). Each taxon of microbes brings uniquely evolved genomes, able to adapt in response to changing environmental stimuli, including nutrient availability, niche occupancy and additional selective pressures. Characterizing how these elements interplay is fundamental in understanding host-microbe and microbe-microbe interactions. The vast majority of the biomass within our microbiota is composed of bacteria averaging a genome size of 3.65 Mb (diCenzo and Finan 2017). Interestingly, genetic diversity could easily be altered through colonization of the intestinal tract via a single protozoan species (average genome size 160 Mb), nematodes (average genome size 40–700 Mb) or fungi (average genome size 9–177 Mb) (Mohanta and Bae 2015; Benchimol et al. 2017; International Helminth Genomes 2019). By itself, colonization with a single non-bacterial species can provide a 50–200 fold increase in surplus genetic material. In consideration of the variety in encoded genes, metabolic and biochemical processes, enrichment of the bacterial microbiome by single representatives of eukaryotic species will significantly impact the intestinal microbial ecosystem. Advancements in experimental procedures and analysis of genetic information at high-resolutions help progress our understanding of the mechanisms utilized by members of distinct microbial taxa that influence each other and the host's physiology. While our understanding of the microbiome has excelled over the last decade, additional questions concerning host-microbe and microbe-microbe interactions remain unanswered. In particular, uncovering complex microbe-microbe interactions that alter host immune responses bare the potential to leave a high impact on our understanding of health and disease. Neglected and understudied non-bacterial microbes are rapidly entering the center stage of these emerging studies.

Protozoa are a Part of the Intestinal Microbiota

Protozoa are a class of unicellular eukaryotes that contain prominent pathogenic members (e.g. *Giardia*, *Toxoplasma*, and *Cryptopsporidium*). While these pathogens are well characterized, far less is known about commensal, symbiotic protozoa. *Blastocystis*, *Entamoeba* and *Trichomonads* are considered "commensal-like" protozoa that can been found in the stool of healthy individuals or asymptomatic during disease (Embree 1998; Parfrey et al. 2014). Interestingly, the analysis of mice across specific pathogen-free (SPF) animal facilities revealed the prevalence of several *Tritrichomonas* spp. with asymptomatic colonization, similar to reports in human patients (Chudnovskiy et al. 2016). Recent literature has acknowledged a higher frequency of these seemingly "harmless" commensals in developing countries and moderate occurrence in closed communities across the developed world (i.e., daycare centers and retirement homes), suggesting a significant presence of protozoan commensals within our microbiome (Chudnovskiy et al. 2016). *Trichomonads* not only colonize mice and humans, but have been identified in the intestinal tracts of dogs, cats, birds and cows, where colonization occurs via the fecal-oral route through contaminated surfaces or food (Maritz et al. 2014). An exception to intestinal *Trichomonads* is the sexually transmitted pathogenic species *Tritrichomonas vaginalis* in humans preferentially colonizing the urogenital tract (Conrad et al. 2013). Zoonotic transmission of *Tritrichomonas* spp. between distinct hosts (e.g. dogs/cats/cows/birds to humans) has additionally been elucidated to, raising the question of host-specificity and pathogenic evolution across interspecies transmission (Maritz et al. 2014).

Morphologically, *Tritrichomonads* are unicellular, flagellated and highly motile organisms of 5–10 μm in size. The primary energy metabolism for *Tritrichomonads* is achieved using the hydrogenosome, an organelle that serves as an alternative to the mitochondria (Johnson et al. 1993). While pathogenic members exist within this family (e.g. *Tritrichomonas vaginalis*), reports have associated intestinal *Pentatrichomonas hominis* colonization with enteric diseases and other less classified *Trichomonads* as opportunistic pathogens (Okamoto et al. 1998; Zalonis et al. 2011; Suzuki et al. 2016). However, the general classification of *Trichomonads* as true pathogens remains under debate, while current notions favors the classification of *Tritrichomonas* spp. as commensal pathobionts.

Dientamoeba fragilis (*D. fragilis*) is one of the closest human relatives to murine *Tritrichomonas* spp. characterized as an intestinal commensal that shares genetic homology to the recently identified murine species *Tritrichomonas musculis* (*T. mu*) (Chudnovskiy et al. 2016; Stark et al. 2016). *D. fragilis* is occasionally associated with intestinal symptoms and its distribution has been widely reported among humans (frequency of ~10%) (Chudnovskiy et al. 2016; Stark et al. 2016). Our findings in mice suggests that *T. mu* colonization occurs at similar, or even higher frequencies across animal facilities in Europe and North America. In line with these observations, our analysis of "wild" pet-store mice, resembling the microbiome of "natural" mice, demonstrates that 100% of all analyzed mice were colonized with *Tritrichomonas* spp. (*Mortha unpublished data*). This suggests that *Tritrichomonas* spp. are permanent and asymptomatic inhabitants of the mouse intestinal tract and could potentially facilitate host regulation as a component of the microbiome.

Host-Microbe Interactions

The intestinal microbiota influences the physiology of our body. A dominant method of host-microbe engagement occurs through the recognition of microbial-derived ligands by host pattern-recognition receptors capable of differentiating between bacteria, fungi, protozoa, worms and viruses (Takeuchi and Akira 2010). Bacteria utilize biochemical pathways to generate metabolites that are mutualistic in supporting bacterial metabolism and dampening host inflammatory responses. A prominent example involves short chain fatty acids (SCFA), produced by intestinal members of the *Clostridiales* group (Rooks and Garrett 2016). These SCFAs drive the differentiation of regulatory T helper cells (Tregs), a critical immunosuppressive lymphocyte subset that strikes immune balance by preventing proinflammatory activity (Smith et al. 2013). Release of bacterial by-products or direct contact of luminal bacteria with the underlying intestinal epithelium initiate the activation of gut-resident innate immune cells that triggers the release of the cytokine interleukin(IL)-22 by group

3 innate lymphoid cells (ILC3) and Retinoic acid-related Orphan Receptor γ (t) (RORγt)-expressing T cells (Sanos et al. 2009). IL-22 in turn signals through the epithelial-expressed IL-22 receptor and initiates the post-translational modification of epithelial-expressed glycoproteins (Goto et al. 2014). The availability of glycosylated proteins on the luminal site of the gut epithelium serve as potential nutrient resource for microbes and foster host-microbe commensalism (Pickard et al. 2014). Intriguingly, the competition for nutrients inside the intestinal tract may additionally be a regulator for microbial gene expression. For example, gene expression in *T. vaginalis* has been reported to show significant adaptations within aerobic culture conditions in comparison to their physiological anaerobic niche. Changes in glucose alterations induced an approximate 30% change in *T. vaginalis* transcriptional activity while iron-deprivation influenced 5% of the expressed genes. *Tritrichomonas* spp., much like bacteria, are capable of generating host modulatory metabolites. Through using their 520 kDa hydrogenosome protein-complex, *Tritrichomonas* spp. produces Adenosine Tri-Phosphate (ATP) from anaerobic fermentation of pyruvate via the reduction of Ferredoxin (Hrdy et al. 2004; Gould et al. 2013). More recent reports suggest that intestinal *Tritrichomonas* spp. in mice increase the accessibility of succinate from dietary fibres (Burrows et al. 2019). ATP and succinate, once released into the extracellular space serve as nutrients for microbes. These metabolites additionally activate the host epithelium and immune cells via the NLRP3 inflammasome, or through taste receptors engagement in myeloid and on Tuft cells respectively (Nadjsombati et al. 2018) (Fig. 10.1). The release of the inflammatory cytokines IL-1 and IL-18, as well as IL-25 were found to be elevated in the intestine of mice colonized with *Tritrichomonas* spp., suggesting the contribution of *Tritrichomonads* to the direct activation of these pathways or indirect modulation through the interactions with the bacterial microbiome (Chudnovskiy et al. 2016; Howitt et al. 2016) (Fig. 10.1).

Collectively, these reports suggest an intimate interplay of protozoa, bacteria and the host immune system within the intestinal ecosystem. While several reports demonstrated that *T. vaginalis* is capable of activating urogenital tract immunity, analysis of intestinal commensal protozoa and their interactions with the host immune system remain limited (Gu et al. 2016). Below, we discuss recent reports highlighting the immune modulatory capacities of murine enteric commensal *Tritrichomonads*.

Tritrichomonas spp. Regulate Intestinal Innate Type 2 Immune Responses

The analysis of the intestinal tract revealed significant changes following *Tritrichomonas* colonization. An increase in goblet cell hyperplasia, epithelial proliferation, cell death and elevated numbers of DCLK1-expressing Tuft cells were reported in mice colonized with *T. muris* and *T. mu* (Chudnovskiy et al. 2016; Escalante et al. 2016; Howitt et al. 2016) (Fig. 10.1). Goblet cell expansion, a consequence

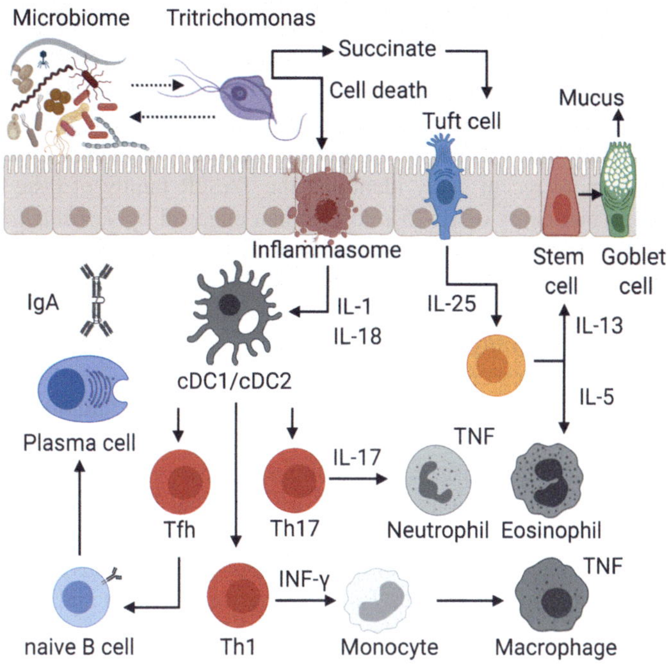

Fig. 10.1 *Tritrichomonas* spp.—mediated intestinal mucosal immune modulation

of ILC2-derived IL-13 stimulation of the epithelium is initiated through the engagement of taste receptors on Tuft cells (Howitt et al. 2016; Schneider et al. 2018) (Fig. 10.1). *Tritrichomonas* spp. or enteric helminths engage the Tuft cell-specific succinate receptor (GPR91) leading to activation of the cation channel TRPM5 on Tuft cells leading to the release of IL-25 (Howitt et al. 2016; Schneider et al. 2018) (Fig. 10.1). IL-25 stimulates the IL-25 receptor on ILC2 and leads to an increase in ILC2 numbers and a boost in ILC2-derived cytokine production. ILC2-derived cytokines IL-5 and IL-13 increased in mice colonized with *Tritrichomonas* spp. and promote Tuft cell and Goblet cell differentiation via the IL-13—STAT6 pathway (Chudnovskiy et al. 2016; Nadjsombati et al. 2018; Schneider et al. 2018). These findings were coupled to a prominent role for the E3-Ubiquitin ligase A20, that negatively regulates ILC2 proliferation and cytokine production downstream of the IL-25R (Schneider et al. 2018). In the absence of A20, higher counts of Tuft cells were observed alongside a significant elongation of the small intestine due to an uncoupled activation of ILC2 (Schneider et al. 2018). Here, the elongation of the intestinal tract may be the result of direct activation of intestinal epithelial stem cells by IL-13 (Zhu et al. 2019).

Dietary metabolites acetone and succinate released by the microbiota engage the GPR91 on Tuft cells and are potent triggers of ILC2 proliferation and cytokine production, which occurs upstream of the A20-mediated inhibition of ILC2s (Nadjsombati et al. 2018; Schneider et al. 2018). A report analyzing GPR91-deficient mice demonstrated that *Tritrichomonas rainier*, a related species to *T. mu* and *T. muris*, initiated GPR91-dependent Tuft cell-ILC2 regulation via IL-25 (Nadjsombati et al. 2018). Strikingly, the authors provided evidence that succinate-mediated activation of Tuft cells by protozoa-colonization triggered a unique activation pathway for Tuft cell-ILC2 crosstalk that was important for anti-helminth responses (Nadjsombati et al. 2018). A more recent study identified the TAS2R family of taste receptors on Tuft cells as mediators of IL-25 release when engaged by *Trichinella* secreted compounds. This mechanism required interactions with TRPM5 and was counterbalanced by inhibitors of G-protein activity and the inositol triphosphate receptor type 2 (Lei et al. 2018). While TAS2Rs showed specific upregulation upon stimulation with *Trichinella spiralis*-derived immunogens, the data collectively suggested cross-kingdom sensory potential by epithelial Tuft cells (Lei et al. 2018). Reactivity to immunogens produced by different kingdoms of enteric microbes including bacteria, protozoa, or helminths can activate the Tuft cell-ILC2 circuit in a taste receptor-dependent manner, triggering host immune alterations and disease exacerbation during microbial infections. However, to extend our limited understanding of the mechanisms exerting these effects further experimental validation is required.

The underappreciated prevalence of species homologous to *T. mu* that colonize the human gastrointestinal tract, such as *Pentatrichomonas hominis* and *Dientamoeba fragilis*, can have implications on the development and severity of human disease. The expansion and release of inflammatory ILC2 (iILC2) from the intestines critically requires Tuft cell-released IL-25, which functions through a S1P-dependent manner (Huang et al. 2018). Therefore, a protozoan-stimulated Tuft cell-ILC2 circuit propagated by IL-25, might promote systemic dissemination of ILC2s. The long-term consequence of these newly distributed ILC2s in tissues is currently unknown, but studies suggest they could retain a persistent epigenetically-poised memory that would allow rapid and robust responses to antigen re-challenge (Schneider et al. 2019). While gut-derived ILC2s do exhibit plasticity and could revert to a non-inflammatory state over time, commensal protozoan colonization was discovered to be chronic during the host's lifetime and therefore could continuously seed distal tissues with newly generated iILC2s (Chudnovskiy et al. 2016; Burrows et al. 2019). The result of ILC2 expansion permits an increased availability in local cytokines. IL-13, as previously mentioned, promotes stem cell expansion which leads to epithelial and goblet cell differentiation. IL-5, another ILC2-released cytokine mediates the recruitment of eosinophils and promotes their activation and survival (Mortha and Burrows 2018). Although protozoan colonization is dualistic in nature by exerting both homeostatic regulation or enteric complications in animal models, it is unknown if colonization may be a contributing factor to human chronic inflammatory conditions and enteric autoimmune disorders such as inflammatory bowel disease.

The microbiota is an established regulator of host immunity with distal effects on tissue immunity influenced by healthy and dysbiotic intestinal microbial composition. Although multiple enteric *Tritrichomonas* spp. have been identified and reported in quick succession, little to no knowledge on their phylogenetic relationships or functional heterogeneity complicates our ability to interpret their contribution to microbiome diversity and host physiology, and highlighting a gap in our understanding of host-microbiome and trans-kingdom interactions (Chudnovskiy et al. 2016; Escalante et al. 2016; Nadjsombati et al. 2018). Future studies should focus on elucidating the full diversity of human commensal protozoan commensals.

In essence, ILC2s remain a key driver of systemic type 2 immunity. The diversity in their development, functional specialization and tissue localization makes these cells pliable moderators for innate immunity. Looking forward, the unique involvement of ILC2s in response to pathogen-driven Tuft cell activation could make them an attractive target for new therapeutic intervention.

Tritrichomonas spp. Shape the Colonic Type 1 and Type 3 Immunity

Innate immune cells bridge the recognition of microbes to the activation of the adaptive immune system. A primary example involves the recognition of microbes through innate pattern recognition receptors, in which subsequent activation of the inflammasome supports $CD4^{+}$ T helper (Th) differentiation (Takeuchi and Akira 2010). The clearance of *Toxoplasma gondii*, a pathogenic intracellular protist was reported to activate the Nlrp1 and Nlrp3 inflammasome for their elimination (Gorfu et al. 2014). Elevated IL-1 and IL-18 production by colonic macrophages and epithelial cells is a feature of mice colonized with *T. mu* (Chudnovskiy et al. 2016; Escalante et al. 2016). Both IL-1 and IL-18 require the activation of the inflammasome for their release and processing into a biologically active form (Chan and Schroder 2019). Moreover, IL-1 is a prominent cytokine that boosts Th17 cell activation, while IL-18 preferentially supports the differentiation of Th1 cells and production of interferon-gamma (IFN-γ) (Okamura et al. 1995; Zielinski et al. 2012) (Fig. 10.1). Analyzing the colonic Th cell compartment in mice colonized with *T. mu* or *T. muris* revealed elevated levels of IFN-γ and IL-17 producing Th1 and Th17 cells, arising within 7–14 days post-colonization. Interestingly, Th1 and Th17 cells remained activated over the course of the animal's life (Chudnovskiy et al. 2016; Escalante et al. 2016). Dendritic cells (DC) including Batf3-dependent cDC1 and Irf4-dependent cDC2 are key migratory sentinels of the intestinal tract and central in determining the effector program of naïve $CD4^{+}$ T cells in tissue draining lymph nodes (Merad et al. 2013). Both DC subsets are specialized in promoting either Th1 (cDC1) or Th17 (cDC2) cell differentiation (Bogunovic et al. 2012). Following colonization with *Tritrichomonas* spp., cDC1 and cDC2 numbers decreased in the lamina propria of the colon, suggesting a possible increase in DC migration to the colon-draining lymph node through a NFκB-dependent upregulation of the chemokine receptor CCR7 on DCs (Chudnovskiy et al. 2016). Conclusively, deficiency in either cDC1, cDC2 or CCR7 eliminated the differentiation of Th1, Th17 or both Th cell subsets *T. mu* colonization (Chudnovskiy et al. 2016) (Fig. 10.1). The major effector functions executed by Th1 and Th17 cells involve the secretion of IFN-γ and IL-17. These cytokines activate macrophages and neutrophils, elevate their release of TNF and contribute to increased anti-microbial immunity, rendering *Tritrichomonas* spp. colonized mice resistant to infections by enteric pathogen Salmonella (Chudnovskiy et al. 2016). However, elevated numbers of Th1 and Th17 cells are also found during intestinal autoimmune inflammation. Using the T cell transfer model of colitis in mice, either free-of, or colonized with *T. mu* or *T. muris*, reveal that protozoan colonization can result in an exacerbated intestinal autoimmune pathology in the absence of immunosuppressive regulatory T cells and B cells (Chudnovskiy et al. 2016; Escalante et al. 2016). However, *Tritrichomonas* colonization in mice competent in a regulatory T cell compartment, did not develop Th1 and Th17 cell-mediated intestinal autoimmunity (Chudnovskiy et al. 2016; Escalante et al. 2016).

Tritrichomonas Musculis Shapes the Peripheral B Cell Response

Th17 and lymph node-resident T follicular helper (T_{fh}) cells are involved in the regulation of mucosal B cells. These T cells drive the terminal differentiation of B cells towards plasma cells that can secrete up to 2–3 g of Immunoglobulin (Ig)A per day into the intestinal lumen of humans. The production of IgA occurs through the B cell-intrinsic process denoted as Class Switch Recombination (CSR) (Muramatsu et al. 2000; Stavnezer et al. Stavnezer et al. 2008; Bemark et al. 2012). During CSR, a selected constant-region segment of the antibody heavy chain (IgH locus) is targeted by the enzyme Activation-induced Cytidine Deaminase (AID). AID deamination initiates double-stranded DNA breaks to allow splicing of the IgA alpha constant region ($C\alpha$) to a successfully re-arranged VDJ segment, resulting in isotype-switched IgA transcripts (Muramatsu et al. 2000; Stavnezer et al. 2008; Bemark et al. 2012). The initiation of CSR is primarily driven by signals from helper T cells (T-dependent, Td) but can also be facilitated in the absence of T cells, through T cell-independent (Ti), soluble factors, released by surrounding tissue or immune cells (Cerutti 2008). During Ti CSR, various molecules such as reactive oxygen species (ROS), retinoic acid (RA), tumor necrosis factor alpha ($TNF\alpha$), transforming growth factor beta ($TGF\beta$), IL-5, IL-6, IL-21, B-cell activating factor (BAFF) and A proliferation-inducing ligand (APRIL) directly or indirectly increase the chromatin accessibility of the IgA locus to promote IgA isotype switching (Tezuka et al. 2011). Within mice, intestinal Ti events partially rely on the peritoneal B-1-cell subset and do not require organized lymphoid tissues or germinal center reactions. Conversely, Td events occur through interactions with $CD4^+$ T_{fh} cells that regulate the production of IgA. Here T_{fh} cells that express CD40L engage the CD40 receptor on B cells (Tezuka et al. 2011; Boulianne et al. 2013). This T cell-B cell interaction is coupled to antigen display by Follicular Dendritic Cells (FDC), providing a template that promotes the selection of high-affinity germinal-center (GC) B cells to ultimately differentiate into long-lived memory B cells or antibody secreting plasma cells (Muramatsu et al. 2000; Stavnezer et al. 2008; Bemark et al. 2012).

Germ-free animals secrete virtually no IgA into the intestinal lumen. Upon colonization with gut microbes, IgA-CSR is induced and IgA will be secreted into the gut lumen, enhancing the mucus barrier and regulating the proliferation, biofilm formation and neutralization of microbes (Macpherson and Slack 2007). Interestingly, colonization of the intestinal tract with *T. mu*, led to elevated levels of IgA in the serum and the fecal matter (Chudnovskiy et al. 2016). Although plasma cells are required for the release of IgA into the gut lumen, they have also been shown to produce immune modulatory cytokines such as IL-10 (Fillatreau et al. 2002). In addition, plasma cells generated in the intestinal tract were reported to be able to exit the intestinal lamina propria and migrate to peripheral tissue including the brain (Rojas et al. 2019). The *Tritrichomonas*-initiated expansion and migration of plasma cells to the brain attenuates disease severity in Experimental Autoimmune Encephalomyelitis (EAE), a mouse model for multiple sclerosis (Rojas et al. 2019). *T. mu* colonization of mice suffering from EAE resulted in reduced cerebral inflammation and demyelination of neurons, indicating *Tritrichomonas musculis* as an intestinal commensal capable of protecting against neuroinflammatory autoimmunity.

Collectively, *Tritrichomonas* spp. unlike their pathogenic protozoan relatives, show signs of symbiosis, similar to commensal microbes that primarily seed the intestinal tract for lifelong colonization. *Tritrichomonas* colonization can potentially provide critical advantages for the host. First, the activation of the epithelium can increase the release of mucins to enforce mucus barrier integrity. Second, elevated levels of activated innate immune cells and Th cells can prime the local gut immune system for improved responses against enteric pathogens. Third, *Tritrichomonas* spp. can incite locally activated immune suppressive cells to migrate from the intestine to the brain and protect against autoimmunity. While these effects resemble a clear advantage to animals colonized with *Tritrichomonas* spp., damaging effects can occur when *Tritrichomonas* spp. colonization supports uncontrolled Th cell activation in the absence of sufficient immune suppression. Instances of pathogenic *Tritrichomonas* spp. have been reported in the development of intestinal inflammation in cats and in humans. While these observations reflect a duality in *Tritrichomonas* spp. nature, more research regarding the heterogeneity of *Tritrichomonas* spp., their interactions with the microbiome and immune system is required to fully grasp the breadth of commensal protozoan impact on the host immune system.

Colonization by *Tritrichomonas* spp. has multiple impacts on the composition and function of the microbiota and immune system. Elevated levels of luminal succinate are recognized by Tuft cell receptors, leading to the production of IL-25 and the release of IL-5 and IL-13 by activated ILC2s. IL-5 facilitates the survival and recruitment of eosinophils, while IL-13 acts on intestinal epithelial stem cells and facilitate their differentiation into goblet cells. *Tritrichomonas* leads to an increase in epithelial death and inflammasome activation. Bioactive IL-1 and IL-18 processed by inflammasome-driven caspase activity mediates cDC1 and cDC2 migration into the mesenteric lymph nodes to prime the differentiation of naïve T cells into Th1 and Th17 effector subsets. Th1 and Th17 cells release IFNγ and IL-17 which promote the differentiation of monocytes into TNF-producing macrophages and the increase neutrophil recruitment. Through the priming of T_{FH} cells, *Tritrichomonas* spp. could mediate intestinal plasma cell differentiation and class-switch recombination into the IgA isotype. In turn, IgA antibodies could modulate the composition of the microbiota. Lastly, expansion and induction of immune regulatory plasma cells that contribute peripheral immune suppression, suggesting a broader role for intestinal *Tritrichomonas* spp. in modulating host immunity.

Acknowledgements E.Y.C. and A.M. would like to thank all members of #theonlylabever for their continued support and discussions. Figure 10.1 was created with BioRender.

Funding A.M. is supported by CIHR-Project Grant (388337) and NSERC-Discovery Grant (RGPIN-2019-04521). A.M. is the Tier 2 Canadian Research Chair in Mucosal Immunology and supported by the Tier 2 CRC-CIHR program.

References

Belkaid, Y., & Hand, T. W. (2014). Role of the microbiota in immunity and inflammation. *Cell, 157*(1), 121–141.

Bemark, M., Boysen, P., & Lycke, N. Y. (2012). Induction of gut IgA production through T cell-dependent and T cell-independent pathways. *Annals of the New York Academy of Sciences, 1247,* 97–116.

Benchimol, M., de Almeida, L. G. P., Vasconcelos, A. T., de Andrade Rosa, I., Reis Bogo, M., Kist, L. W., & de Souza, W. (2017). Draft genome sequence of tritrichomonas foetus strain K. *Genome Announc, 5*(16).

Bogunovic, M., Mortha, A., Muller, P. A., & Merad, M. (2012). Mononuclear phagocyte diversity in the intestine. *Immunologic Research, 54*(1–3), 37–49.

Boulianne, B., Le, M. X., Ward, L. A., Meng, L., Haddad, D., Li, C., et al. (2013). AID-expressing germinal center B cells cluster normally within lymph node follicles in the absence of FDC-$M1^+$ $CD35^+$ follicular dendritic cells but dissipate prematurely. *Journal Immunology, 191*(9), 4521–4530.

Burrows, K., Ngai, L., Wong, F., Won, D., & Mortha, A. (2019). ILC2 activation by protozoan commensal microbes. *International Journal of Molecular Sciences, 20*(19).

Cerutti, A. (2008). The regulation of IgA class switching. *Nature Reviews Immunology, 8*(6), 421–434.

Chan, A. H., & Schroder, K. (2019). Inflammasome signaling and regulation of interleukin-1 family cytokines. *Journal of Experimental Medicine.*

Chudnovskiy, A., Mortha, A., Kana, V., Kennard, A., Ramirez, J. D., Rahman, A., et al. (2016). Host-protozoan interactions protect from mucosal infections through activation of the inflammasome. *Cell, 167*(2), 444–456, e414.

Conrad, M. D., Bradic, M., Warring, S. D., Gorman, A. W., & Carlton, J. M. (2013). Getting Trichy: Tools and approaches to interrogating *Trichomonas vaginalis* in a post-genome world. *Trends in Parasitology, 29*(1), 17–25.

diCenzo, G. C., & Finan, T. M. (2017). The divided bacterial genome: Structure, function, and evolution. *Microbiology and Molecular Biology Reviews, 81*(3).

Dinan, T. G., & Cryan, J. F. (2017). The microbiome-gut-brain axis in health and disease. *Gastroenterology Clinics of North America, 46*(1), 77–89.

Embree, J. E. (1998). Dientamoeba fragilis: A harmless commensal or a mild pathogen? *Paediatrics & Child Health, 3*(2), 81–82.

Escalante, N. K., Lemire, P., Cruz Tleugabulova, M., Prescott, D., Mortha, A., Streutker, C. J., et al. (2016). The common mouse protozoa *Tritrichomonas muris* alters mucosal T cell homeostasis and colitis susceptibility. *Journal of Experimental Medicine, 213*(13), 2841–2850.

Fillatreau, S., Sweenie, C. H., McGeachy, M. J., Gray, D., & Anderton, S. M. (2002). B cells regulate autoimmunity by provision of IL-10. *Nature Immunology, 3*(10), 944–950.

Filyk, H. A., & Osborne, L. C. (2016). The multibiome: The intestinal ecosystem's influence on immune homeostasis, health, and disease. *EBioMedicine, 13,* 46–54.

Gilbert, J. A., Blaser, M. J., Caporaso, J. G., Jansson, J. K., Lynch, S. V., & Knight, R. (2018). Current understanding of the human microbiome. *Nature Medicine, 24*(4), 392–400.

Gorfu, G., Cirelli, K. M., Melo, M. B., Mayer-Barber, K., Crown, D., Koller, B. H., et al. (2014). Dual role for inflammasome sensors NLRP1 and NLRP3 in murine resistance to Toxoplasma gondii. *MBio, 5*(1).

Goto, Y., Obata, T., Kunisawa, J., Sato, S., Ivanov, A., II, Lamichhane, N., et al. (2014). Innate lymphoid cells regulate intestinal epithelial cell glycosylation. *Science, 345*(6202), 1254009.

Gould, A. L., Zhang, V., Lamberti, L., Jones, E. W., Obadia, B., Korasidis, N., et al. (2018). Microbiome interactions shape host fitness. *Proc Natl Acad Sci U S A, 115*(51), E11951–E11960.

Gould, S. B., Woehle, C., Kusdian, G., Landan, G., Tachezy, J., Zimorski, V., et al. (2013). Deep sequencing of Trichomonas vaginalis during the early infection of vaginal epithelial cells and amoeboid transition. *International Journal for Parasitology, 43*(9), 707–719.

Gu, N. Y., Kim, J. H., Han, I. H., Im, S. J., Seo, M. Y., Chung, Y. H., et al. (2016). Trichomonas vaginalis induces IL-1beta production in a human prostate epithelial cell line by activating the NLRP3 inflammasome via reactive oxygen species and potassium ion efflux. *Prostate, 76*(10), 885–896.

Hernandez, C. J., Guss, J. D., Luna, M., & Goldring, S. R. (2016). Links between the microbiome and bone. *Journal of Bone and Mineral Research, 31*(9), 1638–1646.

Howitt, M. R., Lavoie, S., Michaud, M., Blum, A. M., Tran, S. V., Weinstock, J. V., et al. (2016). Tuft cells, taste-chemosensory cells, orchestrate parasite type 2 immunity in the gut. *Science, 351*(6279), 1329–1333.

Hrdy, I., Hirt, R. P., Dolezal, P., Bardonova, L., Foster, P. G., Tachezy, J., et al. (2004). Trichomonas hydrogenosomes contain the NADH dehydrogenase module of mitochondrial complex I. *Nature, 432*(7017), 618–622.

Huang, Y., Mao, K., Chen, X., Sun, M. A., Kawabe, T., Li, W., et al. (2018). S1P-dependent interorgan trafficking of group 2 innate lymphoid cells supports host defense. *Science, 359*(6371), 114–119.

International Helminth Genomes, C. (2019). Comparative genomics of the major parasitic worms. *Nature Genetics, 51*(1), 163–174.

Johnson, P. J., Lahti, C. J., & Bradley, P. J. (1993). Biogenesis of the hydrogenosome in the anaerobic protist *Trichomonas vaginalis*. *Journal of Parasitology, 79*(5), 664–670.

Lei, W., Ren, W., Ohmoto, M., Urban, J. F., Jr., Matsumoto, I., Margolskee, R. F., et al. (2018). Activation of intestinal tuft cell-expressed Sucnr1 triggers type 2 immunity in the mouse small intestine. *Proceedings of the National Academy of Sciences of the United States of America, 115*(21), 5552–5557.

Li, X. V., Leonardi, I., & Iliev, I. D. (2019). Gut mycobiota in immunity and inflammatory disease. *Immunity, 50*(6), 1365–1379.

Macpherson, A. J., & Slack, E. (2007). The functional interactions of commensal bacteria with intestinal secretory IgA. *Current Opinion in Gastroenterology, 23*(6), 673–678.

Maritz, J. M., Land, K. M., Carlton, J. M., & Hirt, R. P. (2014). What is the importance of zoonotic trichomonads for human health? *Trends in Parasitology, 30*(7), 333–341.

Maruvada, P., Leone, V., Kaplan, L. M., & Chang, E. B. (2017). The human microbiome and obesity: Moving beyond associations. *Cell Host & Microbe, 22*(5), 589–599.

Merad, M., Sathe, P., Helft, J., Miller, J., & Mortha, A. (2013). The dendritic cell lineage: Ontogeny and function of dendritic cells and their subsets in the steady state and the inflamed setting. *Annual Review of Immunology, 31,* 563–604.

Mohanta, T. K., & Bae, H. (2015). The diversity of fungal genome. *Biological Procedures Online, 17,* 8.

Mortha, A., & Burrows, K. (2018). Cytokine networks between innate lymphoid cells and myeloid cells. *Frontiers in Immunology, 9,* 191.

Muramatsu, M., Kinoshita, K., Fagarasan, S., Yamada, S., Shinkai, Y., & Honjo, T. (2000). Class switch recombination and hypermutation require activation-induced cytidine deaminase (AID), a potential RNA editing enzyme. *Cell, 102*(5), 553–563.

Nadjsombati, M. S., McGinty, J. W., Lyons-Cohen, M. R., Jaffe, J. B., DiPeso, L., Schneider, C., et al. (2018). Detection of succinate by intestinal tuft cells triggers a type 2 innate immune circuit. *Immunity, 49*(1), 33–41, e37.

Okamoto, S., Wakui, M., Kobayashi, H., Sato, N., Ishida, A., Tanabe, M., et al. (1998). Trichomonas foetus meningoencephalitis after allogeneic peripheral blood stem cell transplantation. *Bone Marrow Transplantation, 21*(1), 89–91.

Okamura, H., Tsutsi, H., Komatsu, T., Yutsudo, M., Hakura, A., Tanimoto, T., et al. (1995). Cloning of a new cytokine that induces IFN-gamma production by T cells. *Nature, 378*(6552), 88–91.

Parfrey, L. W., Walters, W. A., Lauber, C. L., Clemente, J. C., Berg-Lyons, D., Teiling, C., et al. (2014). Communities of microbial eukaryotes in the mammalian gut within the context of environmental eukaryotic diversity. *Frontier in Microbiology, 5,* 298.

Pickard, J. M., Maurice, C. F., Kinnebrew, M. A., Abt, M. C., Schenten, D., Golovkina, T. V., et al. (2014). Rapid fucosylation of intestinal epithelium sustains host-commensal symbiosis in sickness. *Nature, 514*(7524), 638–641.

Rojas, O. L., Probstel, A. K., Porfilio, E. A., Wang, A. A., Charabati, M., Sun, T., et al. (2019). Recirculating intestinal IgA-producing cells regulate neuroinflammation via IL-10. *Cell, 176*(3), 610–624, e618.

Rooks, M. G., & Garrett, W. S. (2016). Gut microbiota, metabolites and host immunity. *Nature Reviews Immunology, 16*(6), 341–352.

Sanos, S. L., Bui, V. L., Mortha, A., Oberle, K., Heners, C., Johner, C., et al. (2009). RORgammat and commensal microflora are required for the differentiation of mucosal interleukin 22-producing NKp46+ cells. *Nature Immunology, 10*(1), 83–91.

Schneider, C., Lee, J., Koga, S., Ricardo-Gonzalez, R. R., Nussbaum, J. C., Smith, L. K., et al. (2019). Tissue-resident group 2 innate lymphoid cells differentiate by layered ontogeny and in situ perinatal priming. *Immunity, 50*(6), 1425–1438, e1425.

Schneider, C., O'Leary, C. E., von Moltke, J., Liang, H. E., Ang, Q. Y., Turnbaugh, P. J., et al. (2018). A metabolite-triggered tuft cell-ILC2 circuit drives small intestinal remodeling. *Cell, 174*(2), 271–284, e214.

Smith, P. M., Howitt, M. R., Panikov, N., Michaud, M., Gallini, C. A., Bohlooly, Y. M., et al. (2013). The microbial metabolites, short-chain fatty acids, regulate colonic Treg cell homeostasis. *Science, 341*(6145), 569–573.

Stark, D., Barratt, J., Chan, D., & Ellis, J. T. (2016). Dientamoeba fragilis, the neglected trichomonad of the human bowel. *Clinical Microbiology Reviews, 29*(3), 553–580.

Stavnezer, J., Guikema, J. E., & Schrader, C. E. (2008). Mechanism and regulation of class switch recombination. *Annual Review of Immunology, 26,* 261–292.

Suzuki, J., Kobayashi, S., Osuka, H., Kawahata, D., Oishi, T., Sekiguchi, K., et al. (2016). Characterization of a human isolate of Tritrichomonas foetus (cattle/swine genotype) infected by a zoonotic opportunistic infection. *Journal of Veterinary Medical Science, 78*(4), 633–640.

Takeuchi, O., & Akira, S. (2010). Pattern recognition receptors and inflammation. *Cell, 140*(6), 805–820.

Tezuka, H., Abe, Y., Asano, J., Sato, T., Liu, J., Iwata, M., et al. (2011). Prominent role for plasmacytoid dendritic cells in mucosal T cell-independent IgA induction. *Immunity, 34*(2), 247–257.

Zalonis, C. A., Pillay, A., Secor, W., Humburg, B., & Aber, R. (2011). Rare case of trichomonal peritonitis. *Emerging Infectious Diseases, 17*(7), 1312–1313.

Zhu, P., Zhu, X., Wu, J., He, L., Lu, T., Wang, Y., et al. (2019). IL-13 secreted by ILC2s promotes the self-renewal of intestinal stem cells through circular RNA circPan3. *Nature Immunology, 20*(2), 183–194.

Zielinski, C. E., Mele, F., Aschenbrenner, D., Jarrossay, D., Ronchi, F., Gattorno, M., et al. (2012). Pathogen-induced human TH17 cells produce IFN-gamma or IL-10 and are regulated by IL-1beta. *Nature, 484*(7395), 514–518.

Chapter 11
Contribution of Host Immunity to the Development of *Entamoeba histolytica*-Induced Liver Abscess

Julie Sellau and Hanna Lotter

Abstract Amebic liver abscess (ALA) is a focal destruction of the liver tissue due to infection with the protozoan parasite *Entamoeba histolytica* (*E. histolytica*). So far, it is generally accepted that pathogenicity factors of the parasite contribute to tissue damage. However, epidemiological studies as well as data that evolve from respective murine models suggest a contribution of host factors as well. Independent from the infection rates, men have a higher risk to develop ALA compared to women. The mouse model for ALA exhibits the same sex difference and based on this model, we found that IFNγ provided by Natural Killer T cells confers the female resistance towards ALA. This specific immune response is clearly modulated by androgens. Hence, female mice which were substituted with testosterone develop larger abscesses and are less able to control *E. histolytica* viability in the liver compared to males. In male mice, on the other side, an enhanced CCL2-dependent recruitment of inflammatory Ly6C^{hi} monocytes via the IL-23/IL-17 immune pathological axis is responsible for tissue destruction during ALA development. Furthermore, TNFα, that exhibits a pivotal role in cytotoxicity, is an effector molecule also responsible for increased liver damage in this case. Interestingly, in human asymptomatically infected with *E. histolytica*, men exhibit higher CCL2 serum levels compared to women, suggesting a similar mechanism to mice underlying the immune response to the parasite. Indeed, there are a variety of phenotypical similarities within the inflammatory monocyte subsets between humans and mice including sex-dependent differences in the expression of the surface receptor for CCL2 as well as a male bias in the production of specific cytokines involved in the recruitment of innate immune cells. In summary, we assume that the host immune response is considerably involved in liver tissue damage due to the strong reactivity of inflammatory monocytes, which is significantly more distinct in male individuals.

Keywords *Entamoeba histolytica* · Amebic liver abscess · Sex difference · Immune protection · Immunopathology · Natural Killer T cells (NKT cells)

J. Sellau · H. Lotter (✉)
Department of Molecular Parasitology, Bernhard Nocht Institute for Tropical Medicine, Bernhard-Nocht-Str. 74, 20359 Hamburg, Germany
e-mail: lotter@bnitm.de

© Springer Nature Switzerland AG 2020

N. Guillen (ed.), *Eukaryome Impact on Human Intestine Homeostasis and Mucosal Immunology*, https://doi.org/10.1007/978-3-030-44826-4_11

Abbreviations

ALA	Amebic liver abscess
E. histolytica	*Entamoeba histolytica*
IFNγ	Interferon gamma
TNFα	Tumor necrosis factor alpha
CCL2	C-C chemokine ligand 2
Ly6C	Lymphocyte antigen 6 complex
IL-23	Interleukin 23
IL-17	Interleukin 17
αGalCer	αGalactosylceramide

Introduction

Humans are the only natural host for *E. histolytica*. An infection occurs with the oral ingestion of environmentally resistant form of the parasite (cysts) via contaminated food or water. In the small intestine, motile trophozoites of *E. histolytica* emerge from these cysts, multiply and usually colonize the large intestine without inducing any symptoms. In only 10% of the cases (Blessmann et al. 2003), tissue invasion and spread of the pathogen occur. Tissue invasion leads to ulcerative hemorrhagic colitis, one of the main symptoms of invasive amebiasis. In case of a further spread via the bloodstream, abscesses may develop in various organs, but mainly in the liver. So far, signals that are responsible to trigger the transition from the asymptomatic stage to the invasive form are not completely understood. On the one side, *E. histolytica* possesses a variety of pathogenicity factors that enables tissue destruction, including amoeba pores and cysteine peptidases (Leippe 1997; Bruchhaus et al. 2003). Overexpression of various cysteine peptidases in low pathogenic amebic cell clones increased their ability to induce amebic liver abscesses ALA in the murine model for this disease (Matthiesen 2013). Along with the previous observation, chemical inhibition of theses virulence factors reduced liver damage upon infection (Stanley et al. 1995). Most recently, new pathogenicity factors of *E. histolytica* affecting liver pathology were identified which include metallopeptidases, C2 domain proteins, alcohol dehydrogenases and hypothetical proteins (Meyer et al. 2016). However, on the other side, epidemiological studies and health surveys suggest that host factors contribute to the development of ALA as well. Unlike the amebic colitis and despite similar infection rates in the study groups, a clear bias in the occurrence of ALA towards male individuals could have been observed (Blessmann et al. 2003; Acuna-Soto et al. 2000). Most interestingly, the risk to develop ALA increases in males after puberty with a peak incidence around the age of 40 years and declines afterwards, In women, ALAs are less frequent and are distributed equally among the different age groups with slight increase in elderly women (Blessmann et al. 2003). This distribution of the occurrence of ALA implicates that two factors, the hormone status of the host and the immune system, which is influenced by sex hormones (Kadel and Kovats 2018), are probably involved in the hepatic form of invasive amebiasis, apart from the pathogenicity factors of the ameba.

Sex Differences in the Immune Response During Amebiasis

In addition to behavioral, genetic, and hormonal factors, differences in the abundance and activation of various types of immune cells could explain some of the observed sexual dimorphisms in infectious diseases (Klein 2004; Markle and Fish 2014). Sex-specific differences in immune responses of men and women could underlie the higher susceptibility of the male sex towards infectious diseases caused by bacteria, viruses and parasites, e.g., tuberculosis, hepatitis B, leishmaniasis and amebiasis (Klein and Flanagan 2016). By contrast, women exhibit more vigorous humoral and cellular immune responses, with the beneficial aspect of a better responsiveness to foreign antigens like vaccines (Klein and Pekosz 2014) but combined with a higher risk for cell-mediated autoimmune diseases (Markle and Fish 2014).

Amebiasis is a prototypical example for a sex-biased disease and initial data support the thesis that women exhibit a stronger humoral immune response that might protect them from the invasive form of the disease (Bernin et al. 2014). Immunological variables from male and female ALA patients, asymptomatic carrier and *E. dispar*-infected endemic residents included *Entamoeba*-specific IgG and the various IgG subclasses, along with IgA. The humoral immune responses to *E. histolytica* showed that the patients with ALA had the highest anti-*E. histolytica* IgG antibody titers. Asymptomatically infected females, however, had significantly higher anti-*E. histolytica* total IgG and IgG1 levels than males. This was also found in female ALA patients, however, due to the low number of ALA cases in women, this result revealed in no significance (Bernin et al. 2014).

IgG1 is known to be a strong complement activator through binding of the complement factor to C1q. The complement system is a major part of the host innate immune defense mechanism against *E. histolytica* trophozoites, which are highly sensitive to complement-mediated lysis (Reed and Gigli 1990). More recently it was shown that serum from women kills *E. histolytica* trophozoites significantly more efficiently than serum from men, and that the mechanisms involve complement-mediated amebic lysis (Snow et al. 2008; Forster et al. 1994; Urban et al. 1996). Another IgG1-related mechanism that may promote *E. histolytica* trophozoite killing is that cells of the innate immune system, e.g. monocytes, macrophages, neutrophils and dendritic cells bear high affinity receptors for IgG1 (and IgG3), which leads to the expression of inducible nitric oxide synthase (iNOS) and thus to an increased cytotoxicity against amoebic trophozoites (Spiegelberg 1989; Seguin et al. 1997). This would suggest that subclinical invasion exist continuously in both women and men, but is controlled in women by a stronger humoral IgG immune response, thereby inhibiting the development of ALA at an early stage.

Another relevant immunological cell type may prevent ALA in female individuals at an early stage, which comprises the non-conventional T cells, especially the invariant Natural Killer T (NKT) cells. These cells bridge the innate and adaptive immune response, recognizing lipids and glycolipids and being early producers of cytokines in response to infection. Furthermore, these cells are highly abundant in the liver (Tupin et al. 2007; Godfrey et al. 2010).

In human, stimulated peripheral blood, with the NKT-ligand α-Galactosylceramide (αGalCer) revealed in a higher expression of inflammatory cytokines, such as TNFα and IL-17 by female-derived NKT cells. (Bernin et al. 2016). Additionally, women exhibit higher NKT cell counts in the peripheral blood compared to men (Sandberg et al. 2003). Our group identified a natural NKT cell ligand in the membrane of *E. histolytica* trophozoites indicating that the parasite can indeed activate these cells (Lotter et al. 2009). Stimulation of murine liver-derived NKT cells with this glycolipid, the *E. histolytica* lipopeptidophosphoglycan (*Eh*LPPG), resulted in an increased production of protective IFNγ, but not IL-4 in NKT cells (Lotter et al. 2009). The importance of NKT cells for the control of ALA in vivo was previously shown by the use of Jα18 knock-out mutant mice, which lack iNKT cells and were more susceptible to ALA development (Lotter et al. 2006). In addition, application of the most potent and specific NKT cell activator, αGalCer reduced the sizes of ALA significantly (Lotter et al. 2009). Similarly, murine models for other infectious diseases like malaria, cryptococcosis or trypanosomiasis have shown that a single αGalCer-treatment stimulated IFNγ production by iNKT cells and reduced the respective pathogen load (Mattner et al. 2005; Gonzalez-Aseguinolaza et al. 2000; Duthie and Kahn 2002) (Fig. 11.1).

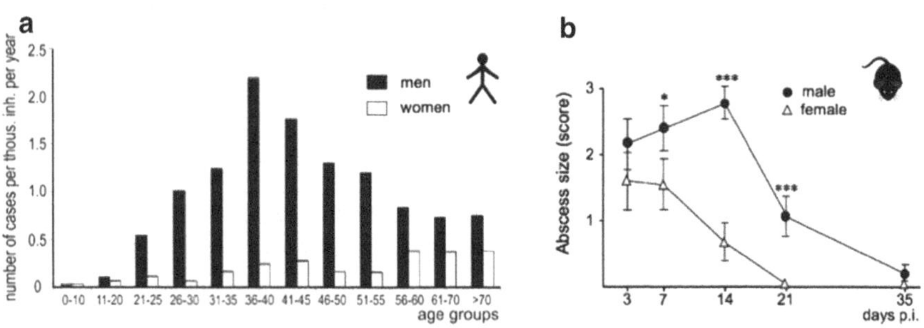

Fig. 11.1 Age and sex dependency in the occurrence of amebic liver abscess in human (Blessmann et al. 2002) (**a**) and mice (Lotter et al. 2006) (**b**)

Influence of Sex Hormones on the Outcome of ALA in the Murine Model

Since invasive amebiasis is only present in humans, attempts to study immune mechanisms underlying the observed sex difference in ALA was hampered by the lack of suitable, immunocompetent animal models. Several rodent species have been successfully used as models to study the development of ALA in the past. Although amebic liver lesions, that resemble those found in humans, could be provoked by direct inoculation of *E. histolytica* trophozoites into the livers or the peritoneum of rabbits, Syrian hamsters, or Mongolian gerbils, respectively (Gil-Barbosa et al. 1972; Roncolato et al. 2015; Chadee and Meerovitch 1984), these models showed substantial limitations. The use of a variety of immunodeficient mouse strains, such as SCID mice, IFNγ- and iNOS knock-out mice revealed in part the contribution of polymorphonuclear leukocytes for the development of ALA since they were found to massively infiltrate the abscess region at early time points after infection (Seydel et al. 1997). In addition, it was found that ALA development was strongly controlled by IFNγ and the production of amebicidal NO (Seydel et al. 2000) which had been shown to be the major effector molecule in the killing of amebic trophozoites in vitro (Lin and Chadee 1992). The first successful liver lesions in mice mentioned, demonstrated a contribution of innate immune cells, e.g. neutrophils and macrophages, but here, no sex differences were not found or analyzed yet (Velazquez et al. 1998; Jarillo-Luna et al. 2000). Attempt to induce considerable abscess lesions upon intrahepatic injection of trophozoites in female C57BL/6 mice, we used their male counterparts as well as an amebic strain which had recently been passaged via the gerbil liver. Most surprisingly, male mice now developed reliable and large abscesses that were larger than those obtained in females and moreover, we found that they were less able to control the parasite load compared to female mice. However, in contrast to human, these mice were able to control the liver lesions within 30 days p.i. (Lotter et al. 2006). Basic immunological analysis of spleen derived lymphocytes isolated during the course of infection already revealed a significantly higher production of protective IFNγ by female-derived splenocytes compared to those isolated from male mice. Vice versa, a non-protective IL-4 production was higher in male-derived splenocytes compared to female derived splenocytes. This model now allowed us to analyze the influence of sex hormones on the outcome of ALA and underlying immune mechanisms. Initially, mice were castrated or female mice were substituted with the opposite sex hormone testosterone (Lotter et al. 2013). These experiments revealed a dominant role for testosterone on the outcome of ALA. Male mice or testosterone substituted female mice developed larger abscesses with higher parasite loads in the liver compared to castrated male mice and female mice (Lotter et al. 2013).

Since previous investigations have shown that NKT cells are critical in the control of amebic liver abscess development, and can be induced by the amebic glycolipid *Eh*LPPG, we assessed whether these cells were activated to produce protective IFNγ in a sex specific manner. Without any hormone substitution, female derived liver

NKT cells produced higher amounts of IFNγ compared to liver NKT cells isolated from male mice which is in line with earlier investigations, demonstrating higher serum levels of IFNγ following administration of αGalCer in female mice compared to male mice (Gourdy et al. 2005). Substitution of female mice with testosterone reduced the production of IFNγ in liver NKT cells from female mice as well as castration of male mice led to an increase in the IFNγ production by liver NKT cells. However, studies on NKT cells from human ALA patients are needed to confirm the results obtained by these experiments in the murine model. More particularly, the frequency of NKT cells in the liver is higher in mice compared to humans and Kenna and colleagues postulated that NKT cells from humans and mice are phenotypically and functionally different from another (Subleski and Ortaldo 2009; Kenna 2003).

Immunopathological Mechanisms Underlying ALA Development

In contrast to females, male individuals in general show a higher susceptibility to infectious diseases, and especially to parasitic diseases due to a more suppressive effect upon testosterone binding to its receptors on immune cells (Klein 2004; Roberts et al. 2001; Bouman et al. 2004; Gubbels Bupp and Jorgensen 2018).

This has been shown for cells of the adaptive immune system, by repressing the pro-inflammatory immunity of T helper cells in inhibiting the production of IL-12 as well as leading to lower numbers of $CD8^+$ T cells (Klein and Flanagan 2016; Kissick et al. 2014). Moreover, we could show that testosterone treatment specifically reduced the production of pro-inflammatory cytokines, e.g. IFNγ by NKT cells in the murine model for ALA (Lotter et al. 2013). The stimulation of peripheral NKT cells from humans with amebic antigens led also to an increased IFNγ release in female derived-samples (Bernin et al. 2016). However, there is one major exception of this overall accepted paradigm, which refers to male-derived monocytes. Stimulation of monocytes from men leads to an increase in the production of pro-inflammatory cytokines IL-12, TNFα and IL-1β compared to monocytes from female individuals (Markle and Fish 2014; Klein and Flanagan 2016; Bouman et al. 2004; Gubbels Bupp and Jorgensen 2018). Monocytes are within the first wave of the immune response towards infections or injury and are crucial in the control of microorganisms, however, if these cells are not properly controlled, they are able to destroy the host tissue (Nathan 2006; Laskin et al. 2011). In contrast to NKT cells, monocytes share large similarities in human and mice. Murine monocytes are characterized by the expression of the surface markers CD11b and Ly6C and can be further subdivided into a classical, inflammatory $Ly6C^{hi}$ and a non-classical, anti-inflammatory $Ly6C^{lo}$ subpopulation (Guilliams et al. 2014; Ziegler-Heitbrock et al. 2010). Based on the expression of CD14 and CD16, three monocyte subpopulations

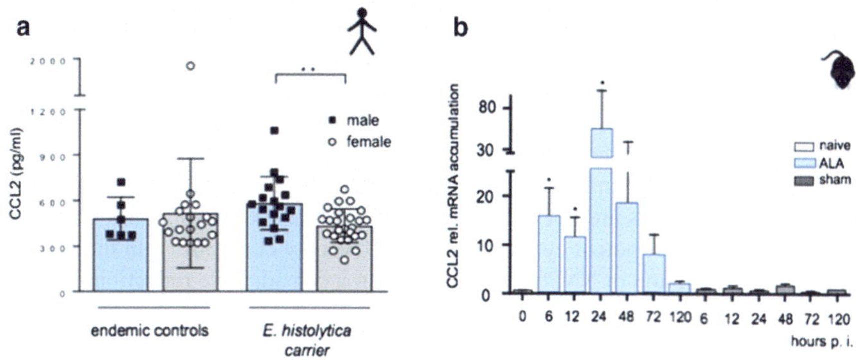

Fig. 11.2 Serum levels of CCL2 in endemic controls and *E. histolytica* infected individuals out of an amebiasis endemic area stratified by the sex of the host (Bernin et al. 2014) (**a**). Increase in CCL2 mRNA accumulation in the liver of male mice during ALA (Helk et al. 2013) (**b**)

are present humans: classical $CD14^{+}CD16^{-}$, intermediate $CD14^{+}CD16^{+}$, and non-classical $CD14^{-}CD16^{+}$ monocytes. In humans and mice, the expression of C-C chemokine receptor 2 (CCR2) and the chemokine receptor CX_3CR1 allows a further subdivision of these monocyte subsets (Zimmermann et al. 2012). Especially relevant for the recruitment of classical monocytes is the expression of CCR2. Binding of CCR2 to its ligand CCL2 is a requirement for the recruitment of $CCR2^{+}$ classical monocytes to the site of infection (Shi and Pamer 2011; Patel et al. 2017; Serbina and Pamer 2006). A previous study has shown that ALA patients, as well as asymptomatically infected patients with *E. histolytica* develop high serum levels of CCL2. However, following stratification of the groups by sex revealed significant higher levels of CCL2 in men compared to women (Bernin et al. 2014) (Fig. 11.2).

Together with the reports of higher circulating monocyte numbers in men (Bouman et al. 2005), one can suggest that male individuals have a higher predisposition to develop monocyte-mediated immunopathology compared to women and indeed, a variety of infection studies in murine infection models using CCR2 knock-out mice revealed a detrimental effect for classical monocytes (Bosschaerts et al. 2010; Terrazas et al. 2017; Dunlap et al. 2018; Gurczynski et al. 2019). The CCR2/CCL2 axis was found to play a deleterious role in the murine model of amebiasis, which is clearly influenced by a male-bias (Helk et al. 2013). CCR2 knock-out mice showed reduced numbers of circulating and liver infiltrating classical inflammatory $Ly6C^{hi}$ monocytes and developed significantly smaller liver lesions compared to wildtype C57BL/6 mice. Accordingly, the substitution of CCR2 knock-out mice with classical monocytes from wildtype mice via adoptive transfer restored the liver pathology after intrahepatic infection with *E. histolytica* trophozoites. However, liver destruction is not only due to the amebic invasion, but mainly dependent on monocytes, neutrophils and liver resident macrophages (Kupffer cells). Depletion of neutrophils using the neutralizing monoclonal antibody Ly6G diminished abscess lesions to the half, whereas the depletion of Kupffer cell by clodronate treatment led to a complete inhibition of abscess development. Kupffer cells are the main producers of NO in the liver and play a relevant role for organ infiltrating monocytes since they provide an inflammatory milieu upon *E. histolytica* invasion and lead to a further polarization to inflammatory monocytes (Geissmann et al. 2003). Substantial amebicidial effects are not only due to NO, as referred above, but also TNFα is considered to exert potent cytotoxic activity (Uversky et al. 2017). In the murine ALA model, depletion of TNFα resulted again in a considerable reduction of ALA lesions supporting its immunopathological role in the context of *E. histolytica* liver infection (Helk et al. 2013).

Finally, the question for the signaling cascade underlying the CCL2-dependent recruitment and activation of monocytes remains open. Upstream of the chemokines and effectors which are crucial for the liver damage by *E. histolytica* is the IL-23/IL-17 immune axis. This immune axis is on the one hand involved in the host protection from microbial pathogens but can be on the other hand responsible for the immune mediated pathology (Gaffen et al. 2014). By this, IL-23 mediates and stabilizes the Th17 immune response via the IL-17 family members IL-17A and IL-17F (Paust et al. 2009). IL-17 production promotes inflammation but also migration and infiltration of monocytes at the site of infection through its induction of CCL2 (Chen et al. 2011; Shahrara et al. 2009; Yang et al. 2008). Indeed, we found an increased expression of IL-23 and IL-17 mRNA within hours post intrahepatic infection with *E. histolytica* and vice versa, respective cytokine knock-out mutant mice upstream of CCL2 were largely resistant to the development of ALA (Noll et al. 2016). Most interestingly, in IL-23 knock-out mutant mice there were no differences in the percentage of $Ly6C^{hi}$ monocytes or neutrophils between wildtype mice and knock-out mice but instead an increase in tissue-repairing $Ly6C^{lo}$ monocytes that produced the anti-inflammatory cytokine IL-13 and arginase 1. As recently shown, the transition of early recruited inflammatory monocytes into regenerative and anti-inflammatory monocytes occurs very fast under certain conditions and depends on the degree of inflammation and the local cytokine milieu (Dal-Secco et al. 2015) (Fig. 11.3).

Taken together, the fact that hepatic amebiasis occurs in a strict sex-dependent manner suggests that not only the parasite, but also host factors contribute to the damage of the liver tissue. To understand immunological mechanisms underlying the female resistance and the male susceptibility both sexes need to be investigated in parallel, as possible accompanied by human samples.

So far, we conclude that the female resistance is mediated by NKT cell-dependent production of pro-inflammatory, amebicidal cytokine like IFNγ leading to an increased NO production by macrophages. Male individuals suffer from a misbalanced increased recruitment of inflammatory, tissue-destroying monocytes. The immune responses in men as well as in women due to *E. histolytica*-induced abscess development is mainly dependent on the sex hormones, especially the presence or the absence of testosterone.

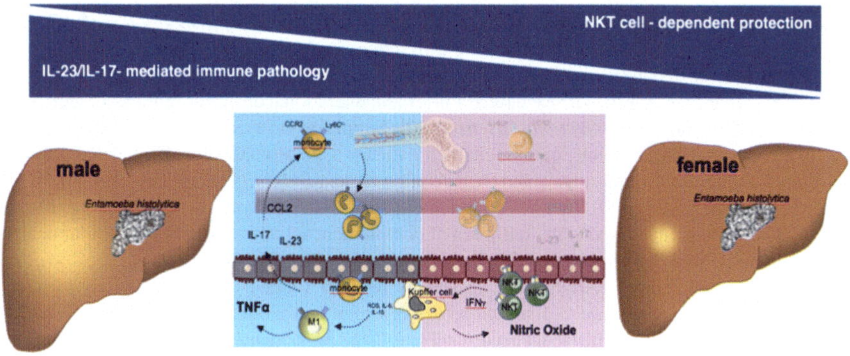

Fig. 11.3 Immune mechanisms underlying the sex dimorphism in hepatic amebiasis. In male mice, infection with *E. histolytica* initiates a pathological IL-23/IL-17 immune axis which in terms leads to the production of CCL2 by various cell types and egress of Ly6C^{hi} inflammatory monocytes from the bone marrow. Monocytes recruited to the site of infection encounter a pro-inflammatory milieu leading to their polarization towards inflammatory M1 macrophages and TNFα-mediated tissue damage. In female mice, protective IFNg-produced by liver NKT cells activates NO-production by macrophages leading to the early control of *E. histolytica* survival

Acknowledgements This work was supported by the German Research Foundation within Collaborative Research Centre 841 and by the State Research Funding Hamburg, LFF-FV45, Hamburg, Germany.

References

Blessmann, J., et al. (2003). Longitudinal study of intestinal *Entamoeba histolytica* infections in asymptomatic adult carriers. *Journal of Clinical Microbiology, 41,* 4745–4750.

Leippe, M. (1997). Amoebapores. *Parasitology Today (Personal ed),* 13, 178–183.

Bruchhaus, I., Loftus, B. J., Hall, N., & Tannich, E. (2003). The intestinal protozoan parasite *Entamoeba histolytica* contains 20 cysteine protease genes, of which only a small subset is expressed during in vitro cultivation. *Eukaryotic Cell, 2,* 501–509.

Matthiesen, J., et al. (2013). Overexpression of specific cysteine peptidases confers pathogenicity to a nonpathogenic *Entamoeba histolytica* clone. *MBio, 4.*

Stanley, S. L., Jr., Zhang, T., Rubin, D., & Li, E. (1995). Role of the *Entamoeba histolytica* cysteine proteinase in amebic liver abscess formation in severe combined immunodeficient mice. *Infection and Immunity, 63,* 1587–1590.

Meyer, M., et al. (2016). Overexpression of differentially expressed genes identified in non-pathogenic and pathogenic *Entamoeba histolytica* clones allow identification of new pathogenicity factors involved in amoebic liver abscess formation. *PLoS Pathogens, 12,* e1005853.

Acuna-Soto, R., Maguire, J. H., & Wirth, D. F. (2000). Gender distribution in asymptomatic and invasive amebiasis. *The American Journal of Gastroenterology, 95,* 1277–1283.

Kadel, S., & Kovats, S. (2018). Sex hormones regulate innate immune cells and promote sex differences in respiratory virus infection. *Frontiers in Immunology, 9,* 1653.

Klein, S. L. (2004). Hormonal and immunological mechanisms mediating sex differences in parasite infection. *Parasite Immunology, 26,* 247–264.

Markle, J. G., & Fish, E. N. (2014). SeXX matters in immunity. *Trends in Immunology, 35,* 97–104.
Klein, S. L., & Flanagan, K. L. (2016). Sex differences in immune responses. *Nature Reviews Immunology, 16,* 626–638.
Klein, S. L., & Pekosz, A. (2014). Sex-based biology and the rational design of influenza vaccination strategies. *The Journal of Infectious Diseases, 209*(Suppl 3), S114–S119.
Bernin, H., et al. (2014). Immune markers characteristic for asymptomatically infected and diseased *Entamoeba histolytica* individuals and their relation to sex. *BMC Infectious Diseases, 14,* 621.
Reed, S. L., & Gigli, I. (1990). Lysis of complement-sensitive *Entamoeba histolytica* by activated terminal complement components. Initiation of complement activation by an extracellular neutral cysteine proteinase. *The Journal of Clinical Investigation,* 86, 1815–1822.
Snow, M., Chen, M., Guo, J., Atkinson, J., & Stanley, S. L., Jr. (2008). Differences in complement-mediated killing of *Entamoeba histolytica* between men and women—An explanation for the increased susceptibility of men to invasive amebiasis? *The American Journal of Tropical Medicine and Hygiene, 78,* 922–923.
Forster, B., Ebert, F., & Horstmann, R. D. (1994). Complement sensitivity of *Entamoeba histolytica* and various nonpathogenic amoeba species. *Tropical Medicine and Parasitology, 45,* 355–356.
Urban, B., Blasig, C., Forster, B., Hamelmann, C., & Horstmann, R. D. (1996). Putative serine/threonine protein kinase expressed in complement-resistant forms of *Entamoeba histolytica. Molecular and Biochemical Parasitology, 80,* 171–178.
Spiegelberg, H. L. (1989). Biological role of different antibody classes. *International Archives of Allergy and Immunology, 90*(Suppl. 1), 22–27.
Seguin, R., Mann, B. J., Keller, K., & Chadee, K. (1997). The tumor necrosis factor alpha-stimulating region of galactose-inhibitable lectin of *Entamoeba histolytica* activates gamma interferon-primed macrophages for amebicidal activity mediated by nitric oxide. *Infection and Immunity, 65,* 2522–2527.
Tupin, E., Kinjo, Y., & Kronenberg, M. (2007). The unique role of natural killer T cells in the response to microorganisms. *Nature Reviews Microbiology, 5,* 405–417.
Godfrey, D. I., Stankovic, S., & Baxter, A. G. (2010). Raising the NKT cell family. *Nature Immunology, 11,* 197–206.
Bernin, H., Fehling, H., Marggraff, C., Tannich, E., & Lotter, H. (2016). The cytokine profile of human NKT cells and PBMCs is dependent on donor sex and stimulus. *Medical Microbiology and Immunology, 205,* 321–332.
Sandberg, J. K., Bhardwaj, N., & Nixon, D. F. (2003). Dominant effector memory characteristics, capacity for dynamic adaptive expansion, and sex bias in the innate Valpha24 NKT cell compartment. *European Journal of Immunology, 33,* 588–596.
Lotter, H., et al. (2009). Natural killer T cells activated by a lipopeptidophosphoglycan from *Entamoeba histolytica* are critically important to control amebic liver abscess. *PLoS Pathogens, 5,* e1000434.
Lotter, H., Jacobs, T., Gaworski, I., & Tannich, E. (2006). Sexual dimorphism in the control of amebic liver abscess in a mouse model of disease. *Infection and Immunity, 74,* 118–124.
Mattner, J., et al. (2005). Exogenous and endogenous glycolipid antigens activate NKT cells during microbial infections. *Nature, 434,* 525–529.
Gonzalez-Aseguinolaza, G., et al. (2000). alpha -galactosylceramide-activated Valpha 14 natural killer T cells mediate protection against murine malaria. *Proceedings of the National Academy of Sciences of the United States of America, 97,* 8461–8466.
Duthie, M. S., & Kahn, S. J. (2002). Treatment with alpha-galactosylceramide before *Trypanosoma cruzi* infection provides protection or induces failure to thrive. *Journal of Immunology (Baltimore, Md: 1950), 168*, 5778–5785.
Blessmann, J., et al. (2002). Epidemiology of amebiasis in a region of high incidence of amebic liver abscess in central Vietnam. *The American Journal of Tropical Medicine and Hygiene, 66,* 578–583.
Gil-Barbosa, M., Fastag de Shor, A., De la Torre, M., & Villegas-Gonzalez, J. (1972). Sequence of amebic hepatic lesions in the rabbit. *Archives of Medical Research (Mex),* 2(Suppl. 2), 349–354.

Roncolato, E. C., Teixeira, J. E., Barbosa, J. E., Zambelli Ramalho, L. N., & Huston, C. D. (2015). Immunization with the *Entamoeba histolytica* surface metalloprotease EhMSP-1 protects hamsters from amebic liver abscess. *Infection and Immunity, 83,* 713–720.

Chadee, K., & Meerovitch, E. (1984). The pathogenesis of experimentally induced amebic liver abscess in the gerbil (Meriones unguiculatus). *The American Journal of Pathology, 117,* 71–80.

Seydel, K. B., Zhang, T., & Stanley, S. L., Jr. (1997). Neutrophils play a critical role in early resistance to amebic liver abscesses in severe combined immunodeficient mice. *Infection and Immunity, 65,* 3951–3953.

Seydel, K. B., Smith, S. J., & Stanley, S. L., Jr. (2000). Innate immunity to amebic liver abscess is dependent on gamma interferon and nitric oxide in a murine model of disease. *Infection and Immunity, 68,* 400–402.

Lin, J. Y., & Chadee, K. (1992). Macrophage cytotoxicity against *Entamoeba histolytica* trophozoites is mediated by nitric oxide from L-arginine. *Journal of Immunology (Baltimore, Md: 1950), 148,* 3999–4005.

Velazquez, C., Shibayama-Salas, M., Aguirre-Garcia, J., Tsutsumi, V., & Calderon, J. (1998). Role of neutrophils in innate resistance to *Entamoeba histolytica* liver infection in mice. *Parasite Immunology, 20,* 255–262.

Jarillo-Luna, R. A., Campos-Rodriguez, R., & Tsutsumi, V. (2000). Participation of neutrophils, macrophages, and endothelial cells in the amebic liver lesion in the mouse. *Archives of Medical Research, 31,* S101–S103.

Lotter, H., et al. (2013). Testosterone increases susceptibility to amebic liver abscess in mice and mediates inhibition of IFNgamma secretion in natural killer T cells. *PLoS ONE, 8,* e55694.

Gourdy, P., et al. (2005). Relevance of sexual dimorphism to regulatory T cells: Estradiol promotes IFN-gamma production by invariant natural killer T cells. *Blood, 105,* 2415–2420.

Subleski, J. J., & Ortaldo, J. R. (2009). Editorial: NKT cells: To suppress or not to suppress, that is the question. *Journal of Leukocyte Biology, 86,* 751–752.

Kenna, T., et al. (2003). NKT cells from normal and tumor-bearing human livers are phenotypically and functionally distinct from murine NKT cells. *Journal of Immunology (Baltimore, Md: 1950), 171,* 1775–1779.

Roberts, C. W., Walker, W., & Alexander, J. (2001). Sex-associated hormones and immunity to protozoan parasites. *Clinical Microbiology Reviews, 14,* 476–488.

Bouman, A., Schipper, M., Heineman, M. J., & Faas, M. M. (2004). Gender difference in the non-specific and specific immune response in humans. *American Journal of Reproductive Immunology (New York, NY: 1989),* 52, 19–26.

Gubbels Bupp, M. R., & Jorgensen, T. N. (2018). Androgen-induced immunosuppression. *Frontiers in Immunology, 9,* 794.

Kissick, H. T., et al. (2014). Androgens alter T-cell immunity by inhibiting T-helper 1 differentiation. *Proceedings of the National Academy of Sciences, 111,* 9887.

Nathan, C. (2006). Neutrophils and immunity: Challenges and opportunities. *Nature Reviews Immunology, 6,* 173–182.

Laskin, D. L., Sunil, V. R., Gardner, C. R., & Laskin, J. D. (2011). Macrophages and tissue injury: Agents of defense or destruction? *Annual Review of Pharmacology and Toxicology, 51,* 267–288.

Guilliams, M., et al. (2014). Dendritic cells, monocytes and macrophages: A unified nomenclature based on ontogeny. *Nature Reviews Immunology, 14,* 571–578.

Ziegler-Heitbrock, L., et al. (2010). Nomenclature of monocytes and dendritic cells in blood. *Blood, 116,* e74–e80.

Zimmermann, H., Trautwein, C., & Tacke, F. (2012). Functional role of monocytes and macrophages for the inflammatory response in acute liver injury. *Frontiers in Physiology 3.*

Shi, C., & Pamer, E. G. (2011). Monocyte recruitment during infection and inflammation. *Nature Reviews Immunology, 11,* 762–774.

Patel, A. A., et al. (2017). The fate and lifespan of human monocyte subsets in steady state and systemic inflammation. *The Journal of Experimental Medicine, 214,* 1913–1923.

Serbina, N. V., & Pamer, E. G. (2006). Monocyte emigration from bone marrow during bacterial infection requires signals mediated by chemokine receptor CCR2. *Nature Immunology, 7,* 311–317.

Helk, E., et al. (2013). TNFalpha-mediated liver destruction by Kupffer cells and Ly6Chi monocytes during *Entamoeba histolytica* infection. *PLoS Pathogens, 9,* e1003096.

Bouman, A., Heineman, M. J., & Faas, M. M. (2005). Sex hormones and the immune response in humans. *Human Reproduction Update, 11,* 411–423.

Bosschaerts, T., et al. (2010). Tip-DC development during parasitic infection is regulated by IL-10 and requires CCL2/CCR2, IFN-gamma and MyD88 signaling. *PLoS Pathogens, 6,* e1001045.

Terrazas, C., et al. (2017). Ly6C(hi) inflammatory monocytes promote susceptibility to *Leishmania donovani* infection. *Scientific Reports, 7,* 14693.

Dunlap, M. D., et al. (2018). A novel role for C-C motif chemokine receptor 2 during infection with hypervirulent *Mycobacterium tuberculosis. Mucosal Immunology, 11,* 1727–1742.

Gurczynski, S. J., et al. (2019). CCR2 mediates increased susceptibility to post-H1N1 bacterial pneumonia by limiting dendritic cell induction of IL-17. *Mucosal Immunology, 12,* 518–530.

Geissmann, F., Jung, S., & Littman, D. R. (2003). Blood monocytes consist of two principal subsets with distinct migratory properties. *Immunity, 19,* 71–82.

Uversky, V. N., et al. (2017). Functionality of intrinsic disorder in tumor necrosis factor-alpha and its receptors. *FEBS Journal, 284,* 3589–3618.

Gaffen, S. L., Jain, R., Garg, A. V., & Cua, D. J. (2014). The IL-23-IL-17 immune axis: From mechanisms to therapeutic testing. *Nature Reviews Immunology, 14,* 585–600.

Paust, H. J., et al. (2009). The IL-23/Th17 axis contributes to renal injury in experimental glomerulonephritis. *Journal of the American Society of Nephrology, 20,* 969–979.

Chen, K., et al. (2011). IL-17RA is required for CCL2 expression, macrophage recruitment, and emphysema in response to cigarette smoke. *PLoS ONE, 6,* e20333.

Shahrara, S., Pickens, S. R., Dorfleutner, A., & Pope, R. M. (2009). IL-17 induces monocyte migration in rheumatoid arthritis. *Journal of Immunology (Baltimore, Md: 1950), 182*, 3884–3891.

Yang, X. O., et al. (2008). Regulation of inflammatory responses by IL-17F. *The Journal of Experimental Medicine, 205,* 1063–1075.

Noll, J., et al. (2016). IL-23 prevents IL-13-dependent tissue repair associated with Ly6C(lo) monocytes in *Entamoeba histolytica*-induced liver damage. *Journal of Hepatology, 64,* 1147–1157.

Dal-Secco, D., et al. (2015). A dynamic spectrum of monocytes arising from the in situ reprogramming of CCR2$^+$ monocytes at a site of sterile injury. *The Journal of Experimental Medicine, 212,* 447–456.

Chapter 12
Oxidative Stress and Heat Stress in Experimental Amoebic Liver Abscess

Alfonso Olivos-García, Emma Saavedra, Sergio Enríquez-Flores, Nancy Guillen, Mario Nequiz-Avendaño, Yoalli Martínez, Azucena Cortes, Edith Mendoza, Julio C. Carrero, Porfirio Zúñiga, Gabriel López-Velázquez, and Ruy Pérez-Tamayo

Abstract Several species belonging to the Genus *Entamoeba* can colonize the mouth or the human gut; however, only *Entamoeba histolytica* is pathogenic to the host, causing amoebiasis. This illness leads to one hundred thousand deaths per year worldwide, affecting mainly underdeveloped countries in areas with poor sanitary conditions. Throughout its life cycle or during the invasion of human tissues, the parasite is constantly subjected to stress conditions. In in vitro culture, this microaerophilic parasite can tolerate up to 50 μM oxygen concentrations; however, during invasion the parasite has to cope with the higher oxygen content found in the blood and well perfused tissues (60–130 μM) and with reactive oxygen and nitrogen species (ROS and NOS, respectively) derived from both the host (as a first line of defense against the infection) and from the oxygen and nitrogen detoxification systems within the parasite. Furthermore, notwithstanding that host body temperatures (36.5–39 °C) observed during amoebic liver infection affect amoebic growth in culture (~50% survival); during tissue invasion the parasite adapts to this stress and proliferates.

A. Olivos-García (✉) · M. Nequiz-Avendaño · Y. Martínez · A. Cortes · E. Mendoza · R. Pérez-Tamayo
Facultad de Medicina, Unidad de Investigación en Medicina Experimental, Universidad Nacional Autónoma de México, Mexico City 04510, Mexico
e-mail: olivosa@yahoo.com

E. Saavedra
Departamento de Bioquímica, Instituto Nacional de Cardiología Ignacio Chávez, Secretaría de Salud, Mexico City 14080, Mexico

S. Enríquez-Flores · G. López-Velázquez
Laboratorio de Errores Innatos del Metabolismo y Tamiz, Grupo de Investigación en Biomoléculas y Salud Infantil, Instituto Nacional de Pediatría, Secretaria de Salud, Mexico City 04530, Mexico

N. Guillen
Institut Pasteur and Centre National de la Recherche Scientifique ERL9195, Paris 75015, France

J. C. Carrero
Departamento de Inmunología, Instituto de Investigaciones Biomédicas, Universidad Nacional Autónoma de México, Mexico City, Mexico

P. Zúñiga
Centro Médico Dalinde, Tuxpan 25, Roma sur, Mexico City 06760, Mexico

© Springer Nature Switzerland AG 2020

N. Guillen (ed.), *Eukaryome Impact on Human Intestine Homeostasis and Mucosal Immunology*, https://doi.org/10.1007/978-3-030-44826-4_12

In the present work, we analyzed the possible mechanisms by which *E. histolytica* resists the physiological stressors present during the host tissue invasion. It is also discussed that during the experimental amoebic liver abscess development in hamsters, several adaptation stages are occurring, including a host liver metabolic reprograming that creates a reductive environment that favors amoebic survival.

Keywords *Entamoeba histolytica* · Pathogenicity · Oxidative stress · Heat shock

Introduction

The human large intestine usually harbors several amoebic species, however only *E. histolytica* is able to invade the gut mucosa and cause tissue damage. The pathogenicity of *E. histolytica* has been attributed to increased levels in a concerted action of several functions such as, proteolysis, cytotoxicity, complement resistance and inflammation (Mirelman et al. 2000; Marie and Petri 2014); if some of them are diminished or missing, virulence is impaired. The higher levels of such functions in *E. histolytica* are in comparison with those observed in other amoebae such as the non pathogenic species *E. dispar* or the non virulent strain *E. histolytica* Rahman, that are able to perform all these functions but at lower extents. Other functions essential for amoebic pathogenicity are resistance to oxidative and heat stresses as well as an efficient stress repairing system held by the heat shock protein (HSP) family (Santos et al. 2015).

Amoebic Resistance to Oxygen and ROS

As anaerobic microorganisms, oxygen is toxic for *Entamoeba* sp. because sensitivity to oxygen and reactive oxygen species (ROS) of Fe^{2+} or Fe–S centers in many proteins (Imlay 2003). Examples of the latter are aldehyde-alcohol dehydrogenase (ADHE) and pyruvate: ferredoxin oxidoreductase (PFOR), enzymes from fermentative glycolysis (Saavedra et al. 2019) whose inhibition by oxygen exposure can compromise the energy supply to the parasite (Pineda et al. 2010, 2013). Also, oxidative stress in *E. histolytica* modifies the transcription of thousands of genes implicated in several functions including those related to pathogenicity (Vicente et al. 2009). In addition, such stress causes protein oxidation, inhibition of protein synthesis and overexpression of general stress response-related proteins HSPs (Shahi et al. 2016a).

Entamoeba can resist the low oxygen concentrations of the human gut (0.1–2.3%) (Ladas et al. 2007) due to an efficient oxygen reduction pathway (ORP). This is composed by (1) a flavodiiron protein that transforms oxygen into water without ROS formation; (2) thioredoxin reductases that reduce oxygen to hydrogen peroxide, and (3) peroxiredoxin and rubrerytrin that transform hydrogen peroxide into water (Olivos-García et al. 2016); the intracellular superoxide produced adventitiously is incorporated to ORP by a FeSOD. In addition, we have recently characterized a recombinant iron-sulfur flavoprotein of *E. histolytica* able to reduce oxygen and hydrogen peroxide to water (unpublished results); although its participation in the amoebic ORP remains to be analyzed. We have proposed that the principal function of the ORP is to protect the glycolytic pathway from oxidants since: (1) ORP proteins are constitutively expressed and maintain an intracellular hypoxic environment; (2) a hypoxic microenvironment allows adequate glycolytic fluxes to produce reduced intermediaries (coenzymes and small thiol proteins) and ATP used by the ORP and the repairing system of stress (HSPs) respectively, and (3) the oxygen-sensitive glycolytic components are similarly localized with ORP proteins at the periplasmic region.

In contrast with *E. dispar* (non-pathogenic *Entamoeba* species that infects humans), *E. histolytica* shows higher expression of the ORP proteins (Santos et al. 2015), very probably enabling the parasite to resist the higher oxygen concentrations of the gut wall (1% near the lumen and 5–10% near the vascularized submucosa or deeper muscle layers; Zheng et al. 2015). Such amoebic exposure to gut tissular oxygen is for ~6 h, after which inflammation promotes ischaemia, thus diminishing the local oxygen concentration. The proposal that ischaemia protects the parasite from complement and oxygen is supported by the observation that during experimental amoebic liver infection in hamsters (EALAH) these amoebae are unable to survive in animals devoid of inflammatory cells (Olivos-García et al. 2004). In addition, the parasite's HSP response is necessary to repair the damage caused by oxidative stress. In this regard, the *E. histolytica* loss of virulence by prolonged culture is principally due to its inability to overexpress HSPs under hyperoxia (Santos et al. 2015); moreover, a specific inhibitor of HSP70 (central to HSP response) drastically diminish the ability of virulent *E. histolytica* to survive in the hamster liver (Santos et al. 2015).

Similar to what happens in the human gut, during the early steps of EALAH (~6 h, before inflammation creates ischaemia; Pérez-Tamayo et al. 1992) amoebae are challenged with physiological tissular oxygen from 4 to 14% (Nauck et al. 1981) (Fig. 12.1). The oxidative stress in the parasite caused by oxygen (Akbar et al. 2004; Santos et al. 2015) along with blood's complement, could be the principal causes of the massive amoebic death observed before 12 h of EALAH (Rigothier et al. 2002). The evidences supporting these statements are that: (1) oxidative stress and complement are present during early EALAH, (2) the in vitro lytic activity of complement increases in *E. histolytica* under oxidative stress, (3) during this infection period, amoebic HSP70 that protects the parasite from oxidative stress is not overexpressed. The amoebic oxidative stress caused by the inflammatory ROS is expected to be marginal, since its production is temporal (~30 min), superoxide has limited diffusion to membranes due its charge, and the hydrogen peroxide amount produced of ~30 μM (Winterbourn and Kettle 2013) is not toxic for the parasite (Santos et al. 2015). In addition, on the late stages of EALAH macrophages are not able to produce ROS due to a prostaglandin E2 (PGE2) delivered by the parasite, among others molecules (Denis and Chadee 1988).

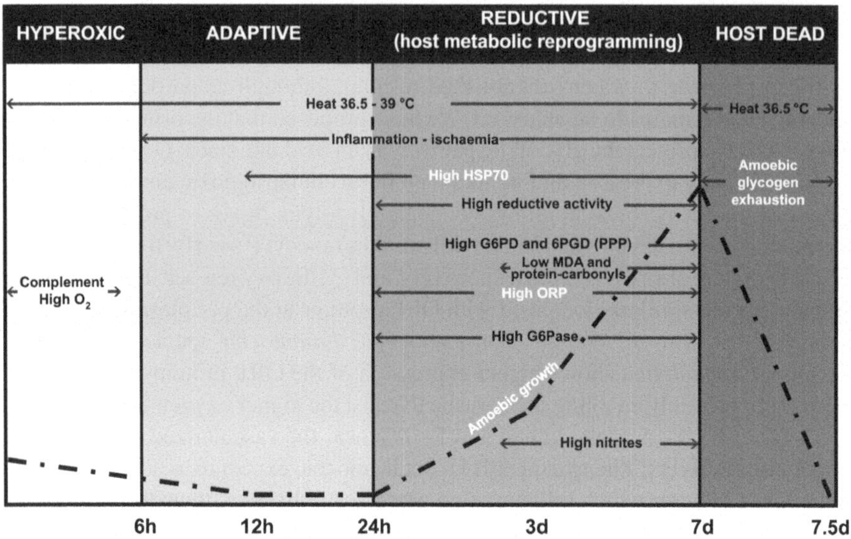

Fig. 12.1 Phases of the experimental amoebic liver abscess in hamster. The amoebic liver infection starts with a hyperoxic phase of 6 h in which the parasite is exposed to heat, oxygen and complement causing ~70% of amoebic death. In the adaptive phase, ischaemia induced by the inflammation process diminish the amoebic contact with oxygen and complement. Heat exposure remains and the parasite does not proliferate but overexpress HSP70 that helps in adaption to the heat stress. Afterwards, in the reductive phase, a metabolic reprogramming of the liver occurs, promoting a reductive environment probably favored by activation of the pentose phosphate pathway (PPP). Also, overexpression of the ORP in the parasite may help to avoid accumulation of reduced intermediaries generated by the glycolysis. The reductive environment, high G6Pase activity that allow glucose delivering and precursors for anabolism generated by PPP (all from liver), and along with overexpression of the amoebic HSP70 (among others) allow high amoebic proliferation. In addition, nitrites from blood circulation are not toxic for the parasite. Finally, after the animal is dead, the parasite disappear 12 h later after glycogen exhaustion

Amoebic Resistance to Nitric Oxide and Nitrites

Nitric oxide (NO·) is an endogenous, diffusible, trans-cellular messenger shown to participate in survival and death pathways; it can alter protein function directly through post-translational modifications (nitration or s-nitrosylation) or indirectly through interactions with oxygen, superoxide, thiols and heavy metals (Uys et al. 2011). When *E. histolytica* inhabits the human colon, reactive nitrogen species (RNS) are other strong oxidants harmful for the parasite; a concentration of 18 ppm of NO· in the gas phase has been detected in the human colon (Kastner et al. 1997). Besides, it has been documented that NO· can also be produced by microorganisms through bacterial nitric oxide synthase (bNOS), nitrate/nitrite reductases, or nitrite reduction under an acidic environment. Although *E. coli* bacteria protects amoebae from ROS in vitro, they do not protect against NO· (Varet el al. 2018).

During amoebic invasion to colonic tissue the parasite may be exposed to sublethal NO· concentrations produced by epithelial cells in response to cytokines or bacterial products. It has been shown that amoebic resistance to in vitro sublethal NO· concentration is related to up-regulation of HSPs, DNA repair, antioxidant and glycolytic genes (Santi-Rocca et al. 2012). Also, the increased nitric oxide levels in blood circulation (determined as nitrites) during the EALAH (Pacheco-Yépez et al. 2001), positively correlates with the parasite proliferation from intermediate to late stages of EALAH (determined by counting well preserved parasites in histological sections of EALAH). In vitro, NO· inhibits cysteine proteinases, PFOR and ADHE (Siman-Tov and Ankri 2003) and induces apoptosis (Ramos et al. 2007) in *E. histolytica*. Furthermore, NO· produced by activated murine macrophages are able to kill the parasite (Lin and Chadee 1992). Therefore a question that needs to be addressed is why amoebae survive and proliferate despite overproduction of NO/nitrites during EALAH? Two possible explanations are that the parasite is resistant to both oxidants and/or they are not accessible to amoebae.

It is well known that NO· is produced by activated inflammatory cells through inducible nitric oxide synthase (iNOS) (Cinelli et al. 2020). Although the parasite can adapt in vitro to toxic levels of NO· delivered by a donor (Shahi et al. 2016a, b), erythrophagocytosis, motility and cytopathic activity essential for pathogenicity are impaired (Trebicz-Geffen et al. 2017). Furthermore, it has been proposed that the parasite may enzymatically produce NO· from L-arginine (Hernández-Campos et al. 2003). However, by using electronic paramagnetic resonance analysis, we were not able to detect NO· in the parasite in the presence of L-arginine or nitrates. Instead we detected NO· production in the presence of nitrites, suggesting that the parasite may produce endogenous NO· by an unknown nitrite reductase or by nitrite reduction in acidic vesicles.

On the other hand, it was previously proposed that during EALAH, NO· may be produced by inflammatory cells present in the liver, since mRNA of iNOS was increased (Ramírez-Emiliano et al. 2005). Similarly to the previous findings in human liver amoebic abscesses (Ventura-Juárez et al. 2003), we detected only at late stages of EALAH several inflammatory cells positive to iNOS in well preserved liver tissue. However, leukocytes bordering the amoebic necrotic foci were negative to iNOS; if a minute amount of this enzyme was induced in these cells, its substrate arginine would be consumed by the parasite's arginase (Elnekave et al. 2003; Malik et al. 2018). In addition, the highest nitrite concentrations observed at late EALAH stages (7 days) were not toxic for amoebae in culture, nor stimulated its cytotoxicity on human leukocytes. These results suggest that during EALAH, the amoebic stress caused by NO· synthesized by iNOS is not occurring and that the parasite is resistant to nitrosative stress caused by nitrites from blood circulation, probably by using the N-acetyl ornithine deacetylase (Shahi et al. 2016b) and/or flavodiiron protein enzyme that is constitutive and overexpressed at late stages of EALAH.

Amoebic Resistance to Heat

It is known that heat shock protein response of almost all cells is essential to adapt and to resist most stresses, not only heat, to prevent accumulation of unfolded proteins in the endoplasmic reticulum by repairing or promoting its degradation by the proteasome system. Several HSPs have been reported in *E. histolytica*: Hsp10, Hsp40, Hsp60, Hsp70 (two cytosolic, two endoplasmic reticulum and one mitosomal isoforms), Hsp90 (three isoforms), Hsp100 and Hsp101 (Olivos-García et al. 2012). A transcription profiling study during heat shock stress (42 °C for 4 h) in *E. histolytica* showed a somewhat specific response of transcription factors, ubiquitin, proteasome 26S, proteases and HSP70 and HSP90 families of genes, according to a classical heat shock response. Also, a significant down-regulation of genes involved in almost all physiologic functions (including related virulence genes) was observed. Such transcriptional changes correspond with a general adapting response to the new environmental conditions (Weber et al. 2006). In addition, during an in vitro heat stress exposure of 45 °C for 1 h in *E. histolytica,* it was observed nuclear overexpression of HSP100, suggesting a nuclear protection function of this HSP during amoebic heat stress (Bernes et al. 2005).

Despite EALAH development no feverish episodes are observed, the animal's body temperature oscillates from 36.5 to 39 °C, which is very similar to the 37–39 °C range observed in humans during invasive amoebic infection (Arellano-Aguilar et al. 2017). However, the heat stress caused in the parasite by culture during three and six days to the same hamster's body temperatures, inhibits parasite replication (~50%). Notwithstanding, during EALAH the parasite adapts and proliferates; probably, the concurrent action of different host stressors stimulates higher HSPs expression and the production of factors promoting amoebae growth.

Furthermore, it is known that HSPs are ATP dependent enzymes and at least cytosolic HSP70 are overexpressed during the adaptation and reductive events of EALAH (after 12 h of infection). During such steps the glycolytic pathway produces ATP (that can be used by HSPs and other physiological processes) as well as reduced coenzymes (NADH) and small dithiol proteins (ferredoxin). Accumulation of NADH and reduced ferredoxin may cause inhibition of the glycolytic flux due to impaired coenzymes' recycling. In order to avoid that, the parasite may use oxygen as the final acceptor or sink of those electrons. In that regard, the high amoebic affinity for oxygen at low tensions observed by Weinbach and Diamond (1974) may be explained by the overexpression of the ORP enzymes during EALAH.

In addition to the high activity of the amoebic functions related with pathogenicity mentioned above, chemotactic molecules to amoebae, low oxygen concentrations, reductive environment and glucose given by living host cells, are necessary conditions for amoebic survival and proliferation during EALAH. The findings supporting that statement are that: (1) Once the animal is euthanized after 7 days of EALAH and preserved at 37 °C in an incubator, amoebae disappear 12 h later showing glycogen exhaustion, (2) on the contrary, amoebae (displaying high glycogen content) survives at the edge of a lobule of 24 h EALAH implanted into the peritoneum of healthy live hamster without infection (personal observation).

Liver Physiological Reprograming During EALAH

After 12 h of infection in EALAH, amoebae enter in an adapting process whose proliferation is maintained from 72 h until the animal's death (Rigothier et al. 2002). Since a reductive environment is a condition for amoebic replication, parasite proliferation may be favored by metabolic reprogramming in the liver cells, from oxidative to a reductive environment. The latter has been observed by the gradual increase of the reductive activity of the liver tissue (determined by nitroblue tetrazolium reduction) along with decreased malondialdehyde levels from three to seven days by an increased in pentose phosphate pathway (PPP) activity, which produces NADPH that is required for the generation of reduced glutathione (GSH), a major ROS scavenger from host (Nathan and Ding 2010). Similarly, histological cryosections of EALAH showed high reductive activity in the liver cells bordering the amoebal necrotic foci, as determined by nitroblue tetrazolium reduction in the presence of NADPH (Pacheco-Yépez et al. 2001). Such liver reductive activity may be stimulated by the anabolic processes required to replace damaged liver tissue. In addition, the reductive intermediaries produced by the activated liver PPP during EALAH (Guerrero et al. 1981) may be incorporated by amoebae, probably through phagocytosis and/or trogocytosis, to satisfy some anabolic demands and to maintain the redox homeostasis in the parasite (Fig. 12.1).

As observed in hepatocellular carcinoma and others tumors (Kowalik et al. 2017), physiological reprograming of the liver during EALAH creates a reductive microenvironment that diminish oxidants alongside with increased glucose-6-phosphatase activity that stimulates glucose deliverance (Fastag de Shor et al. 1972). Both conditions favor parasite replication.

REDOX and HSP Proteins as Potential Antiamoebic Targets

Several parasite molecules involved in some functions related with pathogenicity have been proposed as potential targets to treat invasive amoebiasis. These include, molecules involved in the thiol-dependent metabolism of *E. histolytica* (Jeelani and Nozaki 2016). For example, the L-cysteine biosynthetic pathway of the parasite, in particular serine O-acetyltransferase 3 and cysteine synthase isoforms, have been suggested as potential drug targets since this pathway is critical for the growth and oxidative stress defenses (Jeelani et al. 2017). In addition, auranofin (a compound that inhibits TrxR activity of the parasite and increase amoebic susceptibility to oxidants), was able to diminish the number of parasites, inflammation and tissue damage when hamsters were treated 4 days after liver infection (Debnath et al. 2012).

After analyzing the stress response of the parasite during the EALAH, we propose the two cytosolic HSP70 isoforms as potential amoebic targets since: (1) they are overexpressed during the adaptive and reductive steps of EALAH; (2) they are necessary for growth and resistance to oxygen and heat (36.5–39 °C), being the last one present during all invasive amoebic infection; (3) the loss of parasite ability to survive and to cause EALAH by prolonged culture is related to its inability to overexpress the heat shock protein gene family, that includes HSP70 isoforms (Santos et al. 2015); (4) they are mainly located in cytoplasm and not confined in substructures that could limit diffusion of potential drugs; (5) at structural level, they have different pockets and two non-related Cys residues exposed to the solvent (149 and 456) that may become their Achilles heel since they are not structurally related with its human counterpart; in particular, Cys 149 is 7.4 Å close to the ATP pocket (Fig. 12.2). In addition, HSP90 of *E. histolytica* may be other important potential target since its inhibition in amoebae leads to cytotoxicity in culture and diminished parasite proliferation in a mouse model of amoebic colitis (Debnath et al. 2014).

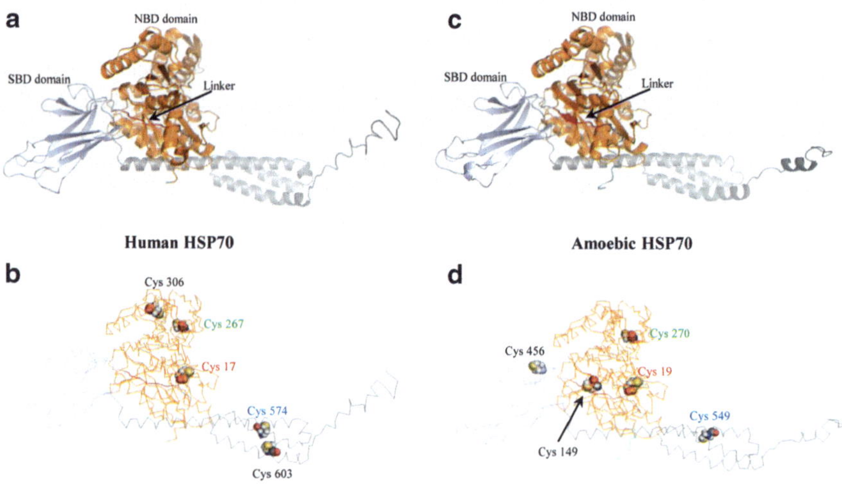

Fig. 12.2 Structural differences of cytosolic EhHSP70 with its human counterpart. Ribbon representations of C_{α} trace of comparative models of EhHSP70-A, and HuHSP70-A1. **a, c** The principal domains are shown as follow: grey, substrate binding domain (SBD); orange, nucleotide binding domain (NBD); red, the linker region. Human HSP70 (**b**) and amoebic HSP70 (**d**) oriented in the same structural position. Human HSP70 and amoebic HSP70 show the Cys residues. As indicated, both HSPs contain 5 Cys/subunit, however amoebic Cys 149 and 456 do not structurally align; in particular, Cyst 149 is 7.4 Å close to the ATP pocket. The figures were prepared with PyMOL (DeLano WL. The PyMOL molecular graphics system. Palo Alto, CA: DeLanoScientific; 2002. Available at: http://www.pymol.org)

Conclusion

The study of the host-parasite interactions is of ultimate importance to understand the molecular mechanisms of *E. histoltytica* pathogenicity. Analysis of the amoebic resistance mechanisms to the stressors present during the amoebic liver infection as well as the metabolic reprograming of the host liver that favors amoebic proliferation, point out the parasite proteins related with HSP response, ORP and thiol-dependent metabolism as potential targets to treat invasive amoebiasis. Further studies are necessary to validate this proposal.

Acknowledgements The research in the author's laboratory is supported by DGAPA Grant IN-214617 and CONACyT 282663. We acknowledge Mr. Marco Gudiño for the informatics and art design.

References

Akbar, M. A., Chatterjee, N. S., Sen, P., et al. (2004). Genes induced by a high-oxygen environment in *Entamoeba histolytica*. *Molecular and Biochemical Parasitology, 133*(2), 187–196.

Arellano-Aguilar, G., Marín-Santillán, E., Castilla-Barajas, J. A., et al. (2017). A brief history of amoebic liver abscess with an illustrative case. *Revista de Gastroenterologia de Mexico, 82*(4), 344–348. https://doi.org/10.1016/j.rgmx.2016.05.007.

Bernes, S., Siman-Tov, R., & Ankri, S. (2005). Epigenetic and classical activation of *Entamoeba histolytica* heat shock protein 100 (EHsp100) expression. *FEBS Letters, 579*(28), 6395–6402.

Cinelli, M. A., Do, H. T., Miley, G. P., et al. (2020). Inducible nitric oxide synthase: Regulation, structure, and inhibition. *Medicinal Research Reviews, 40*(1), 158–189. https://doi.org/10.1002/med.21599.

Debnath, A., Parsonage, D., Andrade, R. M., et al. (2012). A high-throughput drug screen for *Entamoeba histolytica* identifies a new lead and target. *Nature Medicine, 18*(6), 956–960. https://doi.org/10.1038/nm.2758.

Debnath, A., Shahinas, D., Bryant, C., et al. (2014). Hsp90 inhibitors as new leads to target parasitic diarrheal diseases. *Antimicrobial Agents and Chemotherapy, 58*(7), 4138–4444. https://doi.org/10.1128/AAC.02576-14.

Denis, M., & Chadee, K. (1988). In vitro and in vivo studies of macrophage functions in amebiasis. *Infection and Immunity, 56*(12), 3126–3131.

Elnekave, K., Siman-Tov, R., & Ankri, S. (2003). Consumption of L-arginine mediated by *Entamoeba histolytica* L-arginase (EhArg) inhibits amoebicidal activity and nitric oxide production by activated macrophages. *Parasite Immunology, 25*(11–12), 597–608.

Fastag de Shor, A., Ramírez-Espinosa, M. E., Gil-Barbosa, M., et al. (1972). Enzymatic activity of *Entamoeba histolytica* in experimental lesions in rabbits. *Archivos de investigacion medica, 2*(Suppl 2), 287–294.

Guerrero, M., Ríos, D., Ramírez, M. E. (1981). Histochemical aspects of liver lesions induced in hamsters by inoculation of *E. histolytica* in axenic culture. II. Evidence of perinecrotic fibrosis. *Revista de gastroenterologia de Mexico, 46*(2), 49–54.

Hernández-Campos, M. E., Campos-Rodríguez, R., Tsutsumi, V., et al. (2003). Nitric oxide synthase in *Entamoeba histolytica*: Its effect on rat aortic rings. *Experimental Parasitology, 104*(3–4), 87–95.

Imlay, J. A. (2003). Pathways of oxidative damage. *Annual Review of Microbiology, 57,* 395–418.

Jeelani, G., & Nozaki, T. (2016). *Entamoeba* thiol-based redox metabolism: A potential target for drug development. *Molecular and Biochemical Parasitology, 206*(1–2), 39–45. https://doi.org/10.1016/j.molbiopara.2016.01.004.

Jeelani, G., Sato, D., Soga, T., et al. (2017). Genetic, metabolomic and transcriptomic analyses of the de novo L-cysteine biosynthetic pathway in the enteric protozoan parasite *Entamoeba histolytica*. *Scientific Report, 7*(1), 15649. https://doi.org/10.1038/s41598-017-15923-3.

Kastner, J., Pfeffel, F., Rajek, A., et al. (1997). Nitric oxide concentration in the gas phase of the gastrointestinal tract in man. *European Journal of Clinical Investigation, 27*(12), 992–996.

Kowalik, M. A., Columbano, A., & Perra, A. (2017). Emerging role of the pentose phosphate pathway in hepatocellular carcinoma. *Frontiers in Oncology, 7*, 87. https://doi.org/10.3389/fonc.2017.00087.

Ladas, S. D., Karamanolism, G., & Ben-Soussan, E. (2007). Colonic gas explosion during therapeutic colonoscopy with electrocautery. *World Journal of Gastroenterology, 13*(40), 5295–5298.

Lin, J. Y., & Chadee, K. (1992). Macrophage cytotoxicity against *Entamoeba histolytica* trophozoites is mediated by nitric oxide from L-arginine. *The Journal of Immunology, 148*(12), 3999–4005.

Malik, A., Singh, H., Pareek, A., et al. (2018). Biochemical and biophysical insights into the metal binding spectrum and bioactivity of arginase of *Entamoeba histolytica*. *Metallomics, 10*(4), 623–638. https://doi.org/10.1039/c8mt00002f.

Marie, C., & Petri, W. A., Jr. (2014). Regulation of virulence of *Entamoeba histolytica*. *Annual Review of Microbiology, 68*, 493–520. https://doi.org/10.1146/annurev-micro-091313-103550.

Mirelman, D., Ankri, S., Katz, U., et al. (2000). Pathogenesis of *Entamoeba histolytica* depends on the concerted action of numerous virulence factors. *Archives of Medical Research, 31*(4 Suppl), S214–S215.

Nathan, C., & Ding, A. (2010). SnapShot: Reactive oxygen intermediates (ROI). *Cell, 140*(6), 951–951.e2. https://doi.org/10.1016/j.cell.2010.03.008.

Nauck, M., Wölfle, D., Katz, N., et al. (1981). Modulation of the glucagon-dependent induction of phosphoenolpyruvate carboxykinase and tyrosine aminotransferase by arterial and venous oxygen concentrations in hepatocyte cultures. *European Journal of Biochemistry, 119*(3), 657–661.

Olivos-García, A., Nequiz-Avendaño, M., Tello, E., et al. (2004). Inflammation, complement, ischemia and amoebic survival in acute experimental amoebic liver abscesses in hamsters. *Experimental and Molecular Pathology, 77*(1), 66–71.

Olivos-García, A., Saavedra, E., Luis-García, E. R., et al. (2012). Stress response in *Entamoeba histolytica*. In J. M. Requena (Ed.), *Stress response in microbiology* (pp. 405–427). Norfolk: Caister Academic Press.

Olivos-García, A., Saavedra, E., Nequiz, M., et al. (2016). The oxygen reduction pathway and heat shock stress response are both required for *Entamoeba histolytica* pathogenicity. *Current Genetics, 62*(2), 295–300. https://doi.org/10.1007/s00294-015-0543-5.

Pacheco-Yépez, J., Campos-Rodríguez, R., Shibayama, M., et al. (2001). *Entamoeba histolytica*: Production of nitric oxide and in situ activity of NADPH diaphorase in amebic liver abscess of hamsters. *Parasitology Research, 87*(1), 49–56.

Pérez-Tamayo, R., Montfort, I., Tello, E., et al. (1992). Ischemia in experimental acute amebic liver abscess in hamsters. *International Journal for Parasitology, 22*(1), 125–129.

Pineda, E., Encalada, R., Rodríguez-Zavala, J. S., et al. (2010). Pyruvate: Ferredoxin oxidoreductase and bifunctional aldehyde-alcohol dehydrogenase are essential for energy metabolism under oxidative stress in *Entamoeba histolytica*. *FEBS Journal, 277*(16), 3382–3395. https://doi.org/10.1111/j.1742-4658.2010.07743.x.

Pineda, E., Encalada, R., Olivos-García, A., et al. (2013). The bifunctional aldehyde-alcohol dehydrogenase controls ethanol and acetate production in *Entamoeba histolytica* under aerobic conditions. *FEBS Letters, 587*(2), 178–184. https://doi.org/10.1016/j.febslet.2012.11.020.

Ramírez-Emiliano, J., González-Hernández, A., & Arias-Negrete, S. (2005). Expression of inducible nitric oxide synthase mRNA and nitric oxide production during the development of liver abscess in hamster inoculated with *Entamoeba histolytica*. *Current Microbiology, 50*(6), 299–308.

Ramos, E., Olivos-García, A., Nequiz, M., et al. (2007). *Entamoeba histolytica*: Apoptosis induced in vitro by nitric oxide species. *Experimental Parasitology, 116*(3), 257–265.

Rigothier, M. C., Khun, H., Tavares, P., et al. (2002). Fate of *Entamoeba histolytica* during establishment of amoebic liver abscess analyzed by quantitative radioimaging and histology. *Infection and Immunity, 70*(6), 3208–3215.

Saavedra, E., Encalada, R., Vázquez, C., et al. (2019). Control and regulation of the pyrophosphate-dependent glucose metabolism in *Entamoeba histolytica*. *Molecular and Biochemical Parasitology, 229,* 75–87. https://doi.org/10.1016/j.molbiopara.2019.02.002.

Santi-Rocca, J., Smith, S., Weber, C., et al. (2012). Endoplasmic reticulum stress-sensing mechanism is activated in *Entamoeba histolytica* upon treatment with nitric oxide. *PLoS ONE, 7*(2), e31777. https://doi.org/10.1371/journal.pone.0031777.

Santos, F., Nequiz, M., Hernández-Cuevas, N. A., et al. (2015). Maintenance of intracellular hypoxia and adequate heat shock response are essential requirements for pathogenicity and virulence of *Entamoeba histolytica*. *Cellular Microbiology, 17*(7), 1037–1051. https://doi.org/10.1111/cmi.12419.

Shahi, P., Trebicz-Geffen, M., Nagaraja, S., et al. (2016a). Proteomic identification of oxidized proteins in *Entamoeba histolytica* by resin-assisted capture: Insights into the role of arginase in resistance to oxidative stress. *PLoS Neglected Tropical Diseases, 10*(1), e0004340. https://doi.org/10.1371/journal.pntd.0004340.

Shahi, P., Trebicz-Geffen, M., Nagaraja, S., et al. (2016b). N-acetyl ornithine deacetylase is a moonlighting protein and is involved in the adaptation of *Entamoeba histolytica* to nitrosative stress. *Scientific Report, 6,* 36323. https://doi.org/10.1038/srep36323.

Siman-Tov, R., & Ankri, S. (2003). Nitric oxide inhibits cysteine proteinases and alcohol dehydrogenase 2 of *Entamoeba histolytica*. *Parasitology Research, 89*(2), 146–149.

Trebicz-Geffen, M., Shahi, P., Nagaraja, S., et al. (2017). Identification of S-Nitrosylated (SNO) proteins in *Entamoeba histolytica* adapted to nitrosative stress: Insights into the role of SNO actin and *in vitro* virulence. *Frontiers in Cellular and Infection Microbiology, 7,* 192. https://doi.org/10.3389/fcimb.2017.00192.

Uys, J. D., Xiong, Y., & Townsend, D. M. (2011). Nitrosative stress-induced S-glutathionylation of protein disulfide isomerase. *Methods in Enzymology, 490,* 321–332.

Varet, H., Shaulov, Y., Sismeiro, O., et al. (2018). Enteric bacteria boost defences against oxidative stress in *Entamoeba histolytica*. *Scientific Report, 8*(1), 9042. https://doi.org/10.1038/s41598-018-27086-w.

Ventura-Juárez, J., Jarillo-Luna, R. A., Fuentes-Aguilar, E., et al. (2003). Human amoebic hepatic abscess: in situ interactions between trophozoites, macrophages, neutrophils and T cells. *Parasite Immunology, 25*(10), 503–511.

Vicente, J. B., Ehrenkaufer, G. M., & Saraiva, L. M. (2009). *Entamoeba histolytica* modulates a complex repertoire of novel genes in response to oxidative and nitrosative stresses: Implications for amebic pathogenesis. *Cellular Microbiology, 11*(1), 51–69. https://doi.org/10.1111/j.1462-5822.2008.01236.x.

Weber, C., Guigon, G., Bouchier, C., et al. (2006). Stress by heat shock induces massive down regulation of genes and allows differential allelic expression of the Gal/GalNAc lectin in *Entamoeba histolytica*. *Eukaryotic Cell, 5*(5), 871–875.

Weinbach, E. C., & Diamond, L. S. (1974). *Entamoeba histolytica*. I. Aerobic metabolism. *Experimental Parasitology, 35*(2), 232–243.

Winterbourn, C. C., & Kettle, A. J. (2013). Redox reactions and microbial killing in the neutrophil phagosome. *Antioxidants & Redox Signaling, 18*(6), 642–660. https://doi.org/10.1089/ars.2012.4827.

Zheng, L., Kelly, C. J., & Colgan, S. P. (2015). Physiologic hypoxia and oxygen homeostasis in the healthy intestine. A review in the theme: Cellular responses to hypoxia. *American Journal of Physiology-Cell Physiology, 309,* C350–C353.

Chapter 13
Role of Extracellular Traps Promoted by Intestinal Parasites. Relationship with Virulence

César Díaz-Godínez, Arony Martínez-Flores, Raúl Argüello-García, Alfonso Olivos-García, Mario Néquiz-Avendaño, and Julio César Carrero

Abstract NETosis is a form of programmed cell death in neutrophils characterized by the release of extracellular DNA traps (NET) composed of DNA associated with histones, granule enzymes and antimicrobial peptides. This mechanism of innate immunity has been linked to cell defense against different pathogens. In case of bacteria, NETosis has been directly associated with virulence, suggesting that its formation could depend on the pathogen's ability to cause tissue damage. Moreover, NET formation can contribute to tissue destruction by promoting local inflammation and coagulation when exacerbated.

Keywords Neutrophil extracellular traps (NETs) · Pathogenicity/virulence · *Entamoeba histolytica* · *Entamoeba dispar* · *Giardia lamblia* · *Blastocystis* spp. · Reactive oxygen species (ROS)

In previous studies we have shown that trophozoites of *Entamoeba histolytica*, the intestinal protozoan parasite that causes amoebiasis, trigger NET formation in vitro, and have characterized the NETosis process showing that it depends on the production of reactive oxygen species. To determine whether NETosis depends on the pathogenicity or virulence of intestinal parasites, we evaluated here the ability of some non-pathogenic or non-invasive intestinal parasites to induce in vitro NETosis. Human neutrophils isolated from peripheral blood from healthy donors were exposed

C. Díaz-Godínez · J. C. Carrero (✉)
Departmento de Inmunología, Instituto de Investigaciones Biomédicas, Universidad Nacional Autónoma de México (UNAM), 04510 Ciudad de México, Mexico
e-mail: carrero@unam.mx

A. Martínez-Flores
Laboratorio de Ecología de Agentes Patógenos, Hospital General Dr. Manuel Gea González, Ciudad de México, Mexico

R. Argüello-García
Depto de Genética y Biología Molecular, Centro de Investigación y Estudios Avanzados del IPN, Ciudad de México, Mexico

A. Olivos-García · M. Néquiz-Avendaño
Unidad de Investigación en Medicina Experimental, Facultad de Medicina, Universidad Nacional Autónoma de México, Ciudad de México, Mexico

© Springer Nature Switzerland AG 2020

N. Guillen (ed.), *Eukaryome Impact on Human Intestine Homeostasis and Mucosal Immunology*, https://doi.org/10.1007/978-3-030-44826-4_13

to axenic parasites at ratios between 100:1 and 10:1 during 4 h at 37 °C, and the DNA released form the neutrophils into the medium was measured in a spectrofluorometer using SYTOX Green. The results show that trophozoites of *Entamoeba dispar*, the non-pathogenic amoeba relative of *E. histolytica*, as well as non-invasive intestinal parasites *Giardia intestinalis* and *Blastocystis* spp. were unable to induce NETosis at any ratio and time point of incubation studied. In contrast, *E. histolytica* trophozoites attenuated in virulence by maintaining parasites in culture media for more than twenty years, were still able to induce NETosis.

These results suggest that NET formation is associated with the pathogenicity of the intestinal parasites, in terms of their ability to invade the intestinal epithelium, cause tissue damage and expose themselves to the bloodstream where they are faced with neutrophils. More detailed studies are necessary to be carried out using highly efficient and reproducible in vivo NET inhibitory systems to determine the precise role of NETosis in parasitic infections.

Introduction

Neutrophil Extracellular Traps (NETs)

NETs consist of extracellular DNA fibres, histones, antimicrobial peptides and proteins from cytoplasmic granules including neutrophil elastase (NE), myeloperoxidase (MPO), cathepsin G, defensins and cathelicidins, among others (Brinkmann et al. 2004). Neutrophils release these structures in a stepwise process called NETosis that proceeds as follows: pathogen recognition, signalling pathway activation, reactive oxygen species (ROS) generation, histone processing through NE and MPO translocation to the nucleus or by peptidylarginine deiminase 4 (PAD4) activation, internal membrane disruption and cytoplasmatic membrane rupture with subsequent DNA extrusion (Fuchs et al. 2007; Wang et al. 2009; Hakkim et al. 2011; Metzler et al. 2014). Since their discovery, different functions attributed to NETs in the defence against pathogens include trapping—and in some cases, killing—bacteria, fungi and viruses as well as degrading virulence factors (Brinkmann et al. 2004; Young et al. 2011; Bruns et al. 2010; Urban et al. 2006; Saitoh et al. 2012). However, they do not only function to protect, as the NETosis mechanism has been implicated in a number of pathological processes such as intravascular coagulation, autoimmune diseases and cancer (McDonald et al. 2017; Gestermann et al. 2018; Albrengues et al. 2018).

NETs in Intestinal Parasitic Infections

Neutrophils have long been known to play an important role in the defence against pathogens, limiting dissemination of microorganisms or directly killing them

(Papayannopoulos 2018). Despite this, the specific role of NET in the protection against intestinal parasites is not completely understood.

Eimeria Species

One of the first descriptions of NETosis in response to an intestinal parasite was performed by Behrendt et al. (2010), who stimulated bovine neutrophils with *Eimeria bovis* sporozoites. *Eimeria* is a genus of protozoan parasite responsible for coccidiosis in a number of animal species, characterized by inappetence, weight loss, fever and diarrhoea, and is lethal in severe cases (Allen and Fetterer 2002). The disease is acquired when a susceptible host ingests the sporulated oocysts, and once in the intestine, the sporozoites are released and penetrate the intestinal cells. At this stage, asexual reproduction of the parasite occurs by merogonia, resulting in the formation of merozoites. After the final merogonia cycle, sexual reproduction begins, forming the gametes that, when fused, will lead to the formation of immature oocysts subsequently expelled in the faeces. Bovine neutrophils rapidly released NETs when co-cultured with *E. bovis* sporozoites (only 30 min after initial interaction) in a dose-dependent manner. The amount of released DNA was higher for viable sporozoites than heat-inactivated parasites. This NETosis mechanism was partially characterized, demonstrating a dependence on NADPH oxidase complex activity. In addition, it was found when sporozoites were co-cultured with neutrophils, their infective capacity on an epithelial cell line was significantly reduced. Another parasite of the same genus, *Eimeria arloingi*, induced NET release in goat neutrophils (Silva et al. 2014). In these fibres were found histone H3, MPO and NE, which have become hallmarks of NETs. DNA was released in a time-dependent manner and required the formation of ROS, as diphenyleneiodonium (DPI)-treated neutrophils significantly reduced the release of DNA. In addition, during neutrophil/sporozoite co-culture, an increase in ROS generation and NE activity was detected. Sporozoites were observed in close contact with NETs and were trapped by them, and while they did not reduce their viability, NETs did decrease their ability to infect bovine umbilical vein endothelial cells susceptible to parasite entry. The presence of NET in animals infected with these parasites was demonstrated by Muñoz-Caro et al. (2016) who observed that leukocytes were in close contact with different parasite stages (macrogamon, macromeronts and oocysts) in intestinal sections from *E. bovis*-infected calf and *E. arloingi*-infected goat. In addition, they detected the presence of extracellular DNA that co-located with NET-associated components such as histones and NE. In the same year, it was found that goat monocytes were capable of releasing NETs when co-cultured with *E. ninakonhlyakimovae* sporozoites (Pérez et al. 2016). The identification of these networks was done by co-localization of DNA with histone H3 and MPO. As in previous studies, this mechanism depended on the generation of ROS by the NADPH oxidase complex.

Neospora caninum

N. caninum is an apicomplexan parasite that affects a wide variety of mammals. The definitive host (mammals of the *Canidae* family) acquires the infection by ingesting tissue cysts, which release bradyzoites in the intestine that through merogonia culminates in the formation of oocysts eventually expelled in the faeces. The oocysts sporulate in the environment and are ingested by the intermediate hosts where the tissue cysts will be formed, thus completing the cycle (Dubey et al. 2007). In both intermediate and definitive hosts, the infection is acquired orally, suggesting an interaction between the intestinal immune response and the parasite. It has been suggested that neutrophils of some intermediate hosts such as cattle, goats and dolphins are capable of releasing extracellular traps when they are exposed to *N. caninum* tachyzoites (Villagra-Blanco et al. 2017a, b, c) through a mechanism that depends on the generation of ROS by the NADPH oxidase complex; only in goat neutrophils was determined that the process was independent of NADPH oxidase activity. In addition, the presence of NETs has been described using neutrophils from the definitive host (canines). In this case, NETs were generated in a time-dose-dependent manner and required the activity of NADPH oxidase, NE, MPO enzymes and the ERK/p38 signalling pathway (Wei et al. 2016). Interestingly, bovine macrophages are also capable of releasing extracellular traps when they are exposed to this parasite (Wei et al. 2018), suggesting that this immunity mechanism is conserved not only in different species but also in different cell lineages.

Cryptosporidium Species

Cryptosporidium species cause Criptosporidiosis, a parasitosis of great importance due to its high prevalence in immunosuppressed people (Wanyiri et al. 2014; Ghafari et al. 2018). This disease can course asymptomatic, however, in some patients, diarrhoea and abdominal pain may occur. The severity is related to the immune state of the patient: immunosuppressed people are more affected, presenting dehydration and wasting (Bouzid et al. 2013). Infection is acquired by ingestion of food or water contaminated with stool containing the infective sporulated oocyst. Similar to other apicomplexan protozoa, *Cryptosporidium parvum* acts as a NET-inducing agent (Muñoz-Caro et al. 2015a). Both bovine and human neutrophils released NETs when co-cultured with *C. parvum* sporozoites. In human neutrophils, NETosis depended on the activity of NADPH oxidase as pre-treatment of polymorphonuclear cells (PMN) using DPI significantly reduced the amount of DNA released. The mechanism was characterized in greater detail using bovine neutrophils: the release of DNA depended on NADPH oxidase complex activity, store-operated calcium entry (SOCE), ERK1/2 and p28 signalling pathways, as well as NE and MPO activities. *C. parvum* sporozoites were trapped by DNA fibres significantly reducing their ability to infect epithelial cells.

Strongyloides stercoralis

The study of NETs has not been limited to protozoan diseases. Some investigators have studied this immune mechanism against intestinal helminths. *S. stercoralis* is a nematode parasite infecting humans. Infective larvae penetrate intact skin and migrate to the small intestine where they develop into adult forms. Female worms deposit eggs that release rhabditiform larvae expelled in stools or transform into an infective stage that penetrates intestinal mucosa and establishes an auto-infective cycle (Page et al. 2018). It has been proposed that infective larvae from *S. stercoralis* lead NET formation in human neutrophils (Bonne-Année et al. 2014). NETs were released in a time-dependent manner and were able to entrap larvae, although DNA fibres failed to kill parasites. The authors observed that human macrophages also released extracellular traps in response to larvae but similar to the previous study, failed to kill worms. However, when neutrophils and macrophages were cultured together with infective larvae, parasite viability was reduced significantly. This evidence, combined with the observation that to treatment of cultures with DNase prevented parasite death, suggests a synergic role between neutrophils, macrophages and extracellular traps to kill parasites. In contrast with human cells, mouse neutrophils release NETs but macrophages from these animals co-cultured with *S. stercolaris* larvae failed to release DNA. This suggests variations in immune response between different animals to the same stimulus and the existence of differences in the infection resolving mechanism.

Haemonchus contortus

H. contortus is an intestinal nematode that infects small ruminants, causing anaemia, subcutaneous oedema and death in severe cases (Besier et al. 2016). Animals acquire this disease after ingesting L3 infective larva. The nematodes are established in the intestine (abomasum) and develop into hematophagous adults. It is proposed that *H. contortus* infective L3 larva induce bovine neutrophils to form NETs containing NE, MPO and histones released independently of parasite viability, as heat-inactivated larvae also triggered NET formation (Muñoz-Caro et al. 2015b). These DNA fibres were able to entrap larvae yet failed to kill parasites. In addition, parasite-triggered NETosis at low pH—similar to that in the abomasum—suggests that this process can occur in vivo. Additionally, the authors observed that ovine eosinophils also released NET when they were co-cultured with the larvae; however, the role of eosinophil extracellular traps in this disease is still unclear.

Entamoeba histolytica

Previously, our research group observed that *E. histolytica* trophozoites induced NETosis in human neutrophils (Ávila et al. 2016). We found that NETs were released in a time- and dose-dependent manner and these structures failed to kill amoebas although cathelicidin (an antimicrobial peptide) was detected in the DNA fibres. The NETosis mechanism was found to be independent of ROS production from neutrophils, NADPH oxidase and PAD4 activities, but nonetheless required extracellular calcium influx, serine protease activity and signalling via Raf/MEK/ERK and NF-κB (Díaz-Godínez et al. 2018; Fonseca et al. 2018).

Relation of NETosis with Virulence

NETosis is a process that must be tightly regulated to prevent pathology. Therefore, NET formation should be conditioned to occur only in those situations in which they are strictly necessary. Such is the case of large microorganisms, which cannot be phagocytosed by neutrophils. However, NETosis has been reported with a wide variety of small bacteria that could easily be phagocytosed. It is noteworthy that most of these bacteria can survive and escape phagosomes, suggesting that NETosis against small microorganisms may be reserved for virulent ones (Papayannopoulos 2018). There are numerous examples of NET induction and killing by NETs as a function of bacterial virulence factors and we describe a few here. One showed that probiotic bacteria *Lactobacillus rhamnosus* strain GG, unlike enteropathogenic bacteria, was not only incapable of forming NETs but also inhibited NET formation induced by phorbol 12-myristate 13-acetate (PMA) and *Staphylococcus aureus* (Vong et al. 2014). This correlated with the potent anti-oxidative activity of *Lactobacillus rhamnosus* strain GG, which dampened the respiratory burst in murine neutrophils, a requisite for NET formation. It has been suggested that this mechanism may contribute to the recognized potential of probiotics to regulate local inflammation in the intestine and immunopathology (Vong et al. 2014). NETosis related to a physical bacterial virulence factor was observed in *Pseudomona aeruginosa*, whose ability to trigger NETs was dependent on the expression of a motile flagellum (Floyd et al. 2016). The flagellum-deficient bacteria neither induced respiratory burst nor NETosis, suggesting that bacteria's ability to invade tissues is associated with its potential to induce ROS, in part due to its functional flagellum (purified flagellin protein did not activate NETosis). Other bacterial virulence factors associated with the induction of NETosis include pore-forming proteins, such as Panton-Valentine leukocidin and leucotoxin Gh of *Staphylococcus aureus* (Pilsczek et al. 2010; Malachowa et al. 2013) through an ROS-independent but pore-forming dependent mechanism, adhesins such as invasin of *Yersinia pseudotuberculosis* through binding to neutrophil β-integrins (Gillenius and Urban 2015) and proteases of cysteine, such as gingipains from *Porphyromonas gingivalis* through protease-activated receptor-2 (PAR-2) signalling (Bryzek et al. 2019). Two of the most common strategies to evade neutrophil phagocytosis: bacterial biofilms and aggregations (such as supragingival biofilms that induce intra-biofilm release of NETs) are also potent inducers of NETosis (Hirschfeld et al. 2015). *Mycobacterium bovis* aggregates were found to trigger NETosis in a size-dependent manner (Branzk et al. 2014). It is noteworthy that despite the relevance of bacterial virulence, to our knowledge, the literature lacks studies on the role of pathogenicity and virulence factors of parasites driving NETosis.

In summary, the formation of NETs in bacterial infections corresponds with their degree of virulence, and particularly, with their ability to escape phagocytosis. Therefore, NETosis could be a very important innate immune defence mechanism for the control of infections due to invasive pathogenic bacteria.

Are the Virulence and Pathogenicity of Intestinal Parasites Involved in Their Ability to Trigger NETosis in vitro?

Recently, we found that trophozoites of the protozoa *Entamoeba dispar*, a commensal amoeba for humans phylogenetically related to the pathogenic amoeba *E. histolytica*, were unable to induce DNA release when they were co-cultured with human neutrophils (Fonseca et al. 2019). This contrasts with the potent ability of *E. histolytica* trophozoites, a pathogenic species causing amoebic dysentery and liver abscesses, to efficiently trigger NETosis (Ávila et al. 2016; Díaz-Godínez et al. 2018). The two amoeba species are so similar that they are indistinguishable by non-molecular methods, and the differences regarding NETosis seem to be based on the differential expression of several virulence factors that condition the invasive capacity of *E. histolytica* (Diamond and Clark 1993). This observation served to pose the hypothesis that non-invasive parasites fail to induce NET release compared to the invasive ones. In order to test this theory, herein we determine the ability of *Giardia lamblia* and *Blastocystis* spp., human parasites that do not penetrate the intestinal mucosa during infection, to trigger NETosis in human neutrophils.

Materials and Methods

Parasite Culture

Virulent and attenuated *E. histolytica* trophozoites (HM1:IMSS strain) were cultured in TYIS-33 medium supplemented with 15% decomplemented adult bovine serum (Microlab) and 1x Diamond vitamin tween solution (Sigma-Aldrich). Cultures were incubated for 72 h at 37 °C and parasites were harvested by ice chilling for 5 min. Trophozoites were centrifuged at 1400 rpm for 5 min at room temperature and the pellet was resuspended in PBS pH 7.4. Amoebas were counted in a hemocytometer and used immediately.

For *G. lamblia* culture, trophozoites were grown in the same medium used for *E. histolytica* with an added of 1 mg/mL of bovine bile (Sigma). Cultures were incubated for 96 h at 37 °C and parasites were harvested mechanically. Trophozoites were centrifuged at 4000 rpm for 2 min at room temperature and the pellet was resuspended in PBS pH 7.4. Cells were counted in a hemocytometer and used immediately.

Blastocystis spp. were purified from stool samples previously identified as positive by direct faecal screening. Briefly, 50 mg of the samples were cultured in 3 mL Jones medium supplemented with 10% decomplemented adult horse serum for 72 h at 37 °C. The large particles were removed from the cultures by washing with PBS and centrifugation at 500 x g for 5 min. For *Blastocystis* maintenance, subcultures were performed every third day transferring 100 μL of sediment into fresh Jones medium and cultured at 37 °C. To 72 h cultures, the supernatant was removed and Iscove Modified Dulbecco Medium (IMDM) supplemented with 10% horse serum and a mixture of antibiotics containing piperacillin/tazobactam 0.1 g/L, cefotaxime 0.1 g/L and vancomycin 0.01 g/L was added. Cultures were maintained in anaerobic chamber for 72 h and then the sediment was seeded in semisolid IMDM-agar with the mixture of antibiotics. After 96 h in anerobic chamber, colonies were obtained which were cultured in supplemented IMDM medium with the mixture of antibiotic for 7 days in anaerobiosis. Cultures of vacuolar forms highly reduced of fecal bacteria but not axenic were obtained. Finally, vacuolar forms were counted in a hemocytometer and used immediately.

Neutrophil Isolation

Neutrophils were purified as described elsewhere (García-García et al. 2013). Briefly, peripheral blood samples were collected from healthy donors and neutrophils were obtained by a Ficoll-Paque gradient (GE Healthcare) and hypertonic shock. Cells were resuspended in PBS pH 7.4, counted in a hemocytometer and used immediately.

NET Quantitation Assay

For NET quantitation, neutrophils (5×10^5) were centrifuged at 4000 rpm for 2 min and resuspended in 500 μL of RPMI medium (Biological Industries) supplemented with 5% foetal bovine serum (Gibco) and 500 nM SytoxGreen (Invitrogen). Cells (1×10^5) were seeded in a 96 well plate, each well containing 100 μL of cell suspension and were stimulated with 1×10^3, 2×10^3, 5×10^3 or 1×10^4 (parasite:neutrophil ratios 1:100, 1:50, 1:20 and 1:10 respectively) of virulent or attenuated *E. histolytica* trophozoites, *G. lamblia* trophozoites or *Blastocystis* spp. vacuolar forms. Co-cultures were incubated at 37 °C for 4 h and fluorescence was read from the well bottom in a spectrofluorometer (Synergy HTX, BioTek) using 485 nm excitation and 528 nm emission filters. PMA (50 nM) and A23187 (10 μM) were used as positive controls for NETosis. All experiments were performed two times in triplicate.

NET Visualization by DNA Staining

Neutrophils (1 × 10^6) were centrifuged at 4000 rpm for 2 min and resuspended in 500 μL of RPMI medium supplemented with 5% foetal bovine serum. Cells (2 × 10^5) were seeded on coverslips using 100 μL of cell suspension and were stimulated with 2 × 10^3, 4 × 10^3, 1 × 10^4 or 2 × 10^4 (parasite:neutrophil ratios 1:100, 1:50, 1:20 and 1:10 respectively) of virulent or attenuated *E. histolytica* trophozoites, *G. lamblia* trophozoites or *Blastocystis* spp. vacuolar forms. PMA (50 nM) and A23187 (10 μM) were used as positive controls for NETosis. Samples were incubated for 4 h at 37 °C and fixed with 3.7% formaldehyde. After fixation, DNA was stained using DAPI (5 μg/mL). Coverslips were mounted with Fluoroshield (Sigma-Aldrich) and fluorescence was visualized using a fluorescence microscope (Olympus BX51).

Detection of ROS Generated by Neutrophils During Interaction with Parasites

Neutrophils were isolated as described before: 5 × 10^5 cells were resuspended in 500 μL of PBS added of 10 μM 2′,7′-Dichlorofluorescin diacetate (H_2DCFDA) and incubated for 30 min at 37 °C in the dark. Neutrophils were then centrifuged at 4000 rpm for 2 min and resuspended in 500 μL of RPMI medium supplemented with 5% foetal bovine serum. Cells (1 × 10^5) were seeded in a 96-well plate, each well containing 100 μL of cell suspension then stimulated with 5 × 10^5 virulent and attenuated *E. histolytica* trophozoites or *G. lamblia* trophozoites. Fluorescence was read 1 h after stimulation from the well bottom in a spectrofluorometer (Synergy HTX, BioTek) using 485 nm excitation and 528 nm emission filters. PMA (50 nM) was used as a positive control for ROS.

Visualization of ROS in Amoebas

For ROS detection in *E. histolytica* and *E. dispar* trophozoites during their interaction with neutrophils, 1 × 10^5 amoebas were resuspended in 1 mL of PBS added of 10 μM H_2DCFDA (Sigma-Aldrich) and were incubated for 30 min in the dark. Amoebas were then washed three times and resuspended in 200 μL PBS. To induce NETosis, 20 μL of trophozoite suspension (1 × 10^4 amoebas) was used to stimulate 2 × 10^5 neutrophils (amoeba:neutrophil of 1:20) previously seeded in 100 μL of RPMI medium supplemented with 5% foetal bovine serum on a coverslip. Samples were cultured for 4 h at 37 °C, fixed with 3.7% formaldehyde and counterstained with DAPI (5 μg/mL). Coverslips were mounted with Fluoroshield (Sigma-Aldrich) and fluorescence was visualized using a fluorescence microscope (Olympus BX51).

Quantification of ROS Generated by Parasites

To determine the generation of ROS from parasites, 5×10^5 *E. histolytica* trophozoites (virulent and attenuated), *G. lamblia* trophozoites or *Blastocystis* spp. vacuolar forms were resuspended in 1 mL PBS added of 100 μM H_2DCFDA and incubated for 1 h at 37 °C in the dark. Parasites were then centrifuged and resuspended in 500 μL RPMI medium supplemented with 5% foetal bovine serum. Cells (1×10^5) were seeded in 96-well plate each well containing 100 μL cell suspension and fluorescence was read immediately from the well bottom using a spectrofluorometer (Synergy HTX, BioTek) with 485 nm excitation and 528 nm emission filters. In selected experiments, virulent and attenuated *E. histolytica* trophozoites were pre-treated with the ROS scavenger pyrocatechol (200 μM) 30 min before H_2DCFDA addition and these pyrocatechol pretreated-amoebas also were used to induce NETosis as described above in an amoeba: neutrophil ratio of 1:100.

Results

Non-invasive Parasites Do not Induce NET Formation

Recently, we found that *E. dispar*, a commensal amoeba of humans closely related phylogenetically with the pathogenic amoeba *E. histolytica*, failed to induce NET release when it was co-cultured with human neutrophils, while *E. histolytica* trophozoites triggered NETosis (Fonseca et al. 2019). This observation led to the hypothesis that non-invasive parasites (defined as the incapacity to penetrate intestinal mucosa) could not induce NET release. To determine this, we evaluated whether *Giardia lamblia* and *Blastocystis* spp., two parasites that do not penetrate the intestinal mucosa during infection, triggered NETosis in human neutrophils.

When neutrophils were stimulated with PMA or A23187 (two well-characterized NETosis inducers) there was a greater amount of extracellular DNA released than in the unstimulated control (Fig. 13.1a). In contrast, *G. lamblia* trophozoites did not induce DNA release at any ratio used in the assay. This observation was verified using DNA staining to detect the presence of extracellular DNA fibres corresponding to NETs. Control neutrophils (Fig. 13.1b) exhibited condensed nuclei whereas PMA- and A23187-treated neutrophils showed decondensed nuclei and extracellular DNA fibres. In neutrophil-*G. lamblia* cocultures DNA fibres were not detected at any ratio used, nevertheless, neutrophil agglomerates were observed surrounding trophozoites (white arrows). In addition, viability of *G. lamblia* was not affected during interaction with neutrophils since the trophozoites remained mobile after 4 h (data not shown). Similar results were observed for *Blastocystis* spp. When neutrophils were cultured in the presence of *Blastocystis* spp. vacuolar forms, there was no increase in the amount of extracellular DNA at any ratio used (Fig. 13.2a), in contrast with PMA- and A23187-treated neutrophils. In addition, fluorescence images showed absence of

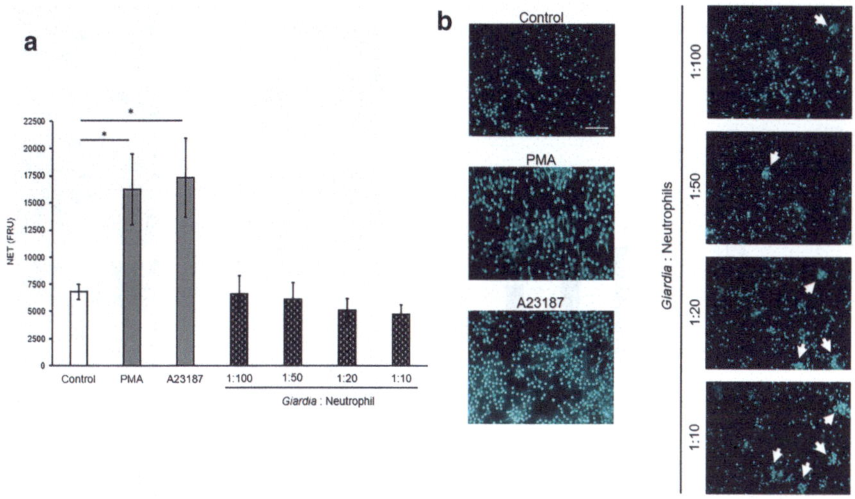

Fig. 13.1 *Giardia lamblia* trophozoites do not trigger NETosis in human neutrophils. **a** Neutrophils (1×10^5) cultured in the presence of 1×10^4, 5×10^3, 2×10^3 or 1×10^3 *G. lamblia* trophozoites (1:10, 1:20, 1:50 and 1:100 trophozoite:neutrophil ratios, respectively) in RPMI medium supplemented with 5% foetal bovine serum and 500 nM Sytox green. Cultures were incubated for 4 h at 37 °C and the fluorescence was measured in the plate reader Synergy HTX using 485 nm excitation and 528 nm emission filters. PMA (50 nM) and A23187 (10 μM) were used as positive controls of NETosis. Graph represent means ± SD of three independent experiments in triplicate. $^*p < 0.001$. **b** Neutrophils (2×10^5) were cultured in RPMI medium supplemented with 5% foetal bovine serum in the presence of PMA (50 nM), A23187 (10 μM), 2×10^4, 1×10^4, 4×10^3 or 2×10^3 *G. lamblia* trophozoites for 4 h at 37 °C. Cultures were formaldehyde-fixed and stained with DAPI (5 μg/mL). Images were taken at 60x magnification and pseudocolor was apply using Image J software. White arrows indicate the presence of neutrophil clusters. Scale bar 100 μm

extracellular DNA fibres (Fig. 13.2b) but, similar to *G.lamblia*, neutrophil clusters were observed surrounding *Blastocystis* spp. parasites. These data suggest that *G. lamblia* and *Blastocystis* spp. do not possess the capability to induce NETosis on human neutrophils.

Triggering NETosis Is Independent of the E. histolytica *Virulence*

Previously, we reported that virulent *E. histolytica* trophozoites (which can cause liver abscess development in a hamster model) induce NET release on human neutrophils (Ávila et al. 2016; Díaz-Godínez et al. 2018). We decided to determine if an attenuated *E. histolytica* strain, which has lost its ability to infect hamsters after being cultivated for 20 years without being inoculated in animals, was able to trigger NETosis. As observed in Fig. 13.3a, as expected, virulent amoebas induced NET

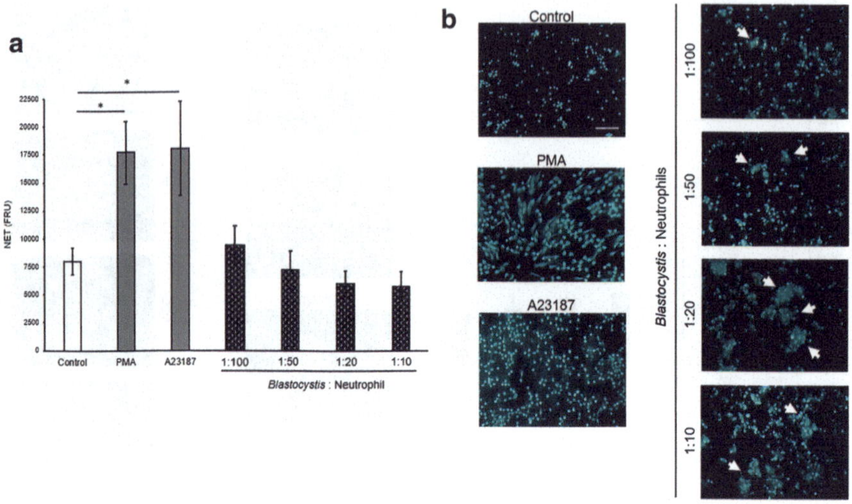

Fig. 13.2 *Blastocystis* spp. vacuolar forms do not trigger NETosis in human neutrophils. **a** Neutrophils (1×10^5) were cultured in the presence of 1×10^4, 5×10^3, 2×10^3 or 1×10^3 *Blastocystis* spp. vacuolar forms (1:10, 1:20, 1:50 and 1:100 *Blastocystis*:neutrophil ratios, respectively) in RPMI medium supplemented with 5% fetal bovine serum and 500 nM Sytox green. Cultures were incubated for 4 h at 37 °C and the fluorescence was measured in the plate reader Synergy HTX using 485 nm excitation and 528 nm emission filters. PMA (50 nM) and A23187 (10 μM) were used as positive controls of NETosis. Graph are means ± SD of three independent experiments in triplicate. *$p < 0.001$. **b** Neutrophils (2×10^5) were cultured in RPMI medium supplemented with 5% fetal bovine serum in the presence of PMA (50 nM), A23187 (10 μM), 2×10^4, 1×10^4, 4×10^3 or 2×10^3 *Blastocystis* spp. vacuolar forms during 4 h at 37 °C. Cultures were formaldehyde fixed and stained with DAPI (5 μg/mL). Images were taken at 60x magnification and pseudocolor was applied using Image J software. White arrows indicate the presence of neutrophil clusters. Scale bar 100 μm

release at all doses, denoted by the increase in extracellular DNA. Furthermore, attenuated *E. histolytica* trophozoites also induced NETosis on human neutrophils at the same magnitude as virulent amoebas (Fig. 13.3a). NET release was also verified by DNA staining in which extracellular DNA fibres were detected using virulent and attenuated *E. histolytica* trophozoites at all ratios (Fig. 13.3b). In both cases, the number of intact neutrophils (condensed nuclei) was reduced when we augmented the amoeba numbers indicating that the process occurs in a dose-dependent manner.

Triggering NETosis Is Related to Pathogen Capacity to Produce ROS

In order to establish what feature makes *E. dispar* trophozoites incapable of inducing NETosis, we tested the participation of ROS in the process. Kenny et al. (2017)

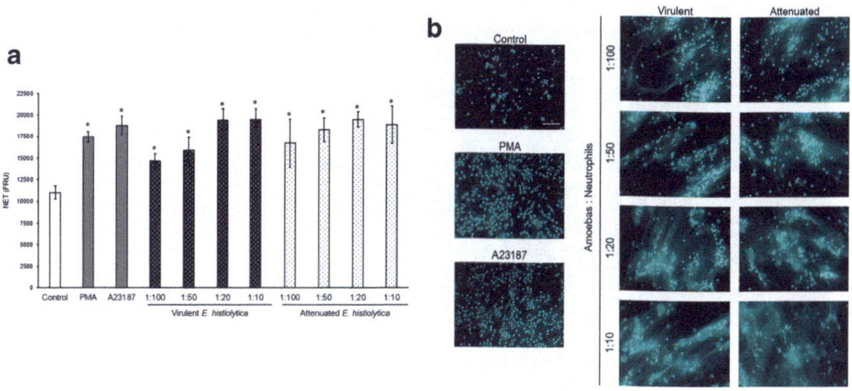

Fig. 13.3 Virulent and attenuated *Entamoeba histolytica* trophozoites induce NET release. **a** Neutrophils (1×10^5) were cultured in the presence of 1×10^4, 5×10^3, 2×10^3 or 1×10^3 virulent and attenuated *E. histolytica* trophozoites (1:10, 1:20, 1:50 and 1:100 trophozoite:neutrophil ratios, respectively) in RPMI medium supplemented with 5% fetal bovine serum and 500 nM Sytox green. Cultures were incubated for 4 h at 37 °C and the fluorescence was measured in plate reader Synergy HTX using 485 nm excitation and 528 nm emission filters. PMA (50 nM) and A23187 (10 μM) were used as positive controls of NETosis. The graph represents means ± SD of three independent experiments in triplicate. $*p < 0.001$. **b** Neutrophils (2×10^5) were cultured in RPMI medium supplemented with 5% fetal bovine serum in the presence of PMA (50 nM), A23187 (10 μM), 2×10^4, 1×10^4, 4×10^3 or 2×10^3 *E. histolytica* trophozoites for 4 h at 37 °C. Cultures were formaldehyde-fixed and stained with DAPI (5 μg/mL). Images were taken at 60x magnification and pseudocolor was apply using Image J software. Scale bar 100 μm

reported that ROS production by *Candida albicans* fungi are responsible for triggering NETosis in neutrophils in chronic granulomatosis disease patients that are unable to generate ROS from NADPH oxidase, suggesting that ROS produced for pathogens are required to trigger NETosis in some cases. In this vein, we analysed the differences in ROS production from *E. histolytica* and *E. dispar* trophozoites. We observed that neutrophils in the absence of amoebas do not release DNA (Fig. 13.4) and maintain their characteristic multilobular nucleus (top panel). The presence of ROS in *E. histolytica* trophozoites is denoted by the intense green fluorescence detected in trophozoites (middle panel) and NET were appreciated as extracellular fibres trapping amoebas. Interestingly, ROS were not detected in *E. dispar* trophozoites since no fluorescence was observed inside amoebas and neutrophils did not lose nuclear morphology despite being in close contact with trophozoites (bottom panel). These data suggest that *E. histolytica* trophozoites are the source of ROS necessary to lead NETosis in human neutrophils as these molecules were not generated by neutrophils during interaction with the parasites tested (Fig. 13.5a) and were only produced when these leukocytes were stimulated with PMA.

To confirm the relevance of ROS produced by *E. histolytica* trophozoites to trigger NETosis, virulent and attenuated amoebas were treated with the ROS scavenger pyrocatechol to decrease ROS level in trophozoites. As shown in Fig. 13.5b, *E. histolytica* trophozoites exhibit similar basal ROS levels that were significantly decreased by

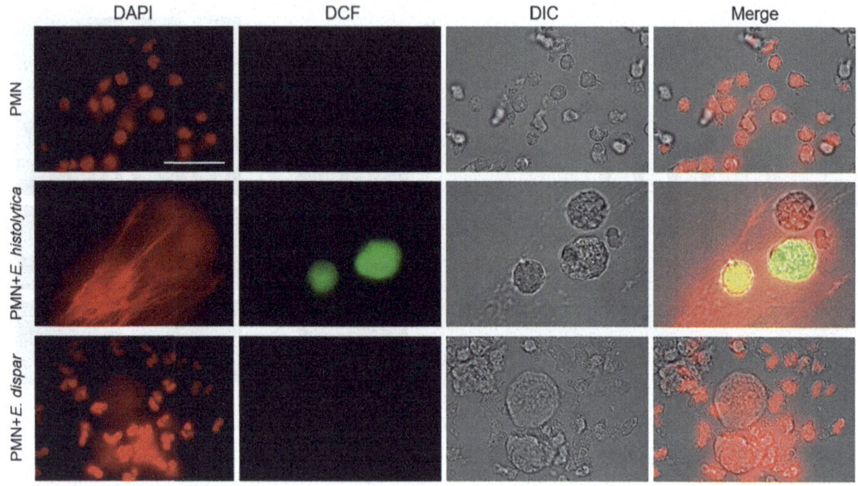

Fig. 13.4 *Entamoeba dispar* does not trigger NETosis and produces less ROS than *Entamoeba histolytica*. *E. histolytica* and *E. dispar* trophozoites (1×10^4) were marked with 10 μM H_2DCFDA for 30 min for ROS detection. After washing, amoebas were cultured with 2×10^5 neutrophils (amoeba:neutrophil ratio 1:20) during 4 h at 37 °C. Samples were fixed with 3.7% formaldehyde and counterstained with 5 μg/mL DAPI. DCF (oxidation product of H_2DCFDA) fluorescence was observed in FITC channel and pseudocolor was applied using Image J software. Images were taken at 100x magnification. The top panel shows non-stimulated neutrophils; middle panel shows neutrophil-*E. histolytica* coculture; and bottom panel shows neutrophil-*E. dispar* coculture. Scale bar 50 μm

pyrocatechol treatment. When virulent and attenuated pyrocatechol-treated amoebas were used to induce NETosis, they induced very little NET formation compared to non-treated trophozoites (Fig. 13.5c, d) and produced neutrophil aggregations similar to neutrophils in contact with *G. lamblia* and *Blastocystis* spp. (Figs. 13.1b and 13.2b), suggesting that ROS produced by *E. histolytica* trophozoites is a key factor leading to NETosis in human neutrophils. In addition, *G. lamblia* trophozoites that lack the capacity to trigger NETosis (Fig. 13.1) did not induce ROS production by neutrophils and produced a lower ROS level compared to *E. histolytica* trophozoites treated with pyrocatechol (Fig. 13.5b) suggesting the importance of an ROS source in the NETosis mechanism. Even though *Blastocystis* spp. did not induce NET release (Fig. 13.2), the ROS amount detected in this sample was even higher than that produced by *E. histolytica* trophozoites. In this case, it seems possible that, given that the sample comes from faecal matter, there were traces of contaminating bacteria.

Discussion

Since the discovery of NETs 15 years ago by the Schilinzky group (Brinkmann et al. 2004), an enormous number of NET studies have been conducted and understanding

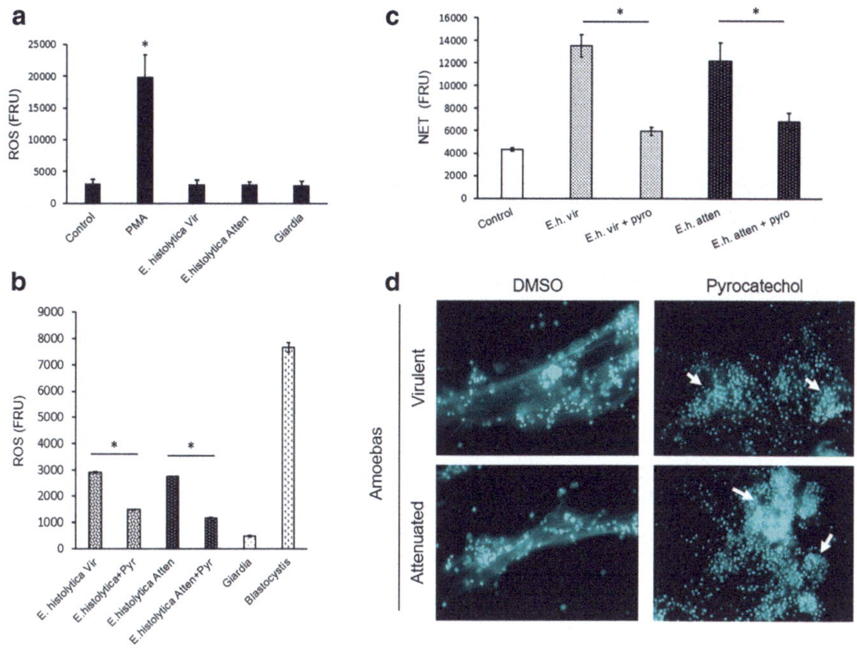

Fig. 13.5 Reactive oxygen species produced by *Entamoeba histolytica* trophozoites are required to trigger NETosis. **a** Human neutrophils (1×10^5) were pre-treated with the ROS indicator H_2DCFDA for 30 min, then resuspended in RPMI medium supplemented with 5% fetal bovine serum and transferred into 96 well plates. Leukocytes were stimulated with 20 nM PMA, 5×10^5 virulent and attenuated *E. histolytica* trophozoites or 5×10^5 *G. lamblia* trophozoites. Neutrophil ROS production was measured at 1 h using a plate reader with 485 nm excitation and 528 nm emission filters. Graph represents means ± SD of three independent experiments in triplicate. *$p < 0.0001$. **b** Virulent and attenuated *E. histolytica* trophozoites, *G. lamblia* trophozoites and *Blastocystis* spp. vacuolar forms were treated with 100 μM H_2DCFDA for 1 h, then resuspended in RPMI medium supplemented with 5% fetal bovine serum and 1×10^5 parasites were cultured in 100 μL of medium. Fluorescence was read using a plate reader with 485 nm excitation and 528 nm emission filters. For some experiments, virulent and attenuated *E. histolytica* trophozoites were pre-treated with 200 μM pyrocatechol for 30 min before H_2DCFDA addition. Graph represents means ± SD of three independent experiments in triplicate. *$p < 0.001$. **c** Neutrophils (1×10^5) were cultured in 100 μL of RPMI medium with 500 nM Sytox green and were stimulated with 1×10^3 virulent and attenuated *E. histolytica* trophozoites pre-treated 1.5 h with 200 μM pyrocatechol or the vehicle DMSO. Cocultures were incubated for 4 h at 37 °C and fluorescence was measured using a plate reader with 485 nm excitation and 528 nm emission filters. Graph represents means ± SD of three independent experiments in triplicate. *$p < 0.001$. **d** Neutrophils (2×10^5) were cultured in 100 μL of RPMI medium and were stimulated with 2×10^3 virulent and attenuated *E. histolytica* trophozoites pre-treated 1.5 h with 200 μM pyrocatechol or the vehicle DMSO. Cocultures were incubated for 4 h at 37 °C, cells were formaldehyde-fixed and stained with DAPI. Images were taken at 60x magnification and pseudocolor was apply using Image J software. White arrows show neutrophil clusters. Scale bar 100 μm

of their function is advancing rapidly. We now know that NETosis is a type of finely regulated programmed neutrophil cell death (Fuchs et al. 2007; Hakkim et al. 2011) which can be triggered by a wide range of stimuli that can lead to the release of these DNA traps by various intracellular pathways (Hakkim et al. 2011; Khan and Palaniyar 2017). This tangle of DNA in the extracellular space can trap pathogens such as viruses, bacteria, fungi and protozoans (Urban et al. 2006; Guimarães-Costa et al. 2009; Young et al. 2011; Saitoh et al. 2012), preventing their spread, and in some cases, killing them. The process has also been identified in macrophages, eosinophils, basophils and mast cells (Yousefi et al. 2008; Lin et al. 2011; Yousefi et al. 2015; Okubo et al. 2018), therefore, the release of DNA traps, globally known as ETosis, has been positioned as a novel effector mechanism of the innate immunity of PMN.

Although at first it was believed that NETs were formed in the presence of a wide range of microorganisms, particularly well-studied bacteria, it soon became clear that not all bacteria had that potential (Pieterse et al. 2016), and that in particular, those capable of doing so were characterized as being pathogenic or highly virulent. Thus, for example, the probiotic bacteria of the human intestine, considered non-pathogenic, as well as variants of some bacterium lacking a well-known virulence factor (e.g. toxins, pore-forming proteins, flagella and adhesins), are unable to activate NETosis or do so very weakly (reviewed in Papayannopoulos 2018). It has become evident then that pathogenicity and/or virulence is a relevant factor during activation of trap release in neutrophils (Malachowa et al. 2013; Rada et al. 2013; Francis et al. 2014). However, in the case of parasites, to our knowledge, this type of information is absent, except for our previous report on the inability of *E. dispar*, a non-pathogenic amoeba, to induce NET release (Fonseca et al. 2019).

In the current study, we found that, like the *E. dispar* trophozoites, the trophozoites of *G. lamblia* and the vacuolar forms of *Blastocystis* spp. were unable to activate human neutrophils for the formation of NETs at ratios between 1:10 and 1:100 parasites/neutrophil. These three intestinal parasites represent three different scenarios in terms of pathogenicity: *E. dispar* is considered a commensal (Diamond and Clark 1993), *Blastocystis* spp. is considered on the boundary between a commensal and a pathogen (reviewed in Agrawal 2017), while *G. lamblia* is considered a pathogen (reviewed in Buret 2008). However, the three infections converge in their inability to invade the intestinal mucosa, and therefore, encounter neutrophils in the bloodstream. We speculate that unlike *E. histolytica* (that can reach the bloodstream) and other protozoan blood parasites in which NETosis has been described such as *T. cruzi* (Sousa-Rocha et al. 2015), *Leishmania* spp. (Guimarães-Costa et al. 2009) and *T. gondii* (Abdallah et al. 2012), intestinal parasites that do not invade extra intestinal tissues have not developed in their evolution the mechanisms (most likely invasive factors) to trigger NETosis, nor have neutrophils developed the mechanisms to recognize them in the context of the formation of NETs.

A constant in most studies is the need for ROS formation for NETosis to occur, either from the neutrophils or from an external source such as the parasite itself who would donate them to the neutrophil during contact (Fuchs et al. 2007; Kenny et al. 2017). In our previous reports on NETosis by *E. histolytica*, we show that neutrophils do not produce ROS derived from NOX2 in the presence of the trophozoites, suggesting an ROS/NOX2-independent process (Díaz-Godínez et al. 2018). We reported a similar result in neutrophils exposed to *E. dispar* (Fonseca et al. 2019) and in this current work, we report the same with *G. lamblia* trophozoites (see Fig. 13.5a). However, the results presented here suggest that *E. histolytica*-induced NETosis depends on the production of ROS by the trophozoites. In this regard, in determining the production of ROS by the three parasites evaluated, *E. dispar*, *G. lamblia* and *Blastocystis* spp., we found that *E. dispar* and *G. lamblia* produced less ROS than *E. histolytica* (see Figs. 13.4 and 13.5), suggesting that their inability to activate NETosis is due, at least in part, to the lack of a respiratory burst that exceeds the threshold necessary for triggering NETosis. In the case of *Blastocystis* spp., we found ROS in the analysed sample, but these could come from some contaminating intestinal bacteria that remained after purification of the parasites from the faeces. The notable observation, in this case, is that, despite the presence of ROS in the sample, NETosis was not activated, suggesting that the ROS of the parasite in close contact with neutrophils, along with other factors such as those involved in its invasiveness, could be necessary for the formation of NETs. However, we cannot rule out that the absence of NETosis by *Blastocystis* spp. could be due to phagocytosis of bacteria in the sample by neutrophils. Cross-inhibition of both processes, NETosis and phagocytosis, is a well-known phenomenon and the underlying mechanism has been reported (reviewed in Manfredi et al. 2018). Since the axenization of *Blastocystis* spp. from stool samples is a long-term and hardly successful process, we decided to use samples enriched in vacuolar forms with a highly reduced content of fecal bacteria; however, the use of axenic cultures will be necessary to use in future studies to determine the real role of ROS from this parasite in the NETosis process.

Oxygen and derived ROS cause the oxidation and inactivation of numerous important enzymes in anaerobic and microaerophilic organism, principally those that possess Fe–S centers (reviewed by Imlay 2006). *E. histolytica* as a microaerophilic protozoon in contact with oxidant agents during the invasive process, possesses a complex oxygen reduction pathway to ensure its survival. Thus, oxygen can be converted to hydrogen peroxide by NADPH:flavin oxidoreductases present in amoebas, explaining the basal ROS level detected in this work, and posteriorly hydrogen peroxide is rapidly converted to water by rubrerythrin or peroxiredoxin (reviewed by Olivos-García et al. 2016; Jeelani and Nozaki 2016). In contrast, *E. dispar* consumes lower levels of oxygen and produces lower amount of hydrogen peroxide under normoxia and hyperoxia conditions (Santos et al. 2015). It could explain why we did not detect ROS in *E. dispar* trophozoites (Santos et al. 2015). In the case of *G. lamblia* trophozoites, a NAD(P)H:menadione oxidoreductase (DT-diaphorase) is responsible to generate superoxide from oxygen which is further converted to hydrogen peroxide (Li and Wang 2006). The DT-diaphorase is probably the only enzyme responsible to detoxify oxygen in this parasite and it might explain the small amount of ROS detected in *G. lamblia* in this work. Although adapted to anaerobic environments, *Blastocystis* spp. seems to respire oxygen using the complex II and producing high amounts of superoxide and hydrogen peroxide (Tsaousis et al. 2018; Quinlan et al. 2012). Therefore, complex II can be the source of ROS detected by us in the enriched cultures of vacuolar forms of *Blastocystis* spp.; nevertheless, as we mention before, the cultures were not axenic and further studies are required to determine the real source of ROS.

Finally, in order to evaluate the role of the virulence of the parasite in the ability to form NETs, we studied NETosis by an attenuated culture of *E. histolytica*, which has been maintained for years without passing through the experimental model of amoebic liver abscesses, unlike the virulent culture that passes through this model once per month. Interestingly, the attenuated trophozoites of *E. histolytica* induced NETosis at the same level as the trophozoites of highly virulent cultures, suggesting that in the case of *E. histolytica*, the ability to induce the formation of NETs is more associated with pathogenicity than with the degree of virulence. Notably, the ability of attenuated cultures to produce ROS is maintained (see Fig. 13.5b), emphasizing the importance of the parasite's respiratory burst to activate NETosis over the decrease in the expression of other factors involved in the capacity of the amoeba to establish tissue infection.

In conclusion, our observations suggest that NETosis in vitro by intestinal protozoan parasites is a process that depends on its pathogenicity, in terms of its ability to invade tissues and therefore, confront neutrophils. NETosis could then be reserved for invasive protozoan parasites such as *E. histolytica*, which activate the formation of NETs through the release of ROS and maintain that capacity despite attenuation. Pathogenic or commensal parasites of the intestine that do not invade may not have the capacity to induce NETosis, which could be due to their inability to produce ROS. More studies with other parasites and their respective attenuated cultures are required to be able to define precisely the participation of the factors associated with pathogenicity and virulence in the activation of NETosis, as well as their role in the definition of NETs as a protective or inmunopathogenic mechanism, significantly affecting the outcome of the infection.

Acknowledgements César Díaz-Godínez is a Ph.D. student of the Programa de Maestría y Doctorado en Ciencias Bioquímicas, UNAM, and received a scholarship from Consejo Nacional de Ciencia y Tecnología (CONACyT), México (432205/596731). We thank Pavel Petrosyan for critical reading of the manuscript.

Interest Disclosure All authors declare they have no conflict of interest.

Funding This work was funded by Consejo Nacional de Ciencia y Tecnología (CONACyT) grant No. 284830 and Programa de Apoyo a Proyectos de Investigación e Innovación Tecnológica (PAPIIT-UNAM) grant No. IN206316, both granted to Julio César Carrero.

References

Abdallah, D. S. A., Lin, C., Ball, C. J., et al. (2012). *Toxoplasma gondii* triggers release of human and mouse neutrophil extracellular traps. *Infection and Immunity, 80*, 768–777. https://doi.org/10.1128/IAI.05730-11.

Agrawal, S. K. (2017). *Blastocystis Hominis* infection pathogenic or commensal: Short review. *International Journal of Research Studies in Microbiology and Biotechnology, 3,* 1–5. https://doi.org/10.20431/2454-9428.0303001.

Albrengues, J., Shields, M. A., Ng, D., et al. (2018). Neutrophil extracellular traps produced during inflammation awaken dormant cancer cells in mice. *Science, 361*(80-). https://doi.org/10.1126/science.aao4227.

Allen, P. C., & Fetterer, R. H. (2002). Recent advances in biology and immunobiology of Eimeria species and in diagnosis and control of infection with these coccidian parasites of poultry. *Clinical microbiology reviews, 15,* 58–65. https://doi.org/10.1128/CMR.15.1.58.

Ávila, E. E., Salaiza, N., Pulido, J., et al. (2016). *Entamoeba histolytica* trophozoites and lipopeptidophosphoglycan trigger human neutrophil extracellular traps. *PLoS ONE, 11,* 1–18. https://doi.org/10.1371/journal.pone.0158979.

Behrendt, J. H., Ruiz, A., Zahner, H., et al. (2010). Neutrophil extracellular trap formation as innate immune reactions against the apicomplexan parasite *Eimeria bovis*. *Veterinary Immunology and Immunopathology, 133,* 1–8. https://doi.org/10.1016/j.vetimm.2009.06.012.

Besier, R. B., Kahn, L. P., Sargison, N. D., Van Wyk, J. A. (2016). *Diagnosis, treatment and management of Haemonchus contortus in small ruminants*. Elsevier Ltd.

Bonne-Année, S., Kerepesi, L. A., Hess, J. A., et al. (2014). Extracellular traps are associated with human and mouse neutrophil and macrophage mediated killing of larval *Strongyloides stercoralis*. *Microbes and Infection, 16,* 502–511. https://doi.org/10.1016/j.micinf.2014.02.012.

Bouzid, M., Hunter, P. R., Chalmers, R. M., & Tyler, K. M. (2013). Cryptosporidium pathogenicity and virulence. *Clinical Microbiology Reviews, 26,* 115–134. https://doi.org/10.1128/CMR.00076-12.

Branzk, N., Lubojemska, A., Hardison, S. E., et al. (2014). Neutrophils sense microbe size and selectively release neutrophil extracellular traps in response to large pathogens. *Nature Immunology, 15,* 1017–1025. https://doi.org/10.1038/ni.2987.

Brinkmann, V., Reichard, U., Goosmann, C., et al. (2004). Neutrophil extracellular traps kill bacteria. *Science, 303*(80-), 1532–1535. https://doi.org/10.1126/science.1092385.

Bruns, S., Kniemeyer, O., Hasenberg, M., et al. (2010). Production of extracellular traps against *Aspergillus fumigatus* in vitro and in infected lung tissue is pependent on invading neutrophils and influenced by hydrophobin RodA. *PLoS Pathogens, 6,* e1000873. https://doi.org/10.1371/journal.ppat.1000873.

Bryzek, D., Ciaston, I., Dobosz, E., et al. (2019). Triggering NETosis via protease-activated receptor (PAR)-2 signaling as a mechanism of hijacking neutrophils function for pathogen benefits. *PLoS Pathogens, 15,* 1–27. https://doi.org/10.1371/journal.ppat.1007773.

Buret, A. G. (2008). Pathophysiology of enteric infections with *Giardia duodenalis*. *Parasite, 15,* 261–265. https://doi.org/10.1051/parasite/2008153261.

Diamond, L. S., & Clark, C. G. (1993). A redescription of *Entamoeba histolytica* Schaudinn, 1903 (Emended Walker, 1911) separating it from *Entamoeba dispar* Brumpt, 1925. *Journal of Eukaryotic Microbiology, 40,* 340–344. https://doi.org/10.1111/j.1550-7408.1993.tb04926.x.

Díaz-Godínez, C., Fonseca, Z., Néquiz, M., et al. (2018). *Entamoeba histolytica* trophozoites induce a rapid non-classical NETosis mechanism independent of NOX2-derived reactive oxygen species and PAD4 activity. *Frontiers in Cellular and Infection Microbiology, 8,* 1–17. https://doi.org/10.3389/fcimb.2018.00184.

Dubey, J. P., Schares, G., & Ortega-Mora, L. M. (2007). Epidemiology and control of neosporosis and *Neospora caninum*. *Clinical Microbiology Reviews, 20,* 323–367. https://doi.org/10.1128/CMR.00031-06.

Floyd, M., Winn, M., Cullen, C., et al. (2016). Swimming motility mediates the formation of neutrophil extracellular traps induced by flagellated *Pseudomonas aeruginosa*. *PLoS Pathogens, 12,* 1–32. https://doi.org/10.1371/journal.ppat.1005987.

Fonseca, Z., Díaz-Godínez, C., Mora, N., et al. (2018). *Entamoeba histolytica* induce signaling via Raf/MEK/ERK for neutrophil extracellular trap (NET) formation. *Frontiers in Cellular and Infection Microbiology, 8*. https://doi.org/10.3389/fcimb.2018.00226.

Fonseca, Z., Uribe-Querol, E., Díaz-Godínez, C., et al. (2019). Pathogenic *Entamoeba histolytica*, but not *Entamoeba dispar*, induce neutrophil extracellular trap (NET) formation. *Journal of Leukocyte Biology, 105,* 1167–1181. https://doi.org/10.1002/JLB.MA0818-309RRR.

Francis, R. J., Butler, R. E., & Stewart, G. R. (2014). *Mycobacterium tuberculosis* ESAT-6 is a leukocidin causing Ca^{2+} influx, necrosis and neutrophil extracellular trap formation. *Cell Death and Disease, 5,* e1474–7. https://doi.org/10.1038/cddis.2014.394.

Fuchs, T. A., Abed, U., Goosmann, C., et al. (2007). Novel cell death program leads to neutrophil extracellular traps. *Journal of Cell Biology, 176,* 231–241. https://doi.org/10.1083/jcb.200606027.

García-García, E., Uribe-Querol, E., Rosales, C. (2013). A simple and efficient method to detect nuclear factor activation in human neutrophils by flow cytometry. *Journal of Visualized Experiments*, 1–7. https://doi.org/10.3791/50410.

Gestermann, N., Di Domizio, J., Lande, R., et al. (2018). Netting neutrophils activate autoreactive B cells in lupus. *The Journal of Immunology, 200,* 3364–3371. https://doi.org/10.4049/jimmunol.1700778.

Ghafari, R., Rafiei, A., Tavalla, M., et al. (2018). Prevalence of Cryptosporidium species isolated from HIV/AIDS patients in southwest of Iran. *Comparative Immunology, Microbiology and Infectious Diseases, 56,* 39–44. https://doi.org/10.1016/j.cimid.2017.12.002.

Gillenius, E., & Urban, C. F. (2015). The adhesive protein invasin of *Yersinia pseudotuberculosis* induces neutrophil extracellular traps via β1 integrins. *Microbes and Infection, 17,* 327–336. https://doi.org/10.1016/j.micinf.2014.12.014.

Guimarães-Costa, A. B., Nascimento, M. T. C., Froment, G. S., et al. (2009). *Leishmania amazonensis* promastigotes induce and are killed by neutrophil extracellular traps. *Proceedings of the National Academy of Sciences of United States of America, 106,* 6748–6753. https://doi.org/10.1073/pnas.0900226106.

Hakkim, A., Fuchs, T. A., Martinez, N. E., et al. (2011). Activation of the Raf-MEK-ERK pathway is required for neutrophil extracellular trap formation. *Nature Chemical Biology, 7,* 75–77. https://doi.org/10.1038/nchembio.496.

Hirschfeld, J., Dommisch, H., Skora, P., et al. (2015). Neutrophil extracellular trap formation in supragingival biofilms. *International Journal of Medical Microbiology, 305,* 453–463. https://doi.org/10.1016/j.ijmm.2015.04.002.

Imlay, J. A. (2006). Iron-sulphur clusters and the problem with oxygen. *Molecular Microbiology, 59,* 1073–1082. https://doi.org/10.1111/j.1365-2958.2005.05028.x.

Jeelani, G., & Nozaki, T. (2016). Entamoeba thiol-based redox metabolism: A potential target for drug development. *Molecular and Biochemical Parasitology, 206,* 39–45. https://doi.org/10.1016/j.molbiopara.2016.01.004.

Kenny, E. F., Muth, A., Mondal, S., et al. (2017). *Diverse stimuli engage different neutrophil extracellular trap pathways*, pp. 1–21. https://doi.org/10.7554/elife.24437.

Khan, M. A., & Palaniyar, N. (2017). Transcriptional firing helps to drive NETosis. *Scientifc Reports, 7,* 1–16. https://doi.org/10.1038/srep41749.

Li, L., & Wang, C. C. (2006). A likely molecular basis of the susceptibility of *Giardia lamblia* towards oxygen. *Molecular Microbiology, 59,* 202–211. https://doi.org/10.1111/j.1365-2958.2005.04896.x.

Lin, A. M., Rubin, C. J., Khandpur, R., et al. (2011). Mast cells and neutrophils release IL-17 through extracellular trap formation in psoriasis. *The Journal of Immunology, 187,* 490–500. https://doi.org/10.4049/jimmunol.1100123.

Malachowa, N., Kobayashi, S. D., Freedman, B., et al. (2013). *Staphylococcus aureus* leukotoxin GH promotes formation of neutrophil extracellular traps. *The Journal of Immunology, 191,* 6022–6029. https://doi.org/10.4049/jimmunol.1301821.

Manfredi, A. A., Ramirez, G. A., Rovere-Querini, P., & Maugeri, N. (2018). The neutrophil's choice: Phagocytose vs make neutrophil extracellular traps. *Frontiers in Immunology, 9,* 9–13. https://doi.org/10.3389/fimmu.2018.00288.

McDonald, B., Davis, R. P., Kim, S. J., et al. (2017). Platelets and neutrophil extracellular traps collaborate to promote intravascular coagulation during sepsis in mice. *Blood, 129,* 1357–1367. https://doi.org/10.1182/blood-2016-09-741298.

Metzler, K. D., Goosmann, C., Lubojemska, A., et al. (2014). Myeloperoxidase-containing complex regulates neutrophil elastase release and actin dynamics during NETosis. *Cell Reports, 8,* 883–896. https://doi.org/10.1016/j.celrep.2014.06.044.

Muñoz-Caro, T., Lendner, M., Daugschies, A., et al. (2015a). NADPH oxidase, MPO, NE, ERK1/2, p38 MAPK and Ca^{2+} influx are essential for *Cryptosporidium parvum*-induced NET formation. *Developmental and Comparative Immunology, 52,* 245–254. https://doi.org/10.1016/j.dci.2015.05.007.

Muñoz-Caro, T., Ribeiro, Machado, da Silva, L., Rentería-Solis, Z., et al. (2016). Neutrophil extracellular traps in the intestinal mucosa of *Eimeria*-infected animals. *Asian Pacific Journal of Tropical Biomedicine, 6,* 301–307. https://doi.org/10.1016/j.apjtb.2016.01.001.

Muñoz-Caro, T., Rubio, R. M. C., Silva, L. M. R., et al. (2015). Leucocyte-derived extracellular trap formation significantly contributes to *Haemonchus contortus* larval entrapment. *Parasites and Vectors, 8,* 1–12. https://doi.org/10.1186/s13071-015-1219-1.

Okubo, K., Kurosawa, M., Kamiya, M., et al. (2018). Macrophage extracellular trap formation promoted by platelet activation is a key mediator of rhabdomyolysis-induced acute kidney injury. *Nature Medicine, 24,* 232–238. https://doi.org/10.1038/nm.4462.

Olivos-García, A., Saavedra, E., Nequiz, M., et al. (2016). The oxygen reduction pathway and heat shock stress response are both required for *Entamoeba histolytica* pathogenicity. *Current Genetics, 62,* 295–300. https://doi.org/10.1007/s00294-015-0543-5.

Page, W., Judd, J., & Bradbury, R. (2018). The unique life cycle of *Strongyloides stercoralis* and implications for public health action. *Tropical Medicine and Infectious Disease, 3,* 53. https://doi.org/10.3390/tropicalmed3020053.

Papayannopoulos, V. (2018). Neutrophil extracellular traps in immunity and disease. *Nature Reviews Immunology, 18,* 134–147. https://doi.org/10.1038/nri.2017.105.

Pérez, D., Muñoz, M. C., Molina, J. M., et al. (2016). *Eimeria ninakohlyakimovae* induces NADPH oxidase-dependent monocyte extracellular trap formation and upregulates IL-12 and TNF-α, IL-6 and CCL2 gene transcription. *Veterinary Parasitology, 227,* 143–150. https://doi.org/10.1016/j.vetpar.2016.07.028.

Pieterse, E., Rother, N., Yanginlar, C., et al. (2016). Neutrophils discriminate between lipopolysaccharides of different bacterial sources and selectively release neutrophil extracellular traps. *Frontiers in Immunology, 7,* 1–13. https://doi.org/10.3389/fimmu.2016.00484.

Pilsczek, F. H., Salina, D., Poon, K. K. H., et al. (2010). A novel mechanism of rapid nuclear neutrophil extracellular trap formation in response to *Staphylococcus aureus*. *The Journal of Immunology, 185,* 7413–7425. https://doi.org/10.4049/jimmunol.1000675.

Quinlan, C. L., Orr, A. L., Perevoshchikova, I. V., et al. (2012). Mitochondrial complex II can generate reactive oxygen species at high rates in both the forward and reverse reactions. *Journal of Biological Chemistry, 287,* 27255–27264. https://doi.org/10.1074/jbc.M112.374629.

Rada, B., Jendrysik, M. A., Pang, L., et al. (2013). Pyocyanin-enhanced neutrophil extracellular trap formation requires the NADPH oxidase. *PLoS One, 8.* https://doi.org/10.1371/journal.pone.0054205.

Saitoh, T., Komano, J., Saitoh, Y., et al. (2012). Neutrophil extracellular traps mediate a host defense response to human immunodeficiency virus-1. *Cell Host & Microbe, 12,* 109–116. https://doi.org/10.1016/j.chom.2012.05.015.

Santos, F., Nequiz, M., Hernández-Cuevas, N. A., et al. (2015). Maintenance of intracellular hypoxia and adequate heat shock response are essential requirements for pathogenicity and virulence of *Entamoeba histolytica*. *Cellular Microbiology, 17,* 1037–1051. https://doi.org/10.1111/cmi.12419.

Silva, L. M. R., Muñoz Caro, T., Gerstberger, R., et al. (2014). The apicomplexan parasite *Eimeria arloingi* induces caprine neutrophil extracellular traps. *Parasitology Research, 113,* 2797–2807. https://doi.org/10.1007/s00436-014-3939-0.

Sousa-Rocha, D., Thomaz-Tobias, M., Diniz, L. F. A., et al. (2015). *Trypanosoma cruzi* and its soluble antigens induce NET release by stimulating toll-like receptors. *PLoS ONE, 10,* 1–16. https://doi.org/10.1371/journal.pone.0139569.

Tsaousis, A. D., Hamblin, K. A., Elliott, C. R., et al. (2018). The human gut colonizer Blastocystis respires using complex ii and alternative oxidase to buffer transient oxygen fluctuations in the gut. *Frontiers in Cellular and Infectious Microbiology, 8,* 371. https://doi.org/10.3389/fcimb.2018.00371.

Urban, C. F., Reichard, U., Brinkmann, V., & Zychlinsky, A. (2006). Neutrophil extracellular traps capture and kill *Candida albicans* and hyphal forms. *Cellular Microbiology, 8,* 668–676. https://doi.org/10.1111/j.1462-5822.2005.00659.x.

Villagra-Blanco, R., Silva, L. M. R., Aguilella-Segura, A., et al. (2017a). Bottlenose dolphins (*Tursiops truncatus*) do also cast neutrophil extracellular traps against the apicomplexan parasite *Neospora caninum. International Journal for Parasitology: Parasites and Wildlife, 6,* 287–294. https://doi.org/10.1016/j.ijppaw.2017.09.002.

Villagra-Blanco, R., Silva, L. M. R., Gärtner, U., et al. (2017b). Molecular analyses on *Neospora caninum*-triggered NETosis in the caprine system. *Developmental and Comparative Immunology, 72,* 119–127. https://doi.org/10.1016/j.dci.2017.02.020.

Villagra-Blanco, R., Silva, L. M. R., Muñoz-Caro, T., et al. (2017c). Bovine polymorphonuclear neutrophils cast neutrophil extracellular traps against the abortive parasite *Neospora caninum. Frontiers in Immunology, 8,* 1–11. https://doi.org/10.3389/fimmu.2017.00606.

Vong, L., Lorentz, R. J., Assa, A., et al. (2014). Probiotic *Lactobacillus rhamnosus* inhibits the formation of neutrophil extracellular traps. *The Journal of Immunology, 192,* 1870–1877. https://doi.org/10.4049/jimmunol.1302286.

Wang, Y., Li, M., Stadler, S., et al. (2009). Histone hypercitrullination mediates chromatin decondensation and neutrophil extracellular trap formation. *Journal of Cell Biology, 184,* 205–213. https://doi.org/10.1083/jcb.200806072.

Wanyiri, J. W., Kanyi, H., Maina, S., et al. (2014). Cryptosporidiosis in HIV/AIDS patients in Kenya: Clinical features, epidemiology, molecular characterization and antibody responses. *American Journal of Tropical Medicine and Hygeine, 91,* 319–328. https://doi.org/10.4269/ajtmh.13-0254.

Wei, Z., Hermosilla, C., Taubert, A., et al. (2016). Canine neutrophil extracellular traps release induced by the apicomplexan parasite *Neospora caninum* in vitro. *Frontiers in Immunology, 7.* https://doi.org/10.3389/fimmu.2016.00436.

Wei, Z., Wang, Y., Zhang, X., et al. (2018). Bovine macrophage-derived extracellular traps act as early effectors against the abortive parasite *Neospora caninum. Veterinary Parasitology, 258,* 1–7. https://doi.org/10.1016/j.vetpar.2018.06.002.

Young, R. L., Malcolm, K. C., Kret, J. E., et al. (2011). Neutrophil extracellular trap (NET)-mediated killing of *Pseudomonas aeruginosa*: Evidence of acquired resistance within the CF airway, independent of CFTR. *PLoS One, 6.* https://doi.org/10.1371/journal.pone.0023637.

Yousefi, S., Morshed, M., Amini, P., et al. (2015). Basophils exhibit antibacterial activity through extracellular trap formation. *Allergy: European Journal Allergy and Clinical Immunology, 70,* 1184–1188. https://doi.org/10.1111/all.12662.

Yousefi, S., Gold, J. A., Andina, N., et al. (2008). Catapult-like release of mitochondrial DNA by eosinophils contributes to antibacterial defense. *Nature Medicine, 14,* 949–953. https://doi.org/10.1038/nm.1855.

Chapter 14
Parasite Secretory Molecules in Host-Parasite Interaction

Koushik Das and Tomoyoshi Nozaki

Abstract A cross-talk between the host and the parasites is fundamentally important to understand pathophysiology of the diseases they cause. During the adaptation-driven co-evolution of both the host and the parasites, the familiar 'arms race' takes place. Over the years, host–parasite interactions have been studied extensively from both the host and parasitic points of view, which led to new insights into novel strategies exploited by the parasites to manipulate host–parasite cross-talk. Such molecular strategies used by different parasites attacking their hosts often share many similarities. Parasites secrete a number of effector molecules to invade and create the habitat for growth inside the host, as well as to evade the host defense mechanisms. An overview is presented on these parasite secretory molecules used by both intra- and extracellular parasites during host-parasite interactions in important parasitic diseases. Questions with regard to new direction of research that are needed to discover the generic and specific molecular strategies used by the parasite to invade, survive, and grow inside their hosts, and to finally discover parasitic molecular mechanism associated with their development are also discussed.

Keywords *Entamoeba histolytica* · Pathogenesis · Virulence · Exosome · Secreted protein · Host-parasite interaction · Host invasion · Host immunomodulation

Abbreviations

Gal	Galactose
GalNAc	N-acetyl-D-galactosamine
STIRPs	Serine-, threonine-, isoleucine-rich
SREHP	Serine rich *E. histolytica* proteins
KERP	Lysine and glutamic acid rich protein
ROM	Rhomboid protease

K. Das · T. Nozaki (✉)
Department of Biomedical Chemistry, Graduate School of Medicine, The University of Tokyo, Tokyo, Japan
e-mail: nozaki@m.u-tokyo.ac.jp

© Springer Nature Switzerland AG 2020

N. Guillen (ed.), *Eukaryome Impact on Human Intestine Homeostasis and Mucosal Immunology*, https://doi.org/10.1007/978-3-030-44826-4_14

CPs	Cysteine proteases
IFN-γ	Interferon γ
TNF-α	Tumor necrosis factor α
GP	Glycoprotein
TIMP	Tissue inhibitor of metalloproteinases
HSP	Heat shock protein

Introduction

Living organisms are constantly exposed to other organisms. In such mutualistic organism-organism, predator-prey, and host–parasite interactions, all living organisms have acquired strategies to deal with other organisms via molecular cross-talk. In case of host-parasite interaction, all parasites have acquired strategies to deceive or scape from the host defenses to survive, thereby enhancing the chances of completion of their life cycle. Such finely tuned host–parasite molecular cross-talk is the result of the molecular arms race between the host and parasites, where the hosts use different molecules and mechanisms to counteract parasite's invasion, while the parasite deploys many effector molecules to manipulate the host behaviors in such ways that ensure the parasite survival and enhance host-to-host transmission. This race has led to the selection of three parasitic lifestyles: (i) extracellular, (ii) obligate and (iii) facultative intracellular. Moreover, such molecular war between the host and parasite often affects the outcome of infection. The parasite can engage a plethora of surface and secretory molecules either to kill or nullify the host cells to access the host nutrients or to evade the host immunity for their prolonged survival. Many of these molecules are involved in triggering specific signaling pathways, both in the parasite and the host cell, that are critical for parasite's entry to the host cells and survival, in case of intracellular parasites. Several important advances have been achieved in identifying factors that are critical to the pathogenesis of the diseases they cause. In this review, we highlight and summarize the present knowledge on host-parasite cross-talk and discuss how the parasite's secretory molecules regulate the host-parasite relationships in some representative parasitic diseases. This will help us to understand the molecular strategies used by parasites for invasion, survival, and growth in host tissues, and ultimately, to unravel parasite's molecular signatures associated with adaptation, parasitism, and virulence to the host.

Parasite Derived and Induced Molecules in Host Invasion

Mammalian tissue and cell membranes represent major barriers to parasite's invasion, dissemination, and access to essential nutrients. Parasites employ a repertoire of surface associated as well as secretory effectors that are essential for invasion and survival within the host. However, some parasites evade the host's immune response by hiding intracellularly, such as *Trypanosoma cruzi*, *Leishmania*, *Toxoplasma* and *Plasmodium* species, while some can evade the cell immune response including extracellular parasites such as *Entamoeba histolytica*, *Giardia lamblia*, and *Trichomonas vaginalis*. Here, we discuss the functional implications of parasite-derived molecules from extracellular and intracellular parasites in parasitism and pathogenesis. A list of these molecules are summarized in Table 1 and their different mode of action have been represented in Fig. 1.

Intracellular Parasites

Trypanosoma cruzi have various surface molecules such as, GP85, GP83, and GP63 (penetrin), which help the parasite to adhere to and enter into the host cells (Marino et al. 2003; Nde et al. 2006; Cardenas et al. 2010). GP57/51 (cysteine protease), matrix metalloprotease-9 (MMP-9), and POPTc80 (serine protease) are secreted proteases, capable of degrading collagen and fibronectin, and regulates host-parasite interaction (Souto-Padron et al. 1990; Murta et al. 1990; Cazzulo 2002; Nogueira et al. 2010; Santana et al. 1997). *Leishmania* promastigotes degrade the extracellular matrix (ECM) by a zinc-dependent metalloprotease GP63 (leishmanolysin) (McGwire et al. 2003). However, it also helps the parasite to evade the host immune response and promote the survival of intracellular amastigotes (Yao 2010). *Toxoplasma gondii* infection promotes the expression and accumulation of MMP-9 (gelatinase B), MT1-MMP (Buache et al. 2007; Seipel et al. 2010), and MMP-12, which were shown to degrade collagen elastin, laminin, and proteoglycans (Sato et al. 2005; Chou and Lai 2011; Geurts et al. 2012). The ability of the malaria parasite to reach and colonize the liver is attributable to two proteins, the circumsporozoite (CS) protein and the thrombospondin-related adhesive protein (TRAP), which recognize the heparan sulfate proteoglycans and thrombospondin in the ECM, respectively (Pradel et al. 2002). Moreover, some host derived enzymes, such as MMP-9 and TIMP-2, were reported to increase in patients with severe malaria (Prato et al. 2005) and suggested to be involved in the pathogenesis of malaria (Dietmann et al. 2008).

Extracellular Parasites

Entamoeba surface proteins including Gal/GalNAc lectin, serine rich *E. histolytica* proteins (SREHP), lysine and glutamic acid rich protein 1 (KERP1) and lipopeptidephosphoglycan (LPPG), serine-, threonine-, isoleucine-rich proteins (STIRPs), rhomboid protease 1 (ROM1), and surface metalloprotease (MSP-1) have been implicated in adherence, signaling, ingestion, and immune evasion at the host-parasite

Table 1 Parasite derived and induced effector molecules, involved in host-parasite interaction

Parasite	Effector molecules	Localization and effect on host cells	References
T. cruzi	GP63	Localized on the cell surface, zinc-dependent metalloprotease, promotes adhesion to heparin, heparan sulfate, and collagen	Ortega-Barria and Pereira (1991), Ma et al. (2011)
	Cruzipain	Localized in lysosomes, reservosomes, cysteine protease, secreted and degrades collagen, fibronectin	Souto-Padron et al. (1990), Cazzulo et al. (1990, 2001)
	Cathepsin B-like protease	Localized in reservosomes, degrades human type I collagen	Garcia et al. (1998), Nóbrega et al. (2004)
	MMP-9 like protease	Secreted, metalloprotease, regulator of infection and pathogenesis	Nogueira et al. (2010), Visse and Nagase (2003)
	POP Tc80	Localized in a vesicular compartment close to the flagellar pocket, serine protease, degrades collagen types I and IV and fibronectin	Santana et al. (1997), Grellier et al. (2001)
Leishmania spp.	GP63 or leishmanolysin	Localized on the cell surface, metalloprotease, secreted, and degrades collagen IV and fibronectin	Kulkarni et al. (2008), McGwire et al. (2003)
T. gondii	MMP-9	Detected on the infected macrophage surface, metalloprotease, secreted, degrades collagen, laminin and elastin	Visse and Nagase (2003), Seipel et al. (2010), Niehus et al. (2012), Nagase et al. (2006)

(continued)

Table 1 (continued)

Parasite	Effector molecules	Localization and effect on host cells	References
	MT1-MMP or MMP-14	Detected on the infected macrophage surface, metalloprotease, degrades collagen, fibronectin and laminin	Visse and Nagase (2003), Buache et al. (2007), Seipel et al. (2010), Sato et al. (2005)
	MMP-2	Metalloprotease, secreted, degrades collagen, fibronectin, laminin and elastin	Visse and Nagase (2003), Nagase et al. (2006), Mũnoz et al. (2009)
	MMP-12	Metalloprotease, secreted, degrades collagen, fibronectin, laminin and elastin	Visse and Nagase (2003), Nagase et al. (2006), Chou and Lai (2011), Shipley et al. (1996)
P. falciparum	MMP-9	Metalloprotease, secreted, degrades collagen, laminin and elastin	Visse and Nagase (2003), Prato et al. (2005)
	MMP-8	Metalloprotease, secreted, degrades collagen	Visse and Nagase (2003), Dietmann et al. (2008)
E. histolytica	Amoebapore A	Membranolytic peptide, implicated in host cell killing before ingestion	Andrä et al. (2003)
	Amoebapore B	Membranolytic peptide, shows bactericidal activity	Andrä et al. (2003)
	Amoebapore C	Membranolytic peptide, cytotoxic to Jurkat and U937 cells	Leippe et al. (1994), Andrä et al. (2003)
	Cysteine proteinases (CPs)	Act on a variety of host substrates such as mucin, villin, laminin, collagen, proteoglycan, and extracellular matrix (ECM), facilitate pathogenesis by mucin (MUC) and ECM degradation	Tillack et al. (2006, 2007), Li et al. (1995), Lidell et al. (2006), Hellberg et al. (2001), Mitra et al. (2007), Nakada-Tsukui et al. (2005), Nowak et al. (2004)

(continued)

Table 1 (continued)

Parasite	Effector molecules	Localization and effect on host cells	References
	Gal/GalNAc lectin	Localized on the cell surface and internal membranes, associated with cytotoxicity and liver abscess formation	Ankri et al. (1999), Sen et al. (2007), Avalos-Padilla et al. (2015)
	*Eh*STIRPs (*Eh* serine, threonine and isoleucine rich proteins)	Localized on the cell surface, associated with adhesion and cytotoxicity	Ankri et al. (1999), Sen et al. (2007)
	*Eh*TMKB1-9 (*Eh* transmembrane kinaseB1-9)	Localized on the cell surface, associated with adhesion and cell killing	Shrimal et al. (2010)
	KERP1 (lysine and glutamic acid rice protein 1)	Secreted, associated with parasite adhesion with host cell	Seigneur et al. (2005), Santi-Rocca et al. (2008), Perdomo et al. (2013, 2016)
	ROM1 (rhomboid proteases)	Intramembrane, involved in adherence, phagocytosis, and motility	Baxt et al. (2010)
	CPADH (CP112+ADH112)	Localized in cytoplasmic vesicles, on the plasma membrane, secreted, involved in adherence, phagocytosis, monolayer destruction, and amoebic liver abscess, degrades collagen, fibronectin, and hemoglobin	Serrano-Luna et al. (2013)
	CPBF1	Localized in the ER and phagosomes, transports cysteine proteases to lysosomes, required for CP5 activity	Nakada-Tsukui et al. (2012)
	CPBF6	Localized in lysosomes and phagosomes, involved in the transport of α-amylase and γ-amylase to phagosomes	Furukawa et al. (2013)

(continued)

Table 1 (continued)

Parasite	Effector molecules	Localization and effect on host cells	References
	CPBF8	Localized in lysosomes and phagosomes, involved in the transport of lysozyme and hexaminidase, and digestion of bacteria in phagosomes and cytotoxicity	Furukawa et al. (2012)
	MSP-1	Localized on the surface, involved in adherence and motility, phagocytosis, and cytotoxicity	Teixeira et al. (2012), Thibeaux et al. (2013)
	β-amylase	Localized on the cell surface and cytoplasmic vesicles, involved in mucin degradation	Thibeaux et al. (2013)
	Lysozyme	Localized on the cell surface and phagosomes, associated with virulence, implicated in amoebic colitis	Gilchrist et al. (2006), Thibeaux et al. (2013)
G. lamblia	66, 45, 30, and 18 kDa proteins	Intracellular cysteine proteases, degrades collagen I	Coradi and Guimaraes (2006)
	145, 96 and 82 kDa proteins	Cysteine proteases, secreted, degrades collagen I	De Carvalho et al. (2008)
T. vaginalis	CP30	Localized on the cell surface, cysteine protease, degrades collagen IV and fibronectin	Hernandez-Gutierrez et al. (2004)
	CP39	Localized on the cell surface, cysteine proteases, degrades collagen I, III, IV, V and fibronectin	Alvarez-Sanchez et al. (2000)
	CP65	Localized on the cell surface, cysteine protease, degrades collagen IV and fibronectin	Huet et al. (1992)

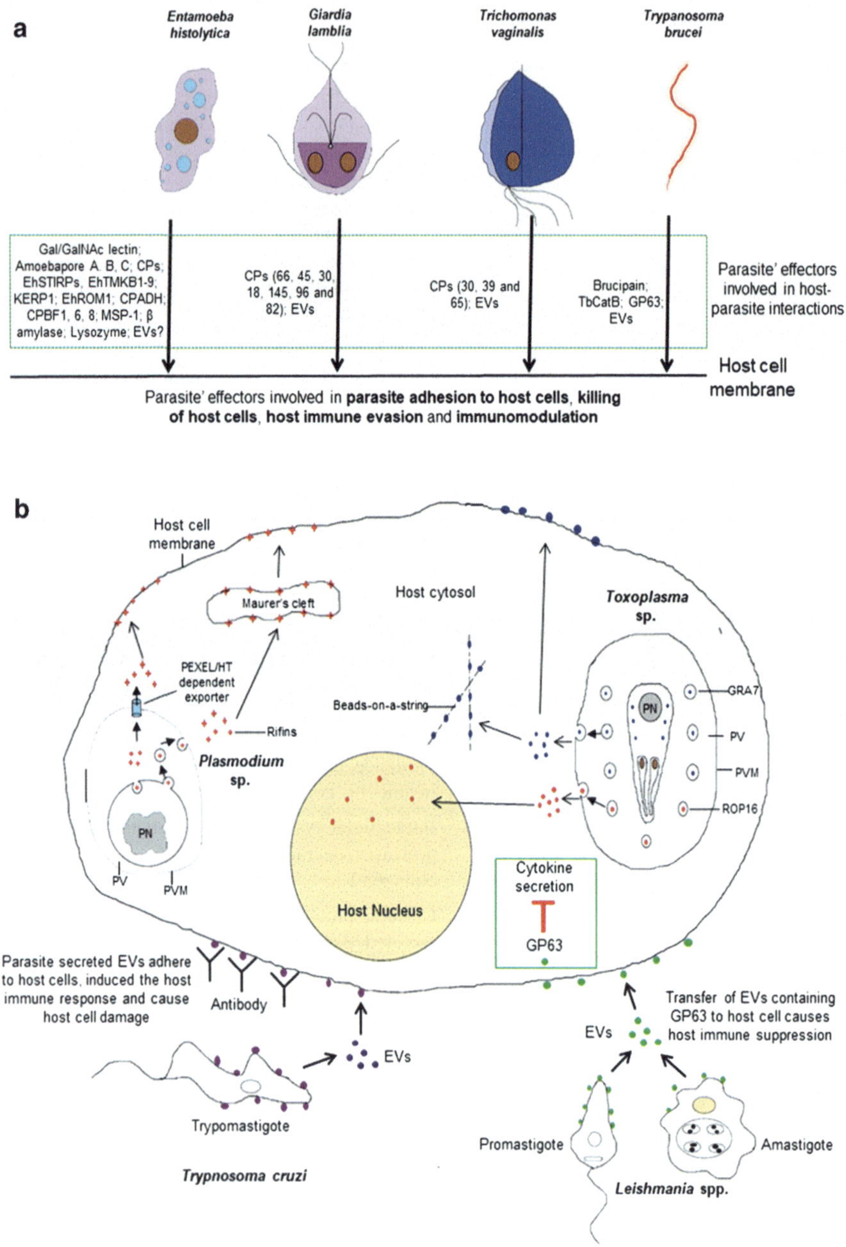
a
Entamoeba histolytica
Giardia lamblia
Trichomonas vaginalis
Trypanosoma brucei
Gal/GalNAc lectin; Amoebapore A, B, C; CPs; EhSTIRPs, EhTMKB1-9; KERP1; EhROM1; CPADH; CPBF1, 6, 8; MSP-1; β amylase; Lysozyme; EVs?
CPs (66, 45, 30, 18, 145, 96 and 82); EVs
CPs (30, 39 and 65); EVs
Brucipain; TbCatB; GP63; EVs
Parasite' effectors involved in host-parasite interactions
Host cell membrane
Parasite' effectors involved in parasite adhesion to host cells, killing of host cells, host immune evasion and immunomodulation
b
Host cell membrane
Maurer's cleft
Host cytosol
Toxoplasma sp.
PEXEL/HT dependent exporter
Rifins
Beads-on-a-string
GRA7
PN
PV
PVM
ROP16
Plasmodium sp.
PN
PV
PVM
Host Nucleus
Cytokine secretion
GP63
Parasite secreted EVs adhere to host cells, induced the host immune response and cause host cell damage
Antibody
EVs
Trypomastigote
Trypnosoma cruzi
EVs
Transfer of EVs containing GP63 to host cell causes host immune suppression
Promastigote
Amastigote
Leishmania spp.

◀**Fig. 1** Parasite' effector molecules involved in host-parasite interaction. **a** Extracellular parasites (such as, *E. histolytica*, *G. lamblia*, *T. vaginalis* and *T. brucei*) employ several effector molecules, involved in parasite' adherence to host cells, lysis of host cells, host immune evasion and immunomodulation. CPs: cysteine proteases; EhSTIRPs: *E. histolytica* serine, threonine and isoleucine rich proteins; EhTMKB1-9: *E. histolytica* transmembrane kinase B1-9; KERP1: lysine and glutamic acid rich protein 1; ROM1: rhomboid protease 1; CPADH: CP112+ ADH112; CPBF: CP binding family protein; MSP-1: metalloprotease 1; TbCatB: *T. brucei* cathepsin B; EVs: extracellular vesicles. **b** Intracellular parasites (such as, *Plasmodium* sp., *Toxoplasma* sp., *Leishmania* spp. and *T. cruzi*) secrete some of effector molecules can translocate to either host cell membrane or to host cell nucleus, can regulates the host immune response. PV: parasitophorous vacuole; PVM: PV membrane; PN: parasite nucleus; HT: host targeting; GRA: granule protein; ROP: rhoptry proteins

interface (Teixeira et al. 2012; Baxt et al. 2010; MacFarlane and Singh 2007). Following adherence, *Entamoeba* trophozoites produce some cytotoxic and bactericidal molecules including proteases (such as cysteine proteases, CPs), lysozymes, glycosidases, amoebapores, and phospholipases (Okada et al. 2006; Serrano-Luna et al. 2013). *E. histolytica* CP-binding proteins (CPBFs), more specifically CPBF1, were identified as the regulator of protease activity and trafficking to phagosomes (Furukawa et al. 2012, 2013; Nakada-Tsukui et al. 2012). Williams and Coombs explored intracellular proteases present in *Giardia* trophozoites and observed collagen degradation by a group of proteases (30–65, 120 kDa) (Williams and Coombs 1995). The hydrolysis of collagen type I by trophozoites lysates was associated with proteases from 18 to >116 kDa (Carvalho et al. 2008). *Trichomonas vaginalis* possess glyceraldehyde 3 phosphate dehydrogenase (GAPDH), which serves as surface-associated fibronectin-binding protein. In addition to 450 in silico inferred proteases in the genome, there are three reports of *T. vaginalis* cysteine proteases (CPs): TvCP30, TvCP39, and TvCP65. TvCP30 is located at the *T. vaginalis* cell surface and in the cytoplasm, involved in parasite adhesion to Hela cells (Meddoza-Lopez et al. 2000).

Parasite Secreted Extracellular Vesicles (EVs) in Host-Parasite Interaction

In parasitic diseases, EVs have emerged as a common mechanism for mediating host-parasite interactions. Three groups of EVs are associated with parasitic infections: (1) EVs directly produced from extracellular parasites, (2) EVs produced by host cells infected by intracellular parasites, and (3) EVs produced by host cells stimulated by parasite derived antigens. Here, we review the present findings regarding EVs (especially exosomes) to recognize the role of EVs in host-parasite interactions in some of important parasitic diseases such as, malaria, leishmaniasis, toxoplasmosis, trypanosomiasis, trichomoniasis, schistosomiasis, and amoebiasis.

Exosome-like vesicles secreted by *Plasmodium falciparum*-infected red blood cells can deliver DNA among themselves, resulting in an increased number of *P.*

falciparum gametocytes and easier transmission to mosquito vectors (Mantel et al. 2013; Regev-Rudzki et al. 2013). Moreover, communication via exosome-like vesicles is beneficial to parasite survival in various situations such as stress and drug pressure. Exosomes from *Leishmania donovani* were shown to facilitate IL-10 production and dampen TNF-α activation. *Leishmania* GP63, present in exosomes, was shown to play an important role in modulating the immune response of macrophages through the regulation of protein tyrosine phosphatases (PTPs) and transcription factors (TFs) (Halle et al. 2009). In addition to proteins, non-coding RNAs (including rRNA, tRNA, and tRNA-derived small RNAs) were also found (Bhattacharyya et al. 2002). *Toxoplasma gondii*-derived exosomes contained several notable protein markers, including HSP70, CD63, and surface marker P30 (Li et al. 2018). The exosomes promoted macrophage activation with increased production of IFN-γ, IL-12, and TNF-α (Li et al. 2018). Moreover, exosomes are enriched in miRNAs and mRNAs of thymosin beta 4, eukaryotic elongation factor-1a (EF-1a), Rab-13, and LLP homolog, which mediate neurologic effects in toxoplasmosis (Pope and Lasser 2013). *Trypanosoma brucei* exosomes functioned as repellents to keep the pathogens away from injured cells and improper environments to influence parasite social motility (Eliaz et al. 2017). *T. cruzi* exosomes contain small RNAs derived from tRNAs and rRNAs. It was reported that tRNA-derived small RNAs (tsRNAs) carried by exosomes can induce changes in the expression profiles of host genes via interaction with host argonaute protein (Garcia-Silva et al. 2014). Exosomes released by highly adherent *Trichomonas vaginalis* were found to promote the adherence of poorly adherent species to epithelial cells. Moreover, it delivers regulatory molecules to the host and promote the modulation of host immune functions by inhibiting IL-8, IL-6, IL-10, and TNF-α production. (Twu et al. 2013; Olmos-Ortiz et al. 2017). Exosome-like vesicles derived from *Schistosoma japonicum* can activate macrophage (termed as M1 macrophage) with increased production of pro-inflammatory factors such as TNF-α, CD16/32, and iNOS (Wang et al. 2015). These EVs also possess parasite miRNA and 403 proteins including, those with catalytic, translation regulatory, and binding activities (Zhu et al. 2016). Only scarce information of EV is available for *Entamoeba histolytica*. However, bioinformatics analysis revealed that *E. histolytica* possess a few homologs of human exosomal markers tetraspanins (unpublished data), which possibly indicates that like other eukaryotic pathogens, *E. histolytica* also can excrete exosomes into extracellular milieu. However, the physiological conditions that trigger exosome secretion as well as the significance of exosomes in *Entamoeba* biology are not yet explored.

Parasite' Secretory Molecules Invade into Host Cells

Some of the proteins secreted by the parasites from their secretory organelles are translocated to different locations in and on the parasite, as well as invaded into host cells. Here, we have discussed the parasite secretory proteins those have ability to invade the host cells in protozoan parasites (*Theileria* sp., *Toxoplasma* sp., *Plasmodium* sp., and *Entamoeba* sp.) and trematode (*Schistosoma* sp.).

Theileria parva and *Theileria annulata* are tick-transmitted parasites that lead to East Coast fever and tropical theileriosis in cattle. To date there are six *Theileria* macroschizont proteins that were shown to be secreted into the host cell cytoplasm. One, a *T. annulata* secretory protein (TaSE), plays an important role in interaction of host and parasite (Schneider et al. 2007). The remaining five *Theileria* proteins are a group of five *T. annulata* schizont (Tash) proteins with mammalian 'AT-hook' DNA-binding domains, which were shown to modulate the myeloid cell differentiation, division, and apoptosis. These proteins contain apparent nuclear localization signals (NLSs), which facilitate their translocation from the host cell cytosol to the host cell nucleus (Swan et al. 1999, 2001). *T. gondii,* when resides in a parasitophorous vacuole (PV), secretes a panel of dense granule proteins, including GRA24, which has the unique ability to trigger prolonged autophosphorylation and nuclear translocation of the host cell (Braun et al. 2013; Hakansson et al. 2001; Saeij et al. 2006; Jacobs et al. 1998). Moreover, protein phosphatase 2C (PP2C) and a rhoptry protein (ROP16) have consensus NLSs, which are recognized by the host nuclear trafficking machinery and help them to localize to the nucleus of infected host cells. (Saeij et al. 2007). *Plasmodium* CS protein is exported into the hepatocyte cytoplasm using the host targeting (HT) signal, and trafficked to the hepatocyte nucleus via NLS (Hamilton et al. 1988; Hollingdale et al. 1983; Singh et al. 2007). CS modulates the host functions and promotes *Plasmodium* growth and development in hepatocytes. CS is also highly immunogenic and recognized by cytotoxic T cells. *E. histolytica* secretes several membrane associated and/or secretory molecules in extracellular milieu during host-parasite interactions as mentioned earlier in Sect. 2.2. However, no such parasite secretory molecules have been identified so far that can potentially infiltrate into the host cell. Seigneur et al. identified two proteins secreted into the extracellular milieu during interaction with the brush-border epithelium of human enterocytes and named them as *E. histolytica* lysine (K)—and glutamic acid (E)-rich protein 1 (EhKERP1) and 2 (EhKERP2), as they were rich in lysine and in glutamic acid (Seigneur et al. 2005; Perdomo et al. 2016). In comparison to EhKERP1, which is a plasma membrane- and vesicle-associated protein, EhKERP2 possesses NLS, which possibly indicates its nuclear targeting with unique implications in host-parasite interactions as discussed earlier. Other than the protozoan parasites, the parasitic trematode *Schistosoma* spp. also secretes interleukin-4-inducing molecule from eggs (IPSE/alpha-1), which infiltrates into host cells, translocates to the host nucleus via C-terminal conserved NLS motifs, and binds to host nuclear DNA, which suggests a possible role in immune modulation of host cells (Pennington et al. 2017).

Parasites' Secretory Molecules in Host Immune Evasion and Immunomodulation

The balance between anti-parasitic immunity by the host versus immunomodulation and evasion by the parasites may result in parasite clearance or chronic infection without severe clinical symptoms, whereas imbalances characterized by excessive parasite growth, exaggerated immune reactions, or a combination of both cause severe pathology and death, which is detrimental for both parasite and host. Here we discuss host-parasite cross-talk in the perspective of parasite and host derived effector molecules involved in immunomodulation and immune evasion mechanisms in major parasitic diseases.

During asexual blood stage developmental, malaria parasites replicate inside red blood cells, which temporarily protects them from being eliminated by immune cells (Podoba and Stevenson 1991; van der Heyde et al. 1993; Horne-Debets et al. 2013; Imai et al. 2015; Safeukui et al. 2015). Several factors produced by the parasite or originating from the infected host cells, including heme, hemozoin, GPI-anchors, nucleic acids, microparticles and exosomes have prominent immunomodulatory effects and substantially determine the outcome of the malaria infection. Heme downregulates anti-inflammatory mechanisms, by inducing the expression of Cu/Zn superoxide dismutase (SOD-1) in peripheral blood mononuclear cells, which inhibits prostaglandin E2 (PGE2) secretion (Andrade et al. 2010). A *Plasmodium* ortholog of macrophage migration inhibitory factor enhances inflammatory cytokine production and induces antigen-experienced CD4 T cells to develop into short-lived effector cells (Sun et al. 2012). The most effective anti-*Leishmania* response of the macrophages is the production of reactive oxygen species and reactive nitrogen species (Iles and Forman 2002; Fang 2004). In response to host defense, *Leishmania* parasites induce macrophages to produce arginase, which competes with iNOS for arginine and thus diminishes the production of parasitotoxic NO. At the same time, arginase produces essential nutrient, ornithine, for the synthesis of polyamines and urea (Gaur et al. 2007). Leishmanial metalloprotease gp63 interferes with macrophage signaling pathways leading to NOX2 and iNOS induction (Olivier et al. 2012). *Leishmania* recruits suppressors of the cytokine signaling family proteins, SOCS-1 and SOCS-3, which negatively regulate TLR2 induced cytokine induction (de Veer et al. 2003). *T. gondii* inhibitor of STAT1-dependent transcription (TgIST) is a dense granule protein, and blocks transcription of IFN-stimulated genes in human foreskin fibroblasts (HFFs) (Gay et al. 2016; Olias et al. 2016). GRA18, another dense granule protein, can reprogram inflammatory responses (Lima and Lodoen 2019). *T. brucei*-derived kinesin heavy chain (TbKHC1) and adenylate cyclase dampen inflammation and promote parasite growth (Stijlemans et al. 2016). *T. brucei* variable surface glycoproteins shield buried invariant surface proteins from recognition by the hosts' immune system and protect bloodstream parasites against complement-mediated lysis. During amebic invasion, *E. histolytica* expresses superoxide dismutase, NADPH: flavin oxidoreductase, arginase and peroxiredoxin, which protect the parasite against reactive oxygen species from early infiltrating neutrophils (Elnekave et al. 2003; Bruchhaus et al.

1998; Sim et al. 2005; Davis et al. 2006). *Entamoeba* arginase enzyme competitively converts L-arginine, the substrate of nitric oxide species (NOS) to L-ornithine, and thus inhibits the NO production by macrophages (Elnekave et al. 2003). Immunoregulatory molecule prostaglandin E2, synthesized by a cyclooxygenase-like enzyme of the parasite (Dey et al. 2003) increases cyclic adenosine monophosphate (cAMP) levels in macrophages (another immune effector) that inhibits Th1 cytokine release, NADPH-mediated oxidative burst, and NO synthesis (Wang and Chadee 1995). *Entamoeba* produces immunosuppressive monocyte locomotion inhibitory factor (MLIF), which shows anti-inflammatory activities and inhibits the NO production (Rico et al. 2003). *E. histolytica* Gal/GalNAc lectin has sequence resemblance and antigenic cross reactivity with the MAC-inhibitory protein, CD59, and inhibits MAC-mediated lysis (Braga et al. 1992). *E. histolytica* secreted cysteine proteases can cleave complement components C3a and C5a (Reed et al. 1995). Neutrophil-inhibiting factor (NIF), produced in hookworm infection inhibits the function of leukocytes, preventing their adhesion and peroxide release (Ali et al. 2001; Loukas and Prociv 2001). *Taenia solium* secretes antigen B (also known as paramyosin), which prevents damage to the parasite by complement fixation on its surface (Laclette et al. 1990, 1992). EgTeg is an anti-inflammatory molecules produced by *Echinococcus granulosus*, favors the parasite survival in the host.

Conclusion

The host-parasite relationship is a complex phenomenon that is mediated by virulence factors of the parasite as well as exacerbated responses of the host. As summarized, the hosts exploit a variety of strategies to win the fight against the parasites. In response, the parasites have also developed counter measures to mislead and nullify the host's defenses and to ensure the completion of their life cycle, to ultimately win the molecular war. New research challenges to understand host–parasite relationship should address the following points: (i) what are the mechanisms of the parasite factors to be activated dependent on host-parasite interactions (ii) what are the mechanisms of the host proteins that are involved in host-parasite cross-talk repressed (or activated)?; (iii) is the repression/activation of host proteins regulated at genetic, epigenetic, transcriptional, translational, or post-translational levels?; (iv) can biomolecules including proteins, lipids, and nucleic acids identified in the secretome or exosome contents explain resistance or susceptibility of hosts to the parasites?; (v) what is the key (or multiple independent) event(s) of the host–parasite cross-talk in evolution that allowed hijacking of host cellular functions by extra- and intra-cellular parasites? Indeed, elucidation of these complex molecular dialogues between hosts and parasites is clearly desirable not only to improve our understanding of parasite virulence, but also to identify parasite-specific host biomarkers, and to define novel therapeutic targets.

Acknowledgements This work is supported in part by Grants-in-Aid for Scientific Research (B) and Challenging Research (Exploratory) (KAKENHI JP26293093, JP17K19416, and JP18H02650 to T.N.) from the Japan Society for the Promotion of Science, Grant for research on emerging and re-emerging infectious diseases from Japan Agency for Medical Research and Development (AMED, JP18fk0108046 to T.N.), Grant from the National Center for Global Health and Medicine Intramural Research Fund (29–2013 to T. N.), and Grant for Science and Technology Research Partnership for Sustainable Development (SATREPS) from AMED and Japan International Cooperation Agency (JICA) (to T. N.).

References

Ali, F., Brown, A., Stanssens, P., et al. (2001). Vaccination with neutrophil inhibitory factor reduces the fecundity of the hookworm *Ancylostoma ceylanicum. Parasite Immunology, 5,* 237–249.

Alvarez-S´anchez, M. E., Avila-Gonz´alez, L., Becerril-Garc´ıa, C., et al. (2000). A novel cystein proteinase (CP65) of *Trichomonas vaginalis* involved in cytotoxicity. *Microbial Pathogenesis, 28,* 193–202.

Andrä, J., Herbst, R., & Leippe, M. (2003). Amoebapores, archaic effector peptides of protozoan origin, are discharged into phagosomes and kill bacteria by permeabilizing their membranes. *Developmental and Comparative Immunology, 27,* 291–304.

Andrade, B. B., Reis-Filho, A., Souza-Neto, S. M., et al. (2010). Severe *Plasmodium vivax* malaria exhibits marked inflammatory imbalance. *Malaria Journal, 9,* 13.

Ankri, S., Padilla-Vaca, F., Stolarsky, T., et al. (1999). Antisense inhibition of expression of the light subunit (35 kDa) of the Gal/GalNac lectin complex inhibits *Entamoeba histolytica* virulence. *Molecular Microbiology, 33,* 327–337.

Avalos-Padilla, Y., Betanzos, A., Javier-Reyna, R., et al. (2015). EhVps32 is a vacuole-associated protein involved in pinocytosis and phagocytosis of *Entamoeba histolytica. PLoS Pathogens, 11,* e1005079.

Baxt, L. A., Rastew, E., Bracha, R., et al. (2010). Downregulation of an *Entamoeba histolytica* rhomboid protease reveals roles in regulating parasite adhesion and phagocytosis. *Eukaryotic Cell, 9,* 1283–1293.

Bhattacharyya, S., Ghosh, S., Dasgupta, B., et al. (2002). Chemokine-induced leishmanicidal activity in murine macrophages via the generation of nitric oxide. *Journal of Infectious Diseases, 185,* 1704–1708.

Braga, L. L., Ninomiya, H., McCoy, J. J., et al. (1992). Inhibition of the complement membrane attack complex by the galactose-specific adhesion of *Entamoeba histolytica. Journal of Clinical Investigation, 90,* 1131.

Braun, L., Brenier-Pinchart, M. P., Yogavel, M., et al. (2013). A toxoplasma dense granule protein, GRA24, modulates the early immune response to infection by promoting a direct and sustained host p38 MAPK activation. *Journal of Experimental Medicine, 210,* 2071–2086.

Bruchhaus, I., Richter, S., & Tannich, E. (1998). Recombinant expression and biochemical characterization of an NADPH: Flavin oxidoreductase from Entamoeba histolytica. *Biochemical Journal, 330,* 1217–1221.

Buache, E., Garnotel, R., Aubert, D., et al. (2007). Reduced secretion and expression of gelatinase profile in *Toxoplasma gondii*-infected human monocytic cells. *Biochemical and Biophysical Research Communications, 359,* 298–303.

Cardenas, T. C., Johnson, C. A., Pratap, S., et al. (2010). Regulation of the extracellular matrix. *Open Parasitol Journal, 4,* 72–76.

Cazzulo, J. J. (2002). Proteinases of *Trypanosoma cruzi*: Potential targets for the chemotherapy of Chagas disease. *Current Topics in Medicinal Chemistry, 2,* 1261–1271.

Cazzulo, J. J., Cazzulo Franke, M. C., Martinez, J., et al. (1990). Some kinetic properties of a cysteine proteinase (cruzipain) from *Trypanosoma cruzi*. *Biochimica et Biophysica Acta, 1037,* 186–191.

Cazzulo, J. J., Stoka, V., & Turk, V. (2001). The major cysteine proteinase of *Trypanosoma cruzi*: A valid target for chemotherapy of Chagas disease. *Current Pharmaceutical Design, 7,* 1143–1156.

Chou, P. H., & Lai, S. C. (2011). Elevated concentrations of matrix metalloproteinase-12 and elastin degradation products in the sera of pregnant women infected with *Toxoplasma gondii*. *Annals of Tropical Medicine and Parasitology, 105,* 225–231.

Coradi, S. T., & Guimar˜aes, S. (2006). *Giardia duodenalis*: Protein substrates degradation by trophozoite proteases. *Parasitology Research, 99,* 131–136.

Davis, P. H., Zhang, X., Guo, J., et al. (2006). Comparative proteomic analysis of two *Entamoeba histolytica* strains with different virulence phenotypes identifies peroxiredox in as an important component of amoebic virulence. *Molecular Microbiology, 61,* 1523–1532.

De Carvalho, T. B., David, E. B., Coradi, S. T., et al. (2008). Protease activity in extracellular products secreted in vitro by trophozoites of *Giardia duodenalis*. *Parasitology Research, 104,* 185–190.

de Veer, M. J., Curtis, J. M., Baldwin, T. M., et al. (2003). MyD88 is essential for clearance of *Leishmania major*: Possible role for lipophosphoglycan and Toll-like receptor 2 signaling. *European Journal of Immunology, 33,* 2822.

Dey, I., Keller, K., Belley, A., et al. (2003). Identification and characterization of a cyclooxygenase-like enzyme from *Entamoeba histolytica*. *Proceedings of the National Academy of Sciences of the United States of America, 100,* 13561–13566.

Dietmann, A., Helbok, R., Lackner, P., et al. (2008). Matrix metalloproteinases and their tissue inhibitors (TIMPs) in *Plasmodium falciparum* malaria: Serum levels of TIMP-1 are associated with disease severity. *Journal of Infectious Diseases, 197,* 1614–1620.

Eliaz, D., Kannan, S., Shaked, H., et al. (2017). Exosome secretion affects social motility in *Trypanosoma brucei*. *PLoS Pathogens, 13,* e1006245.

Elnekave, K., Siman-Tov, R., & Ankri, S. (2003). Consumption of L-arginine mediated by *Entamoeba histolytica* L-arginase (EhArg) inhibits amoebicidal activity and nitric oxide production by activated macrophages. *Parasite Immunology, 25,* 597–608.

Fang, F. C. (2004). Antimicrobial reactive oxygen and nitrogen species: Concepts and controversies. *Nature Reviews Microbiology, 2,* 820.

Furukawa, A., Nakada-Tsukui, K., & Nozaki, T. (2012). Novel transmembrane receptor involved in phagosome transport of lysozymes and β-hexosaminidase in the enteric protozoan *Entamoeba histolytica*. *PLoS Pathogens, 8,* e1002539.

Furukawa, A., Nakada-Tsukui, K., & Nozaki, T. (2013). Cysteine protease-binding protein family 6 mediates the trafficking of amylases to phagosomes in the enteric protozoan *Entamoeba histolytica*. *Infection and Immunity, 81,* 1820–1829.

Garcia, M. P., Nobrega, O. T., Teixeira, A. R. L., et al. (1998). Characterisation of a *Trypanosoma cruzi* acidic 30 kDa cysteine protease. *Molecular and Biochemical Parasitology, 91,* 263–272.

Garcia-Silva, M. R., Cabrera-Cabrera, F., Das, N. R., et al. (2014). Gene expression changes induced by *Trypanosoma cruzi* shed microvesicles in mammalian host cells: Relevance of tRNA-derived halves. *BioMed Research International, 2014,* 305239.

Gaur, U., Roberts, S. C., Dalvi, R. P., et al. (2007). An effect of parasite-encoded arginase on the outcome of murine cutaneous leishmaniasis. *Journal of Immunology, 179,* 8446.

Gay, G., Braun, L., Brenier-Pinchart, M. P., et al. (2016). *Toxoplasma gondii* TgIST co-opts host chromatin repressors dampening STAT1-dependent gene regulation and IFN-γ-mediated host defenses. *Journal of Experimental Medicine, 213,* 1779–1798.

Geurts, N., Opdenakker, G., & Van Den Steen, P. E. (2012). Matrix metalloproteinases as therapeutic targets in protozoan parasitic infections. *Pharmacology & Therapeutics, 133,* 257–279.

Gilchrist, C. A., Houpt, E., Trapaidze, N., et al. (2006). Impact of intestinal colonization and invasion on the *Entamoeba histolytica* transcriptome. *Molecular and Biochemical Parasitology, 147,* 163–176.

Grellier, P., Vendeville, S., Joyeau, R., et al. (2001). *Trypanosoma cruzi* prolyl oligopeptidase Tc80 Is involved in nonphagocytic mammalian cell invasion by trypomastigotes. *Journal of Biological Chemistry, 276,* 47078–47086.

Hakansson, S., Charron, A. J., & Sibley, L. D. (2001). Toxoplasma evacuoles: A twostep process of secretion and fusion forms the parasitophorous vacuole. *EMBO Journal, 20,* 3132–3144.

Halle, M., Gomez, M. A., Stuible, M., et al. (2009). The *Leishmania* surface protease GP63 cleaves multiple intracellular proteins and actively participates in p38 mitogen activated protein kinase inactivation. *Journal of Biological Chemistry, 284,* 6893–6908.

Hamilton, A. J., Suhrbier, A., Nicholas, J., et al. (1988). Immunoelectron microscopic localization of circumsporozoite antigen in the differentiating exoerythrocytic trophozoite of *Plasmodium berghei. Cell Biology International Reports, 12,* 123–129.

Hellberg, A., Nickel, R., Lotter, H., et al. (2001). Overexpression of cysteine proteinase 2 in *Entamoeba histolytica* or *Entamoeba dispar* increases amoeba-induced monolayer destruction *invitro* but does not augment amoebic liver abscess formation in gerbils. *Cellular Microbiology, 3,* 13–20.

Hernandez-Gutierrez, R., Avila-Gonz´alez, L., Ortega-L´opez, J., et al. (2004). *Trichomonas vaginalis*: Characterization of a 39-kDa cysteine proteinase found in patient vaginal secretions. *Experimental Parasitology, 107,* 125–135.

Hollingdale, M. R., Leland, P., Leef, J. L., et al. (1983). Serological reactivity of in vitro cultured exoerythrocytic stages of *Plasmodium berghei* in indirect immunofluorescent or immunoperoxidase antibody tests. *American Journal of Tropical Medicine and Hygeine, 32,* 24–30.

Horne-Debets, J. M., Faleiro, R., Karunarathne, D. S., et al. (2013). PD-1 dependent exhaustion of CD8+ T cells drives chronic malaria. *Cell Reports, 5,* 1204–1213.

Huet, G., Richet, C., Demeyer, D., et al. (1992). Characterization of different proteolytic activities in *Trypanosoma brucei brucei. Biochimica et Biophysica Acta, 1138,* 213–221.

Iles, K. E., & Forman, H. J. (2002). Macrophage signaling and respiratory burst. *Immunologic Research, 26,* 95.

Imai, T., Ishida, H., Suzue, K., et al. (2015). Cytotoxic activities of CD8 (+) T cells collaborate with macrophages to protect against blood stage murine malaria. *Elife, 4,* e04232.

Jacobs, D., Dubremetz, J. F., Loyens, A., et al. (1998). Identification and heterologous expression of a new dense granule protein (GRA7) from *Toxoplasma gondii. Molecular and Biochemical Parasitology, 91,* 237–249.

Kulkarni, M. M., Jones, E. A., McMaster, W. R., et al. (2008). Fibronectin binding and proteolytic degradation by *Leishmania* and effects on macrophage activation. *Infection and Immunity, 76,* 1738–1747.

Laclette, J. P., Alagón, A., Willms, K., et al. (1990). Purification of antigen B from *Taenia solium* cysticerci by affinity to mammalian collagen. *Journal of Parasitology, 76,* 273–275.

Laclette, J. P., Shoemaker, C. B., Richter, D., et al. (1992). Paramyosin inhibits complement C1. *Journal of Immunology, 148,* 124–128.

Leippe, M., Andra, J., & Müller-Eberhard, H. J. (1994). Cytolytic and antibacterial activity of synthetic peptides derived from amoebapore, the pore- forming peptide of *Entamoeba histolytica. Proceedings of the National Academy of Sciences of the United States of America, 91,* 2602–2606.

Li, E., Yang, W. G., Zhang, T., et al. (1995). Interaction of laminin with *Entamoeba histolytica* cysteine proteinases and its effect on amebic pathogenesis. *Infection and Immunity, 63,* 4150–4153.

Li, Y., Liu, Y., Xiu, F., et al. (2018). Characterization of exosomes derived from *Toxoplasma gondii* and their functions in modulating immune responses. *International Journal of Nanomedicine, 13,* 467–477.

Lidell, M. E., Moncada, D. M., Chadee, K., et al. (2006). *Entamoeba histolytica* cysteine proteases cleave the MUC2 mucin in its C-terminal domain and dissolve the protective colonic mucus gel. *Proceedings of the National Academy of Sciences of the United States of America, 103,* 9298–9303.

Lima, T. S., & Lodoen, M. B. (2019). Mechanisms of human innate immune evasion by *Toxoplasma gondii. Frontiers in Cellular and Infection Microbiology, 9,* 103.

Loukas, A., & Prociv, P. (2001). Immune responses in hookworm infections. *Clinical Microbiology Reviews, 4,* 689–703.
Ma, L., Chen, K., Meng, Q., et al. (2011). An evolutionary analysis of trypanosomatid GP63 proteases. *Parasitology Research, 109,* 1075–1084.
MacFarlane, R. C., & Singh, U. (2007). Identification of an *Entamoeba histolytica* serine-, threonine-, and isoleucine-rich protein with roles in adhesion and cytotoxicity. *Eukaryotic Cell, 6,* 2139–2146.
Mantel, P. Y., Hoang, A. N., Goldowitz, I., et al. (2013). Malaria-infected erythrocyte-derived microvesicles mediate cellular communication within the parasite population and with the host immune system. *Cell Host & Microbe, 13,* 521–534.
Marino, A. P., Silva, A. A., Pinho, R. T., et al. (2003). *Trypanosoma cruzi* infection: A continuous invader host cell cross talk with participation of extracellular matrix and adhesion and chemoattractant molecules. *Brazilian Journal of Medical and Biological Research, 36,* 1121–1133.
McGwire, B. S., Chang, K. P., & Engman, D. M. (2003). Migration through the extracellular matrix by the parasitic protozoan *Leishmania* is enhanced by surface metalloprotease gp63. *Infection and Immunity, 71,* 1008–1010.
Mendoza-Lopez, M. R., Becerril-Garcia, C., Fattel- Facenda, L. V., et al. (2000). CP30, a cysteine proteinase involved in *Trichomonas vaginalis* cytoadherence. *Infection and Immunity, 68,* 4907–4912.
Mitra, B. N., Saito-Nakano, Y., Nakada-Tsukui, K., et al. (2007). Rab11B small GTPase regulates secretion of cysteine proteases in the enteric protozoan parasite *Entamoeba histolytica. Cellular Microbiology, 9,* 2112–2125.
Mũnoz, M., Heimesaat, M. M., Danker, K., et al. (2009). Interleukin (IL)-23 mediates *Toxoplasma gondii*-induced immunopathology in the gut via matrix metalloproteinase-2 and IL-22 but independent of IL-17. *Journal of Experimental Medicine, 206,* 3047–3059.
Murta, A. C. M., Persechini, P. M., De Souto, Padron T., et al. (1990). Structural and functional identification of GP57/51 antigen of *Trypanosoma cruzi* as a cysteine proteinase. *Molecular and Biochemical Parasitology, 43,* 27–38.
Nagase, H., Visse, R., & Murphy, G. (2006). Structure and function of matrix metalloproteinases and TIMPs. *Cardiovascular Research, 69,* 562–573.
Nakada-Tsukui, K., Saito-Nakano, Y., Ali, V., et al. (2005). A retromerlike complex is a novel Rab7 effector that is involved in the transport of the virulence factor cysteine protease in the enteric protozoan parasite *Entamoeba histolytica. Molecular Biology of the Cell, 11,* 5294–5303.
Nakada-Tsukui, K., Tsuboi, K., Furukawa, A., et al. (2012). A novel class of cysteine protease receptors that mediate lysosomal transport. *Cellular Microbiology, 14,* 1299–1317.
Nde, P. N., Simmons, K. J., Kleshchenko, Y. Y., et al. (2006). Silencing of the larninin γ-1 gene blocks *Trypanosoma cruzi* infection. *Infection and Immunity, 74,* 1643–1648.
Niehus, S., Elass, E., Coddeville, B., et al. (2012). Glycosylphosphatidylinositols of *Toxoplasma gondii* induce matrix metalloproteinase-9 production and degradation of galectin-3. *Immunobiology, 217,* 61–64.
Nóbrega, O. T., Santana, J. M., Sturm, N. R., et al. (2004). HygR and PurR plasmid vectors for episomal transfection of *Trypanosoma cruzi. Memórias do Instituto Oswaldo Cruz, 99,* 513–516.
Nogueira de Melo, A. C., de Souza, E. P., Elias, C. G. R., et al. (2010). Detection of matrix metallopeptidase-9-like proteins in *Trypanosoma cruzi. Experimental Parasitology, 125,* 256–263.
Nowak, N., Lotter, H., Tannich, E., et al. (2004). Resistance of *Entamoeba histolytica* to the cysteine proteinase inhibitor E64 is associated with secretion of pro enzymes and reduced pathogenicity. *Journal of Biological Chemistry, 37,* 38260–38266.
Okada, M., Huston, C. D., Oue, M., et al. (2006). Kinetics and strain variation of phagosome proteins of *Entamoeba histolytica* by proteomic analysis. *Molecular and Biochemical Parasitology, 145,* 171–183.

Olias, P., Etheridge, R. D., Zhang, Y., et al. (2016). *Toxoplasma* effector recruits the Mi-2/NuRD complex to repress STAT1 transcription and block IFN-g-dependent gene expression. *Cell Host & Microbe, 20,* 72–82.

Olivier, M., Atayde, V. D., Isnard, A., et al. (2012). *Leishmania* virulence factors: Focus on the metalloprotease GP63. *Microbes and Infection, 14,* 1377.

Olmos-Ortiz, L. M., Barajas-Mendiola, M. A., Barrios-Rodiles, M., et al. (2017). *Trichomonas vaginalis* exosome-like vesicles modify the cytokine profile and reduce inflammation in parasite infected mice. *Parasite Immunology, 39,* e12426.

Ortega-Barria, E., & Pereira, M. E. A. (1991). A novel *T. cruzi* heparin-binding protein promotes fibroblast adhesion and penetration of engineered bacteria and trypanosomes into mammalian cells. *Cell, 67,* 411–421.

Pennington, L. F., Alouffi, A., Mbanefo, E. C., et al. (2017). H-IPSE is a pathogen secreted host nucleus-infiltrating protein (infiltrin) expressed exclusively by the *Schistosoma haematobium* egg stage. *Infection and Immunity, 12,* e00301–e00317.

Perdomo, D., Baron, B., Rojo-Domínguez, A., et al. (2013). The α helical regions of KERP1 are important in *Entamoeba histolytica* adherence to human Cells. *Scientific Reports, 3,* 1171–1181.

Perdomo, D., Manich, M., Syan, S., et al. (2016). Intracellular traffic of the lysine and glutamic acid rich protein KERP1 reveals features of endomembrane organization in *Entamoeba histolytica. Cellular Microbiology, 8,* 1134–1152.

Podoba, J. E., & Stevenson, M. M. (1991). CD4+ and CD8+ T lymphocytes both contribute to acquired immunity to blood-stage *Plasmodium chabaudi* AS. *Infection and Immunity, 59,* 51–58.

Pope, S. M., & Lasser, C. (2013). *Toxoplasma gondii* infection of fibroblasts causes the production of exosome-like vesicles containing a unique array of mRNA and miRNA transcripts compared to serum starvation. *Journal of Extracellular Vesicles, 2,* 22484.

Pradel, G., Garapaty, S., & Frevert, U. (2002). Proteoglycans mediate malaria sporozoite targeting to the liver. *Molecular Microbiology, 45,* 637–651.

Prato, M., Giribaldi, G., Polimeni, M., et al. (2005). Phagocytosis of hemozoin enhances matrix metalloproteinase-9 activity and TNF-*α* production in human monocytes: Role of matrix metalloproteinases in the pathogenesis of falciparum malaria. *Journal of Immunology, 175,* 6436–6442.

Reed, S. L., Ember, J. A., Herdman, D. S., et al. (1995). The extracellular neutral cysteine proteinase of *Entamoeba histolytica* degrades anaphylatoxins C3a and C5a. *Journal of Immunology, 155,* 266–274.

Regev-Rudzki, N., Wilson, D. W., Carvalho, T. G., et al. (2013). Cell-cell communication between malaria-infected red blood cells via exosome-like vesicles. *Cell, 153,* 1120–1133.

Rico, G., Leandro, E., Rojas, S., et al. (2003). The monocyte locomotion inhibitory factor produced by *Entamoeba histolytica* inhibits induced nitric oxide production in human leukocytes. *Parasitology Research, 90,* 264–267.

Saeij, J. P., Boyle, J. P., Coller, S., et al. (2006). Polymorphic secreted kinases are key virulence factors in toxoplasmosis. *Science, 314,* 1780–1783.

Saeij, J. P., Coller, S., Boyle, J. P., et al. (2007). *Toxoplasma* co-opts host gene expression by injection of a polymorphic kinase homologue. *Nature, 445,* 324–327.

Safeukui, I., Gomez, N. D., Adelani, A. A., et al. (2015). Malaria induces anemia through CD8+ T cell-dependent parasite clearance and erythrocyte removal in the spleen. *MBio, 6,* e02493–e02514.

Santana, J. M., Grellier, P., Schr´evel, J., et al. (1997). A *Trypanosoma cruzi*-secreted 80 kDa proteinase with specificity for human collagen types I and IV. *Biochemical Journal, 325,* 129–137.

Santi-Rocca, J., Weber, C., Guigon, G., et al. (2008). The lysine- and glutamic acid rich protein KERP1 plays a role in *Entamoeba histolytica* liver abscess pathogenesis. *Cellular Microbiology, 10,* 201–217.

Sato, H., Takino, T., & Miyamori, H. (2005). Roles of membrane type matrix metalloproteinase-1 in tumor invasion and metastasis. *Cancer Science, 96,* 212–217.

Schneider, I., Haller, D., Kullmann, B., et al. (2007). Identification, molecular characterization and subcellular localization of a *Theileria annulata* parasite protein secreted into the host cell cytoplasm. *Parasitology Research, 101,* 1471–1482.
Seigneur, M., Mounier, J., Prevost, M. C., et al. (2005). A lysine- and glutamic acid-rich protein, KERP1, from *Entamoeba histolytica* binds to human enterocytes. *Cellular Microbiology, 7,* 569–579.
Seipel, D., de Lima, Oliveira B. C., Resende, T. L., et al. (2010). *Toxoplasma gondii* infection positively modulates the macrophages migratory molecular complex by increasing matrix metalloproteinases, CD44 and alpha v beta3 integrin. *Veterinary Parasitology, 169,* 312–319.
Sen, A., Chatterjee, N. S., Akbar, M. A., et al. (2007). The 29-kilodalton thiol-dependent peroxidase of *Entamoeba histolytica* is a factor involved in pathogenesis and survival of the parasite during oxidative stress. *Eukaryotic Cell, 6,* 664–673.
Serrano-Luna, J., Pi˜na-V´azquez, C., Reyes-L´opez, M., et al. (2013). Proteases from *Entamoeba* spp. and pathogenic free-living amoebae as virulence factors. *Journal of Tropical Medicine, 2013,* 890603.
Shipley, J. M., Wesselschmidt, R. L., Kobayashi, D. K., et al. (1996). Metalloelastase is required for macrophage-mediated proteolysis and matrix invasion in mice. *Proceedings of the National Academy of Sciences of the United States of America, 93,* 3942–3946.
Shrimal, S., Bhattacharya, S., & Bhattacharya, A. (2010). Serum-dependent selective expression of EhTMKB1-9, a member of *Entamoeba histolytica* B1 family of transmembrane kinases. *PLoS Pathogens, 6,* e1000929.
Sim, S., Yong, T. S., Park, S. J., et al. (2005). NADPH oxidase-derived reactive oxygen species-mediated activation of ERK1/2 is required for apoptosis of human neutrophils induced by *Entamoeba histolytica. Journal of Immunology, 174,* 4279–4288.
Singh, A. P., Buscaglia, C. A., Wang, Q., et al. (2007). *Plasmodium* circumsporozoite protein promotes the development of the liver stages of the parasite. *Cell, 131,* 492–504.
Souto-Padron, T., Campetella, O. E., Cazzulo, J. J., et al. (1990). Cysteine proteinase in *Trypanosoma cruzi*: Immunocytochemical localization and involvement in parasite-host cell interaction. *Journal of Cell Science, 96,* 485–490.
Stijlemans, B., Caljon, G., Van Den Abbeele, J., et al. (2016). Immune evasion strategies of *Trypanosoma brucei* within the mammalian host: Progression to pathogenicity. *Frontiers in Immunology, 7,* 233.
Sun, T., Holowka, T., Song, Y., et al. (2012). A *Plasmodium*-encoded cytokine suppresses T-cell immunity during malaria. *Proceedings of the National Academy of Sciences of the USA, 31,* 2117–2126.
Swan, D. G., Phillips, K., Tait, A., et al. (1999). Evidence for localization of a *Theileria* parasite AT hook DNA-binding protein to the nucleus of immortalized bovine host cells. *Molecular and Biochemical Parasitology, 101,* 117–129.
Swan, D. G., Stern, R., McKellar, S., et al. (2001). Characterisation of a cluster of genes encoding *Theileria annulata* AT hook DNA-binding proteins and evidence for localisation to the host cell nucleus. *Journal of Cell Science, 114,* 2747–2754.
Teixeira, J. E., Sateriale, A., Bessoff, K. E., et al. (2012). Control of *Entamoeba histolytica* adherence involves metallosurface protease 1, an M8 family surface metalloprotease with homology to leishmanolysin. *Infection and Immunity, 80,* 2165–2176.
Thibeaux, R., Weber, C., Hon, C. C., et al. (2013). Identification of the virulence landscape essential for *Entamoeba histolytica* invasion of the human colon. *PLoS Pathogens, 12,* e1003824.
Tillack, M., Nowak, N., Lotter, H., et al. (2006). Increased expression of the major cysteine proteinases by stable episomal transfection underlines the important role of EhCP5 for the pathogenicity of *Entamoeba histolytica. Molecular and Biochemical Parasitology, 149,* 58–64.
Tillack, M., Biller, L., Irmer, H., et al. (2007). The *Entamoeba histolytica* genome: Primary structure and expression of proteolytic enzymes. *BMC Genomics, 8,* 170.
Twu, O., de Miguel, N., Lustig, G., et al. (2013). *Trichomonas vaginalis* exosomes deliver cargo to host cells and mediate host ratio parasite interactions. *PLoS Pathogens, 9,* e1003482.

van der Heyde, H. C., Manning, D. D., & Weidanz, W. P. (1993). Role of CD4+ T cells in the expansion of the CD4-, CD8-gamma delta T cell subset in the spleens of mice during blood-stage malaria. *Journal of Immunology, 151,* 6311–6317.

Visse, R., & Nagase, H. (2003). Matrix metalloproteinases and tissue inhibitors of metalloproteinases: Structure, function, and biochemistry. *Circulation Research, 92,* 827–839.

Wang, W., & Chadee, K. (1995). *Entamoeba histolytica* suppresses gamma interferon-induced macrophage class II major histocompatibility complex Ia molecule and I-A beta mRNA expression by a prostaglandin E2-dependent mechanism. *Infection and Immunity, 63,* 1089–1094.

Wang, L., Li, Z., Shen, J., et al. (2015). Exosome-like vesicles derived by *Schistosoma japonicum* adult worms mediates M1 type immune- activity of macrophage. *Parasitology Research, 114,* 1865–1873.

Williams, A. G., & Coombs, G. H. (1995). Multiple protease activities in *Giardia intestinalis* trophozoites. *International Journal for Parasitology, 25,* 771–778.

Yao, C. (2010). Major surface protease of trypanosomatids: One size fits all? Infect. *Immunology, 78,* 22–31.

Zhu, L., Liu, J., Dao, J., et al. (2016). Molecular characterization of *S. japonicum* exosome-like vesicles reveals their regulatory roles in parasite-host interactions. *Scientific Reports, 6,* 25885.

Chapter 15
The Role of Host PKCα During Intracellular *Cryptosporidium* Infection

Sayo McCowin and Chelsea Marie

Abstract Protein Kinase C-α (PKCα) is a serine/threonine kinase that has recently been associated with increased susceptibility in several enteric infections through regulation of host cell actin cytoskeleton. In the scope of *Cryptosporidium* infection, changes in PKCα expression and activity have been reported in vitro. Past studies have shown *Cryptosporidium* requires host cell actin polymerization during infection with the identification of some important mediators, [e.g. c-Src, PI3K, Cdc42, N-WASP, Arp 2/3 complex], however a potential mechanism between PKCα and regulation of actin has not been established. Hence, the aim of this review is to present an overview of the progress made thus far attempting to understand changes in host cell characteristics during *Cryptosporidium* infection and to propose the role of PKCα during this process. We report on findings in the field using techniques such as immunofluorescent imaging, RNA-seq, and transmission electron microscopy (TEM). Although the data linking PKCα and *Cryptosporidium* is limited in the field, the mechanism outlined is promising for PKCα as a potential target for treatment of cryptosporidiosis.

Keywords Cryptosporidiosis · *Cryptosporidium* · Protein kinase C · Host-pathogen interactions · Intracellular invasion · Actin remodeling · Host targeted therapeutics

Abbreviations

PKC	Protein Kinase C
PKCα	Protein Kinase C alpha-protein
CDPK1	Calcium-dependent protein kinase 1
PKD3	Protein Kinase D3
RNA-seq	(RNA-sequencing

S. McCowin · C. Marie (✉)
Department of Medicine: Infectious Diseases and International Health, University of Virginia School of Medicine, Charlottesville, VA, USA
e-mail: csm8r@virginia.edu

© Springer Nature Switzerland AG 2020

N. Guillen (ed.), *Eukaryome Impact on Human Intestine Homeostasis and Mucosal Immunology*, https://doi.org/10.1007/978-3-030-44826-4_15

TEM	Transmission electron microscopy
PRKCA	Protein Kinase C alpha-gene
BKI	Bumped-kinase Inhibitors
c-Src	Proto-oncogene tyrosine-protein kinase Src
PI3K	Phosphoinositide 3-kinase
Cdc42	Cell Division Cycle 42
N-WASP	Neural Wiskott-Aldrich syndrome protein
Arp 2/3	Actin-related protein 2/3
C. parvum	*Cryptosporidium parvum*
C. hominis	*Cryptosporidium hominis*
PKCβ	Protein Kinase C beta
JAK2	Janus Kinase 2
Chk1/2	Checkpoint Kinase 1/2
SI	Small intestinal
PKCε	Protein Kinase C epsilon
F-actin	Filamentous actin
GTPase	Guanosine-5′-triphosphate hydrolase enzyme
GEF	Guanine nucleotide exchange factor
Aqp1	Aquaporin 1
Sglt1	Sodium/Glucose transporter 1
HTT	Host-targeted therapies
FDA	Food and drug administration
IEC	Intestinal epithelial cell
EC_{50}	Effective concentration of inhibitor that reduced infection by half
IC_{50}	Inhibitory concentration where enzyme binding is reduced by half

Host Factors in Susceptibility to *Cryptosporidium*

As an obligate intracellular parasite, *Cryptosporidium* depends on host cell machinery to complete its lifecycle. Enteric infection is initiated by ingestion of oocysts and subsequent excystation of sporozoites in the small intestine. Sporozoites use gliding motility to reach and adhere to epithelial cells of the brush border. Parasite adherence induces rapid uptake and encapsulation of parasites in a dual membrane parasitophorous vacuole on the surface of the infected epithelial cells. The host-parasite interface is marked by a dense band of polymerized actin with an adjacent filamentous network (Bonnin et al. 1999; Elliott et al. 2001) creating an extracytoplasmic yet intramembranous niche for intracellular parasites within infected epithelial cells. To date, the host cell signaling cascades and proteins that are usurped by parasites during invasion and replication are not well understood. To identify novel host factors involved in *Cryptosporidium* infection we compared the genotypes of children living in an endemically exposed area of Bangladesh who contracted cryptosporidiosis to those that did not (177 cases, 866 controls). This genome-wide study identified an intronic region in the gene encoding host protein kinase C alpha (*PRKCA*) that was significantly associated with susceptibility to cryptosporidiosis in the first year of life (Wojcik et al. 2019). Multiple lines of evidence from the literature support the biological plausibility of this association (Hashim et al. 2006; Love et al. 2017; Liu et al. 2018). However, the direct role of host PKCα during *Cryptosporidium* infection remains to be definitively determined. Here we review past experimental evidence supporting the importance of host PKCα in cryptosporidiosis.

Protein Kinase C: Inhibitor Studies and High-Throughput Drug Screens

While the role of host cell PKCα has not been directly investigated, several lines of evidence suggest that PKCα may also be a critical host factor for the establishment of intracellular *Cryptosporidium* infection. Two studies have independently characterized the anticryptosporidial activity of the PKCα inhibitor Gö6976. Gö6976 is a potent inhibitor of protein kinase C (isotypes α and β) as well as the tyrosine kinases JAK2 and the checkpoint kinases Chk1/2 (Grandage et al. 2006). One study characterized the effects of a variety of kinase inhibitors on *Cryptosporidium* infection of HCT-8 and primary epithelial cells (Hashim et al. 2006). Staurosporine, a general protein kinase inhibitor, completely abolished entry of *C. parvum* and *C. hominis* into both HCT-8 cells and primary epithelial cells (Hashim et al. 2006). In primary cells, dual inhibitor experiments found that staurosporine acted primarily via inhibition of PKC signaling. However, in HCT-8 cells the strong inhibitory effects of staurosporine were only partly recapitulated by other PKC inhibitors and appeared to also involve the induction of apoptosis. The more specific PKC α/β inhibitor, Gö6976, reduced invasion of *C. hominis* and *C. parvum* invasion by ~50% for both HCT 8 and pri mary human intestinal epithelial cells (Table 15.1). Interestingly, none of the other PKC inhibitors tested had such significant effects on parasite invasion. The kinase inhibitors calphostin C and chelerythrine chloride had inhibitory effects on invasion of HCT-8 cells by *C. hominis* and *C. parvum* respectively, but neither had activity in primary epithelial cells. These results suggest that distinct host signaling pathways may be activated depending on the *Cryptosporidium* species and host cell-type.

The observation that pharmacologic inhibitors of host PKC have anticryptosporidial activity was recently independently reproduced in a high throughput phenotypic screen (Love et al. 2017). The authors identified 12 of 78,942 compounds tested with potent sub-micromolar anticryptosporidial activity, including the PKC α/β inhibitor Gö6976. Similar to the results by Hashim et al. Gö6976 strongly inhibited both *C. parvum* and *C. hominis* infection of HCT-8 cells with similar half-maximal effective concentrations (*C. parvum* $EC_{50} = 2.5$ nM $\pm$ 1.8 nM and *C. hominis* (EC_{50} $= 5$ nM $\pm$ 2.6 nM). In contrast to earlier work, Love et al. experimentally determined the EC_{50} and demonstrated it to be 10-fold lower than the single concentration tested previously. The precise determination of EC_{50} in this study further suggests that

Table 15.1 Anticryptosporidial activity of PKC inhibitors

Inhibitor	Mode of action	Inhibitory conc.	Parasite and cell type
Staurosporine		2.5 μM	100% inhibition of *C. hominis* and *C. parvum* infection of HCT-8[a] Reduced *C. hominis* invasion of primary IECs from 95.2% ± 0.2% to 60.0% ± 0.2%[a] Reduced *C. parvum* primary IECs 97.3% ± 0.2% to 60.4% ± 0.4%[a]
Gö6976	Ca^{2+}-dependent PKCα	2.5 nM ± 1.8 nM 5.0 nM ± 2.6 nM 80 nM	EC_{50} *C. parvum* HCT-8[b] EC_{50} *C. hominis* HCT-8[b] Reduced *C. hominis* infection of HCT-8 from ~55% to ~35%[a] Reduced *C. parvum* infection of HCT-8 from 91.4% ± 0.5% to 59.9% ± 2.6%[a] Reduced *C. hominis* infection of primary IECs from 95.2% ± 0.2% to 60.0% ± 0.2%[a] Reduced *C. parvum* infection of primary IECs from 97.4% ± 0.2% to 49.6% ± 1.6%[a]
Chelerythrine chloride	Inhibits the catalytic domain of PKC	660 nM	Reduced *C. parvum* infection of HCT-8 cells from 91.4% to ~46%[a] No effect on *C. hominis* infection of HCT-8
Calphostin C	Competes at the binding site of diacylglycerol and phorbol esters of PKC	5 μM 0.5 μM	*Reduced C. hominis* infection of HCT-8 from ~57% to ~38%[a] No effect on *C. parvum* infection of HCT-8 cells (tested at 10-fold lower conc.)[a]

[a]Hashim et al. (2006)
[b]Love et al. (2017)

Gö6976 may act via inhibition of PKCα which is inhibited at lower concentrations than PKCβ (PKCα IC_{50} = 2.3 nM, PKCβ IC_{50} = 6.2 nM) (Martiny-Baron et al. 1993) (Table 15.1).

The fundamental limitation of pharmacologic inhibitors studies is that it is not possible to differentiate between inhibition of host and *Cryptosporidium* targets, though the study by Hashim et al. removed inhibitors from host cells prior to infection, suggesting the mechanism of action is related to a host protein (Hashim et al. 2006). A further limitation of these studies is that many of the PKC inhibitors, including Gö6976, act on multiple targets making it difficult to discern if their anticryptosporidial activity is due to specific inhibition of host PKCα or to another similar kinase.

This point is illustrated by the recent identification of the *Cryptosporidium* calcium-dependent protein kinase 1 (CDPK1) as an attractive drug target that must also be considered in the interpretation of inhibitor studies. Bumped-kinase inhibitors (BKI) with high activity to purified CDPK1 as well as in in vitro and in vivo *Cryptosporidium* infection have been well-characterized (Castellanos-Gonzalez et al. 2016; Hulverson et al. 2017). Nonetheless, the efficacy of BKIs for their intended target has been questioned. Some BKIs have been shown to have efficacy against mammalian c-Src, which has known activity in *Cryptosporidium* invasion, further supporting the potential for off-target effects (Hulverson et al. 2017). A recent analysis found that many BKI analogs with potent activity toward purified CDPK1 had little or no effect on parasite growth in HCT-8 cells despite favorable absorption in mammalian cells. Furthermore, the IC_{50} of BKIs for purified CDPK1 had a negative correlation with the EC_{50} for blocking parasite growth suggesting these compounds may have an alternative target in parasite or host cells (Kuhlenschmidt et al. 2015).

Compound optimization has led to the development of BKIs that are reportedly >200-fold more active against parasite CDPK1 than 20 representative human kinases in vitro (Van Voorhis et al. 2017). However, the reported data show that even these optimized BKIs had the highest off-target effects on protein kinase D3 (PKD3), a member of the protein kinase D family, that is related to protein kinase C. No results for conventional PKC isoforms were reported. While both BKIs had 60–1000-fold great activity toward purified CDPK1 than purified PKD3, the EC_{50} in cell-based assays for *Cryptosporidium* was >10-fold higher than the mammalian PKD3 IC_{50} (BKI 1553) and only 2-fold lower than that of BKI 1517 (Table 15.2). This finding suggests that the efficacy of BKIs may be at least partially mediated via inhibition of host cell kinases, including PKC. A recent study found that CDPK1 knockdown decreased parasite invasion and growth by ~50% in vitro, which was comparable to BKI 1517 activity alone (Castellanos-Gonzalez et al. 2016). Combining CDPK1 knockdown with BKI 1517 treatment decreased parasite growth by ~95%. The authors interpret this finding as confirmation of the specificity of BK 1517 for parasite CDPK1, however these data do not exclude the possibility that BKI 1517 acts in an additive fashion with CDPK1 knockdown via off-target effects on host cell kinases.

Table 15.2 BKI efficacy on target CDPK1 and host PKD3

Assay	1517 Compound 39	1553 Compound 32
CDPK1 enzyme inhibition IC_{50}	1 nM (Schaefer et al. 2016)	2 nM (Schaefer et al. 2016)
Cell based assay EC_{50}	600 nM (Schaefer et al. 2016)	1600 nM (Schaefer et al. 2016)
PKD3 enzyme inhibition IC_{50}	1110 nM (Vidadala et al. 2016)	120 nM (Zhang et al. 2013)

Downregulation of *PRKCA* Expression by *Cryptosporidium*

In a more recent study, *PRKCA* was identified as the most significantly down-regulated gene in a transcriptomic analysis of HCT-8 cells infected with *C. parvum* (Liu et al. 2018), providing an independent line of evidence for the involvement of PKCα in *Cryptosporidium* infection. In this study, HCT-8 cells were infected with *C. parvum* and differentially expressed genes were identified by RNA-seq 24 h post-infection. In addition to *PRKCA*, other differentially regulated genes in this analysis corresponded to inflammation, anti-apoptosis, and initiation and regulation of mucosal response pathways. As inhibitor studies have established a compelling link between blockade of host PKCα and decreased parasite invasion, this may be a protective response to prevent further infection. We hypothesize that down-regulation of *PRKCA* decreases parasite-induced actin rearrangements that are necessary for invasion and intracellular survival. An important note, the study showed opposite trends of mRNA expression during infection of two different *C. parvum* subtypes highlighting different pathogenic mechanisms between species and subspecies, however down-regulation of *PRKCA* remained consistent.

A recent study assessed the transcriptomic landscape of epithelial human organoids infected with *C. parvum* which support the full life cycle of the parasite up to 72 h post-infection (Heo et al. 2018). In small intestinal (SI) organoids at 72 h post-infection, there is significant upregulation of several genes which were not observed at 24 h post-infection. Numerous changes in host cell transcriptome in response to infection were described, in addition to differential regulation at 24- and 72-h post-infection, supporting distinct host transcriptional responses at various stages of infection. While *PRKCA* was not among the significantly differentially expressed genes in this analysis, a significant enrichment of upregulated genes related to 'cytoskeleton' and 'cell mobility' was identified at 24 h post-infection, of which PKCα is a key regulator (Heo et al. 2018). A paired analysis of *C. parvum* gene expression over the course on intracellular infection found an enrichment of transcripts for parasite kinases later in infection (24 h = 0%, 72 h = 5%) suggesting that parasite kinases may be necessary for progeny invasion or replication in the organoid model. It is notable that the authors did not report a significant down-regulation of *PRKCA* as observed in Liu et al. Similarly, no significant changes in expression of

PRKCA interacting proteins were observed. This may be a consequence of a low efficiency of *Cryptosporidium* infection in the stem cell-derived organoids compared to HCT-8 intestinal epithelial cells. In the context of a low burden of infection, the uninfected-dominant population of cells may dilute or confound the identification of differential gene expression due to infection. To address this challenge, Yang et al. analyzed the transcriptome of uninfected and infected cells separately (Yang et al. 2010). In this study, the authors sorted human intestinal epithelial cells for presence of *C. parvum* infection to identify differentially transcribed genes as a direct consequence of infection (Yang et al. 2010). The authors observed 31 down-regulated genes associated with the functional category termed 'cytoskeleton'. Interestingly, protein kinase C epsilon (PKCε) was significantly downregulated in infected cells compared to uninfected controls which is known to have opposing effects to PKCα on F-actin stability in the intestinal epithelium (Song et al. 2002).

Host Cytoskeleton Remodeling in *Cryptosporidium* Infection

Numerous studies have documented the requirement for host cell actin during cellular invasion by *Cryptosporidium*, which is directly regulated by PKCα (Song et al. 2002; Nakashima 2002). Both *C. parvum* and *C. hominis* colocalized with host cell actin in primary human intestinal cells during infection (Hashim et al. 2006) and induced host cell actin rearrangement in immortalized cell lines (Hashim et al. 2004). Supporting the importance of actin rearrangements, pretreatment of host cells with cytoskeletal inhibitors cytochalasin B and cytochalasin D blocked parasite invasion (Hashim et al. 2006). Mechanistic studies of *C. parvum*-induced cytoskeleton remodeling have shown that parasites rapidly trigger tyrosine phosphorylation at the site of attachment (Forney et al. 1999). Further studies identified activation of c-Src, a membrane-associated protein tyrosine kinase and phosphatidylinositol 3-kinase (PI3K) leading to subsequent recruitment of host Rho-GTPase Cdc42 and the Cdc42-associated GEF, frabin (Chen et al. 2004a). This phosphorylation cascade results in N-WASP activation and subsequent recruitment and activation of the actin nucleation and branching complex of proteins, Arp2/3 (Chen et al. 2004a, b). Inhibition of Arp2/3 (Elliott et al. 2001), PI3K (Forney et al. 1999; Chen et al. 2003), Cdc42, and N-WASP (Chen et al. 2004a) blocked parasite infection suggesting that host actin polymerization is essential for initial parasite invasion. Parasites have also been shown to induce volume increases via activation of host aquaporin 1 (Aqp1) and the sodium/glucose transporter (Sglt1) at the site of host cell attachment which is thought to additionally contribute to actin-dependent host cell invasion (Borowski et al. 2008; Chen et al. 2005). In a separate study, it was shown that silencing *AQP1* gene expression results in an impairment in the organization host cell actin in human endothelial and melanoma cell lines (Monzani et al. 2009). Based on the prior evidence we have developed a model of PKCα-actin-mediated *Cryptosporidium* invasion of host cells (Fig. 15.1).

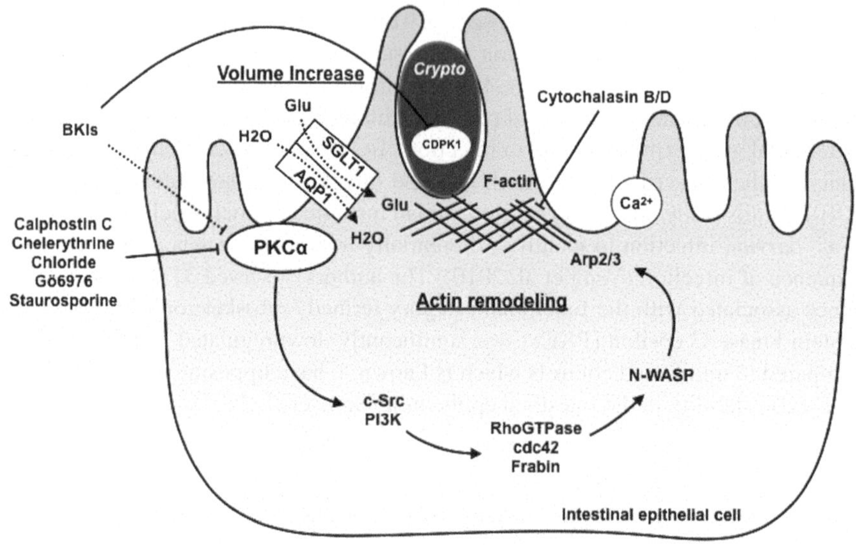

Fig. 15.1 PKCα mediated *Cryptosporidium* invasion of host cells. Contact of *Cryptosporidium* causes recruitment of host factors necessary for invasion. In our model, parasite invasion results in activation and translocation of PKCα to the site of infection at the plasma membrane. Active PKCα initiates a phosphorylation cascade in the host cytoplasm causing activation of the Arp2/3 complex and subsequent remodeling of the actin cytoskeleton. Concurrently, host cell contact with *Cryptosporidium* leads to activation of Sglt1/Aqp1 and a volume increase, from co-transport of glucose and water, further contributing to the actin remodeling required for infection. Pharmaceutical inhibition of PKCα by several drugs has been reported to decrease parasite invasion significantly. Additionally, inhibition of the recently identified CDPK1 by BKIs has been observed to significantly decrease parasite invasion potentially through off-target effects (e.g. PKCα)

PKCα has been shown to play an important role in the invasion of human endothelial cells by *Escherichia coli*. Like *Cryptosporidium*, *E. coli* induces host actin condensation at the site of host cell invasion and immunocytochemical studies indicated that activated PKCα co-localized with actin condensation at the bacterial entry site (Sukumaran et al. 2002). A similar mechanism was described in *Pseudomonas aeruginosa* invasion of middle ear epithelial cells (Mittal et al. 2016). In this study, the bacterial effector *oprF* was necessary for phosphorylation of host PKCα which translocated to the plasma membrane at the site of bacterial invasion and actin condensation. Altogether these findings support a role for *PRKCA* gene expression and resultant PKCα protein activity potentially acting via cytoskeleton remodeling in diverse enteric pathogenesis.

The finding that genetic variability in PKCα increased susceptibility to cryptosporidiosis in infants in combination with the evidence from pharmacologic and transcriptomic studies indicate the potential importance of host PKCα in infection. As an intracellular parasite, *Cryptosporidium* has minimal metabolic and energetic pathways, lacking drug targets of other parasites. An additional challenge is that drugs must penetrate both host and parasite membranes without toxicity to the host cell. Host-targeted therapies (HTTs) are a promising new class of compounds for the treatment of infectious diseases, particularly for pathogens with limited drug targets such as *Cryptosporidium,* or high levels of antimicrobial resistance. The PKC family of enzymes has been a focus of drug discovery effort due to their role in a variety of pathogenic conditions including diabetes, cancer, heart disease, Parkinson's disease, Alzheimer's disease, bipolar disease, psoriasis, and inflammatory bowel disease leading to numerous FDA-approved PKCα modulators (Yang and Yan 2014), though to our knowledge none have been tested for efficacy against *Cryptosporidium.* Our finding that genetic variation in the *PRKCA* is important in susceptibility to cryptosporidiosis provides strong foundational evidence for further exploring host PKCα as a potential target for the treatment of cryptosporidiosis.

Compliance with Ethical Standards Funding This study was funded by Gates Cryptosporidium Grant #: OPP1160655 and Cryptosporidium Field Studies Grant#: R01AI043596.

Disclosure of Interests All authors declare they have no conflicts of interest.

Human and animal studies discussed in this paper have been previously published. Informed consent for participation and publication was obtained from all individuals.

References

Bonnin, A., Lapillonne, A., Petrella, T., Lopez, J., Chaponnier, C., Gabbiani, G., et al. (1999). Immunodetection of the microvillous cytoskeleton molecules Villin and Ezrin in the parasitophorous vacuole wall of Cryptosporidium Parvum (Protozoa: Apicomplexa). *European Journal of Cell Biology, 78*(11), 794–801.

Borowski, H., Clode, P. L., & Andrew Thompson, R. C. (2008). Active invasion and/or encapsulation? A reappraisal of host-cell parasitism by cryptosporidium. *Trends in Parasitology, 24*(11), 509–516.

Castellanos-Gonzalez, A., Sparks, H., Nava, S., Huang, W., Zhang, Z., Rivas, K., et al. (2016). A novel calcium-dependent kinase inhibitor, bumped kinase inhibitor 1517, Cures Cryptosporidiosis in immunosuppressed mice. *The Journal of Infectious Diseases, 214*(12), 1850–1855.

Chen, X.-M., Huang, B. Q., Splinter, P. L., Cao, H., Zhu, G., Mcniven, M. A., et al. (2003). Cryptosporidium Parvum invasion of biliary epithelia requires host cell tyrosine phosphorylation of cortactin via C-Src. *Gastroenterology, 125*(1), 216–228.

Chen, X.-M., Huang, B. Q., Splinter, P. L., Orth, J. D., Billadeau, D. D., McNiven, M. A., et al. (2004a). Cdc42 and the actin-related protein/neural Wiskott-Aldrich syndrome protein network mediate cellular invasion by cryptosporidium parvum. *Infection and Immunity, 72*(5), 3011.

Chen, X.-M., Splinter, P. L., Tietz, P. S., Huang, B. Q., Billadeau, D. D., & LaRusso, N. F. (2004b). Phosphatidylinositol 3-kinase and frabin mediate cryptosporidium parvum cellular invasion via activation of Cdc42. *Journal of Biological Chemistry, 279*(30), 31671–31678.

Chen, X.-M., O'Hara, S. P., Huang, B. Q., Splinter, P. L., Nelson, J. B., & LaRusso, N. F. (2005). Localized glucose and water influx facilitates Cryptosporidium Parvum cellular invasion by means of modulation of host-cell membrane protrusion. *Proceedings of the National Academy of Sciences of the United States of America, 102*(18), 6338–6343.

Elliott, D. A., Coleman, D. J., Lane, M. A., May, R. C., Machesky, L. M., & Clark, D. P. (2001). Cryptosporidium parvum infection requires host cell actin polymerization. *Infection and Immunity, 69*(9), 5940–5942.

Forney, J. R., DeWald, D. B., Yang, S., Speer, C. A., & Healey, M. C. (1999). A role for host phosphoinositide 3-kinase and cytoskeletal remodeling during cryptosporidium parvum infection. *Infection and Immunity, 67*(2), 844–852.

Grandage, V. L., Everington, T., Linch, D. C., & Khwaja, A. (2006). Gö6976 is a potent inhibitor of the JAK 2 and FLT3 Tyrosine kinases with significant activity in primary acute myeloid leukaemia cells. *British Journal of Haematology, 135*(3), 303–316.

Hashim, A., Clyne, M., Mulcahy, G., Akiyoshi, D., Chalmers, R., & Bourke, B. (2004). Host cell tropism underlies species restriction of human and bovine cryptosporidium parvum genotypes. *Infection and Immunity, 72*(10), 6125–6131.

Hashim, A., Mulcahy, G., Bourke, B., & Clyne, M. (2006). Interaction of cryptosporidium hominis and cryptosporidium parvum with primary human and bovine intestinal cells. *Infection and Immunity, 74*(1), 99–107.

Heo, I., Dutta, D., Schaefer, D. A., Iakobachvili, N., Artegiani, B., Sachs, N., et al. (2018). Modeling cryptosporidium infection in human small intestinal and lung organoids. *Nature Microbiology, 3*(7), 814–823.

Hulverson, M. A., Vinayak, S., Choi, R., Schaefer, D. A., Castellanos-Gonzalez, A., Vidadala, Rama S. R., et al. (2017). Bumped-kinase inhibitors for cryptosporidiosis therapy. *The Journal of Infectious Diseases, 215*(8), 1275–1284.

Kuhlenschmidt, T. B., Rutaganira, F. U., Long, S., Tang, K., Shokat, K. M., Kuhlenschmidt, M. S., et al. (2015). Inhibition of calcium-dependent protein kinase 1 (CDPK1) in vitro by pyrazolopyrimidine derivatives does not correlate with sensitivity of cryptosporidium parvum growth in cell culture. *Antimicrobial Agents and Chemotherapy, 60*(1), 570–579.

Liu, T.-L., Fan, X.-C., Li, Y.-H., Yuan, Y.-J., Yin, Y.-L., Wang, X.-T., et al. (2018). Expression profiles of MRNA and LncRNA in HCT-8 cells infected with Cryptosporidium Parvum IId subtype. *Frontiers in Microbiology, 9,* 1409.

Love, M. S., Beasley, F. C., Jumani, R. S., Wright, T. M., Chatterjee, A. K., Huston, C. D., Schultz, P. G., & McNamara, C. W. (2017) A high-throughput phenotypic screen identifies clofazimine as a potential treatment for cryptosporidiosis. *PLoS Neglected Tropical Diseases, 11*(2).

Martiny-Baron, G., Kazanietz, M. G., Mischak, H., Blumberg, P. M., Kochs, G., Hug, H., et al. (1993). Selective inhibition of protein kinase C isozymes by the indolocarbazole Gö 6976. *Journal of Biological Chemistry, 268*(13), 9194–9197.

Mittal, R., Grati, M., Yan, D., & Liu, X. Z. (2016). Pseudomonas Aeruginosa activates PKC-alpha to invade middle ear epithelial cells. *Frontiers in Microbiology, 7.*

Monzani, E., Bazzotti, R., Perego, C., & La Porta, C. A. M. (2009). AQP1 is not only a water channel: It contributes to cell migration through Lin7/Beta-Catenin. *PLoS ONE, 4,* (7).

Nakashima, S. (2002). Protein kinase Cα (PKCα): Regulation and biological function. *The Journal of Biochemistry, 132*(5), 669–675.

Schaefer, D. A., Betzer, D. P., Smith, K. D., Millman, Z. G., Michalski, H. C., Menchaca, S. E., et al. (2016). Novel bumped kinase inhibitors are safe and effective therapeutics in the calf clinical model for cryptosporidiosis. *The Journal of Infectious Diseases, 214*(12), 1856–1864.

Song, J. C., Rangachari, P. K., & Matthews, J. B. (2002). Opposing effects of PKCα and PKCε on basolateral membrane dynamics in intestinal epithelia. *American Journal of Physiology-Cell Physiology, 283*(5), C1548–C1556.

Sukumaran, S. K., Quon, M. J., & Prasadarao, N. V. (2002). Escherichia coli K1 internalization via caveolae requires Caveolin-1 and protein kinase calpha interaction in human brain microvascular endothelial cells. *The Journal of biological chemistry, 277*(52), 50716–50724.

Voorhis, V., Wesley, C., Stone Doggett, J., Parsons, M., Hulverson, M. A., Choi, R., et al. (2017). Extended-spectrum antiprotozoal bumped kinase inhibitors: A review. *Experimental Parasitology, 180*(September), 71–83.

Vidadala, R. S., Rao, K. L., Rivas, K. K., Ojo, M. A., Hulverson, J. A., Zambriski, I. B., et al. (2016). Development of an orally available and Central Nervous System (CNS)-penetrant Toxoplasma

Gondii Calcium-Dependent Protein Kinase 1 (TgCDPK1) inhibitor with minimal Human Ether-à-Go-Go-Related Gene (HERG) activity for the treatment of toxoplasmosis. *Journal of Medicinal Chemistry, 59*(13), 6531–6546.

Wojcik, G. L., Korpe, P., Marie, C., Mychaleckyj, J., Kirkpatrick, B. D., Rich, S. S., Concannon, P., et al. (2019). Genome-wide association study of cryptosporidiosis in infants implicates PRKCA. *BioRxiv*, October, 819052.

Yang, L., & Yan, Y. (2014). Protein kinases are potential targets to treat inflammatory bowel disease. *World Journal of Gastrointestinal Pharmacology and Therapeutics, 5*(4), 209–217.

Yang, Y.-L., Buck, G. A., & Widmer, G. (2010). Cell sorting-assisted microarray profiling of host cell response to cryptosporidium parvum infection. *Infection and Immunity, 78*(3), 1040–1048.

Zhang, Z., Ojo, K. K., Vidadala, R., Huang, W., Geiger, J. A., Scheele, S., et al. (2013). Potent and selective inhibitors of CDPK1 from T. Gondii and C. Parvum based on a 5-Aminopyrazole-4-Carboxamide Scaffold. *ACS Medicinal Chemistry Letters, 5*(1), 40–44.

Chapter 16
Human Immune Response Triggered by *Entamoeba histolytica* in a 3D-Intestinal Model

Silvia Castellanos-Castro, Arturo Aguilar-Rojas, Mariette Matondo, Quentin Gian Gianetto, Hugo Varet, Odile Sismeiro, Rachel Legendre, Julien Fernandes, David Hardy, Jean Yves Coppée, Jean Christophe Olivo-Marin, and Nancy Guillen

Abstract *Entamoeba histolytica*, the agent of amoebiasis, colonizes the human colon and can invade the lining of the colon to disseminate in the deep layers of the intestine. Amoebiasis mainly affects poor people in developing countries, where the barriers between human feces and food or water are inadequate. Humans are the only reservoir of *E. histolytica* and are the sole target organism of the development of the disease, which limits our knowledge of the crosstalk between the colon and the parasite, especially during the acute phase of infection. In the present work, we

---

Castellanos-Castro and Aguilar-Rojas: These authors have equally contributed to the work.

S. Castellanos-Castro · A. Aguilar-Rojas · J. C. Olivo-Marin
Institut Pasteur, Bioimage Analysis Unit, 75015 Paris, France

S. Castellanos-Castro
Universidad Autónoma de la Ciudad de México, Colegio de Ciencias y Humanidades, Plantel Casa Libertad, Ciudad de México, Mexico

A. Aguilar-Rojas
Instituto Mexicano del Seguro Social, Unidad de Investigación Médica en Medicina Reproductiva, UMAE no. 4, Ciudad de México, Mexico

M. Matondo · Q. Gian Gianetto
Institut Pasteur, Unité de Spectrométrie de Masse pour la Biologie (MSBio), Plateforme Protéomique, Centre de Ressources et Recherches Technologiques (C2RT), USR 2000 CNRS, Paris, France

H. Varet · O. Sismeiro · R. Legendre · J. Y. Coppée
Institut Pasteur, Plateforme Transcriptome et EpiGenome, Biomics, Centre de Ressources et Recherches Technologiques (C2RT), Paris, France

Q. Gian Gianetto · H. Varet · R. Legendre
Institut Pasteur, Département de Biologie Computationnelle (USR 3756 IP CNRS), Hub Bioinformatique et Biostatistique, Paris, France

J. Fernandes
Institut Pasteur, UTechS PBI, Centre de Ressources et Recherches Technologiques (C2RT), Paris, France

D. Hardy
Institut Pasteur, Experimental Neuropathology Unit, 75015 Paris, France

© Springer Nature Switzerland AG 2020

N. Guillen (ed.), *Eukaryome Impact on Human Intestine Homeostasis and Mucosal Immunology*, https://doi.org/10.1007/978-3-030-44826-4_16

constructed an in vitro model of intestinal epithelium that includes an immune component to mimic the immune response against pathogenic microorganisms such as *E. histolytica*. Using this model and leading-edge technologies, including tissue and cell imaging, transcriptomics, proteomics and ELISA, we investigated the early stages of amoebic infection, in particular, the early immune response. The data obtained highlight the importance of several previously showed virulence markers in patients and experimental models. In addition, we underscored the involvement of other factors that appear to be key regulators of gene expression in the cellular stress responses against amoebiasis and we found novel regulatory mechanisms used by this parasite to modulate the immune response and survive within the human intestine.

Keywords *Entamoeba histolytica* · 3D-intestinal model · Cytokines · Mucin · Transcriptomics · Proteomics

Abbreviations

3D	Three-dimensional
ECM	Extracellular matrix
IL-8	Interleukin-8
TNF-α	Tumor Necrosis Factor alpha
IL-1β	Interleukin-1beta
IL-6	Interleukin-6
PMA	Phorbol 12-myristate 13-acetate
TJ	Tight junctions
AJ	Adherence junctions
PMA	Phorbol 12-myristate 13-acetate
MUC2	Mucin 2
NF-kβ	Nuclear factor-kβ
EhCP-A5	Amoebic cysteine protease A5
AP-1	Activating protein-1
IFN-γ	Interferon-γ
GM-CSF	Macrophage-colony stimulating factor
ELISA	Enzyme-linked immunosorbent assay
PAMPs	Pathogen-associated molecular patterns
LC-MS/MS proteomics	Liquid chromatography coupled to mass spectrometry
PI3K	Phosphoinositide 3-kinase
CSP	Surface amoebic protein
MMPs	Metalloproteinases

N. Guillen (✉)
Institut Pasteur, Centre National de la Recherche Scientifique-ERL9195, 25 Rue du Dr Roux, 75015 Paris, France
e-mail: nguillen@pasteur.f

MIF	Macrophage migration inhibitory factor
SHG	Second Harmonic Generation signal

Introduction

Tissue engineering is an interdisciplinary field that applies the principles of engineering and the life science towards the development of biological three-dimensional (3D) systems. Tissue substitutes are emerging as promising tools for understanding human diseases by mean of the development of reliable models of experimentation (Dutton et al. 2018). In this point, the application of a "synthetic" intestine in research work is a technical challenge due the complexity shown by this tissue; however, several research groups have been working to build in vitro models that represent specific technical advantages keeping important features of native intestine. Examples are the 3D scaffolds that mimic the small intestine constructed with different proportions of enterocytic-like cells and macrophages, as well as different extracellular matrix (ECM) compositions (Huang et al. 2014). The physiological intestinal peristaltic motion has been introduced in the "gut-on-a-chip" model by applying microfluidic techniques, (Kim et al. 2016; Shin and Kim 2018; Jalili-Firoozinezhad et al. 2019). Likewise, the structure of colon crypts has recently been recapitulated by the cultivation of human intestinal epithelium on a collagen scaffold micro fabricated with an array of crypt-like invaginations (Wang et al. 2018) and, using human-derived colonic epithelial stem cells, a thick layer of mucus has been produced in vitro (Wang et al. 2019). Although all of these substitutes for intestinal tissue are emerging as promising tools for understanding human tissues behavior in health and disease conditions, their implementation in daily laboratory work has been complex.

Entamoeba histolytica, is a human intestinal parasite which is the causative agent of amoebiasis. In humans, the motile form of the parasite known as trophozoite, is able to cause two kinds of infections. A silent infection occurs in the majority of patients. In these cases, trophozoites live in the intestine as commensals and there are no symptoms (Blessmann et al. 2003). In contrast, in 10% of infected persons, trophozoites cause intestinal disease such as colitis or colon and/or liver abscess (Marie and Petri 2014; Shirley et al. 2018). Tissue destruction and invasion of human colon by *E. histolytica* begins with the crossing of the mucus layer by trophozoites. They become then able to bind, kill and ingest epithelial cells with the concomitant release of immunological mediators (Nakada-Tsukui and Nozaki 2016). In response, goblet cells secrete mucin 2 (MUC2) via vesicular exocytosis (Cornick et al. 2017). After this step, amoebas can invade the lamina propria with the subsequent activation of an acute human inflammatory response (Ghosh et al. 2019). As a result, macrophages and neutrophils are recruited into the damaged area to secrete reactive oxygen species and nitric oxide attempting to clear the infection (Nagaraja and Ankri 2018). Among the immunological factors secreted by human cells are, Interleukin-8 (IL-8), Tumor Necrosis Factor alpha (TNF-α), Interleukin-1beta (IL-1β), and Interleukin-6 (IL-6)

(Moonah et al. 2014; Bansal et al. 2009). In response, *E. histolytica* is capable of damaging the host tissue by the secretion of virulence factors such as amoebic proteases necessary to disrupt cell-cell tight junctions (TJ) (Leroy et al. 2000; Hernández-Nava et al. 2017) and to promote cytolytic effects and apoptosis of human cells (Huston et al. 2000; Ralston and Petri 2011). Although the infection process has been studied in several in vitro models, such as cell cultures (Li et al. 1994), human tissue explants (Bansal et al. 2009), or human intestine xenografts in mice (Seydel et al. 1997), the understanding of parasite behaviors during early steps of amoebic intestinal infection and the molecules involved during human-parasite interaction is scarce, principally due to the lack of models similar to in vivo conditions.

To overcome these methodological limitations, our research group has developed an intestinal 3D model with human cells that is able to reproduce important features of the native intestine in vitro. This system is easily reproducible under conventional laboratory conditions and is suitable for evaluating different intestinal processes. Using this 3D-intestinal model, we studied the early steps of interaction between *E. histolytica* and human cells (Aguilar Rojas et al. 2020) and here, we discuss the molecules linked to the immune responses of human cells confronted to an amoebic infection.

Results

Construction of the Intestinal Model

The 3D-intestinal model consists of a basal ECM made with collagen type I from rat tail containing the non-transformed human colon fibroblast, CCD-18Co and human monocyte cell line THP-1 differentiated into macrophages (standardized ratio of both cell lines of 1:2.2) (Fig. 16.1a, b). THP-1 differentiation was performed using phorbol 12-myristate 13-acetate (PMA) and it was validated using morphology and viability indicators (Chitra et al. 2014) and CD11b surface marker (Mittar et al. 2011) (data not shown). Over the ECM was added an epithelial layer assembled by the co-culture of human cell line, Caco-2/TC7 as enterocyte-like cells and the human adenocarcinoma HT29-MTX cells as mucus-producing cells (standardized ratio of both cells: 4:1, respectively) (Fig. 16.1a). Those cell lines were selected by their specific characteristics. In the case of Caco-2/TC7 clone, they are able to express all the characteristics of Caco2 cells and in addition, they exhibit the expression of functional properties of fully differentiated intestinal enterocytes (Sambuy et al. 2005). HT29-MTX cells, were adapted from the parental cell line HT29 to acquire the morphological and mucin-producing characteristics of goblet cells (Pontier et al. 2001). Functional properties of 3D-intestinal model were evaluated after 21 days of culture. The model was able to promote epithelial polarization as was demonstrated by the immunolocalization of villin in the apical side of the epithelial cells (0–15 μm). Likewise, below the apical zone, were localized the TJ as was revealed

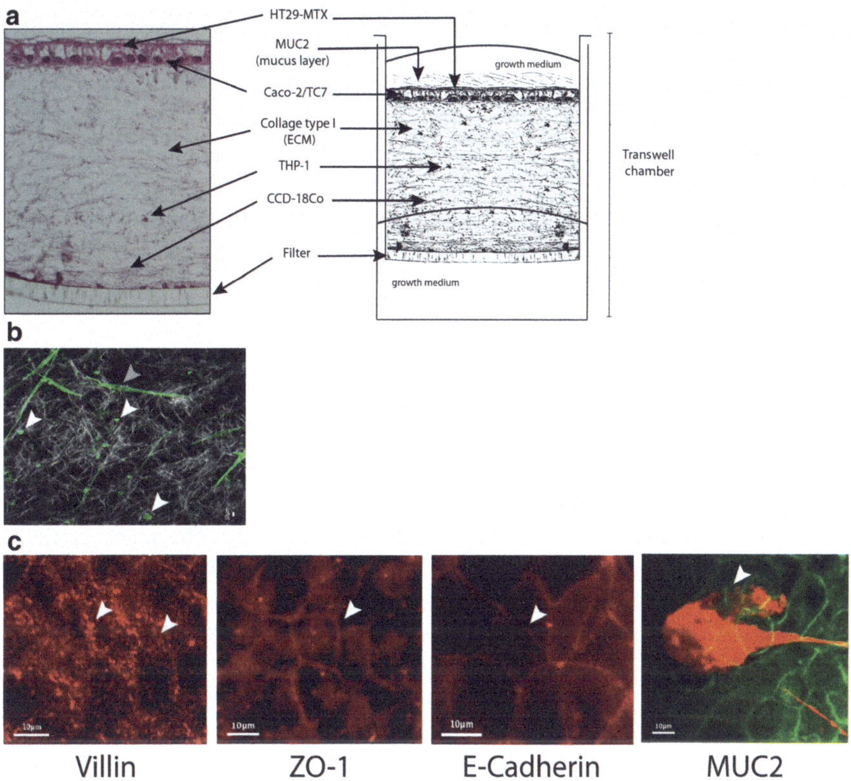

Fig. 16.1 **a** Histology of 3D-intestinal model (*left*). Cells were visualized by light microscopy. Hematoxylin and eosin (HE) were employed to stain the cells. General view of 3D-intestinal model (*right*). The device was built on the upper layer of a transwell chamber (0.4-μm pore in 24 mm well plate). Epithelial layer was composed of enterocyte-like cells (Caco-2/TC7) as epithelial cells and human colorectal cell line (HT29-MTX) as mucus-producing cells. A thin mucus layer was produced above the epithelium. In the basal compartment, lamina propria was constituted of collagen type I as extracellular matrix (ECM). **b** Macrophages [THP-1 cells, (*white arrowhead*) and fibroblasts CCD 18 Co cells (*gray arrowheads*)] were grown within the ECM. Cells were stained with fluorescent phalloidin decorating actin filaments (*green*). Collagen was detected by the Second Harmonic Generation (SHG) signal (*gray*). **c** Maturation markers. After 21 days of culture, villin (*white arrowheads*), tight junction protein ZO-1 (*white arrowheads*), and E-cadherin (*white arrowheads*), were visualized by immunofluorescence using confocal laser microscopy (*red*). The mucus layer was evaluated by immunolocalization of mucin 2 (MUC2) (*red, white arrowheads*), by confocal microscopy. F-actin decoration was revealed with fluorescent phalloidin (*green*) by confocal microscopy

by the immunolocalization of the tight junction protein ZO-1 protein (15 μm). Down TJ, adherence junctions (AJ) were found as demonstrated by the immunolocalization of E-Cadherin (25–30 μm) (Fig. 16.1c). Finally, a layer of mucus was secreted on the apical side of the epithelial-like cells as shown by confocal microscopy analysis of

MUC2 staining (Fig. 16.1c). After several experiments, the data showed that the 3D-intestinal model is reproducible, standardizable under regular laboratory conditions, and suitable for studying different intestinal processes. The model was used to perform *E. histolytica* infection assays and proteomics to determine the proteins secreted by human cells to the medium. The changes in gene expression of human cells were analyzed by transcriptomics (Aguilar Rojas et al. 2020). In addition, here were quantified the pro-inflammatory cytokines secreted to the interstitial and luminal area of the 3D-intestinal model and we also discuss the effects on the mucus barrier and the immune responses triggered by *E. histolytica* when infected the human 3D-intestinal model.

Mucus Barrier Defense

During infection, when *E. histolytica* reaches the intestinal parenchyma, it releases various enzymes into the extracellular milieu to break the tissue. The mucus barrier is critical for gut homeostasis and during amoebiasis, it is the first layer of defense which avoids the adherence and cytotoxicity of the parasites to the epithelial cells (Chadee et al. 1987). Amoeba secretes some cysteine proteases, glycosidases and β-amylase to degrade the mucus barrier (Bansal et al. 2009; Thibeaux et al. 2013). Mucus is mainly composed of different mucins, in particular MUC2, which is enriched in hydroxy amino acids that attach to the O-glycans forming a dense net-like polymer. It is also enriched in lysozymes and β-defensins secreted by Paneth cells (Cobo et al. 2018). In murine models, Goblet cells activate the secretion of MUC2 via vesicular exocytosis when *E. histolytica* is present. Furthermore, mice MUC2-/-present an exacerbated infection and colonic damage (Cornick et al. 2017).

During 3D-intestinal model infection by *E. histolytica*, using confocal images and anti-MUC2 signal, we have observed that mucin layer is diminished after two hours of infection (Fig. 16.2a), although the number of vesicles secreted by goblet cells augmented. We suggest that when trophozoites are in contact with human cells of the 3D model, they activate mucus production as defense response, and at the same time, parasite destroy it to reach epithelial cells. Furthermore, we observed many trophozoites with mucus signal-containing vesicles ingested, suggesting that amoebas also shallow pieces of the mucus layer.

Immune Response of the 3D-Intestinal Model in Contact with E. histolytica

Activation of the immune system in the intestine, by loss of tolerance for non-pathogenic commensal microbes or by pathogen infections, leads to inflammation and contributes to the development of intestinal disorders such as amoebiasis.

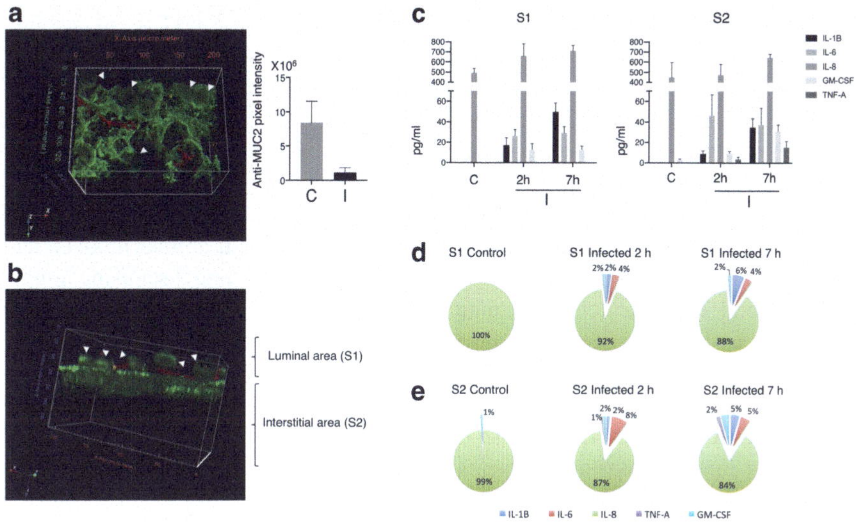

Fig. 16.2 **a** Mucus layer destruction and invasion of colon 3D model by *E. histolytica*. The micrograph was obtained by two-photon microscopy after 2 h of infection. The parasite adheres (*white arrow heads*) to cells and induces cytotoxic and cytolytic effects destroying part of the mucus and epithelial barriers. Cells were stained (*green*) with fluorescent phalloidin decorating actin filaments. The graph shows the quantification of anti-Muc2 antibody in *red*. 3D-intestinal models without parasites were used as controls (*C*), infected samples (*I*). **b** 3D reconstruction from a scanned section of the 3D-instestinal model incubated 2 h with trophozoites (*white arrow heads*). **c** Quantification and **d** relative abundance of pro-inflammatory cytokines secreted in both compartments of the device. 3D-intestinal models were incubated with trophozoites during 2 and 7 h, then the culture medium on top (*S1 fraction*) was collected. The remaining cell device was centrifuged for 5 min at 1600 g to get the medium inside (*S2 fraction*). Cytokines were quantified in both fractions by ELISA assays (Qiagen multi-analyte kit) for human IL-1α, IL-1β, IL-2, IL-6, IL-8, IL-10, IL-12, 1L-17α, INF-γ, TNF-α and GM-CSF. Experiments were made in triplicate

Immune response during infection is characterized by secretion of pro-inflammatory mediators, which in turn induce the activation of blood complement, the production of reactive nitrogen and oxygen radicals (by macrophages and neutrophils) and the cell death due to stresses (Marie and Petri 2014). Understanding the biology of the host defense responses to *E. histolytica* is important to develop new treatment strategies.

Cell-Mediated Immunity

During the invasion of the human colon, trophozoites directly and indirectly interact with human cells activating the immune response by the secretion of immunomodulatory proteins (Yu and Chadee 1997). The gut inflammatory response contributes to the infiltration of the colon by neutrophils and macrophages, which promotes tissue destruction by trophozoites. The direct contact of epithelial cells and parasites occurs through the Gal/GalNAc lectin via toll-like receptor-2/4 TLR-2/4, leading to nuclear factor-kβ (NF-kβ) signaling pathway activation and the production of pro-inflammatory cytokines, all of them important to modulate the participation of different immune cells in inflammatory responses (Moonah et al. 2014). For instance, the amoebic cysteine protease A5 (EhCP-A5) is involved in the increase of cytokine production, such as IL-1β and IL-8.

The inflammatory response induced after cell-parasite interactions have been reported in vitro using cell monolayers cultures, in vivo with animal models, and human colon explants. Taking advantage of the 3D-intestinal model set using human epithelial and immune cells, the major changes in the pro-inflammatory tissue response at early stages of amoeba infection were determined.

Using transcriptomic approaches, we highlighted the activation of different signalling pathways, such as NF-kβ, which regulates cytokine production in conjunction with other transcriptional factors also upregulated such as activating protein-1 (AP-1), Fos and Jun. In addition, the genes related to p53 transcriptional factor, which is involved in immune response during viral infection, were also found upregulated (Aguilar Rojas et al. 2020). In contrast, other signaling pathways were downregulated including interferon-γ (IFN-γ) via JAK-STAT signaling, which regulates genes related to inflammation via inflammasome, the production of oxidative response (Kopitar-Jerala 2017) and anti-microbial molecules, and those genes involved in lysosomes functions (MacMicking 2012). Then, the presence of pro-inflammatory cytokines including IL-1α, IL-1β, IL-2, IL-6, IL-8, IL-10, IL-12, 1L-17α, INF-γ, TNF-α and the granulocyte macrophage-colony stimulating factor (GM-CSF) was examined by enzyme-linked immunosorbent assay (ELISA), in the interstitial and luminal areas of the 3D-intestinal model after 2 and 7 h of infection (Fig. 16.b–e). 3D-models without infection were taken as control. In these experiments, IL-1α, IL-2, IL-10, IL-12, IL-17α and IFN-γ were not detected.

Immune Response Turned on in the 3D-Intestinal Model in Contact with *E. histolytica*

We observed that IL-1β was secreted in the interstitial area of cell layers and in the luminal part of the human 3D-intestinal model in similar concentrations and it increased from 2 to 7 h after infection. IL-1β is a pro-inflammatory cytokine associated to cell proliferation, differentiation and apoptosis. During amoebiasis, it is activated when amoebic cysteine proteases, secreted in the milieu, cause cytopathic effects on the host cells. IL-1β is also produced by macrophages when in direct contact with *E. histolytica.* Parasite-macrophage contact triggers cytoskeletal reorganization and the EhCP-A5 protease binds to α5β1 integrin of macrophages inducing the secretion of IL-1β, which in mice activates the NLRP3 inflammasome (Mortimer et al. 2015). Inflammasomes are multiprotein signaling complexes that are able to activate the inflammatory caspases and to maturate the IL-1β and IL-18. Activation of these factors is induced by Toll-like receptors, NF- kβ pathway and some characteristic pathogen-associated molecular patterns (PAMPs). During infection of the

3D-intestinal model we found some downregulated inflammasome related genes. Alterations on inflammasome activation are associated with certain autoimmune and chronic inflammatory diseases in humans (Jo et al. 2016). Thus, downregulation of the inflammasome by *E. histolytica* may be necessary for functions preventing a potentially harmful reaction for the host at early stage of infection.

Epithelial cells produce IL-6 during colonic infection and it is also known as an intestine inflammatory marker (Zavala et al. 2018). In our experiments, IL-6 was found in similar concentrations in interstitial and luminal areas of the 3D-intestinal model during infection. This finding correlates with previous reports in animals and human cells (Meurens et al. 2009; Moonah et al. 2014; Bansal et al. 2009). IL-8 was present in control and infected sample and increased during infection in interstitial and luminal areas of the 3D-intestinal model. We also found the presence of IL-8 in the secretome [by liquid chromatography coupled to mass spectrometry (LC-MS/MS proteomics)] in control and infected samples. Epithelial cells express IL-8 constitutively and they are stimulated for IL-8 increases during infection. IL-8 is a chemoattractant for neutrophils and it has been identified in high concentrations during severe diseases (Dickson-Gonzalez et al. 2009). IL-8 has other effects on endothelial cells, macrophages and mast cells (Lee et al. 2014). IL-8 also attracts the trophozoites inducing specific rearrangements of the trophozoites actin-myosin cytoskeleton components (Diaz-Valencia et al. 2015).

TNF-α was secreted only in the interstitial part of the 3D-intestinal model and it increases during time of infection. In general, TNF-α is produced by neutrophils, eosinophils, mast cells and the main source are the activated macrophages. TNF-α is a potent chemoattractant for circulating monocytes and also for *E. histolytica* (Blazquez et al. 2006) where it induces signaling through the phosphoinositide 3-kinase (PI3K) and the Gal-GalNAc lectin leading to cytoskeleton reorganization (Blazquez et al. 2008). The chemotaxis of trophozoites towards TNF-α involves a surface amoebic protein (CSP), which is enriched in uropods and sense TNF-α (Silvestre et al. 2015). In human tissues, TNF-α promotes the degradation of ECM via the metalloproteinases (MMPs) activity, a fact that facilitates the immune cells migration to the site of infection and is also essential for the tissue damage processes leading to amoebiasis (Thibeaux et al. 2014). TNF-α production by macrophages is enhanced by the homolog of the pro-inflammatory cytokine macrophage migration inhibitory factor (MIF) (Moonah et al. 2014), which is present in the 3D-intestinal model. Another cytokine found in the experiments here discussed was GM-CSF. It was secreted in the interstitial part of the control and increased levels appeared over time during infection in both sections of the 3D-model. GM-CSF is a chemokine recruiting monocytes and is involved in modulation of immune response. The presence of GM-CSF in high levels has been previously reported during porcine epithelial jejunal IPI-21 cells interacting with *E. histolytica* (Meurens et al. 2009).

Cell motility of both *E. histolytica* and immune cells is essential for tissue damage processes leading to amoebiasis. During parasite invasion, intestinal cells can recruit circulating monocytes and neutrophils by sending distress signals following secretion of pro-inflammatory cytokines (e.g. GM-CSF, IL-8 and mainly TNF-α). These cytokines allow cell infiltration and may activate MMPs to remodel the

ECM, an essential feature for parasite migration within the lamina propria. This early inflammatory response in the tissue can be considered as a signal enhancing parasite motility.

Immune Response Turned off in the 3D-Intestinal Model in Contact with *E. histolytica*

An important cytokine previously reported during amoebiasis is IFN-γ, which has been associated with the clearance of infection. IFN-γ is produced by peripheral mononuclear cells and activates neutrophils and macrophages. Natural killers T cells also secrete IFN-γ with protective roles during amoebic liver abscesses (Lotter et al. 2009). The JAK-STAT1 signaling pathway activates IFN-γ production. Intriguingly, during infection of the 3D-intestinal model, we did not find IFN-γ by ELISA and IFN-γ signaling pathways were downregulated in the transcriptome profile (Aguilar Rojas et al. 2020). IFN-γ signaling pathways are involved in the β-defensins antimicrobial program needed to control infection. Defensins are constitutively expressed in the intestine and have been considered as a part of innate immune responses according to their ability to destroy the microbe's membranes (Ulm et al. 2012). IFN-γ is also involved in the production of oxidative and nitrosative radicals in macrophages. These molecules generate high stresses during infection; host cells and trophozoites die due to free radical damage contributing to the immunopathology of amoebic dysentery (Lin and Chadee 1992). The downregulation of genes encoding β-defensins and those encoding factors triggering stress, may contribute to parasite evasion from the immune response during the early steps of infection. Furthermore, in addition to the modulation of stress triggered by macrophages via IFN-γ, we have found that the gene encoding the transcription factor TFEC is downregulated during infection of the 3D-intestinal model. TFEC, restricted to macrophages, promotes the expression of lysosomal and oxidative metabolism genes (Ploper and De Robertis 2015); the reduction of these functions implicates a limitation for human cells to fulfill their metabolism leading to slow down human cells defenses during amoebic infection. *E. histolytica* can also resist intestinal oxidative stress by several ways (Jeelani and Nozaki 2016; Varet et al. 2018; Nagaraja and Ankri 2018) and thus survive in a stressful oxidative environment (Pineda and Perdomo 2017).

By proteomics analysis of the secreted components in the 3D-intestinal model other downregulated proteins were identified. These are related to macrophage's functions, such as CAPG protein, which participates in actin filaments capping and, in this condition, can alter macrophages motility. Furthermore, other chemokines were present in low amounts during infection, such as CCL15, GDF-15 and attractin. In addition, proteins related to lysosomal function were also diminished confirming the transcriptome features above discussed (Aguilar Rojas et al. 2020). Since we did not find IFN-γ after 7 h of infection in 3D-intestinal model, neither the transcription of genes whose products are involved in oxidative or nitrosative attack nor microbe degradation via lysosomes, we suggest that *E. histolytica* may complete its strategies to evade the innate immune response by modulating IFN-γ signaling pathway and thus attempt to survive inside the tissue during the early stages of infection. The experimental exploration of the multiple routes used by *E. histolytica* to turn off the macrophage defense activity during amoebiasis is a matter of further studies.

Conclusions

The 3D-intestinal model allows us to determine and to quantify the changes of certain components of the immune response at the very early stages of an amoebic intestinal infection. The virulence factors secreted by *E. histolytica*, as well as its interaction with host cells, induce a potential inflammatory response leading to the secretion of chemokines and cytokines that can recruit immune cells by sending distress signals. These compounds also recruit amoeba promoting tissue invasion. Some of them like IL-1β, IL-6, IL-8, TNF-α and GM-CSF have been found in murine models or in patients infected with *E. histolytica* and confirmed here. In addition, our data suggest that trophozoites decrease the signalling pathways regulated by IFN-γ including the β-defensins antimicrobial responses, oxidative attack and macrophages activation. Our working hypothesis suggests that the strategies used by trophozoites to invade the intestinal tissue combine the absence of IFN-γ (promoting parasite survival at early stages) and the exploitation of pro-inflammatory mechanisms to activate MMPs leading to tissue invasion.

Acknowledgements The authors gratefully acknowledge Professor Jost Enninga from the Dynamics of Host Pathogen Interactions Unit, Institut Pasteur, Paris, France, for his kind supply of the cell lines, Caco-2/TC7 and HT29-MTX. Special thanks are due to Maria Manich, BIA Unit-Institut Pasteur, for her support in laboratory organization. The authors express their gratitude to all members of Icy software group from BIA Unit-Institut Pasteur, Paris, France, for their help in image analysis. We acknowledge the generous support of "DIM Ile de France" for the support in the acquisition of the two-photon microscope.

Funding The project has received funding from the European ERA-NET Infect-ERA program AMOEBAC (French National Agency for Research (ANR) grants ANR-14-IFEC-0001-01 and ANR-14-IFEC-0001-02). Biomics Platform (Institute Pasteur) is supported by France Génomique (ANR-10-INBS-09-09) and IBISA. AAR is a recipient of Research Career Development Awards from Fundación-IMSS, México. SCC received a postdoctoral fellowship with number 173697 from the program S190-Conacyt, 2017, Mexico.

Conflict of Interest All the authors declare no potential conflicts of interest.

References

Aguilar Rojas, A., Castellanos Castro, S., Matondo, M., Giai Gianetto, Q., Varet, H., Sismeiro, O., et al. (2020). Insights into amebiasis using a human 3D-intestinal model. *Cellular Microbiology, 16*, e13203. https://doi.org/10.1111/cmi.13203.

Bansal, D., Ave, P., Kerneis, S., Frileux, P., Boché, O., Baglin, A. C., et al. (2009). An ex-vivo human intestinal model to study Entamoeba histolytica pathogenesis. *PLOS Neglected Tropical Diseases, 3,* e551.

Blazquez, S., Guigon, G., Weber, C., Syan, S., Sismeiro, O., Coppée, J. Y., et al. (2008). Chemotaxis of Entamoeba histolytica towards the pro-inflammatory cytokine TNF is based on PI3K signalling, cytoskeleton reorganization and the Galactose/N-acetylgalactosamine lectin activity. *Cellular Microbiology, 10,* 1676–1686.

Blazquez, S., Zimmer, C., Guigon, G., Olivo-Marin, J. C., Guillén, N., & Labruyère, E. (2006). Human tumor necrosis factor is a chemoattractant for the parasite Entamoeba histolytica. *Infection and Immunity, 74,* 1407–1411.

Blessmann, J., Ali, I. K., Nu, P. A., Dinh, B. T., Viet, T. Q., Van, A. L., et al. (2003). Longitudinal study of intestinal Entamoeba histolytica infections in asymptomatic adult carriers. *Journal of Clinical Microbiology, 41,* 4745–4750.

Chadee, K., Petri, W. A., Innes, D. J., & Ravdin, J. I. (1987). Rat and human colonic mucins bind to and inhibit adherence lectin of Entamoeba histolytica. *Journal of Clinical Investigation, 80,* 1245–1254.

Chitra, S., Ganesan, N., & Lokeswari, T. S. (2014). Comparison of differentiation to macrophages in isolated monocytes from human peripheral blood and THP1 cells. *Sri Ramachandra Journal of Medicine, 7*(1).

Cobo, E. R., Holani, R., Moreau, F., Nakamura, K., Ayabe, T., Mastroianni, J. R., et al. (2018). Entamoeba histolytica alters ileal Paneth cell functions in intact and Muc2 mucin deficiency. *Infection and Immunity, 86,* e00208–e00218.

Cornick, S., Moreau, F., Gaisano, H. Y., & Chadee, K. (2017). Entamoeba histolytica-induced mucin exocytosis is mediated by VAMP8 and is critical in mucosal innate host defense. *MBio, 8,* e01317–e01323.

Diaz-Valencia, J. D., Pérez-Yépez, E. A., Ayala-Sumuano, J. T., Franco, E., & Meza, I. (2015). A surface membrane protein of Entamoeba histolytica functions as a receptor for human chemokine IL-8: its role in the attraction of trophozoites to inflammation sites. *International Journal for Parasitology, 45,* 915–923.

Dickson-Gonzalez, S. M., de Uribe, M. L., & Rodriguez-Morales, A. J. (2009). Polymorphonuclear neutrophil infiltration intensity as consequence of Entamoeba histolytica density in amebic colitis. *Surgical Infections, 10,* 91–97.

Dutton, J. S., Hinman, S. S., Kim, R., Wang, Y., & Allbritton, N. L. (2018) Primary cell-derived intestinal models: Recapitulating physiology. *Trends in Biotechnology.*

Ghosh, S., Padalia, J., & Moonah, S. (2019). Tissue destruction caused by Entamoeba histolytica parasite: Cell death, inflammation, invasion, and the gut microbiome. *Current Clinical Microbiology Reports, 6,* 51–57.

Hernández-Nava, E., Cuellar, P., Nava, P., Chávez-Munguía, B., Schnoor, M., Orozco, E., et al. (2017). Adherens junctions and desmosomes are damaged by Entamoeba histolytica: Participation of EhCPADH complex and EhCP112 protease. *Cellular Microbiology, 19,* e12761.

Huang, Z., Wang, Z., Long, S., Jiang, H., Chen, J., Zhang, J., et al. (2014). A 3-D artificial colon tissue mimic for the evaluation of nanoparticle–based drug delivery system. *Molecular Pharmaceutics, 11,* 2051–2061.

Huston, C. D., Houpt, E. R., Mann, B. J., Hahn, C. S., & Petri, W. A. (2000). Caspase 3-dependent killing of host cells by the parasite Entamoeba histolytica. *Cellular Microbiology, 2,* 617–625.

Jalili-Firoozinezhad, S., Gazzaniga, F. S., Calamari, E. L., Camacho, D. M., Fadel, C. W., Bein, A., et al. (2019). A complex human gut microbiome cultured in an anaerobic intestine-on-a-chip. *Nature Biomedical Engineering, 3,* 520–531.

Jeelani, G., & Nozaki, T. (2016). Entamoeba thiol-based redox metabolism: A potential target for drug development. *Molecular and Biochemical Parasitology, 206,* 39–45.

Jo, E. K., Kim, J. K., Shin, D. M., & Sasakawa, C. (2016). Molecular mechanisms regulating NLRP3 inflammasome activation. *Cellular & Molecular Immunology, 13,* 148–159.

Kim, H. J., Li, H., Collins, J. J., & Ingber, D. E. (2016). Contributions of microbiome and mechanical deformation to intestinal bacterial overgrowth and inflammation in a human gut-on-a-chip. *Proceedings of the National Academy of Sciences of the United States of America, 113,* E7–15.
Kopitar-Jerala, N. (2017). The role of interferons in inflammation and inflammasome activation. *Frontiers in Immunology, 8,* 873.
Lee, Y. A., Nam, Y. H., Min, A., Kim, K. A., Nozaki, T., Saito-Nakano, Y., et al. (2014). Entamoeba histolytica-secreted cysteine proteases induce IL-8 production in human mast cells via a PAR2-independent mechanism. *Parasite, 21,* 1.
Leroy, A., Lauwaet, T., De Bruyne, G., Cornelissen, M., & Mareel, M. (2000). Entamoeba histolytica disturbs the tight junction complex in human enteric T84 cell layers. *The FASEB Journal, 14,* 1139–1146.
Li, E., Stenson, W. F., Kunz-Jenkins, C., Swanson, P. E., Duncan, R., & Stanley, S. L. (1994). Entamoeba histolytica interactions with polarized human intestinal Caco-2 epithelial cells. *Infection and Immunity, 62,* 5112–5119.
Lin, J. Y., & Chadee, K. (1992). Macrophage cytotoxicity against Entamoeba histolytica trophozoites is mediated by nitric oxide from L-arginine. *Journal of Immunology, 148,* 3999–4005.
Lotter, H., González-Roldán, N., Lindner, B., Winau, F., Isibasi, A., Moreno-Lafont, M., et al. (2009). Natural killer T cells activated by a lipopeptidophosphoglycan from Entamoeba histolytica are critically important to control amebic liver abscess. *PLoS Pathogens, 5,* e1000434.
MacMicking, J. D. (2012). Interferon-inducible effector mechanisms in cell-autonomous immunity. *Nature Reviews Immunology, 12,* 367–382.
Marie, C., & Petri, W. A. (2014). Regulation of virulence of Entamoeba histolytica. *Annual Review of Microbiology, 68,* 493–520.
Meurens, F., Girard-Misguich, F., Melo, S., Grave, A., Salmon, H., & Guillén, N. (2009). Broad early immune response of porcine epithelial jejunal IPI-2I cells to Entamoeba histolytica. *Molecular Immunology, 46,* 927–936.
Mittar, D., Paramban, R.,& McIntyre, C. (2011) Flow cytometry and high-content imaging to identify markers of monocyte-macrophage differentiation. *BD Biosciences, 20.*
Moonah, S. N., Abhyankar, M. M., Haque, R., & Petri, W. A. (2014). The macrophage migration inhibitory factor homolog of Entamoeba histolytica binds to and immunomodulates host macrophages. *Infection and Immunity, 82,* 3523–3530.
Mortimer, L., Moreau, F., Cornick, S., & Chadee, K. (2015). The NLRP3 inflammasome is a pathogen sensor for invasive Entamoeba histolytica via activation of α5β1 integrin at the macrophage-amebae intercellular junction. *PLoS Pathogens, 11,* e1004887.
Nagaraja, S., & Ankri, S. (2018). Utilization of different omic approaches to unravel stress response mechanisms in the parasite Entamoeba histolytica. *Frontiers in Cellular and Infection Microbiology, 8,* 19.
Nakada-Tsukui, K., & Nozaki, T. (2016). Immune response of amebiasis and immune evasion by Entamoeba histolytica. *Frontiers in Immunology, 7,* 175.
Pineda, E., & Perdomo, D. (2017). Entamoeba histolytica under oxidative stress: What countermeasure mechanisms are in place. *Cells, 6.*
Ploper, D., & De Robertis, E. M. (2015). The MITF family of transcription factors: Role in endolysosomal biogenesis, Wnt signaling, and oncogenesis. *Pharmacological Research, 99,* 36–43.
Pontier, C., Pachot, J., Botham, R., Lenfant, B., & Arnaud, P. (2001). HT29-MTX and Caco-2/TC7 monolayers as predictive models for human intestinal absorption: Role of the mucus layer. *Journal of Pharmaceutical Sciences, 90,* 1608–1619.
Ralston, K. S., & Petri, W. A. (2011). Tissue destruction and invasion by Entamoeba histolytica. *Trends in Parasitology, 27,* 254–263.
Sambuy, Y., De Angelis, I., Ranaldi, G., Scarino, M. L., Stammati, A., & Zucco, F. (2005). The Caco-2 cell line as a model of the intestinal barrier: influence of cell and culture-related factors on Caco-2 cell functional characteristics. *Cell Biology and Toxicology, 21,* 1–26.

Seydel, K. B., Li, E., Swanson, P. E., & Stanley, S. L. (1997). Human intestinal epithelial cells produce proinflammatory cytokines in response to infection in a SCID mouse-human intestinal xenograft model of amebiasis. *Infection and Immunity, 65,* 1631–1639.

Shin, W., & Kim, H. J. (2018). Intestinal barrier dysfunction orchestrates the onset of inflammatory host-microbiome cross-talk in a human gut inflammation-on-a-chip. *Proceedings of the National Academy of Sciences of the United States of America, 115,* E10539–E10547.

Shirley, D. T., Farr, L., Watanabe, K., & Moonah, S. (2018). A review of the global burden, new diagnostics, and current therapeutics for amebiasis. *Open Forum Infectious Disease, 5,* ofy161.

Silvestre, A., Plaze, A., Berthon, P., Thibeaux, R., Guillen, N., & Labruyère, E. (2015). Entamoeba histolytica, a BspA family protein is required for chemotaxis toward tumour necrosis factor. *Microbial Cell, 2,* 235–246.

Thibeaux, R., Avé, P., Bernier, M., Morcelet, M., Frileux, P., Guillén, N., et al. (2014). The parasite Entamoeba histolytica exploits the activities of human matrix metalloproteinases to invade colonic tissue. *Nature Communications, 5,* 5142.

Thibeaux, R., Weber, C., Hon, C. C., Dillies, M. A., Avé, P., Coppée, J. Y., et al. (2013). Identification of the virulence landscape essential for Entamoeba histolytica invasion of the human colon. *PLoS Pathogens, 9,* e1003824.

Ulm, H., Wilmes, M., Shai, Y., & Sahl, H. G. (2012). Antimicrobial host defensins—specific antibiotic activities and innate defense modulation. *Frontiers in Immunology, 3,* 249.

Varet, H., Shaulov, Y., Sismeiro, O., Trebicz-Geffen, M., Legendre, R., Coppée, J. Y., et al. (2018). Enteric bacteria boost defences against oxidative stress in Entamoeba histolytica. *Scientific Reports, 8,* 9042.

Wang, Y., Kim, R., Gunasekara, D. B., Reed, M. I., DiSalvo, M., Nguyen, D. L., et al. (2018). Formation of human colonic crypt array by application of chemical gradients across a shaped epithelial monolayer. *Cellular and Molecular Gastroenterology Hepatology, 5,* 113–130.

Wang, Y., Kim, R., Sims, C. E., & Allbritton, N. L. (2019). Building a thick mucus hydrogel layer to improve the physiological relevance of in vitro primary colonic epithelial models. *Cellular and Molecular Gastroenterology Hepatology, 8,* 653–655.e5.

Yu, Y., & Chadee, K. (1997). Entamoeba histolytica stimulates interleukin 8 from human colonic epithelial cells without parasite-enterocyte contact. *Gastroenterology, 112,* 1536–1547.

Zavala, G. A., García, O. P., Camacho, M., Ronquillo, D., Campos-Ponce, M., Doak, C., et al. (2018). Intestinal parasites: Associations with intestinal and systemic inflammation. *Parasite Immunology, 40,* e12518.

Chapter 17
Receptors for Phagocytosis and Trogocytosis in *Entamoeba histolytica*

Kumiko Nakada-Tsukui and Tomoyoshi Nozaki

Abstract In the protozoan parasite *Entamoeba histolytica*, phagocytosis is considered to be indispensable for its parasitic lifestyle, and centrally involved in the pathogenesis. Currently, two modes of target internalization that can be differentiated at phenomenal and molecular levels have been demonstrated in *E. histolytica*: phagocytosis and trogocytosis. In phagocytosis, *E. histolytica* ingests a dead mammalian cell or an undeformable microorganism such as bacteria and fungi, as a whole. In contrast, in trogocytosis, live cells are nibbled (chewed) and ingested as fragments. It has been reported in *E. histolytica* that trogocytosis, not phagocytosis, is involved in immune evasion from complement attack via presenting cell surface membrane proteins acquired from host cells onto the amebic surface. Mechanistic differences at molecular levels between phagocytosis and trogocytosis have been only partially demonstrated. Only proteins known to differentiate these processes are a set of kinases, AGCK1 and AGCK2. However, what triggers phagocytosis versus trogocytosis remains largely unknown. In model organisms, it is well established that receptor for the prey is the key that regulates phagocytic internalization and the following signaling events defined by the prey. To get insights into receptor candidates for phagocytosis and trogocytosis, here we summarize previously identified potential receptors and surface molecules involved in adhesion and phagocytosis/trogocytosis in *E. histolytica*.

Keywords *Entamoeba histolytica* · Phagocytosis · Trogocytosis · Receptor

---

K. Nakada-Tsukui (✉)
Department of Parasitology, National Institute of Infectious Diseases, Tokyo, Japan
e-mail: kumiko@nih.go.jp

T. Nozaki (✉)
Department of Biomedical Chemistry, Graduate School of Medicine, The University of Tokyo, Tokyo, Japan
e-mail: nozaki@m.u-tokyo.ac.jp

© Springer Nature Switzerland AG 2020

N. Guillen (ed.), *Eukaryome Impact on Human Intestine Homeostasis and Mucosal Immunology*, https://doi.org/10.1007/978-3-030-44826-4_17

Abbreviations

AGCK	AGC family kinase
Arial	Asparagine-rich *E. histolytica* antigen
CRT	Calreticulin
Gal	Galactose
GalNAc	*N*-acetyl-D-galactosamine
LRP1	LDL receptor-related protein 1
SREHP	Serine rich *E. histolytica* protein
TMK	Transmembrane kinase

Introduction

Ancestral eukaryotes are believed to have originated from an archaeal organism which engulfed and enslaved α-proteobacterium, followed by successful endosymbiosis. Such phagocytosis-associated endosymbiosis was thus the key biological event that have led to further evolution of all eukaryotes. Phagocytosis is a fundamental process shared by majority of eukaryotic cells. It is currently well accepted that not only professional phagocytes such as neutrophils, dendritic cells, and macrophages, but somatic cells are in general capable of ingesting damaged cells for homeostasis in multicellular organisms, thus such activity is indispensable for a majority of, if not all, eukaryotes (Arandjelovic and Ravichandran 2015). Phagocytosis is a multistep process composed of: recognition of a prey, signal transduction leading to cytoskeletal rearrangement, internalization of the prey, closure of the phagosome, maturation of nascent phagosomes by fusion of lysosomes, and degradation of the prey. Several different downstream signaling pathways are initiated during internalization of different preys such as microbial pathogens, cancer cells, apoptotic and necroptotic cells (Pincetic et al. 2014; Betancur et al. 2017; Fadok et al. 1998; Zargarian et al. 2017). When the macrophage encounters opsonized pathogenic microbes, the prey is recognized by Fcγ receptor, degraded in phagosomes, and then elicits inflammatory responses. On the contrary, when the macrophage finds apoptotic cells, the prey is recognized by phosphatidylserine receptors, which leads to anti-inflammatory response. To regulate such diversified and often opposite downstream pathways, the ligand receptors that recognize the preys have a pivotal role.

In the protozoan parasite *Entamoeba histolytica*, phagocytosis is indispensable for its parasitic lifestyle, and it is also considered to be centrally involved in the pathogenesis (Orozco et al. 1983). Two modes of internalization of the preys that are differentiated at phenomenal and molecular levels have been demonstrated in *E. histolytica*: phagocytosis and trogocytosis. In phagocytosis, *E. histolytica* ingests a dead mammalian cell or an undeformable microorganism such as bacteria and fungi, as a whole. In contrast, in trogocytosis, live cells are nibbled (chewed) and ingested in fragments. It has been recently reported in *E. histolytica* that trogocytosis is also involved in immune evasion from complement attack via presenting cell surface membrane proteins acquired from host cells onto the amebic surface (Miller et al. 2019). Mechanistic differences at molecular levels between phagocytosis and trogocytosis have been only partially demonstrated. Only molecule known to differentiate these processes is a different set of kinases, AGCK1 and AGCK2 (Somlata et al. 2017). However, what triggers phagocytosis versus trogocytosis remains largely unknown. Since phagocytosis and trogocytosis were not carefully separated in most of experimental designs in the past (and often only described as "phagocytosis" all together) before identification of trogocytosis-specific AGCK1, it is possible that surface receptors and their ligands involved in target-specific recognition of phagocytosis and trogocytosis have been overlooked or misidentified. To get insights into receptor candidates for phagocytosis and trogocytosis, here we summarize previously identified potential receptors and surface molecules involved in adhesion and phagocytosis/trogocytosis in *E. histolytica*. For the existing excellent review on the molecular mechanisms of host cell killing and phagocytosis of apoptotic cells, see Marion and Guillén (2006), and Sateriale and Huston (2011); and for trogocytosis, see Ralston (2015).

Gal/GalNAc Specific Lectins

During the intestinal colonization, trophozoites bind to colonic mucin and host tissue to make their residing niche. Galactose (Gal) and N-acetyl-D-galactosamine (GalNAc) specific lectin (Gal/GalNAc lectin hereinafter) is the major surface glycoprotein and has high affinity to Gal and GalNAc, and thus was suggested to have a pivotal role in adhesion to colonic mucin, mammalian cells, erythrocytes, and some bacteria. The Gal/GalNAc lectin was originally identified by asialoorosomucoid affinity chromatography and further confirmed by affinity purification by using anti-lectin monoclonal antibodies (Fig. 17.1; Petri et al. 1987, 1989; Tannich et al. 1991). Asialoorosomucoid is a glycoprotein which has a galactose moiety at the end and was shown to efficiently inhibit *E. histolytica* adhesion to CHO cells (Petri et al. 1987). The dependence of the binding of the lectin purified from the parasite to the terminal sugars was demonstrated by using Chinese Ovary Cell (CHO)-derived glycosylation mutant strains (Ravdin et al. 1989). The Gal/GalNAc lectin is a 260-kDa heterodimeric complex composed of disulfide-linked 170-kDa heavy and 35-kDa light subunits. The complex is also noncovalently associated with a 150-kDa intermediate subunit, which was independently identified as surface antigen and then identified as a component of the Gal/GalNAc lectin (Cheng et al. 1998). Among the three subunits, the heavy subunit has a single transmembrane domain, whereas other two subunits are associated with membranes via a glycophosphatidylinositol (GPI) anchor. The purified 260-kDa heterodimer retains Gal/GalNAc-dependent binding activity to cells (Petri et al. 1990) and neoglycoproteins (Adler et al. 1995). Since the heavy subunit contains a defined carbohydrate recognition domain (CRD) (Dodson

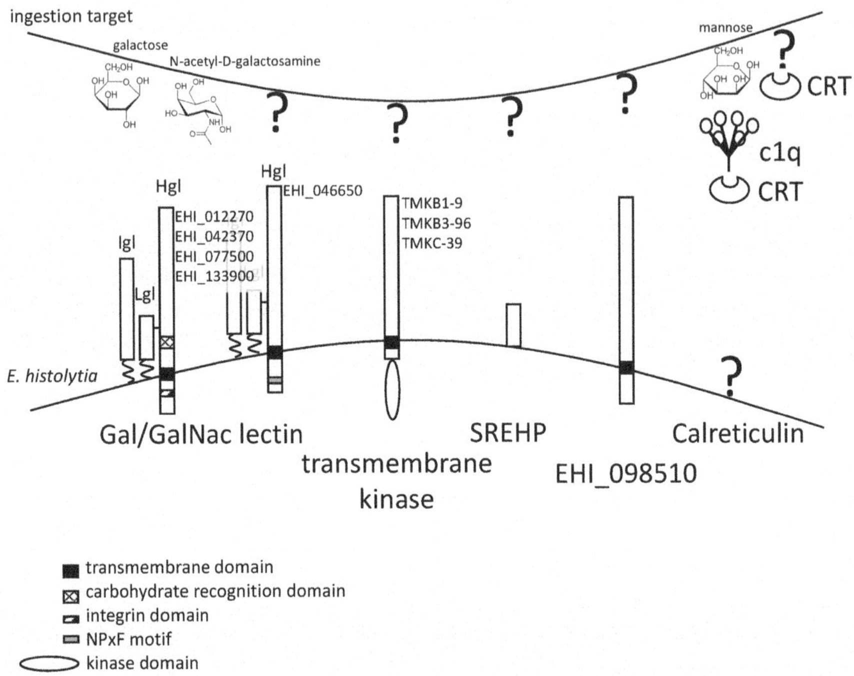

Fig. 17.1 Schematic diagram of known and predicted ligand and receptor candidates involved in adherence and internalization. Phagocytosis/trogocytosis receptor candidates described in the text are shown. Abbreviations are: Hgl—Gal/GalNAc lectin heavy subunit; Igl—intermediate subunit; Lgl—light subunit; TMK-transmembrane kinase; SREHP-serine rich *E. histolytica* protein; CRT—calreticulin

et al. 1999) and the cytosolic domain after the single transmembrane domain, unlike other two subunits, the heavy subunit was presumed to have a major role for ligand recognition and downstream signaling. In fact, expression of only the intercellular domain of the heavy subunit caused dominant negative effect, i.e., reduction of adherence to and cytotoxicity against CHO cells (Vines et al. 1998), and reduction of adherence to enterocytes, Caco-2 cells (Tavares et al. 2005); the effect was reversed by mutation in the integrin motif in the cytoplasmic domain (Vines et al. 1998). In addition, expression of an amino terminus truncated dominant negative mutant of the light subunit, or antisense-mediated repression of the light chain caused inhibition of adhesion, cytopathic effect, and erythrophagocytosis (Katz et al. 2002), which could be interpreted as indirect inhibition of the function, i.e., ligand recognition, of the heavy subunit of Gal/GalNAc lectin.

The specific role of the Gal/GalNAc lectin in adhesion versus internalization does not seem to be unequivocally demonstrated until now despite several important findings (Ravdin and Guerrant 1981; Ravdin et al. 1989; Vines et al. 1998; Huston et al. 2003; Boettner et al. 2005). It was puzzling that adhesion to CHO cells was inhibited by addition of Gal and GalNAc; however, phagocytosis of live CHO cells (or maybe mixture of trogocytosis of live cells and phatocytosis of dead cells, as explained above) was not (Ravdin and Guerrant 1981). In Similar experiments conducted using human erythrocytes and T-cell lymphoma, Jurkat cells, adhesion and phagocytosis of intact cells were inhibited by Gal, whereas only adhesion, but not internalization, was reduced when calcium ionophore-treated erythrocytes and apoptotic Jurkat cells were given to the trophozoites in the presence of Gal (Huston et al. 2003; Boettner et al. 2005). Also, as mentioned above, expression of the cytoplasmic domain of the heavy chain caused reduction of adherence to CHO cells but no effects were seen for erythrophagocytosis (Vines et al. 1998). It was also demonstrated that trogocytosis of live Jurkat cells was inhibited by monoclonal antibody against CRD in the heavy subunit (clone 3F4), suggesting its important role on the initiation of trogoytosis (Ralston et al. 2014; Ralston 2015). However, it should be noted that the monoclonal antibody was previously described as an adhesion enhancing antibody (Petri et al. 2002).

Moreover, a previous study by Heron et al. (2011) indicates that internalization of ovalbumin-coated beads is dependent on sugar(s) other than Gal. Hence, different ligands are recognized by different receptors on the *E. histolytica* surface. Overexpression of the cytoplasmic domain of the heavy subunit of Gal/GalNAc lectin reduced adherence to CHO cells, and the intergrin-like domain in the cytoplasmic tail seems to play necessary role (Vines et al. 1998). These inconsistent data on the involvement of Gal/GalNAc lectin in internalization are likely due to lack of clear understanding of the receptors on the ameba and the ligands on live and dead human cells needed for the process. However, Gal/GalNAc lectins surely have important role in adhesion to the target cells.

Our gene survey indicate that there are five genes encoding the lectin heavy subunit of Gal/GalNAc lectins in *E. histolytica* genome: EHI_012270, EHI_042370, EHI_046650, EHI_077500, and EHI_133900 (Fig. 17.1). These proteins show 31–97% mutual identities at the amino acid levels. Four proteins other than EHI_046650 are very similar to one another, showing 89–97% identities. In contrast, EHI_046650 shows only 31% identity to the other isotypes, and apparently lacks the CRD (Fig. 17.1; Dodson et al. 1999). Furthermore, EHI_046650 lacks the integrin motif, but has NPxF motif in the cytosolic tail. The NPxF motif is known as an adaptor binding site implicated for receptor internalization. Since no adaptors that are known to recognize the NPxF motif are conserved in *E. histolytica*, it is not at all understood whether EHI_046650 is and, if so, how its motif is involved in receptor internalization during phagocytosis and trogocytosis. It is worth mentioning that in a previous study where the substrate of Rhomboid protease, EhRhom1 was identified (Baxt et al. 2008), all five members of Gal/GalNAc lectin heavy subunits were ectopically expressed in COS cells. EHI_046650, but not four other members, was aggregated in the endoplasmic reticulum (ER), possibly indicating that this protein is under post-translational modifications and/or trafficking distinct from those of the other members.

Transmembrane Kinases

Over ninety transmembrane kinases (TMKs) are encoded in the *E. histolytica* genome (Beck et al. 2005). Amebic TMKs typically have the amino-terminal signal peptide, the extracellular domain, the transmembrane domain, and the intracellular kinase domain (Beck et al. 2005). Amebic TMKs have been categorized into nine subgroups (A, B1-3, C, D1-2, E and F) based on signature motifs in the kinase domain and the number of cysteine motifs in the extracellular domain (Beck et al. 2005; Sateriale and Huston 2011). This huge expansion of TMK is unique in *E. histolytica* among protozoan parasites (Naula et al. 2005). At least three TMKs, TMKB3-96 (PATMK), TMKC-39, and TMKB1-9, were shown to be involved in phagocytosis (possibly both phagocytosis and trogocytosis in a strict sense) (Fig. 17.1; Boettner et al. 2008; Buss et al. 2010; Abhyankar et al. 2012). TMKE-54 was shown to be involved in membrane targeting of Gal/GalNAc lectin (Buss et al. 2010).

TMKB3-96 was first identified from purified phagosomes containing carboxylated magnetic beads in an attempt to identify receptors involved in phagocytosis of apoptotic cells. Down regulation of the expression by short hairpin RNA or expression of a dominant negative mutant lacking the cytoplasmic kinase domain resulted in reduction of phagocytosis of calcium-treated apoptotic erythrocytes (Boettner et al. 2008). TMKC-39 was detected from phagosomes containing carboxylated latex beads (Okada et al. 2006). Overexpression of a dominant negative TMKC-39 mutant lacking the kinase domain caused reduction of phagocytosis of carboxylated beads and apoptotic Jurkat cells, but no effects were observed on phagocytosis of calcium-treated erythrocytes and endocytosis of the fluid-phase marker, dextran (Buss et al. 2010). A dominant negative form of TMKE-54 caused reduction in the amount of Gal/GalNAc lectin on the cell surface, as well as growth defect, which is consistent with the notion that TMKE-54 is involved, as mentioned above, in recycling of Gal/GalNAc lectin, and thus may be indirectly involved in both phagocytosis and trogocytosis. However, its role in cell adhesion and phago/trogocytosis was not directly examined (Buss et al. 2010).

TMKB1-9 was initially identified as a serum-responsive TMK, and its downregulation by antisense or dominant negative expression caused reduction of cell growth (Shrimal et al. 2010). It was reported that both expression of a dominant negative mutant of TMKB1-9, in which the kinase domain was deleted and repression of TMKB1-9 expression by antisense caused reduction in erythrophagocytosis, adhesion to fixed CHO cells, and killing of live cells (Shrimal et al. 2010; Abhyankar et al. 2012). TMKB1-9 was considered to be unlikely the primary receptor for erythrophagocytosis because the reduction of erythrophagocytosis by inhibition of TMKB1-9 was smaller than that of TMKB3-96 (Abhyankar et al. 2012). Fluid-phase endocytosis was also reduced by transcriptional repression; however, expression of the dominant negative mutant lacking cytoplasmic kinase domain showed no effect. It may reflect the importance of the extracellular domain of TMKB1-9 in fluid- phase endocytosis.

Altogether, only three TMKs were shown to be involved in phagocytosis: TMKB3-96, TMKB1-9, and TMKC-36, among which the ligand specificity seems to vary. TMKB3-96 and TMKB1-9 are involved in erythrophagocytosis, while TMKC-36 is involved in phagocytosis of carboxylated beads and apoptotic Jurkat cells. However, no TMKs have been shown to be specifically involved in trogocytosis, like AGCK1. As TMKs have unique extracellular domain architecture and may serve as scavenger receptors (Beck et al. 2005; Sateriale and Huston 2011), while some TMK members might interact only with dead cells uniquely marked with oxidized or nitrosylated substances. Furthermore, as *E. histolytica* trophozoites express multiple TMKs, it is also conceivable that a combination of multiple TMKs may diversify ligand specificities (Mehra et al. 2006).

SREHP

Serine-rich *E. histolytica* protein (SREHP, EHI_116360) was first discovered as surface antigen (Stanley et al. 1990, 1995), and later identified as a receptor for phagocytosis of an apoptotic cell by screening 43 monoclonal antibodies raised against cell membranes for ability to inhibit phagocytosis of an apoptotic Jurkat cell (Teixeira and Huston 2008; Sateriale and Huston 2011). SREHP, which was recognized by such antibody, 10D11, has an amino-terminal signal sequence and is associated with the plasma membrane via the carboxyl terminal hydrophobic stretch, as demonstrated by immunofluorescence and subcellular fractionation (Stanley et al. 1995; Teixeira and Huston 2008). Since no apparent transmembrane and cytoplasmic regions are present in SREHP, it may be primarily involved in adhesion to the apoptotic host cells as adhesin (Fig. 17.1). Asparagine-rich *E. histolytica* antigen surface immunogenic proteins (Ariel) also have similar domain architectures including the amino-terminal signal sequence and the carboxy-terminal hydrophobic stretch, and the plasma membrane localization (Mai and Samuelson 1998). Ariel was shown to be internalized during phagocytosis of bacteria and localized in phagosomes. There is no direct evidence for involvement of SREHP and Ariel in phagocytosis pe se, but they are potential candidates of phagocytic and trogocytic receptors.

EHI_098510 Hypothetical Protein

A gene encoding EHI_098510 was initially identified as one of six genes that were up-regulated in the trophozoites that ingested C1q-coated beads (Sateriale et al. 2012). Among the six encoded proteins analyzed, EHI_098510 is the only protein that harbors the amino-terminal signal sequence and the transmembrane domain, and thus was presumed to serve as a surface receptor (Fig. 17.1). EHI_098510 was also detected on the plasma membrane surface by proteomics, which agrees to a premise that it plays a role as a receptor for C1q-decorated apoptotic cell to initiate internalization (Biller et al. 2014). Gene silencing of EHI_098510 selectively caused reduction in phagocytosis of an apoptotic Jurkat cell, but not trogocytosis of a live Jurkat cell, adhesion to CHO cells in monolayer and Jurkat cells in suspension, and cytotoxicity to Jurkat cells (Sateriale et al. 2016), which may indicate EHI_098510 specifically involved in phagocytosis, but not trogocytosis. In Amoeba DB (release 45), there are two proteins, EHI_185450 and EHI_049650, which show high similarity to EHI_098510 (*E*-value over 6×10^{-87}), but remain uncharacterized.

Calreticulin

Calreticulin (CRT) is an ER luminal chaperone involved in quality control of glycoproteins. CRT is known to translocate to cell surface and serve as eat-me signal during apoptosis. CRT translocates to the surface of the apoptotic cell, and is subsequently recognized by LRP1 on the phagocyte (Gardai et al. 2005). CRT is a C1q binding protein, which is a complement component and also belongs to the collectin family of mannose-binding lectins (Fig. 17.1). When CRT is translocated to the surface, C1q binds to CRT, which enhances phagocytosis of the opsonized prey. As explained above, *E. histolytica* phagocytoses an apoptotic Jurkat cell in a C1q-dependent manner (Teixeira et al. 2008). CRT was shown to be concentrated at the interface between *E. histolytica* and Jurkat cell and also on the phagocytic cup of a trophozoite ingesting an erythrocyte (Vaithilingam et al. 2012). Overexpression of CRT caused enhancement of phagocytosis of apoptotic Jurkat cells and calcium ionophore-treated erythrocytes; however, no effect was observed for adhesion and cytopathy to CHO cell monolayer (Vaithilingam et al. 2012). It is to be noted that, *E. histolytica* CRT binds to human C1q, but also directly bind to apoptotic Jurkat cells independent of C1q (Fig. 17.1). Further studies are needed to clarify whether CRT and C1q function in coordination or independently during phagocytosis of apoptotic cells. So far, no evidence has been shown for the involvement of CRT in trogocytosis. Altogether, CRT and C1q are not the receptor for phagocytosis per se; however, they likely form a bridge between the receptor on *E. histolytica* and a not-yet-identified marker on dead prey (Fig. 17.1).

Conclusion

To initialize commitment toward phagocytosis versus trogocytosis, recognition of the specific marker(s) on dead or live target cells is necessary. Several candidate proteins are likely to serve as a potential receptor or an opsonin-like coreceptor, like CRT, which specifically recognize phagocytosis- and/or trogocytosis-specific ligands. Although Gal/GalNAc lectin has been implicated for adhesion, it has not been clearly shown how precisely Gal/GalNAc lectin is involved in phago- and trogocytosis, due to the complexity of the recognition and internalization mechanisms in the course of phago- and trogocytosis. It will be of interest to find if the most divergent heavy subunit of Gal/GalNAc lectin that contains the NPxF motif, as well as other candidates summarized here are selectively involved in trogocytosis, as current methods focus on screening molecules involved in phagocytosis for their role in trogocytosis. At last, we do not currently know what molecules on the live cells are responsible to transduce signal that leads to a signaling cascade for trogocytosis. Further studies are needed to identify such "nibble-me" signal(s) on the live cell and trogocytosis-specific receptor on the *E. histolytica* surface.

References

Abhyankar, M. M., Shrimal, S., Gilchrist, C. A., Bhattacharya, A., & Petri, W. A., Jr. (2012). The *Entamoeba histolytica* serum-inducible transmembrane kinase EhTMKB1-9 is involved in intestinal amebiasis. *International Journal for Parasitology: Drugs and Drug Resistance, 2,* 243–248.

Adler, P., Wood, S. J., Lee, Y. C., Lee, R. T., Petri, W. A., Jr., & Schnaar, R. L. (1995). High affinity binding of the *Entamoeba histolytica* lectin to polyvalent *N*-acetylgalactosaminides. *Journal of Biological Chemistry, 270,* 5164–5171.

Arandjelovic, S., & Ravichandran, K. S. (2015). Phagocytosis of apoptotic cells in homeostasis. *Nature Immunology, 16,* 907–917.

Baxt, L. A., Baker, R. P., Singh, U., & Urban, S. (2008). An Entamoeba histolytica rhomboid protease with atypical specificity cleaves a surface lectin involved in phagocytosis and immune evasion. *Genes & Development, 22,* 1636–1646.

Beck, D. L., Boettner, D. R., Dragulev, B., Ready, K., Nozaki, T., & Petri, W. A., Jr. (2005). Identification and gene expression analysis of a large family of transmembrane kinases related to the Gal/GalNAc lectin in *Entamoeba histolytica*. *Eukaryotic Cell, 4,* 722–732.

Betancur, P. A., Abraham, B. J., Yiu, Y. Y., Willingham, S. B., Khameneh, F., Zarnegar, M., et al. (2017). A CD47 associated super enhancer links pro inflammatory signalling to CD47 upregulation in breast cancer. *Nature Communications, 8,* 14802.

Biller, L., Matthiesen, J., Kühne, V., Lotter, H., Handal, G., Nozaki, T., Saito-Nakano, Y., et al. (2014). The cell surface proteome of *Entamoeba histolytica*. *Molecular & Cellular Proteomics, 13,* 132–144.

Boettner, D. R., Huston, C. D., Linford, A. S., Buss, S. N., Houpt, E., Sherman, N. E., & Petri, W. A. Jr. (2008). *Entamoeba histolytica* phagocytosis of human erythrocytes involves PATMK, a member of the transmembrane kinase family. PLoS Pathogens, e8.

Boettner, D. R., Huston, C. D., Sullivan, J. A., & Petri, W. A., Jr. (2005). *Entamoeba histolytica* and *Entamoeba dispar* utilize externalized phosphatidylserine for recognition and phagocytosis of erythrocytes. *Infection and Immunity, 73,* 3422–3430.

Buss, S. N., Hamano, S., Vidrich, A., Evans, C., Zhang, Y., Crasta, O. R., et al. (2010). Members of the *Entamoeba histolytica* transmembrane kinase family play non-redundant roles in growth and phagocytosis. *International Journal for Parasitology, 40,* 833–843.

Cheng, X. J., Tsukamoto, H., Kaneda, Y., & Tachibana, H. (1998). Identification of the 150-kDa surface antigen of *Entamoeba histolytica* as a galactose- and *N*-acetyl-D-galactosamine-inhibitable lectin. *Parasitology Research, 84,* 632–639.

Dodson, J. M., Lenkowski, P. W., Jr., Eubanks, A. C., Jackson, T. F., Napodano, J., Lyerly, D. M., et al. (1999). Infection and immunity mediated by the carbohydrate recognition domain of the *Entamoeba histolytica* Gal/GalNAc lectin. *Journal of Infectious Diseases, 179,* 460–466.

Fadok, V. A., Bratton, D. L., Konowal, A., Freed, P. W., Westcott, J. Y., & Henson, P. M. (1998). Macrophages that have ingested apoptotic cells in vitro inhibit proinflammatory cytokine production through autocrine/paracrine mechanisms involving TGF-beta, PGE2, and PAF. *Journal of Clinical Investigation, 101,* 890–898.

Gardai, S. J., McPhillips, K. A., Frasch, S. C., Janssen, W. J., Starefeldt, A., Murphy-Ullrich, J. E., Bratton, D. L., et al. (2005). Cell-surface calreticulin initiates clearance of viable or apoptotic cells through trans-activation of LRP on the phagocyte. *Cell, 123,* 321–334.

Heron, B. T., Sateriale, A., Teixeira, J. E., & Huston, C. D. (2011). Evidence for a novel *Entamoeba histolytica* lectin activity that recognises carbohydrates present on ovalbumin. *International Journal for Parasitology, 41,* 137–144.

Huston, C. D., Boettner, D. R., Miller-Sims, V., & Petri, W. A., Jr. (2003). Apoptotic killing and phagocytosis of host cells by the parasite *Entamoeba histolytica. Infection and Immunity, 71,* 964–972.

Katz, U., Ankri, S., Stolarsky, T., Nuchamowitz, Y., & Mirelman, D. (2002). *Entamoeba histolytica* expressing a dominant negative N-truncated light subunit of its gal-lectin are less virulent. *Molecular Biology of the Cell, 13,* 4256–4265.

Mai, Z., & Samuelson, J. (1998). A new gene family (ariel) encodes asparagine-rich *Entamoeba histolytica* antigens, which resemble the amebic vaccine candidate serine-rich *E. histolytica* protein. *Infection and Immunity, 66,* 353–355.

Marion, S., & Guillén, N. (2006). Genomic and proteomic approaches highlight phagocytosis of living and apoptotic human cells by the parasite *Entamoeba histolytica. International Journal for Parasitology, 36,* 131–139.

Mehra, A., Fredrick, J., Petri, W. A., Jr., Bhattacharya, S., & Bhattacharya, A. (2006). Expression and function of a family of transmembrane kinases from the protozoan parasite *Entamoeba histolytica. Infection and Immunity, 74,* 5341–5351.

Miller, H. W., Suleiman, R. L., & Ralston, K. S. (2019). Trogocytosis by *Entamoeba histolytica* mediates acquisition and display of human cell membrane proteins and evasion of lysis by human serum. *mBio, 10*(2). pii: e00068-19.

Naula, C., Parsons, M., & Mottram, J. C. (2005). Protein kinases as drug targets in trypanosomes and Leishmania. *Biochimica et Biophysica Acta, 1754,* 151–159.

Okada, M., Huston, C. D., Oue, M., Mann, B. J., Petri, W. A., Jr., Kita, K., et al. (2006). Kinetics and strain variation of phagosome proteins of *Entamoeba histolytica* by proteomic analysis. *Molecular and Biochemical Parasitology, 145,* 171–183.

Orozco, E., Guarneros, G., Martinez-Palomo, A., & Sánchez, T. (1983). *Entamoeba histolytica.* Phagocytosis as a virulence factor. *Journal of Experimental Medicine, 158,* 1511–1521.

Petri, W. A., Jr., Chapman, M. D., Snodgrass, T., Mann, B. J., Broman, J., & Ravdin, J. I. (1989). Subunit structure of the galactose and N-acetyl-D-galactosamine-inhibitable adherence lectin of *Entamoeba histolytica. Journal of Biological Chemistry, 264,* 3007–3012.

Petri, W. A., Jr., Haque, R., & Mann, B. J. (2002). The bittersweet interface of parasite and host: lectin-carbohydrate interactions during human invasion by the parasite *Entamoeba histolytica. Annual Review of Microbiology, 56,* 39–64.

Petri, W. A., Jr., Smith, R. D., Schlesinger, P. H., Murphy, C. F., & Ravdin, J. I. (1987). Isolation of the galactose-binding lectin that mediates the in vitro adherence of *Entamoeba histolytica. Journal of Clinical Investigation, 80,* 1238–1244.

Petri, W. A., Jr., Snodgrass, T. L., Jackson, T. F., Gathiram, V., Simjee, A. E., Chadee, K., et al. (1990). Monoclonal antibodies directed against the galactose-binding lectin of *Entamoeba histolytica* enhance adherence. *Journal of Immunology, 144,* 4803–4809.

Pincetic, A., Bournazos, S., DiLillo, D. J., Maamary, J., Wang, T. T., Dahan, R., et al. (2014). Type I and type II Fc receptors regulate innate and adaptive immunity. *Nature Immunology, 15,* 707–716.

Ralston, K. S. (2015). Taking a bite: Amoebic trogocytosis in *Entamoeba histolytica* and beyond. *Current Opinion in Microbiology, 28,* 26–35.

Ralston, K. S., Solga, M. D., Mackey-Lawrence, N. M., Somlata, Bhattacharya A., & Petri, W. A., Jr. (2014). Trogocytosis by *Entamoeba histolytica* contributes to cell killing and tissue invasion. *Nature, 508,* 526–530.

Ravdin, J. I., & Guerrant, R. L. (1981). Role of adherence in cytopathogenic mechanisms of *Entamoeba histolytica*. Study with mammalian tissue culture cells and human erythrocytes. *Journal of Clinical Investigation, 68,* 1305–1313.

Ravdin, J. I., Stanley, P., Murphy, C. F., & Petri, W. A., Jr. (1989). Characterization of cell surface carbohydrate receptors for *Entamoeba histolytica* adherence lectin. *Infection and Immunity, 57,* 2179–2186.

Sateriale, A., & Huston, C. D. (2011). A sequential model of host cell killing and phagocytosis by Entamoeba histolytica. *Journal of Parasitology Research*. https://doi.org/10.1155/2011/926706. Epub 2011 Jan 20. PubMed PMID: 21331284; PubMed Central PMCID: PMC3038552.

Sateriale, A., Miller, P., Huston, C. D. (2016, March 24). Knockdown of five genes encoding uncharacterized proteins inhibits entamoeba histolytica phagocytosis of dead host cells. *Infection and Immunity, 84*(4), 1045–1053. https://doi.org/10.1128/IAI.01325-15. Print 2016 Apr. PubMed PMID: 26810036; PubMed Central PMCID: PMC4807468.

Sateriale, A., Vaithilingam, A., Donnelly, L., Miller, P., & Huston, C. D. (2012). Feed-forward regulation of phagocytosis by *Entamoeba histolytica*. *Infection and Immunity, 2012*(80), 4456–4462.

Shrimal, S., Bhattacharya, S., & Bhattacharya, A. (2010). Serum-dependent selective expression of EhTMKB1-9, a member of *Entamoeba histolytica* B1 family of transmembrane kinases. *PLoS Pathogens, 6*(6), e1000929.

Somlata, Nakada-Tsukui K., & Nozaki, T. (2017). AGC family kinase 1 participates in trogocytosis but not in phagocytosis in *Entamoeba histolytica*. *Nature Communications, 8*(1), 101.

Stanley, S. L. Jr., Becker, A., Kunz-Jenkins, C., Foster, L., & Li, E. (1990, July). Cloning and expression of a membrane antigen of Entamoeba histolytica possessing multiple tandem repeats. *Proceedings of the National Academy of Sciences of the United States of America, 87*(13), 4976–4980. PubMed PMID: 1695007; PubMed Central PMCID: PMC54244.

Stanley, S. L. Jr., Tian, K., Koester, J. P., & Li, E. (1995, February 24). The serine-rich Entamoeba histolytica protein is a phosphorylated membrane protein containing O-linked terminal Nacetylglucosamine residues. *Journal of Biological Chemistry, 270*(8), 4121–4126. PubMed PMID: 7876162.

Tannich, E., Ebert, F., & Horstmann, R. D. (1991). Primary structure of the 170-kDa surface lectin of pathogenic *Entamoeba histolytica*. *Proceedings of the National Academy of Sciences of the United States of America, 88,* 1849–1853.

Tavares, P., Rigothier, M. C., Khun, H., Roux, P., Huerre, M., & Guillén, N. (2005). Roles of cell adhesion and cytoskeleton activity in *Entamoeba histolytica* pathogenesis: a delicate balance. *Infection and Immunity, 73,* 1771–1778.

Teixeira, J. E., Heron, B. T., & Huston, C. D. (2008). C1q- and collectin-dependent phagocytosis of apoptotic host cells by the intestinal protozoan *Entamoeba histolytica*. *Journal of Infectious Diseases, 198,* 1062–1070.

Teixeira, J. E., & Huston, C. D. (2008). Participation of the serine-rich *Entamoeba histolytica* protein in amebic phagocytosis of apoptotic host cells. *Infection and Immunity, 76,* 959–966.

Vaithilingam, A., Teixeira, J. E., Miller, P. J., Heron, B. T., & Huston, C. D. (2012). *Entamoeba histolytica* cell surface calreticulin binds human c1q and functions in amebic phagocytosis of host cells. *Infection and Immunity, 80,* 2008–2018.

Vines, R. R., Ramakrishnan, G., Rogers, J. B., Lockhart, L. A., Mann, B. J., & Petri, W. A., Jr. (1998). Regulation of adherence and virulence by the *Entamoeba histolytica* lectin cytoplasmic domain, which contains a beta2 integrin motif. *Molecular Biology of the Cell, 9,* 2069–2079.

Zargarian, S., Shlomovitz, I., Erlich, Z., Hourizadeh, A., Ofir-Birin, Y., Croker, B. A., et al. (2017). Phosphatidylserine externalization, "necroptotic bodies" release, and phagocytosis during necroptosis. *PLoS Biology, 15*(6), e2002711.

Chapter 18
PtdIns(3,4,5)P_3 Binding Protein Screening Reveals Unique Molecules Involved in Endocytic Processes

Somlata, Ravi Bharadwaj, and Tomoyoshi Nozaki

Abstract *Entamoeba histolytica* is a protozoan parasite which is highly motile and phagocytic in nature. Almost 1% of total world's population is infected by the parasite and an estimated 100,000 deaths occur annually (Ben Ayed and Sabbahi 2017). Previously, it was believed that *E. histolytica* destroys host cells only through contact dependent secretion of toxic molecules and then phagocytosing the dead cells. But recently, it was established that amoeba also ingests host cells by a trogocytosis-like process which involves ingestion of fragments of live host cells resulting in target cell death. *E. histolytica* displays various forms of endocytosis like phagocytosis, trogocytosis and pinocytosis that are important for its survival and virulence within host systems. As phosphatidylinositol (3,4,5)-triphosphate (PtdIns(3,4,5)P_3) is an important signalling molecule in endocytic processes, the affinity enrichment of PtdIns(3,4,5)P_3-binding proteins followed by mass spectroscopy revealed several novel PH (Pleckstrin Homology) domain containing proteins from amoebic lysate. A novel PH domain containing molecule was observed with only 17% identity to known GEF (guanine nucleotide exchange factor) sequences and the function of this protein could not be annotated solely on the basis of conserved domains. However, experimental analysis including immunoprecipitation, intracellular localisation, knockdown, and florescence spectroscopy indicate it to be a Rho-specific GEF involved in actin-dependent endocytic processes.

Keywords Guanine exchange factors · Motility · Pathogenesis · Ptd(3,4,5)P3 binding proteins · Rho

---

Somlata (✉)
Multidisciplinary Centre for Advanced Research and Studies, Jamia MIllia Islamia, New Delhi 110025, India
e-mail: somlata83@gmail.com

R. Bharadwaj
Division of Infectious Diseases, Department of Medicine, UMass Medical School, Worcester, MA 01655, USA

T. Nozaki
Department of Biomedical Chemistry, Graduate School of Medicine, The University of Tokyo, 7-3-1 Hongo, Bunkyo-ku, Tokyo 113-0033, Japan

© Springer Nature Switzerland AG 2020

N. Guillen (ed.), *Eukaryome Impact on Human Intestine Homeostasis and Mucosal Immunology*, https://doi.org/10.1007/978-3-030-44826-4_18

Introduction

E. histolytica is a protozoan parasite which is a major cause of morbidity and mortality in children residing in developing countries. This parasite causes 55000 deaths annually and is among the top reasons for diarrhea in infants under two years of age in Asian and African region, where diarrhea remains third leading cause of deaths in infants (Shirley et al. 2018). The infection is transmitted through the oral fecal route, when food or water contaminated with amebic cysts is consumed. The trophozoites emerge from the cyst in the small intestine from where they migrate to large intestine. The parasite is highly motile and endocytic in nature. Phagocytosis of a wide range of host cells including erythrocytes, lymphocytes, intestinal epithelial cells as well as intestinal bacteria has been shown to be important for the survival and pathogenesis of the parasite (Seigneur et al. 2005; Ralston et al. 2014; Somlata et al. 2017). The invasive form of disease causes mild to severe tissue inflammation and destruction resulting in symptoms ranging from diarrhea, dysentery, colitis to abscess formation mostly in liver and rarely in lungs and brain (Ryan and Ray 2004). The tissue destruction by the parasite involves three steps adherence, cytolysis and phagocytosis. Apart from the conventional pathway a new process called amoebic trogocytosis has been shown to contribute to tissue destruction (Ralston et al. 2014). This process is specific to live cells and has been found in other protozoans like *Hartmannella* and *Naegleria fowleri* (Brown 1979). The process involves nibbling of live host cells by the trophozoite leading to destabilization of plasma membrane and loss of organelles, which eventually causes death of the host cell. Although many details about the mechanism of phagocytosis operating in *E. histolytica* are known but detailed mechanism is yet not known and in addition trogocytosis also has been found to be largely overlapping with existing phagocytic pathway (Ralston et al. 2014). Any interference in actin remodeling, GalNac lectin and phosphoinositide signaling affects both the processes.

Overall Process of Host Cell Killing and Phagocytosis in *E. histolytica*

The host cell killing by trophozoites is a contact dependent mechanism and involves attachment through the amebic Gal/GalNac lectin to the host cells (Petri Jr. et al. 1989). The interactions of Gal/GalNac lectins are thought to induce signals which initiate host cell killing (Saffer and Petri Jr. 1991). Another *E. histolytica* surface protein KERP1 (lysine and glutamic acid rich protein) has been shown to be involved in binding to enterocytes (Seigneur et al. 2005). Similarly, serine-rich *E. histolytica* protein (SREHP) that can recognize apoptotic host cells is also assumed to be sensory molecules on surface of trophozoites (Stanley Jr. et al. 1990). It has been shown that a live trophozoite is necessary for killing action but precise mechanism is not known. The contact with parasite leads to elevated cytosolic Ca^{2+} and global tyrosine dephosphorylation in host cells (Teixeira and Mann 2002). Although there are genes homologous to mammalian membrane-permeabilizing proteins NK-lysin and granulysin present in amoeba (Amoebapore A, B and C), direct evidence of their role in cell killing is missing (Zhang et al. 2004; Leippe et al. 1994) though epigenetic silencing of Amoebapore A gene has shown to reduce the abscess formation by the parasite (Zhang et al, 2004). The *E. histolytica* genome codes for about 50 cysteine proteases and few of the proteases are secreted but the these are more likely to contribute to tissue invasion and damage through degradation of extracellular matrix (Tillack et al. 2006). Amoeba ingests dead cells via phagocytosis while live host cells undergo trogocytosis. Currently, an outline of the phagocytosis pathway is known involving some unique molecules like EhC2PK, which is a C2 domain protein kinase and localizes to the site of phagocytosis during initiation of the process (Somlata et al. 2011). The EhC2PK kinase recruits EhCaBP1, calcium binding protein which also binds actin and provides scaffold for recruitment of other proteins involved in the process like EhAK1 (Mansuri et al. 2014). Recently, Sharma et al. showed that out of the two types of PIPK family, type I and II that synthesize PtdIns(4,5)P_2, only type I PIPK (EhPIPKI) is found in *E. histolytica* and phosphorylates PtdIns(4)P generating PtdIns(4,5)P_2. The mechanism by which PtdIns(4,5)P_2 regulate actin dynamics is not very well understood, but a homologue of WASP could not be identified in *E. histolytica* genome. It is possible that EhPIPK1 interacts with subunits of Arp2/3 complex to promote actin assembly at the site of phagocytosis (Sharma et al. 2019). Overall, the described proteins play an important role in actin polymerization and remodeling during parasite endocytic processes. Furthermore, the combined use of biochemical and genetic approaches revealed that trogocytosis requires several molecules like actin, PI3K, EhC2PK which are also involved in phagocytosis (Ralston et al. 2014). To date, no unique signaling pathway for trogocytosis has been found in any organism and it has been only possible to screen molecules involved in phagocytosis for their role in trogocytosis as well. Nevertheless, the role of phosphoinositides (PI) signaling and actin remodeling is definitive in both the processes.

Role of Phosphoinsitides in Endocytosis

It is widely accepted that PI play an important role in remodeling actin cytoskeleton during micropinocytosis, phagocytosis and trogocytosis, but the signaling cascade linking these processes has not been explored in *E. histolytica* (Sharma et al. 2019). PtdIns(3,4,5)P_3 is important molecule known to be involved in regulation of dynamic processes like motility and endocytosis (Di Paol and De Camilli 2006). In mammalian system, it is well established that class I PI3K-mediated generation of PtdIns(3,4,5)P_3 is responsible for recruitment of Vav1, ARF and RhoGEFs at the site of phagocytosis, which leads to assembly of proteins involved in actin remodeling required during the process (Marshall et al. 2001). Recently, in silico analysis of PI kinases and phosphatases of *E. histolytica* by Nakada-Tsukui et al. reveal that the genome of

E. histolytica encodes 10 PI kinase genes and 23 PI phosphatase genes, which are sufficient to generate all seven species of PIs. Also, multiple copies of homologs of class I PI 3-kinases and Phosphatase and tensin homolog (PTEN) exist in genome which indicates the presence of signaling cascades mediated by PtdIns(3,4,5)P_3 in *E. histolytica* (Nakada-Tsukui et al. 2019). The precursor for PtdIns(3,4,5)P_3 is generated by EhPIPK1, which also interacts with actin in vitro. This property is unique to amoebic type I PIPK and has been not reported before (Sharma et al. 2019). Furthermore, it has been shown that phosphatidylinositol 3-phosphate (PtdIns3P), PtdIns(4,5)P_2, and PtdIns(3,4,5)P_3 localize to phagosome/phagocytic cups and plasma membrane but very little information is available about the PtdIns(3,4,5)P_3 mediated signaling in *E. histolytica* (Powell et al. 2006; Nakada-Tsukui et al. 2009; Byekova et al. 2010; Koushik et al. 2014). As the importance of PtdIns(3,4,5)P_3 in various endocytic processes is well proven (Yin and Janmey 2003), it can be hypothesized that affinity enrichment of PtdIns(3,4,5)P_3 binding proteins from cell free extracts might lead to identification of molecules previously unknown to be involved in any endocytic processes of *E. histolytica*. These experiments may also lead to the discovery of Eh proteins uniquely associated with phagocytosis or trogocytosis. As per our observation, most of the proteins involved in initiation and progression of phagocytic cups leave the site of internalization before it completes or immediately after it completes (Jain et al. 2008; Somlata et al. 2011, 2012). Hence this experiment is also expected to reveal molecule/s which may be highly dynamic during the process and leave the site before or just after the completion of the process as their localization is dependent on the local concentration of PtdIns(3,4,5)P_3. Overall, the experiment may lead to adding more details to existing knowledge about the pathway or discovering molecules unique to endocytic processes and enhances our understanding.

The cell free extracts of *E. histolytica* were enriched for PtdIns(3,4,5)P_3 binding proteins by affinity-based interaction with a PtdIns(3,4,5)P_3 coupled resin. The experiment also included resin coupled to PtdIns(4,5)P_2 and PtdIns3P along with resin alone to compare the specificity of binding. The proteins obtained through affinity enrichment of PtdIns(3,4,5)P_3 interaction were identified by mass spectrometry. The data analysis of mass spectrometry results led to identification of 18 specific PtdIns(3,4,5)P_3 binding proteins. Most of the proteins identified were hypothetical and were not reported to be involved in any amoebic cellular process. In silico analysis revealed the putative domain composition and motifs in the identified proteins, which is given in Table 18.1. Most of the proteins identified contained a predicted PH domain in the amino acid sequence, which is known to bind PtdIns(3,4,5)P_3. The proteins were also searched for their presence in various phagosome proteomes submitted by other research groups, but only a few were found in the mass spectrometry results of phagosome proteomes as mentioned in the Table 18.1. The absence of some of these proteins in phagosome proteome data sets might be attributed to transient and dynamic protein localization at the site phagocytosis, which is difficult to be analyzed through mass spectrometry of enriched phagosome preparation. Although the total number of proteins identified was 18, but proteins for which more than 10% of the total sequence was covered by the mass spectrometry results were further analyzed. Out of 14 proteins (which had more than 10% sequence coverage in mass

Table 18.1 PtdIns(3,4,5)P_3 binding proteins identified through mass spectrometry. Most of the proteins are predicted to have PH domain responsible for the interaction with PtdIns(3,4,5)P_3. The proteins present in the phagosome proteome reports are also mentioned

S. No.	Identified proteins	Mol. wt. (kDa)	Name assigned	Control	Experiment	% coverage	Domain compostion	Proteome reports
1.	EHI_008090	54	$PIP3BP_1$	0	40.473	42	PH domain, may be guanine nucleotide exchange protein	
2.	EHI_168500	47		0	32.378	39	PH domain, may be guanine nucleotide exchange protein	
3.	EHI_095940	45	$PIP3BP_2$	0	37.775	61	Protein kinase domain and weak similarity to 3-phophoinositide-dependent protein kinase 1, weak PH-like domain	
4.	EHI_188930	50	$PIP3BP_3$	0	31.029	46	Catalytic domain of AGC family serine/threonine kinases, PH domain	
5.	EHI_042150	51		0	20.911	30	Catalytic domain of AGC family serine/threonine kinases, Akt like, PH domain	
6.	EHI_040420	56		0	30.355	31	PH domain containing protein	Marion et al. (2005)
7.	EHI_010510	39	$PIP3BP_4$	0	15.515	17	PH domain, SH3 domain, uncharacterized sequence between PH and SH3 domain	Marion et al. (2005), Boettner et al. (2008)
8.	EHI_138540	53		0	13.491	14	PH domain containing protein kinase, May belong to AGC/Rac family	

(continued)

Table 18.1 (continued)

S. No.	Identified proteins	Mol. wt. (kDa)	Name assigned	Control	Experiment	% coverage	Domain compostion	Proteome reports
9.	EHI_053040	50	$PIP3BP_5$	0	12.142	25	PH domain, belongs to AKT-like protein kinase family AGC-3	
10.	EHI_125820	51		0	10.793	17	PH domain containing kinase/may belong to AGC family	
11.	EHI_197790	44		0	6.7455	27	PH domain containing protein kinase	
12.	EHI_091510	76	$PIP3BP_6$	0	5.3964	11	PH domain containing protein, weakly related to Golgin subfamily member A protein	
13.	EHI_127250	65		0	4.7218	7.1	PH domain containing protein	
14.	EHI_103720	54		0	4.0473	15	PH domain containing protein	Marion et al. (2005), Boettner et al. (2008)
15.	EHI_009390	57		0	3.3727	8.7	PH domain like containing protein	
16.	EHI_097620	51		0	2.6982	6.4	PH domain, Rac Family Ser thr kinase	
17.	EHI_192460	135		0	2.0236	3.1	Formin homology 2 familyprotein	Ujang et al. (2016), Marion et al. (2005)
18.	EHI_156210	89		0	2.0236	5.1	Ras GTPase-activating protein, putative	

spectrometry), 6 could be expressed as full length N-terminally HA-tagged and GFP-tagged protein in trophozoites under Neomycin selection method. The expression of tagged proteins was verified by western blotting technique and integrity of ectopically expressed proteins was ensured under experimental conditions. Further, the lipid binding capability of the selected proteins was further confirmed by lipid overlay assay. The assay was performed with cell free extracts of *E. histolytica* cells which were transfected with the HA-tagged protein coding plasmid. The results as shown in Fig. 18.1 indicated that all the selected proteins bound to PtdIns(3,4,5)P_3 spots on lipid array, with some cross reactivity towards PtdIns(3,4)P_2. This cross reactivity is well known for PH domains. We also checked the lipid binding with only HA-tagged PH domain of PIP3BP5 protein and it was found to bind the PtdIns(3,4,5)P_3 in same manner as full-length protein. Hence, it was confirmed that proteins identified through affinity enrichment indeed bind PtdIns(3,4,5)P_3 in vitro as well.

The expression of these PH domain containing proteins was further probed in normal *E. histolytica* trophozoites by immunolocalisation with antibodies against HA tag followed by confocal microcopy. The localization experiment in normal trophozoites was carried to reveal, if any of the identified protein involved in pinocytic, macropinocytic or actin-dependent endocytic processes. Figure 18.2 shows the immunolocalization of the all HA-tagged expressed proteins in trophozoite with actin labelled by FITC-phalloidin.

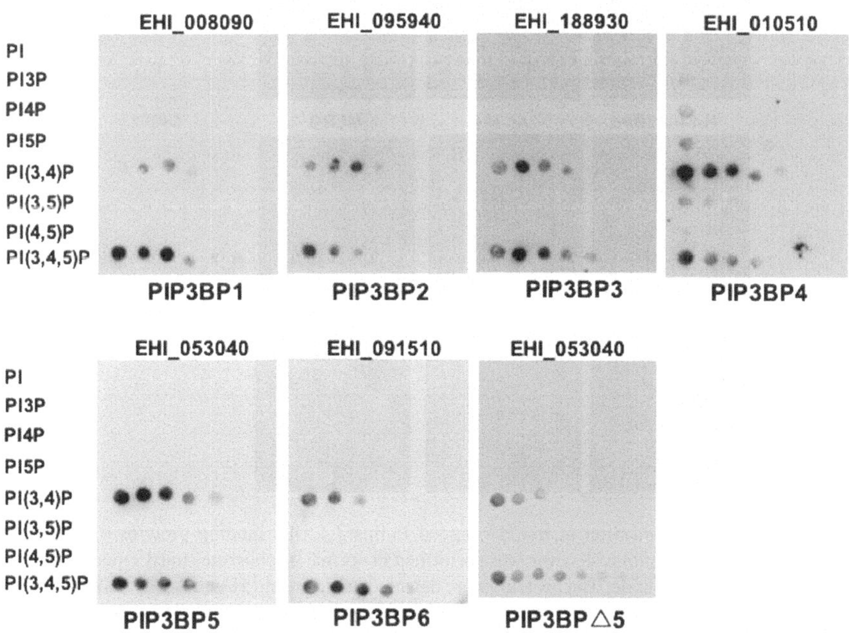

Fig. 18.1 Lipid overlay assay performed with HA-tagged proteins expressed in trophozoites and probed with HA-antibody. PIP3BP△5 is a truncated HA-tagged PH domain of the full-length protein (PIP3BP5)

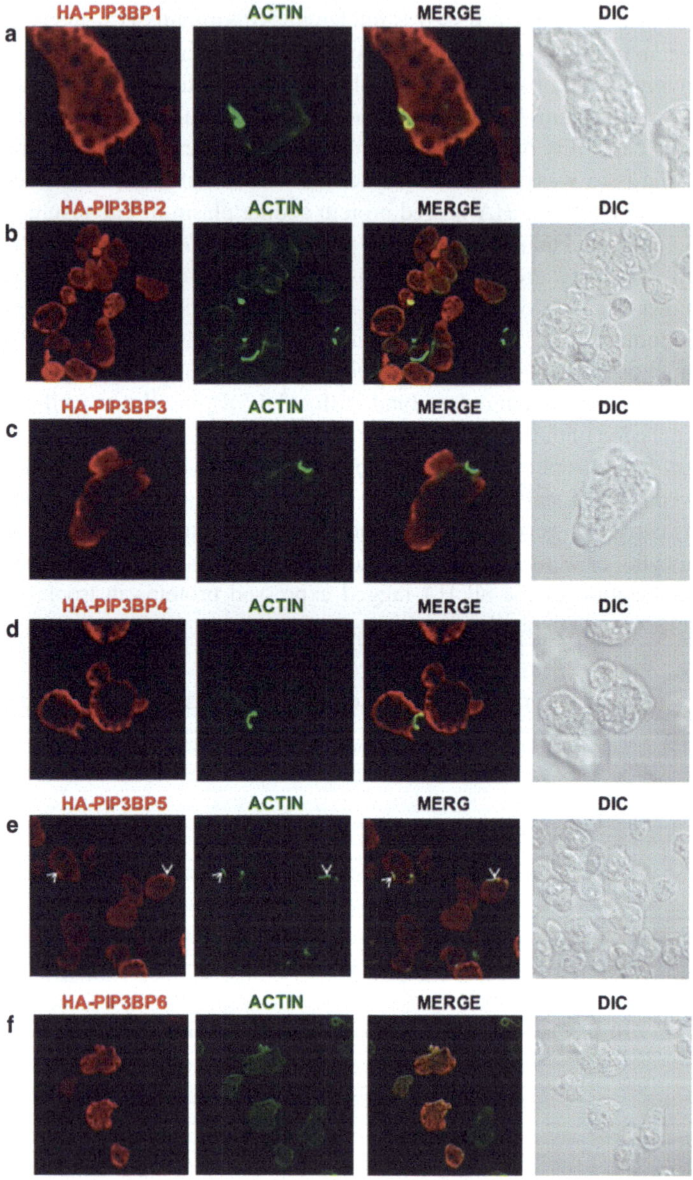

Fig. 18.2 Immunolocalisation of the HA-tagged PtdIns(3,4,5)P_3 binding proteins expressed in trophozoites. The immunolocalisation was performed in normal trophozoites in BIS medium which were fixed by parafoldehyde. The proteins were detected using an anti HA-antibody followed by an Alexaflour-555 labeled secondary antibody and actin is labeled with FITC-phalloidin. The Panel **a** and **c** shows trophozoite showing actin positive endocytic structure at larger magnification to appreciate that the localization of this protein at endocytic structures could not be observed in fixed trophozoites, indicating that protein might be transiently present during the process. While in panel **e** localization of HA-PIP3BP5 was observed at actin rich sites in several cells

The localization of proteins showed higher intensity in regions close to the membrane, which might be due to presence of the PH domain. Overall, only PIP3BP5 showed the localization at the site where actin was also enriched in the plasma membrane of the parasite as shown by arrow (Fig. 18.2, Panel e) in fixed trophozoites. As actin-dependent endocytic processes, which include micropinocytosis, phagocytosis and trogocytosis, are highly dynamic rapid changes in protein localization may not be detectable in fixed cells. To address this challenge, we used live imaging of GFP-tagged proteins in live *E. histolytica* trophozoites. Through this microscopy technique PIP3BP3 (EhAGCK1) was observed to be involved in trogocytosis only while PIP3BP5 (EhAGCK2) actively participating in the actin-dependent endocytic processes like trogocytosis, macropinocytosis and phagocytosis (Somlata et al. 2017). Both PIP3BP3 and 5 are PH-domain containing kinases but PIP3BP3 participates in trogocytosis exclusively and not in other endocytic processes. The role of PIP3BP3 in trogocytosis has been demonstrated with several evidences (Somlata et al. 2017), while PIP3BP5 is involved in all actin-dependent endoytic processes. This study identified the first PH domain containing kinase involved in trogocytosis specifically indicating that this process may be differentiated from phagocytosis. Furthermore, these molecules seem to be different from host kinases of a similar class hence could be developed as a good drug target for treatment of amebiasis. To add to the growing details of endocytic processes, another molecule PIP3BP1 was also considered for further characterization as the protein was also found in the mass spectrometry analysis of EhRho1 binding proteins which is known to be involved in phagocytosis and blebbing in *E. histolytica* (Bharadwaj et al. 2017, 2018). This small GTPase has been shown to regulate actin dynamics in *E. histolytica* through actin nucleating proteins, EhFormin1 and EhProfilin1. It recruits the actin nucleating machinery to site of phagocytosis and promotes actin filament formation to facilitate the process. In order to identify the cognate GEF for EhRho1 an immunoprecipitation was carried out in presence of non-hydrolysable GTP analogue from *E. histolytica* cell extracts. The eluted proteins were identified through mass spectrometry and one of the proteins identified was EhPIP3BP1 (unpublished data). This protein sequentially has very low similarity of known GEFs but our structure predictions indicate that this protein might be functional as GEF in vivo. Further experiments are ongoing to confirm the activity and role of the protein in amoebic cell biology and pathogenesis.

Conclusion

E. histolytica is a protozoan parasite which is highly motile and endocytic in nature. The virulence of the parasite and pathogenesis depends on motility and endocytic capabilities of the parasite. This parasite displays diverse endocytic processes including pinocytosis, micropinocytosis, phagocytosis, and trogocytosis. Most of the processes are actin-dependent and have molecules which are common between the pathway. The actin cytoskeleton plays an important role in the motility and endocytosis and it is regulated by phosphoinositides present in the plasma membrane. The signaling cascade coupling phosphoinositide and actin dynamics remains to be explored. PtdIns(3,4,5)P_3 binding protein profiling in *E. histolytica* has revealed that endocytic pathways also have molecules uniquely associated with a particular pathway, like EhAGCK1 in trogocytosis. Also, a sequentially diverged PH domain-containing protein has been identified with properties of a GEF. Further, characterization of the PtdIns(3,4,5)P_3 binding proteins identified through mass spectrometry will add to the details regarding signaling pathway mediated by PtdIns(3,4,5)P_3 in this parasite which may be useful in developing drug targets for treatment of amebiasis.

References

Aslam, S., Bhattacharya, S., & Bhattacharya, A. (2012). The calmodulin-like calcium binding protein EhCaBP3 of *Entamoeba histolytica* regulates phagocytosis and is involved in actin dynamics. *PLoS Pathogens, 8*, e1003055.

Ben Ayed, L., & Sabbahi, S. (2017). *Entamoeba histolytica*. In J. B. Rose & B. Jiménez-Cisneros (Eds.), *Global Water Pathogen Project*.

Bharadwaj, R., Arya, R., Shahid Mansuri, M., Bhattacharya, S., & Bhattacharya, A. (2017). EhRho1 regulates plasma membrane blebbing through PI3 kinase in *Entamoeba histolytica*. *Cell Microbiology, 19*(10). https://doi.org/10.1111/cmi.12751.

Bharadwaj, R., Sharma, S., Janhawi, Arya, R., Bhattacharya, S., & Bhattacharya, A. (2018). EhRho1 regulates phagocytosis by modulating actin dynamics through EhFormin1 and EhProfilin1 in *Entamoeba histolytica*. *Cell Microbiology, 20*(9), e12851. https://doi.org/10.1111/cmi.1285.

Boettner, D. R., Huston, C. D., Linford, A. S., Buss, S. N., Houpt, E., Sherman, N. E., & Petri, W. A., Jr. (2008). *Entamoeba histolytica* phagocytosis of human erythrocytes involves PATMK, a member of the transmembrane kinase family. *PLoS Pathogens, 4*(1), e8.

Brown, T. (1979). Observations by immunofluorescence microscopy and electron microscopy on the cytopathogenicity of *Naegleria fowleri* in mouse embryo-cell cultures. *Journal of Medical Microbiology, 12*, 363–371.

Byekova, Y. A., Powell, R. R., Welter, B. H., & Temesvari, L. A. (2010). Localization of phosphatidylinositol (3,4,5)-trisphosphate to phagosomes in *Entamoeba histolytica* achieved using glutathione S-transferase- and green fluorescent protein-tagged lipid biosensors. *Infection and Immunity, 78*, 125–137. https://doi.org/10.1128/IAI.00719-09.

Di Paol, G., & De Camilli, P. (2006). Phosphoinositides in cell regulation and membrane dynamics. *Nature*, *443*, 651–657. https://doi.org/10.1038/nature05185.

Enomoto, A., Murakami, H., Asai, N., Morone, N., Watanabe, T., Kawai, K., Murakumo, Y., et al. (2005). Akt/PKB regulates actin organization and cell motility via Girdin/APE. Dev. *Cell, 9*, 389–402.

Jain, R., Santi-Rocca, J., Padhan, N., Bhattacharya, S., Guillen, N., & Bhattacharya, A. (2008). Calcium-binding protein 1 of *Entamoeba histolytica* transiently associates with phagocytic cups in a calcium-independent manner. *Cellular Microbiology, 10*(6), 1373–1389.

Koushik, A. B., Welter, B. H., Rock, M. L., & Temesvari, L. A. (2014). A genome wide overexpression screen identifies genes involved in the phosphatidylinositol 3-kinase pathway in the human protozoan parasite *Entamoeba histolytica*. *Eukaryotic Cell, 13*, 401–411. https://doi.org/10.1128/EC.00329-13.

Leippe, M., Andrä, J., & Müller-Eberhard, H. J. (1994). Cytolytic and antibacterial activity of synthetic peptides derived from amoebapore, the pore-forming peptide of *Entamoeba histolytica*. *Proceedings of the National Academy of Sciences of the United States of America, 91*, 2602–2606.

Mansuri, M. S., Bhattacharya, S., & Bhattacharya, A. (2014). A novel alpha kinase EhAK1 phosphorylates actin and regulates phagocytosis in *Entamoeba histolytica*. *PLoS Pathogens, 10*, e1004411.

Marion, S., Marion, S., Laurent, C., & Guillén, N. (2005). Signalization and cytoskeleton activity through myosin IB during the early steps of phagocytosis in *Entamoeba histolytica*: A proteomic approach. *Cellular Microbiology, 7*(10), 1504–1518.

Marshall, J. G., Booth, J. W., Stambolic, V., Mak, T., Balla, T., Schreiber, A. D., Meyer, T., et al. (2001). Restricted accumulation of phosphatidylinositol 3-kinase products in a plasmalemmal subdomains during Fcγ receptor-mediated phagocytosis. *Journal of Cell Biology, 153*, 1369–1380.

Nakada-Tsukui, K., Okada, H., Mitra, B. N., & Nozaki, T. (2009). Phosphatidylinositol-phosphates mediate cytoskeletal reorganization during phagocytosis via a unique modular protein consisting of RhoGEF/DH and FYVE domains in the parasitic protozoon Entamoeba histolytica. *Cellular Microbiology, 11*, 1471–1491. https://doi.org/10.1111/j.1462-5822.2009.01341.x.

Nakada-Tsukui, K., Watanabe, N., Maehama, T., & Nozaki, T. (2019). Phosphatidylinositol kinases and phosphatases in *Entamoeba histolytica*. *Frontiers in Cellular and Infection Microbiology, 9*, 150. https://doi.org/10.3389/fcimb.2019.00150.

Petri, W. A., Jr., Chapman, M. D., Snodgrass, T. L., & Mann, B. J. (1989). Subunit structure of the galactose and *N*-Acetyl-D-galactosamine inhibitable adherence lectin of *Entamoeba histolytica*. *Journal of Biological Chemistry, 264*, 3007–3012.

Powell, R. R., Welter, B. H., Hwu, R., Bowersox, B., Attaway, C., & Temesvari, L. A. (2006). Entamoeba histolytica: FYVE-finger domains, phosphatidylinositol 3-phosphate biosensors, associate with phagosomes but not fluid filled endosomes. *Experimental Parasitology, 112*, 221–231. https://doi.org/10.1016/j.exppara.2005.11.013.

Ralston, K. S., Solga, M. D., Mackey-Lawrence, M., Somlata, Bhattacharya, A., & Petri, W.A. Jr. (2014). Trogocytosis by Entamoeba histolytica contributes to cell killing and tissue invasion. *Nature 508*, 526–530.

Ryan, K. J., & Ray, C. G. (Eds.). (2004). *Sherris medical microbiology* (4th ed., pp. 733–738). New York: McGraw Hill. ISBN 978-0-8385-8529-0.

Saffer, L. D., & Petri, W. A. Jr. (1991) Role of the galactose lectin of *Entamoeba histolytica* in adherence-dependent killing of mammalian cells. *Infection and Immunity, 59*, 4681–4683.

Seigneur, M., Mounier, J., Prevost, M. C., & Guillén, N. (2005). A lysine- and glutamic acid-rich protein, KERP1, from *Entamoeba histolytica* binds to human enterocytes. *Cellular Microbiology, 7*(4), 569–579.

Sharma, S., Agarwal, S., Bharadwaj, R., Somlata, Bhattacharya, S., & Bhattacharya, A. (2019). Novel regulatory roles of PtdIns(4,5)P_2 generating enzyme EhPIPKI in actin dynamics and phagocytosis of *Entamoeba histolytica*. *Cellular Microbiology*. https://doi.org/10.1111/cmi.13087.

Shirley, D. T., Farr, L., Watanabe, K., & Moonah, S. (2018). A review of the global burden, new diagnostics, and current therapeutics for amebiasis. *Open Forum Infectious Disseases, 5*(7). https://doi.org/10.1093/ofid/ofy161.

Somlata, Bhattacharya, S., & Bhattacharya, A. (2011). A C2 domain protein kinase initiates phagocytosis in the protozoan parasite *Entamoeba histolytica*. *Nature Communications, 2*, 230.

Somlata, Kamanna, S., Agrahari, M., Babuta, M., Bhattacharya, S., & Bhattacharya, A. (2012). Autophosphorylation of Ser428 of EhC2PK plays a critical role in regulating erythrophagocytosis in the protozoan parasite *Entamoeba histolytica*. *Journal of Biological Chemistry, 287*(14), 10844–10852. https://doi.org/10.1074/jbc.M111.308874.

Somlata, Nakada-Tsukui, K., Nozaki T. (2017). AGC family kKinase1 participates in trogocytosis but not in phagocytosis in *Entamoeba histolytica*. *Nature Communications, 8*, 101.

Stanley, S. L., Jr., Becker, A., Kunz-Jenkins, C., Foster, L., & Li, E. (1990). Cloning and expression of a membrane antigen of *Entamoeba histolytica* possessing multiple tandem repeats. *Proceedings of the National Academy of Sciences of the United States of America, 87*, 4976–4980.

Teixeira, J. E., & Mann, B. J. (2002). *Entamoeba histolytica*-induced dephosphorylation in host cells. *Infection and Immunity, 70,* 1816–1823.

Tillack, M., Nowak, N., Lotter, H., Bracha, R., Mirelman, D., Tannich, E., & Bruchhaus, I. (2006). Increased expression of the major cysteine proteinases by stable episomal transfection underlines the important role of EhCP5 for the pathogenicity of *Entamoeba histolytica. Molecular and Biochemical Parasitology, 149*, 58–64.

Ujang, J. A., Kwan, S. H., Ismail, M. N., Lim, B. H., Noordin, R., & Othman, N. (2016). Proteome analysis of excretory-secretory proteins of *Entamoeba histolytica* HM1:IMSS via LC–ESI–MS/MS and LC–MALDI–TOF/TOF. *Clinical Proteomics,13*, 33.

Yin, H. L., & Janmey, P. A. (2003). Phosphoinositide regulation of actin cytoskeleton. *Annual Review of Physiology, 65*, 761–789.

Zhang, X., Zhang, Z., Alexander, D., Bracha, R., Mirelman, D., Stanley, S. L. Jr. (2004). Expression of amoebapores is required for full expression of *Entamoeba histolytica* virulence in amebic liver abscess but is not necessary for the induction of inflammation or tissue damage in amebic colitis. *Infection and Immunity, 72*, 678–683.

Chapter 19
Molecular Insights into *E. histolytica* Mediated Host Tissue Invasion

Megha Jain, Aashutosh Tripathi, Merlyn Emmanuel, Kuldeep Verma, and Sunando Datta

Abstract Pathogenic amoeba, *Entamoeba histolytica*, is a human intestinal parasite which causes dysentery and in severe cases leads to amoebic colitis. Amoebic trophozoites have extraordinary motile ability to contribute to the tissue invasion, which even leads to the crossing of the barrier of host intestinal tissues. The amoebic invasion process requires adherence to host cell, release or activation of proteases, cell/tissue destruction and finally engulfment and clearance of the live and dead target host cells. In this review, we summarize the important molecular players of these cellular processes which are essential for the parasite's survival and virulence.

Keywords *Entamoeba histolytica* · Tissue invasion · Rab GTPase · Phagocytosis · Adhesion

Introduction

Entamoeba histolytica is a protozoan parasite responsible for dysentery and invasive amoebiasis in humans. Its infection has variable outcomes- manifesting asymptomatic colonization, diarrhoea, amoebic colitis, liver abscesses or metastatic infection. The infection occurs worldwide, although the major burden of the disease is associated with the developing world including Latin America, Africa and South Asia

M. Jain · A. Tripathi · M. Emmanuel · K. Verma · S. Datta (✉)
Department of Biological Sciences, Indian Institute of Science Education and Research, Bhopal, Bhopal Bypass Road, Bhauri 462066, India
e-mail: sunando@iiserb.ac.in

K. Verma
e-mail: kuldeep.verma@nirmauni.ac.in

M. Emmanuel
Department of Cell and Developmental Biology, Vanderbilt University School of Medicine Nashville, Nashville, TN 37232-6840, USA

K. Verma
Institute of Science, Nirma University, Sarkhej-Gandhinagar Highway, Ahmedabad, Gujarat 382481, India

© Springer Nature Switzerland AG 2020

N. Guillen (ed.), *Eukaryome Impact on Human Intestine Homeostasis and Mucosal Immunology*, https://doi.org/10.1007/978-3-030-44826-4_19

(Gunther et al. 2011; Shirley et al. 2018). Several studies have indicated that intestinal gut microenvironment, particularly the gut microbiome plays a crucial role in the establishment of infection and invasive behaviour of the trophozoite (Gilchrist et al. 2016; Iyer et al. 2019; Morton et al. 2015). There is also an emerging understanding that parasite genetics contribute significantly to its virulence, and in particular, not all strains are capable of causing the invasive pathologies (Ralston and Petri 2011).

The amoebic invasion begins primarily with the contact-dependent attachment of the parasite to host epithelial cells and extracellular matrix (ECM). The trophozoite's first contact with the host is followed by its penetration into the intestinal mucosa (Cornick and Chadee 2017), degrades the ECM by releasing the proteases and eventually inflict devastating tissue destruction (Marie and Petri 2014; Thibeaux et al. 2014). Tissue destruction is a hallmark of invasive amebiasis. Effective ECM degradation and subsequently phagocytosis or trogocytosis of the human cells with direct contact are the ultimate key features contributing to the pathogenesis of *E. histolytica* (Ralston 2015).

Although many factors contribute to the amoebic invasion, a few of them include motility and chemotaxis, cytotoxic effectors (such as Caspases, Amoebapores, etc.), parasite-induced apoptosis, proteases, pro-inflammatory host-tissue immune reactions (Ralston and Petri 2011). However, the purpose of this review is to discuss the trophozoite mediated adherence, invasion and clearance of host cell/tissue with brief molecular details. In addition, we are also highlighting some new areas of research which will help in improving the understanding of the amoebic invasion.

Adherence

Amoebic trophozoites largely rely on contact-dependent adhesion and cytotoxicity against the host cell. Adhesion is one of the first steps in which trophozoites interact with the host cells and induce cell killing and tissue destruction by multiple mechanisms. The D-Galactose/N-Acetyl-D-Galactosamine (Gal/GalNAc) lectin is one of the most extensively characterized adhesion molecules in the amoebic trophozoites. Gal/GalNAc lectin is composed of three subunits–light, heavy and intermediate chain (Petri et al. 2002). It appears that the heavy chain of Gal/GalNAc lectin (HGL) is a major contributor for host cell adhesion through its carbohydrate recognizing domain (CRD) (Ravdin and Guerrant 1981). CRD of the amoebic Gal/GalNAc lectin has been specifically shown to be important for host cell adhesion (colonic epithelial cells, CHO cells, neutrophils, macrophages, and T lymphocytes) via its ability to interact with Gal/GalNAc (Ravdin et al. 1985; Vines et al. 1998; Yadav et al. 2016). In addition, the cytosolic domain of HGL was also reported for in-out signalling, and these events are thus important for the adherence of the trophozoite to the host cells (Ramakrishnan et al. 2000). Further, it was established that Gal/GalNAc lectin also have a strong affinity for host intestinal mucin MUC-2 (Chadee et al. 1988; Lidell et al. 2006).

In addition to Gal/GalNAc lectin, few other molecules including EhCPADH, EhROM1, EhRhoGAPnc, EhTMK1-9, EhKERP1 and EhSTIRP, have also been reported to be critical for the process of adhesion. Amoebic CPADH is a heterodimeric complex composed of cysteine protease (50 kDa) and EhADH112 adhesin (75 kDa) (García-Rivera et al. 1999). The importance of CPADH is highlighted by the monoclonal antibody against EhADH (mAbAdh), which significantly inhibits the adherence of the amoebic trophozoites on the human erythrocytes (Bañuelos et al. 2005). The rhomboid protease, ROM1, is another membrane protease which is involved in the adhesion with live CHO cells. It has been observed that silencing of ROM1 in amoebic trophozoites significantly decrease the adhesion with CHO cells (Baxt et al. 2010).

Overexpression of a nucleocytoplasmic EhRhoGAPnc in *E. histolytica* has also been reported to inhibit adhesion plate formation, migration, adhesion of *E. histolytica* to MDCK cells, and consequently leads to an impairment of the cytopathic activity (Hernandez-Flores et al. 2016). Meanwhile, a large family of transmembrane kinases, EhTMK1-9, were shown to be important in serum induced cellular responses and their expression levels also correlated with the adherence capacity of the trophozoites. It should also be noted that the expression of TMK9-DN showed impairment in multiple virulence functions including proliferation, adhesion, phagocytosis and target cell destruction (Shrimal et al. 2010).

KERP1, a lysine- and glutamic acid-rich protein is yet another important but unique surface adhesion factor in *E. histolytica* (Perdomo et al. 2013). Interestingly, expression of the KERP1 coiled-coiled domains in the trophozoite significantly reduced its adherence to human cells. Finally, STIRP (serine-threonine-isoleucine rich protein), remains an attractive potential adhesin which is exclusively expressed in *E. histotytica* and is involved in the cytotoxicity against the host cell (MacFarlane and Singh 2007). The downregulation of STIRP in amoebic trophozoites showed a 30% decrease in adhesion to target host cells.

In the current literature, a multitude of adhesion receptors have been reported to be critical for the process of adherence, yet the absence of knowledge on the specificity for their ligands remain unexplored. Further characterization of these molecules will thus provide novel insights into the mechanism underlying adherence of the trophozoite to the host cells and subsequent cytotoxic activities.

Tissue Destruction and Clearance

The interaction of trophozoites through its fibronectin receptor (FNR) with host cell-extracellular matrix component fibronectin (FN) marks the beginning of the destruction of the host tissue. FN is also reported to be degraded and internalized by the amoebic trophozoites (Talamas-Rohana and Meza 1988). On the other hand, host FN also appears to elicit the intracellular signalling cascade in trophozoites which leads to the activation of cytoskeleton rearrangements (Emmanuel et al. 2015). One of these cytoskeleton rearrangement responses involves biogenesis of actin-rich

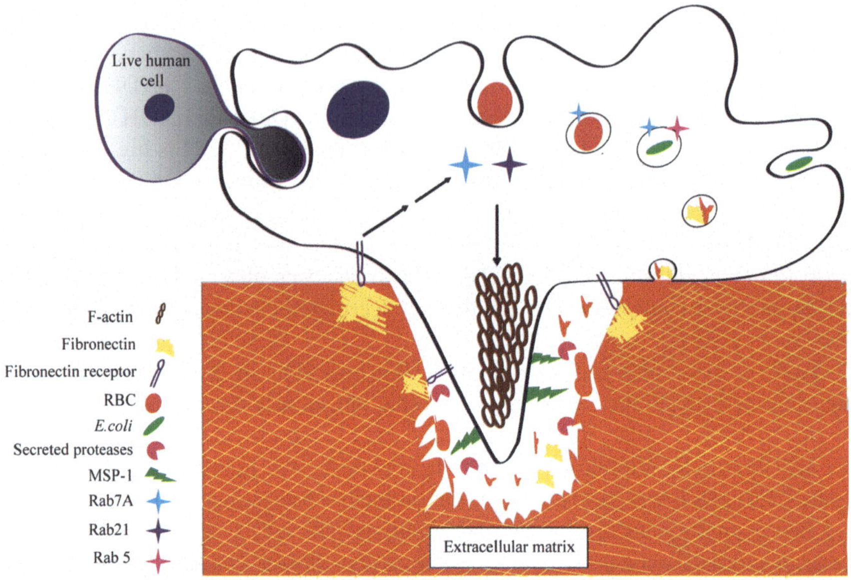

Fig. 19.1 Representative model of host issue destruction and clearance by *E. histolytica*. The amoebic invadosomes are elicited by the interaction with human ECM components, in particular fibronectin. The amoebic GTPases, Rab21 and Rab7A interaction with other components then perhaps lead to activation of actin polymerization and biogenesis of the amoebic invadosomes. Release of the various secretory proteases at the invadosome sites leads to degradation of ECM. Membrane proteases (for e.g. metalloprotease, MSP1) have also been implicated in these pathways through probable regulation of actin dynamics or proteolytic activities. Degraded FN is internalized by the trophozoites. *E.histolytica* also uptake human epithelial cells as well as RBCs by phagocytosis/trogocytosis. These amoebic trophozoites can also phagocytose *E.coli* which are otherwise well known to be a part of the gut microbiota of the host

foci known as “amoebic invadosomes” (Fig. 19.1), (Emmanuel et al. 2015). These invadosomes share the similarity with human podosomes (Linder 2007). Actin dots were first identified upon contact with the human fibronectin and other signalling molecules and were initially characterized in terms of adhesion with host tissue or ECM (Talamas-Rohana and Rios 2000). Further, actin dots were characterized as protease secretion apparatus using amoebic cell surface proteases (Hasan et al. 2018). Several studies have also concluded that cysteine proteases (CP) contribute to cytotoxicity (Hellberg et al. 2001; Mitra et al. 2007; Tillack et al. 2006). Cysteine proteases have also been reported to be secreted by the trophozoites which lead to host mucin degradation, and the resulting degradation products are thus less efficient at preventing the trophozoites’ adherence to target cells (Moncada et al. 2003) The EhMSP1 has also been studied in the regulation of amoebic invadosomes, although it is not yet clear if EhMSP1 directly take part in ECM degradation (Hasan et al. 2018).

Recently, it has also been demonstrated that the ectopic expression of Rab21, Rab7A, and Rab5 upregulated the biogenesis of amoebic invadosomes (Emmanuel et al. 2015). Host fibronectin interacts with trophozoite fibronectin receptor (FNR) and is mobilized through amoebic Rab7 compartments (Javier-Reyna et al. 2012), leading to the regulation of actin dynamics in amoebic trophozoites. In addition, it was shown that Rab21CA (constitutively active) GTP bound version of GTPase activates an array of the signalling cascade, which regulates the amoebic invadosome formation and thus increase the proteolytic capacity of the trophozoites (Emmanuel et al. 2015). However, it is not yet clear how EhRab21 activates the polymerization of F-actin at localized sites. Nevertheless, it is well known that Rab GTPases alternate between Rab GTP-bound (active) and Rab GDP-bound (inactive) forms modulated by guanine nucleotide exchange factors (GEFs) and GTPase activating proteins (GAPs), respectively (Müller and Goody 2018). It is also well established that Rabs in their active form serve as key players in cellular processes by coordinating with their effectors, leading to downstream functions (Rai et al. 2019). Thus, to explore the underlying gap between the Rabs and formation of actin-rich invadosomes, it becomes necessary to explore the possible contribution of the regulators and effectors of the amoebic Rab GTPases.

Phagocytosis/trogocytosis are two other processes that provide a major contribution in virulence of *E. histolytica* and are also considered as markers for invasive amoebiasis. It has been observed that although trogocytosis shares many common features with phagocytosis, they are distinct in nature of engulfment. Trogocytosis refers to the ingestion of 'bites' of the whole target cells, whereas the phagocytosis involves the internalization of the entire target host cells (Ralston et al. 2014). Ralston had also previously reported that the trophozoites appear to specifically approach and ingest live host cells by amoebic trogocytosis, while the pre-killed host cells were ingested whole (Ralston 2015). However, one of the most common features that have been identified between phagocytosis and trogocytosis include the trophozoites membrane receptors which are being recognized by host cell that triggers signalling events leading to phagocytic/trogocytic cup formation through actin polymerization (Meza et al. 2006).

Actin polymerization and phagocytic cup formation during the phagocytosis have also been well established in relation to calcium (Ca^{2+}) signalling in higher eukaryotic cells (Lusche et al. 2009). Similarly, Ca^{2+} is a key player in trogocytosis since it leads to intracellular Ca^{2+} elevation and loss of membrane integrity in the host cells, ultimately causing cell death (Ralston et al. 2014). In the eukaryotic cells, the Ca^{2+} modulates the protein function directly or indirectly. These proteins include kinases, phosphates and transcription factors, Ca^{2+}/calmodulin-binding proteins, and C2 domain proteins (Yáñez et al. 2012). *E. histolytica* genome encodes several numbers of Ca^{2+} binding proteins such as EhCaBP1-EhCaBP27, and multiple C2 domain proteins (Bhattacharya et al. 2006). C2 domain superfamily proteins have been previously known to display this remarkable property of interaction with the variants of phospholipids and inositol polyphosphates as a consequence of calcium-binding (Corbalan-Garcia and Gómez-Fernández 2014). In a recent study, it has been demonstrated that amoebic C2 domain-containing protein kinase (EhC2PK) (Somlata et al. 2011), Gal/GalNAc lectin, and phosphatidylinositol 3-kinase (PI3K) signalling cascade play a central role in trogosome biogenesis (Ralston et al. 2014). In addition,

AGC kinase 2 localizes in close proximity with the trogocytic cup during ingestion of live host cells (Somlata et al. 2017). The molecular mechanism of phagocytic cup formation has been well elucidated in the amoebic trophozoites by probing the role of calcium-sensing machinery. The phagocytic cup formation signalling cascade is initiated upon the cargo (RBCs/CHO cells) interaction with trophozoite leading to an alteration in the local Ca^{2+} concentrations. The local changes in Ca^{2+} recruit the EhC2PK, which is a Ca^{2+}-binding protein (Somlata et al., 2011). Subsequently, another calcium-binding protein EhCaBP1 along with alpha kinase EhAK1 (Mansuri et al. 2014) and Arp2/3 complex protein (EhARPC1) are recruited at cargo binding site (Babuta et al. 2015). Calcium-binding protein EhCaBP3 is also recruited to the phagocytic cup by Arp2/3 complex protein (EhARPC2) and is mediated in the recruitment of myosin 1B (Marion et al. 2004), leading to the formation of the phagosome (Babuta et al. 2018).

Rab GTPases are also prime regulators of intracellular membrane trafficking, through their specific interaction with effector molecules, which coordinate the early and late stages of phagocytosis. Amoebic trophozoites harbor more than 90 Rab GTPases (Saito-Nakano et al. 2004) and the function of some of these GTPases have been established (Chávez-Munguía et al. 2018; Hanadate et al. 2016; Saito-Nakano et al. 2004, 2007; Verma et al. 2015, 2016; Verma and Datta 2017; Welter et al. 2005). Among them, as discussed previously, the functional role of amoebic Rab21 is known in the biogenesis of actin dots (known as amoebic invadosomes) and destruction of host ECM (Emmanuel et al. 2015). Recently, amoebic Rab21 was found to be localized onto lysosome like compartments in a late stage of phagosome progression, suggesting it's possible role in the trafficking of cargos from Golgi apparatus to the lysosome (Constantino-Jonapa et al. 2018). In addition, it has been shown that amoebic Rab5 regulates the biogenesis of pre-phagosomal vacuoles upon interaction with human RBCs (Saito-Nakano et al. 2004). Furthermore, time-lapse confocal microscopy has also revealed that the amoebic Rab35 is translocated to the phagocytic cup upon contact with human RBCs and is subsequently, important for phagosome maturation (Verma and Datta 2017). It has also been shown that during the phagosome maturation, amoebic Rab7A also plays an indispensable part in the formation of phagolysosomes (Saito-Nakano et al. 2007; Verma et al. 2016). Till date, effectors and binding proteins have been identified only for amoebic Rab7A and Rab8A GTPases. Biochemically, it has been shown that amoebic Rab7A interact with retromer component Vps26 and play an important role in the transport of acid hydrolase to the amoebic lysosome (Nakada-Tsukui 2005). Recently, cdc50 was identified as Rab8A interacting partner and was shown to be involved in the trafficking of lipid flippase from the endoplasmic reticulum (ER) to the cell periphery (Hanadate et al. 2016). Thus although, a large number of amoebic Rab GTPases have been tested in multiple cellular pathways, knowledge of their effectors remains very limited.

Future Directions

A large number of existing studies in the broader literature, spreading over the decades, have examined the role of cytoskeleton dynamics and Ca^{2+} signaling in phagocytosis. Although, a large number of authors have shed light on a multitude of molecules playing a role in these processes independently, yet a closer look of the literature on the interplay between these processes leading to amoebic invasion, however, reveals several gaps and shortcomings. Furthermore, the literature on amoebic trogocytosis itself currently stands at a nascent stage. The knowledge of interconnections of the network involving all of the above pathways remains unmapped.

In the mammalian cells, however, Rho GTPases are well known to lay out a network of signalling pathways leading to actin-cytoskeleton rearrangements (Ridley 2006). These networks contribute to a large number of activities, including actin-nucleation, capping, scaffolding. However, in the case of *E. histolytica*, the Rho GTPases and their effectors, as well as regulators, remain largely unexplored. Similarly, Rab GTPases in *E. histolytica* are well known in vesicular trafficking and have been reported to be contributing either in the uptake of the target cells or in the transport of the ingested cargoes towards phagolysosomes during phagocytosis (Saito-Nakano et al. 2004, 2007; Verma et al. 2015, 2016; Verma and Datta 2017). It is also known that cytoskeleton remodeling plays an important role during vesicle transport via their motor protein activities, yet very few molecules have been reported that serve as direct links between these Rabs and cytoskeleton regulating proteins in the amoeba. Hence, in an effort to bridge the looming gaps, it becomes indispensable to identify the intermediate regulatory and effector molecules. This would thus help in better understanding and exposure of the complexity of amoebic invasion–a sophisticated, tunable, and carefully choreographed sequence of events. Through this review, we thus hope to bring the focus of the scientific community on the necessity of the research areas related to invasive behaviour of *E. histolytica* that remain largely unravelled.

Acknowledgements KV acknowledges the DBT-IISc Postdoctoral Fellowship Award, Department of Biotechnology, New Delhi, India and Science and Engineering Research Board (SERB) grant number (SRG/2019/000884 and EEQ/2019/000747), New Delhi, India. SD acknowledges the Science and Engineering Research Board (SERB) grant number (EMR/2016/000340), New Delhi, India, for financial support.

References

Babuta, M., Kumar, S., Gourinath, S., Bhattacharya, S., & Bhattacharya, A. (2018). Calcium-binding protein EhCaBP3 is recruited to the phagocytic complex of *Entamoeba histolytica* by interacting with Arp2/3 complex subunit 2. *Cellular Microbiology, 20*(12), 1–14. https://doi.org/10.1111/cmi.12942.

Babuta, M., Mansuri, M. S., Bhattacharya, S., & Bhattacharya, A. (2015). The *Entamoeba histolytica*, Arp2/3 complex is recruited to phagocytic cups through an atypical kinase EhAK1. *PLoS Pathogens*. https://doi.org/10.1371/journal.ppat.1005310.

Bañuelos, C., García-Rivera, G., López-Reyes, I., & Orozco, E. (2005). Functional characterization of EhADH112: An *Entamoeba histolytica* Bro1 domain-containing protein. *Experimental Parasitology*. https://doi.org/10.1016/j.exppara.2005.03.005.

Baxt, L. A., Rastew, E., Bracha, R., Mirelman, D., & Singh, U. (2010). Downregulation of an *Entamoeba histolytica* rhomboid protease reveals roles in regulating parasite adhesion and phagocytosis. *Eukaryotic Cell*. https://doi.org/10.1128/EC.00015-10.

Bhattacharya, A., Padhan, N., Jain, R., & Bhattacharya, S. (2006). Calcium-binding proteins of *Entamoeba histolytica*. *Archives of Medical Research, 37*(2), 221–225. https://doi.org/10.1016/j.arcmed.2005.10.002.

Bhattacharya, S., & Bhattacharya, A. (2011). A C2 domain protein kinase initiates phagocytosis in the protozoan parasite *Entamoeba histolytica*. *Nature communications, 2*(1), 1–10. https://doi.org/10.1038/ncomms1199.

Chadee, K., Johnson, M. L., Orozco, E., Petri, W. A., & Ravdin, J. I. (1988). Binding and internalization of rat colonic mucins by the galactose/N-acetyl-D-galactosamine adherence lectin of *Entamoeba histolytica*. *Journal of Infectious Diseases*. https://doi.org/10.1093/infdis/158.2.398.

Chávez-Munguía, B., Valle-Solis, M., Rodríguez, M. A., Huerta, M., García-Rivera, G., Bolaños, J., et al. (2018). A calcium/cation exchanger participates in the programmed cell death and in vitro virulence of *Entamoeba histolytica*. *Frontiers in Cellular and Infection Microbiology, 8*(October), 1–16. https://doi.org/10.3389/fcimb.2018.00342.

Constantino-Jonapa, L. A., Hernández-Ramírez, V. I., Osorio-Trujillo, C., & Talamás-Rohana, P. (2018). EhRab21 mobilization during erythrophagocytosis in *Entamoeba histolytica*. *Microscopy Research and Technique*. https://doi.org/10.1002/jemt.23069.

Corbalan-Garcia, S., & Gómez-Fernández, J. C. (2014). Signaling through C2 domains: More than one lipid target. *Biochimica et Biophysica Acta, Biomembranes*. https://doi.org/10.1016/j.bbamem.2014.01.008.

Cornick, S., & Chadee, K. (2017). *Entamoeba histolytica*: Host parasite interactions at the colonic epithelium. *Tissue Barriers*. https://doi.org/10.1080/21688370.2017.1283386.

Emmanuel, M., Nakano, Y. S., Nozaki, T., & Datta, S. (2015). Small GTPase Rab21 mediates fibronectin induced actin reorganization in *Entamoeba histolytica*: implications in pathogen invasion. *PLoS Pathogens*. https://doi.org/10.1371/journal.ppat.1004666.

García-Rivera, G., Rodríguez, M. A., Ocádiz, R., Martínez-López, M. C., Arroyo, R., González-Robles, A., et al. (1999). *Entamoeba histolytica*: A novel cysteine protease and an adhesin form the 112 kDa surface protein. *Molecular Microbiology*. https://doi.org/10.1046/j.1365-2958.1999.01500.x.

Gilchrist, C. A., Petri, S. E., Schneider, B. N., Reichman, D. J., Jiang, N., Begum, S., et al. (2016). Role of the gut microbiota of children in diarrhea due to the protozoan parasite *Entamoeba histolytica*. *Journal of Infectious Diseases*. https://doi.org/10.1093/infdis/jiv772.

Gunther, J., Shafir, S., Bristow, B., & Sorvillo, F. (2011). Short report: Amebiasis-related mortality among United States residents, 1990–2007. *American Journal of Tropical Medicine and Hygiene*. https://doi.org/10.4269/ajtmh.2011.11-0288.

Hanadate, Y., Saito-Nakano, Y., Nakada-Tsukui, K., & Nozaki, T. (2016). Endoplasmic reticulum-resident Rab8A GTPase is involved in phagocytosis in the protozoan parasite *Entamoeba histolytica*. *Cellular Microbiology*. https://doi.org/10.1111/cmi.12570.

Hasan, M. M., Teixeira, J. E., & Huston, & C. D. (2018). Invadosome-mediated human extracellular matrix degradation by *Entamoeba histolytica*. *Infection and Immunity*, *86*(9). https://doi.org/10.1128/iai.00287-18.

Hellberg, A., Nickel, R., Lotter, H., Tannich, E., & Bruchhaus, I. (2001). Overexpression of cysteine proteinase 2 in *Entamoeba histolytica* or Entamoeba dispar increases amoeba-induced monolayer destruction in vitro but does not augment amoebic liver abscess formation in gerbils. *Cellular Microbiology*. https://doi.org/10.1046/j.1462-5822.2001.00086.x.

Hernandez-Flores, A., Almaraz-Barrera, M. D. J., Lozano-Amado, D., Correa-Basurto, J., Rojo-Dominguez, A., Luna-Rivera, E., ... & Vargas, M. (2016). A new nucleocytoplasmic RhoGAP protein contributes to control the pathogenicity of Entamoeba histolytica by regulating EhRacC and EhRacD activity. *Cellular microbiology, 18*(11), 1653–1672. https://doi.org/10.1111/cmi.12603.

Iyer, L. R., Verma, A. K., Paul, J., & Bhattacharya, A. (2019). Phagocytosis of gut bacteria by *Entamoeba histolytica*. *Frontiers in Cellular and Infection Microbiology*. https://doi.org/10.3389/fcimb.2019.00034.

Javier-Reyna, R., Hernández-Ramírez, V. I., González-Robles, A., Galván-Mendoza, I., Osorio-Trujillo, C., & Talamás-Rohana, P. (2012). Rab7 and actin cytoskeleton participate during mobilization of β1EHFNR in fibronectin-stimulated Entamoeba histolyticatrophozoites. *Microscopy Research and Technique*. https://doi.org/10.1002/jemt.21056.

Lidell, M. E., Moncada, D. M., Chadee, K., & Hansson, G. C. (2006). *Entamoeba histolytica* cysteine protease cleave the MUC2 mucin in its C-terminal domain and dissolve the protective colonic mucus gel. *Proceedings of the National Academy of Sciences of the United States of America*. https://doi.org/10.1073/pnas.0600623103.

Linder, S. (2007). The matrix corroded: Podosomes and invadopodia in extracellular matrix degradation. *Trends in Cell Biology*. https://doi.org/10.1016/j.tcb.2007.01.002.

Lusche, D. F., Wessels, D., & Soll, D. R. (2009). The effects of extracellular calcium on motility, pseudopod and uropod formation, chemotaxis, and the cortical localization of myosin II in Dictyostelium discoideum. *Cell Motility and the Cytoskeleton*. https://doi.org/10.1002/cm.20367.

MacFarlane, R. C., & Singh, U. (2007). Identification of an *Entamoeba histolytica* serine-, threonine-, and isoleucine-rich protein with roles in adhesion and cytotoxicity. *Eukaryotic Cell*. https://doi.org/10.1128/EC.00174-07.

Mansuri, M. S., Bhattacharya, S., & Bhattacharya, A. (2014). A novel alpha kinase EhAK1 phosphorylates actin and regulates phagocytosis in *Entamoeba histolytica*. *PLoS Pathogens*. https://doi.org/10.1371/journal.ppat.1004411.

Marie, C., & Petri, W. A. (2014). Regulation of virulence of *Entamoeba histolytica*. *Annual Review of Microbiology*. https://doi.org/10.1146/annurev-micro-091313-103550.

Marion, S., Wilhelm, C., Voigt, H., Bacri, J. C., & Guillén, N. (2004). Overexpression of myosin IB in living *Entamoeba histolytica* enhances cytoplasm viscosity and reduces phagocytosis. *Journal of Cell Science*. https://doi.org/10.1242/jcs.01178.

Meza, I., Talamás-Rohana, P., & Vargas, M. A. (2006). The cytoskeleton of *Entamoeba histolytica*: Structure, function, and regulation by signaling pathways. *Archives of Medical Research*. https://doi.org/10.1016/j.arcmed.2005.09.008.

Mitra, B. N., Saito-Nakano, Y., Nakada-Tsukui, K., Sato, D., & Nozaki, T. (2007). Rab11B small GTPase regulates secretion of cysteine proteases in the enteric protozoan parasite Entamoeba histolytica. *Cellular microbiology, 9*(9), 2112–2125. https://doi.org/10.1111/j.1462-5822.2007.00941.x.

Moncada, D., Keller, K., & Chadee, K. (2003). *Entamoeba histolytica* cysteine proteinases disrupt the polymeric structure of colonic mucin and alter its protective function. *Infection and Immunity*. https://doi.org/10.1128/IAI.71.2.838-844.2003.

Morton, E. R., Lynch, J., Froment, A., Lafosse, S., Heyer, E., Przeworski, M., et al. (2015). Variation in rural African gut microbiota is strongly correlated with colonization by Entamoeba and subsistence. *PLoS Genetics*. https://doi.org/10.1371/journal.pgen.1005658.

Müller, M. P., & Goody, R. S. (2018). *Molecular control of Rab activity by GEFs*. GAPs and GDI: In Small GTPases. https://doi.org/10.1080/21541248.2016.1276999.

Nakada-Tsukui, K. (2005). A retromerlike complex is a novel Rab7 effector that is involved in the transport of the virulence factor cysteine protease in the enteric protozoan parasite *Entamoeba histolytica*. *Molecular Biology of the Cell*. https://doi.org/10.1091/mbc.e05-04-0283.

Perdomo, D., Baron, B., Rojo-Domínguez, A., Raynal, B., England, P., & Guillén, N. (2013). The α-helical regions of KERP1 are important in *Entamoeba histolytica* adherence to human cells. *Scientific Reports*. https://doi.org/10.1038/srep01171.

Petri, W. A., Haque, R., & Mann, B. J. (2002). The bittersweet interface of parasite and host: lectin-carbohydrate interactions during human invasion by the parasite *Entamoeba histolytica*. *Annual Review of Microbiology*. https://doi.org/10.1146/annurev.micro.56.012302.160959.

Rai, A., Goody, R. S., & Müller, M. P. (2019). Multivalency in Rab effector interactions. *Small GTPases*. https://doi.org/10.1080/21541248.2016.1265700.

Ralston, K. S. (2015). Chew on this: Amoebic trogocytosis and host cell killing by *Entamoeba histolytica*. *Trends in Parasitology* . https://doi.org/10.1016/j.pt.2015.05.003.

Ralston, K. S., & Petri, W. A. (2011). Tissue destruction and invasion by *Entamoeba histolytica*. *Trends in Parasitology* . https://doi.org/10.1016/j.pt.2011.02.006.

Ralston, K. S., Solga, M. D., Mackey-Lawrence, N. M., Bhattacharya, A., & Petri, W. A., Jr. (2014). Trogocytosis by *Entamoeba histolytica* contributes to cell killing and tissue invasion. *Nature, 508*(7497), 526–530. https://doi.org/10.1038/nature13242.

Ramakrishnan, G., Lee, S., Mann, B. J., & Petri, W. A. (2000). *Entamoeba histolytica*: Deletion of the GPI anchor signal sequence on the Gal/GalNAc lectin light subunit prevents its assembly into the lectin heterodimer. *Experimental Parasitology*. https://doi.org/10.1006/expr.2000.4543.

Ravdin, J. I., & Guerrant, R. L. (1981). Role of adherence in cytopathogenic mechanisms of *Entamoeba histolytica*. Study with mammalian tissue culture cells and human erythrocytes. *Journal of Clinical Investigation*. https://doi.org/10.1172/JCI110377.

Ravdin, Jonathan I., Murphy, C. F., Salata, R. A., Guerrant, R. L., & Hewlett, E. L. (1985). N-acetyl-d-galactosamine-inhibitable adherence lectin of *Entamoeba histolytica* I. Partial purification and relation to amoebic virulence in vitro. *Journal of Infectious Diseases*. https://doi.org/10.1093/infdis/151.5.804.

Ridley, A. J. (2006). Rho GTPases and actin dynamics in membrane protrusions and vesicle trafficking. *Trends in Cell Biology*. https://doi.org/10.1016/j.tcb.2006.08.006.

Saito-Nakano, Y., Yasuda, T., Nakada-Tsukui, K., Leippe, M., & Nozaki, T. (2004). Rab5-associated vacuoles play a unique role in phagocytosis of the enteric protozoan parasite *Entamoeba histolytica*. *Journal of Biological Chemistry*. https://doi.org/10.1074/jbc.M403987200.

Saito-Nakano, Y., Mitra, B. N., Nakada-tsukui, K., Sato, D., & Nozaki, T. (2007). Two Rab7 isotypes, Eh Rab7A and Eh Rab7B, play distinct roles in biogenesis of lysosomes and phagosomes in the enteric protozoan parasite *Entamoeba histolytica*. *Cellular Microbiology*. https://doi.org/10.1111/j.1462-5822.2007.00915.x.

Shirley, D. A. T., Farr, L., Watanabe, K., & Moonah, S. (2018). A review of the global burden, new diagnostics, and current Therapeutics for amebiasis. *Open Forum Infectious Diseases*. https://doi.org/10.1093/ofid/ofy161.

Shrimal, S., Bhattacharya, S., & Bhattacharya, A. (2010). Serum-dependent selective expression of EhTMKB1-9, a member of *Entamoeba histolytica* B1 family of transmembrane kinases. *PLoS Pathogens*. https://doi.org/10.1371/journal.ppat.1000929.

Somlata, Nakada-Tsukui, K., & Nozaki, T. (2017). AGC family kinase 1 participates in trogocytosis but not in phagocytosis in Entamoeba histolytica. *Nature Communications*. https://doi.org/10.1038/s41467-017-00199-y.

Talamas-Rohana, P., & Meza, I. (1988). Interaction between pathogenic amebas and fibronectin: Substrate degradation and changes in cytoskeleton organization. *Journal of Cell Biology*. https://doi.org/10.1083/jcb.106.5.1787.

Talamas-Rohana, P., & Rios, A. (2000). Actin stress fibers in *Entamoeba histolytica* induced by fibronectin. *Archives of Medical Research*. https://doi.org/10.1016/S0188-4409(00)00190-9.

Thibeaux, R., Avé, P., Bernier, M., Morcelet, M., Frileux, P., Guillén, N., et al. (2014). The parasite *Entamoeba histolytica* exploits the activities of human matrix metalloproteinases to invade colonic tissue. *Nature Communications*. https://doi.org/10.1038/ncomms6142.

Tillack, M., Nowak, N., Lotter, H., Bracha, R., Mirelman, D., Tannich, E., et al. (2006). Increased expression of the major cysteine proteinases by stable episomal transfection underlines the important role of EhCP5 for the pathogenicity of *Entamoeba histolytica*. *Molecular and Biochemical Parasitology*. https://doi.org/10.1016/j.molbiopara.2006.04.009.

Verma, K., & Datta, S. (2017). The monomeric GTPase Rab35 regulates phagocytic cup formation and phagosomal maturation in *Entamoeba histolytica*. *Journal of Biological Chemistry*. https://doi.org/10.1074/jbc.M117.775007.

Verma, K., Nozaki, T., & Datta, S. (2016). Role of EhRab7A in phagocytosis of type 1 fimbriated *E. coli* by *Entamoeba histolytica*. *Molecular Microbiology*, *102*(6), 1043–1061. https://doi.org/10.1111/mmi.13533.

Verma, K., Saito-Nakano, Y., Nozaki, T., & Datta, S. (2015). Insights into endosomal maturation of human holo-transferrin in the enteric parasite *Entamoeba histolytica*: Essential roles of Rab7A and Rab5 in biogenesis of giant early endocytic vacuoles. *Cellular Microbiology*. https://doi.org/10.1111/cmi.12470.

Vines, R. R., Ramakrishnan, G., Rogers, J. B., Lockhart, L. A., Mann, B. J., & Petri, W. A. (1998). Regulation of adherence and virulence by the *Entamoeba histolytica* lectin cytoplasmic domain, which contains a β2 integrin motif. *Molecular Biology of the Cell*. https://doi.org/10.1091/mbc.9.8.2069.

Welter, B. H., Powell, R. R., Leo, M., Smith, C. M., & Temesvari, L. A. (2005). A unique Rab GTPase, EhRabA, is involved in motility and polarization of *Entamoeba histolytica* cells. *Molecular and Biochemical Parasitology*. https://doi.org/10.1016/j.molbiopara.2004.12.011.

Yadav, R., Verma, K., Chandra, M., Mukherjee, M., & Datta, S. (2016). Biophysical studies on calcium and carbohydrate binding to carbohydrate recognition domain of Gal/GalNAc lectin from *Entamoeba histolytica*: Insights into host cell adhesion. *Journal of Biochemistry*. https://doi.org/10.1093/jb/mvw024.

Yáñez, M., Gil-Longo, J., & Campos-Toimil, M. (2012). Calcium binding proteins. *Advances in Experimental Medicine and Biology*. https://doi.org/10.1007/978-94-007-2888-2_19.

Chapter 20
Vesicular Trafficking in *Entamoeba histolytica* is Essential for its Virulence

E. Orozco, A. Betanzos, C. Bañuelos, R. Javier-Reyna, and G. García-Rivera

Abstract Vesicular trafficking is based on membrane remodelling that produces vesicles with distinct composition. This allows the transport of nutrients and molecules, and the communication among intracellular organelles. Vesicular trafficking is one of the main mechanisms for pathogens virulence. Besides, multiple factors elicit parasite resistance to host defences, changing the environment during tissue invasion. Virulence factors are linked to the secretory (exocytosis) and importing (endocytosis) pathways. In this paper, we discuss the relevance of vesicular trafficking in the virulence of the protozoan *Entamoeba histolytica*. An efficient function of virulence factors during adherence to and phagocytosis of target cells and tissue invasion by trophozoites, requires the transport of specific molecules. The EhGal/GalNAc lectin, the EhCPADH complex, phosphoinositides, KERP1, EhRabs, EhCPs, LBPA and cholesterol are some of the molecules involved in the uptake, movement and digestion of the prey. Moreover, phagocytosis is an event that depends on active membranes fusion and fission, finely controlled by rearrangements of the actin cytoskeleton. During this process, a diversity of signalling transducers and effector molecules are implicated in endosomes and phagosomes formation and maturation, including the ESCRT machinery. Furthermore, some of these molecules participate in the exocytosis and others are secreted by the parasite and reach the target cell to initiate the invasion. For instance, proteins forming the EhCPADH complex, involved in adherence and phagocytosis, are secreted. EhCP112 and EhADH reach target cells and penetrate them by caveolae or clathrin-coated vesicles. These proteins destroy the intercellular junctions and provoke cell detachment and the impairment of epithelium homeostasis, eventually resulting in symptomatic manifestations of amoebiasis. In summary, vesicular trafficking results determinant for the transport of virulence factors during the whole *E. histolytica* pathogenesis process.

E. Orozco (✉) · A. Betanzos · C. Bañuelos · R. Javier-Reyna · G. García-Rivera
Centro de Investigación y de Estudios Avanzados del IPN, Ave. IPN no. 2508, Mexico City, Mexico
e-mail: esther@cinvestav.mx

A. Betanzos
Conacyt, Mexico City, Mexico

© Springer Nature Switzerland AG 2020

N. Guillen (ed.), *Eukaryome Impact on Human Intestine Homeostasis and Mucosal Immunology*, https://doi.org/10.1007/978-3-030-44826-4_20

Keywords Amoebiasis · Endocytosis · Host invasion · Vesicular trafficking · Phagocytosis

Abbreviations

ER	Endoplasmic reticulum
GA	Golgi apparatus
SNAREs	Soluble N-ethylmaleimide-sensitive factor attachment protein receptors
ESCRT	Endosomal sorting complexes required for transport
Vps	Vacuolar protein sorting
ES	Endomembrane system
PPV	Pre-phagosomal vacuoles
ILVs	Intraluminal vesicles
LBPA	Lysobisphosphatidic acid
MVBs	Multivesicular bodies
EhSPs	Serine proteases of *E. histolytica*
EhCPs	Cysteine proteases of *E. histolytica*
EhADH	75 kDa adhesin of *E. histolytica*
EhCPADH	124 kDa complex (EhCP112-EhADH)
TJs	Tight junctions
AJs	Adherens junctions
DSMs	Desmosomes

Introduction

Trophozoites, the invasive form of the protozoan *Entamoeba histolytica,* present an intense movement by pseudopodia emission toward nutrients and target cells. Their cytoplasm contains abundant vesicles of distinct size and chemical composition. Vesicle generation involves a dynamic activity of membranes, by fusion and fission events (Fig. 20.1). Vesicles transport molecules that perform specific functions inside or outside the cell. The uneasy detection of organelles, other than the nucleus, has limited the identification and monitoring of molecules that are moved from one place to another inside the parasite. The endoplasmic reticulum (ER) and Golgi apparatus (GA), for instance, appear as simple membranous structures and vesicles, which have been identified only by markers described for other systems (Ghosh et al. 1999; Martinez-Higuera et al. 2013).

Vesicular trafficking is essential during trophozoites attack to target cells. The mucus layer penetration by trophozoites to reach the intestine is facilitated by secreting proteases (Bruchhaus et al. 2003; Cuellar et al. 2017; Debnath et al. 2007). Then, trophozoites adhere to epithelial cells through specific molecules that are mobilized from the parasite cytoplasm to the plasma membrane (Christy and Petri 2011;

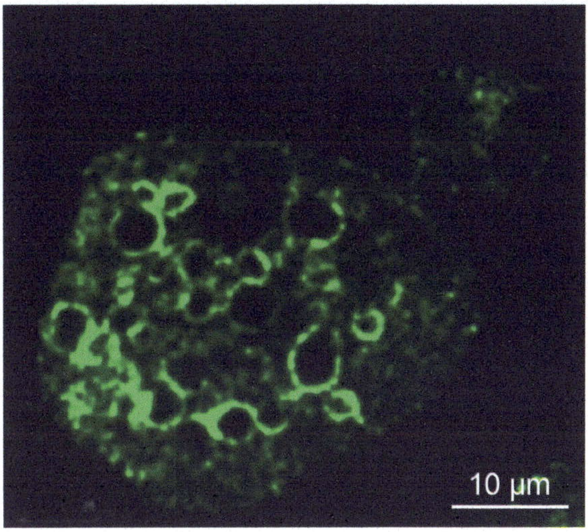

Fig. 20.1 Confocal microscopy image showing a trophozoite with multiple vesicles and vacuoles of distinct size, labelled by the anti-EhRab7 antibody

García-Rivera et al. 1999; Ravdin and Guerrant 1981). Besides, trophozoites also secret proteins that reach and damage target cells (Biller et al. 2014; Cuellar et al. 2017; Ujang et al. 2016). In the intestine, some virulence factors make contact with intercellular junctions of epithelial cells and are even able to disrupt them. Later, cells are detached and phagocytosed (Betanzos et al. 2019). In all these events, vesicular trafficking is a main player.

Since decades ago, it is well known that in eukaryotes, secretory proteins are transported in a vectorial fashion from the ER to the GA, and then, directed to secretory granules to be released to the environment, or led to the corresponding organelles. During the process, donor vesicles bud their content to an acceptor one, providing the transport pathway within the cell. Further work, identified the machinery that promotes vesicular fusion, remarkably, the soluble N-ethylmaleimide-sensitive factor attachment protein receptors (SNAREs) (Han et al. 2017). Zippering of SNAREs provokes membrane fusion, allowing the specific passage of proteins from one compartment to another, directing the cargo to the proper cellular place (Yavuz et al. 2018). However, there are limited studies on SNAREs proteins in *E. histolytica* (Cornick et al. 2017; Dacks and Doolittle 2004).

Later, the endosomal sorting complexes required for transport (ESCRT) machinery and the retromer were identified in eukaryotes (Frankel and Audhya 2018; Seaman 2012). *E. histolytica* possesses both, but with particular features (López-Reyes et al. 2011; Nakada-Tsukui et al. 2005). ESCRT complexes participate in phagocytosis (Avalos-Padilla et al. 2018) and, apparently, also the retromer does (Nakada-Tsukui et al. 2005). The ESCRT machinery is composed by vacuolar protein sorting (Vps) proteins involved in the delivery of soluble molecules carried by vesicles. The study of the *E. histolytica* ESCRT and retromer complexes has provided insights about their evolution. Moreover, it has helped to elucidate some of the molecular basis of the voracious phagocytosis and highly dynamic motility of the parasite. In this paper, we review some of the events involved in vesicular trafficking in the trophozoites and its implications in parasite virulence, including phagocytosis and epithelial invasion.

Participation of EhRab Proteins in *E. histolytica* Vesicular Trafficking

In *E. histolytica,* the endomembrane system (ES) is not well defined; however, it has an essential function for vesicular trafficking and virulence expression (Chavez-Munguía et al. 2000; Ghosh et al. 1999). Small GTPases help to direct proteins through the ES toward different cellular compartments. They act as ubiquitous molecular switches that, in addition to intracellular trafficking, are involved in a variety of cellular processes, including cell proliferation and cytoskeleton assembly (Takai et al. 2001). This superfamily is grossly classified into Ras, Rho/Rac, Rab, Sar/Arf, and Ran families (Godbold et al. 2002; Takai et al. 2001).

So far, approximately 100 EhRab proteins have been identified in *E. histolytica* (Saito-Nakano et al. 2005), and all of them conserve the GTP-binding consensus sequences (GDXXVGKT, DTAGQE, GNKXD, and SAK) (Pereira-Leal and Seabra 2000). Bioinformatics analyses revealed that 22 EhRabs exhibit more than 40% sequence identity to yeast and human Rabs, whereas 39 EhRabs (EhRabA, EhRabB and EhRabH, and EhRabX1 to EhRabX36) show lower homology to them and even to other EhRabs (Saito-Nakano et al. 2005). The EhRabs most studied in this parasite are those involved in phagocytosis; for instance, EhRab5 is located in small vesicular structures, in basal conditions and during erythrophagocytosis. It is also present in pre-phagosomal vacuoles (PPV) at early times of phagocytosis. In trophozoites overexpressing EhRab5, the rate of phagocytosis increases. Additionally, the transport of amoebapores, which are pore-forming membrane proteins, also augments, strengthening the hypothesis that EhRab5 is involved in their carriage. Conversely, mutations in EhRab5 impair PPV formation and phagocytosis (Saito-Nakano et al. 2004).

EhRab7 has nine isoforms (Rab7A-7I). EhRab7A participates in the PPV biogenesis and hydrolases processing, storage and transport to the phagosomes; and it is found in early and late endosomes (Saito-Nakano et al. 2007). In addition, EhRab7 members, together with the actin cytoskeleton, transport the fibronectin receptor from the plasma membrane to the cytoplasm and *vice versa* (Javier-Reyna et al. 2012).

EhRab8A is a key regulator of membrane trafficking in the ER and is also involved in the transport of proteins to the plasma membrane (Hanadate et al. 2016).

Besides, *E. histolytica* presents four homologues to Rab11 (EhRab11A-11D). EhRab11A-C have important roles in encystment (Herrera-Martínez et al. 2013; Saito-Nakano et al. 2005). EhRab11B is associated with non-acidified vesicles, considered as recycling compartments, and regulates the secretion of EhCP1, EhCP2, and EhCP5 (Mitra et al. 2007); hence, playing a central role in the transport and secretion of pathogenic factors.

EhRab21 is involved in migration of and invasion by trophozoites (Emmanuel et al. 2015). It is re-localized to lysosomes during erythrophagocytosis and is also found in the GA (Constantino-Jonapa et al. 2018), pointing out that it may regulate protein trafficking between these organelles.

Other protein involved in phagocytosis is EhRab35, which is located in phagocytic cups and phagolysosomes (Verma and Datta 2017).

EhRabA is found in the cytoplasm in basal conditions; however, in EhRabA mutants, the protein is visualized in tubular structures that are recognized by ER-specific antibodies. Furthermore, these mutants show a reduced rate of erythrophagocytosis (Welter and Temesvari 2009).

Finally, Rodríguez et al. (2000) discovered an atypical EhRabB protein that participates in phagocytosis and in the regulation of vesicle docking and fusion (Juárez-Hernández et al. 2013; Rodríguez et al. 2000). Furthermore, we have reported that EhRabB and actin cytoskeleton are key elements for the EhCPADH mobilization during erythrophagocytosis (Javier-Reyna et al. 2019). All these data strongly support the participation of EhRab proteins in vesicular trafficking during virulence processes of *E. histolytica.*

Contribution of the ESCRT Machinery in *E. histolytica* Vesicular Trafficking

In eukaryotes, multivesicular bodies (MVBs) biogenesis is a key step to direct particles and proteins towards lysosomal degradation and recycling (Hurley 2008). These processes entail membrane-remodelling and molecules (cargo) recognition, finely orchestrated by the ESCRT. The ESCRT machinery drives cargo sorting, membrane deformation and vesicle scission, to generate cargo-enriched intraluminal vesicles (ILVs). Then, fusion of late endosomes or MVBs-containing ILVs with lysosomes results in cargo degradation (Campsteijn et al. 2016). During the last decade, our group has shown evidence on ESCRT participation in *E. histolytica*'s phagocytosis. This parasite has the majority of genes encoding for Vps proteins forming the ESCRT-0, -I, -II and -III complexes, as well as the genes encoding for the ESCRT-accessory proteins (Avalos-Padilla et al. 2015, 2018; Bañuelos et al. 2012; López-Reyes et al. 2010).

By *in silico* analyses, we have identified representative ESCRT components in *E. histolytica*: EhTom1 (our unpublished data), EhHse1 and EhVps27, corresponding to ESCRT-0; EhVps23 and EhVps37, related to ESCRT-I; EhVps22, EhVps25 and EhVps36, forming ESCRT-II; and EhVps2, EhVps20, EhVps24 and EhVps32, belonging to ESCRT-III (López-Reyes et al. 2010). Furthermore, we have also characterized the ESCRT-accessory proteins, EhADH and EhVps4 (Bañuelos et al. 2012; López-Reyes et al. 2010). So far, little is known about the function of members of the ESCRT-0, -I and -II complexes in *E. histolytica*. Nevertheless, they are currently under study by our group, since all display the typical features reported for their orthologues in yeast and human.

The ESCRT-III has been mainly associated to membrane deformation and ILVs scission during MVBs formation and is the complex most widely characterized in this parasite (Avalos-Padilla et al. 2018). EhVps32 has a role in pinocytosis and phagocytosis. It takes part of the scission apparatus involved in membrane invagination and ILVs formation (Avalos-Padilla et al. 2015). In their inactive form, EhVps2, EhVps20, EhVps24 and EhVps32 are located in cytoplasm. However, during phagocytosis, they activate and then, assemble around and inside erythrocytes-containing phagosomes and MVBs. This points out the relevance of the ESCRT-III complex in vesicles membrane remodelling during target cell engulfment and protein degradation. Particularly, EhVps2 and EhVps32 co-localize at phagocytic cups, and EhVps32 is frequently present in nascent ILVs. Interestingly, the knock-down of *Ehvps20* and *Ehvps24* genes diminishes 60–70% the rate of phagocytosis. By giant unilamellar vesicles, we demonstrated that the sequential assembly of ESCRT-III proteins is necessary for ILVs formation. EhVps20 binds to membranes and recruits EhVps32, promoting membrane invaginations. Then, EhVps24 provokes detachment of nascent vesicles and ILVs biogenesis. Finally, EhVps2 modulates ILVs size (Avalos-Padilla et al. 2018).

A defining feature for ESCRT components turnover, is the requirement of an energy-dependent mechanism promoting ESCRT-III disassembly from the endosomal membrane. In other systems, Vps4 dissociates ESCRT-III, allowing new rounds for vesicles formation and cargo transferring from one compartment to another. Accordingly, EhVps4 localizes in the cytosol and displays ATPase activity (López-Reyes et al. 2010), thus, probably contributing to the ESCRT-III disassembly. While trophozoites overexpressing EhVps4 do not display any modification in their rate of erythrophagocytosis, amoebae overexpressing a mutant version of this protein, display a reduced phagocytic activity (López-Reyes et al. 2010).

On the other hand, previous work has evidenced the contribution of EhADH, alone or forming part of the EhCPADH complex, at early and late stages of trophozoites interaction with host cells. EhADH participates in adherence to and phagocytosis of target cells, due to an adherence epitope present at its C-terminus (García-Rivera et al. 1999). Moreover, EhADH displays a N-terminal Bro1 domain, a hallmark of the ALIX family, which mediates protein sorting and trafficking along the MVBs pathway. When the EhADH-Bro1 domain is overexpressed, trophozoites diminish their rate of phagocytosis and exhibit aberrant accumulations of endogenous EhADH in MVBs, resulting in vesicular trafficking alterations. The Bro1 domain is the EhADH interaction site for EhVps32 (Bañuelos et al. 2012). Molecular docking analyses predicted that this domain also interacts with the lysobisphosphatidic acid (LBPA), a lipid predominantly found in acidic vesicles. LBPA, together with cholesterol, phosphatidylinositol 3-phosphate and phosphatidylinositol 3,5-bisphosphate, confers specific characteristics to membranes for being remodelled during endocytosis (Castellanos-Castro et al. 2016; Nakada-Tsukui et al. 2009; Welter et al. 2011). Experimental evidence demonstrated the EhADH and LBPA interaction inside phagolysosomes and MVBs. Moreover, EhADH also binds to EhNPC1 and EhNPC2, both orthologues to the human Niemann-Pick disease type C proteins, involved in cholesterol uptake (Bolaños et al. 2016).

Besides protein degradation, endosomal retrieval and recycling are essential for cell homeostasis. These processes are achieved by the evolutionary conserved retromer complex, which regulates the presence of integral membrane proteins in the cell surface and mediates the recycling of cargo from endosomes to the trans-Golgi network and lysosomes (Elkin et al. 2016). In *E. histolytica*, a retromer-like complex, formed by EhVps26, EhVps29 and EhVps35, has been recently involved in EhCPs transport (Nakada-Tsukui et al. 2005; Srivastava et al. 2017).

Altogether, evidence on the role of the ESCRT machinery and retromer complex in vesicular trafficking of critical virulence factors, strongly suggest a molecular selective process, tightly regulated across the *E. histolytica* ES.

Vesicular Trafficking of *E. histolytica* Proteins During Host Invasion

One of the first steps during trophozoites invasion to the intestine, is the mucus layer penetration, with a remarkably participation of EhCPs (Cuellar et al. 2017; Mitra et al. 2007; Nozaki and Nakada-Tsukui 2006). EhCP5 and EhCP112 digest mucin-2 (Cuellar et al. 2017; Lidell et al. 2006). Besides, EhCPs present proteolytic activities on other epithelial components: (i) EhCP1 digests collagen, the C3 complement factor, pro-IL-18, pro-IL-1β and villin (Serrano-Luna et al. 2013); (ii) EhCP2 cleaves collagen, IgA, IgG, chemokines, C3 and pro-IL-18 (Irmer et al. 2009; Pertuz Belloso et al. 2004); (iii) EhCP5 cuts collagen, IgA, pro-IL-18, haemoglobin, fibrinogen and human pro-matrix metalloproteinase 3 (Cornick et al. 2016; Thibeaux et al. 2014); and (iv) EhCP112 digests gelatine, collagen, fibronectin, haemoglobin and substrates for L-cathepsins (Cuellar et al. 2017; Ocádiz et al. 2005). Moreover, the *Ehcp1*, *Ehcp4* and *Ehcp6* genes are upregulated in trophozoites recovered from mice colon (Gilchrist et al. 2006). These findings denote the EhCPs participation in both, epithelial invasion and extracellular matrix degradation. Therefore, a dynamic vesicular trafficking of EhCPs for their mobilization and secretion must occur. EhCPs trafficking is regulated by: (i) EhRabs (Hanadate et al. 2016; Mitra et al. 2007; Saito-Nakano et al. 2004); (ii) the retromer-like complex (Nakada-Tsukui et al. 2005); (iii) EhCPs inhibitors (Sato et al. 2006); and (iv) the cysteine protease binding family protein 1 (Furukawa et al. 2012; Marumo et al. 2014).

Once trophozoites reach the epithelium, they adhere to the cells by the EhGal/GalNAc lectin, EhADH, EhCPADH and other molecules transported from the cytoplasm to the parasite membrane (Betanzos et al. 2013; García-Rivera et al. 1999; Goplen et al. 2013; Petri et al. 2002). The lectin is composed by the light (Lgl), intermediate (Igl) and heavy (Hgl) subunits (Petri et al. 2002). In basal conditions, Igl resides in raft domains, whereas the Hgl-Lgl dimer is in different compartments. Cholesterol uptake and other events stimulate the recruitment of Hgl-Lgl dimers into raft domains, transforming the lectin into a functional adhesin, thus, increasing the ability of the parasite for adhering to host cells (Mittal et al. 2008; Welter et al. 2011).

Trophozoites pose on the apical part of the intercellular space of epithelial cells, disrupting tight junctions (TJs) and affecting the intestinal permeability (Betanzos et al. 2013; Cuellar et al. 2017; Lauwaet et al. 2004; Martinez-Palomo et al. 1985). TJs are composed by integral proteins as claudins, occludin and junctional adhesion molecules (JAMs), which are bound to actin cytoskeleton by scaffold molecules like *zonula occludens* (ZOs) proteins. The EhCPADH complex and its separated proteins impair the epithelial permeability, reflected as a drop in the transepithelial electrical resistance (TEER) (Betanzos et al. 2013, 2018; Cuellar et al. 2017). Accordingly, these parasite proteins interact with occludin, claudin-1, claudin-2, ZO-1 and ZO-2, dissociating them from TJs. In addition, these junctional proteins are also degraded. Moreover, EhCP112 and EhADH could be endocytosed by host cells through caveolin or clathrin-coated vesicles, suggesting that they can also act from inside the cell (Betanzos et al. 2018; Hernández-Nava et a1. 2017).

Other intercellular junctions like adherens junctions (AJs) and desmosomes (DSMs) are reached and destroyed by trophozoites, with the participation of the EhCPADH complex and EhCP112 (Hernández-Nava et al. 2017). AJs and DSMs are composed by integral proteins of the cadherins family, such as E-cadherin and desmoglein, and their scaffold proteins are catenins and desmoplakin, respectively (Capaldo et al. 2014; Getsios et al. 2004). EhCP112 degrades E-cadherin and desmoglein-2, destabilizing β-catenin and desmoplakin I/II (Hernández-Nava et al. 2017).

Other EhCPs and serine proteases (EhSPs) also participate in TJs disruption. EhCP1, EhCP2 and EhCP5 drop TEER; however, EhCP112 has a stronger effect (Cuellar et al. 2017). Furthermore, trophozoites with low expression of EhCPs activity are unable to damage the intestinal barrier (Zhang et al. 2000). Moreover, EhCPs and EhSPs inhibitors avoid the ZO-1 and ZO-2 proteolysis (Lauwaet et al. 2004).

Other molecules participate in TJs disturbance, particularly, the prostaglandin E_2 ($EhPGE_2$) and a parasite occludin-like protein. Through the EP4 receptor in epithelial cells, $EhPGE_2$ alters the paracellular permeability, probably dissociating claudin-4 from TJs (Lejeune et al. 2011). Instead, the occludin-like protein is also involved in the TEER dropping of human colorectal cells (Goplen et al. 2013).

The lysine glutamic acid rich protein 1 (EhKERP1) and lipopeptidophosphoglycans (EhLPPGs) are also involved in tissue invasion. EhKERP1 is present at the cell surface and intracellular vesicles, whose trafficking is independent of the ER–GA anterograde transport. It is also present in externalized vesicles deposited on the surface of human cells (Perdomo et al. 2016). On the other hand, mice immunization with anti-EhLPPGs antibodies confers them protection against invasive amoebiasis (Marinets et al. 1997; Zhang et al. 2002). Besides, these antibodies decrease trophozoites adherence to target cells (Stanley et al. 1992).

The epithelium homeostasis is also impaired by the cytolytic ability of *E. histolytica*, as revealed by the swelling and massive blebbing of the epithelial surface in contact with trophozoites (Martinez-Palomo et al. 1985; Sateriale and Huston 2011). Parasite molecules participating in cytolysis include EhCPs and amoebapores that are also conducted to the trophozoite surface and the extracellular medium by vesicular trafficking (Irmer et al. 2009; Leippe 1997; Ralston and Petri 2011a, b). Finally, the damage to the epithelium results in cell death by apoptosis and autophagy (Betanzos et al. 2013; Huston et al. 2003; Ralston and Petri 2011a, b).

As it has been referred, throughout all events allowing *E. histolytica* vesicular trafficking, molecules crosstalk and epithelial destruction result crucial for host invasion.

E. histolytica and the Host Microbiota

Even though the relationship among *E. histolytica* with the host microbiota is poorly studied, it is predictable their intimate connection through the crosstalk of molecules transported by vesicles. In the intestine, *E. histolytica* interacts with bacteria and even with other protozoa (Di Genova and Tonelli 2016; Espinosa and Paz-y-Miño-C 2019; Gilchrist et al. 2006; Varet et al. 2018), provoking changes; by one side, on the *E. histolytica* virulence, and by the other, causing dysbiosis and alterations in the host immune response (Cornick and Chadee 2017; Verma et al. 2012).

Early studies demonstrated that the interaction of trophozoites with several Gram-negative bacteria, increases their ability to destroy epithelial cells (Bracha and Mirelman 1984). In addition, certain *Enterobacteriaceae* members help *E. histolytica* to resist oxidative stress and survive in the intestinal mucosa by affecting parasite genes involved in glycolysis and proteolytic activity (Fernández-López et al. 2019; Varet et al. 2018). Particularly, the interaction between *E. histolytica* and enteropathogenic *Escherichia coli*, increases the *Ehcp1-5*, *Ehhgl*, *Ehapa* (encoding for amoebapore A), and *Ehcox-1* (encoding for cyclooxygenase-like 1) mRNA expression and also augments the EhCP1-5 proteolytic activity (Fernández-López et al. 2019). Moreover, bacteria favour the trophozoites infection in gnotobiotic animals (Fernández-López et al. 2019; Phillips and Gorstein 1966). Conversely, the introduction of certain commensal bacteria reduces the susceptibility of mice to develop colitis by *E. histolytica* infection (Burgess and Petri 2016). These results indicate the influence of gut microbiota on parasite infection, where the vesicular trafficking of pivotal virulence factors, such as proteases, could be affected.

The interdependence between *E. histolytica* and the microbiota has been also observed in human populations. A cohort study in South West Cameroon showed that the presence of this protozoa, significantly correlates with the gut microbiota composition and diversity across diet, geographic location, and genetic ancestry (Morton et al. 2015). Patients of Northern India infected with *E. histolytica* presented dysbiosis; at the same time, bacteria as *Bifidobacterium* spp., *Ruminococcus* spp., *Lactobacillus* spp., *Clostridium leptum* and *Eubacterium* spp., appeared reduced in stool samples (Verma et al. 2012), suggesting the effect of *E. histolytica* on intestinal microbiota.

Information regarding the association between *E. histolytica* with other protozoa is limited. In a Southern Côte d'Ivoire population, protists such as *E. histolytica*,

Giardia duodenalis, *Cryptosporidium* spp. and *Blastocystis hominus* induced significant changes in the intestinal microbiota (Burgess et al. 2017). Furthermore, mixed infections of *E. histolytica / E. dispar* and *G. duodenalis* were detected in a vulnerable population of Ethiopia (Mardu et al. 2019), suggesting that the presence of some parasites could favour the infection by other pathogens; however, further research are required to determine this.

Although it is estimated that 3.5 billion people worldwide are infected with parasites (WHO 2015), studies assessing their role in shaping gut microbiota remain scarce. Throughout evolution, bacteria and protozoa have co-inhabited the human gastrointestinal tract. Therefore, implications determined by past and current interactions among *E. histolytica* with the microbiota will provide insights about parasite adaptation, its invasion mechanisms and the disturbance produced on the host immune system. Here, the crosstalk among molecules from *E. histolytica* and bacteria or other protozoa facilitated by vesicular trafficking, could be relevant for the virulence properties expression. However, further studies should be performed to go deeper in this hypothesis.

Concluding Remarks

Many molecules participate during vesicular trafficking to allow the host damage produced by *E. histolytica* (Fig. 20.2). Here, we highlighted the role of: (i) EhRabs acting as effector switches, which in their majority are involved in phagocytosis.

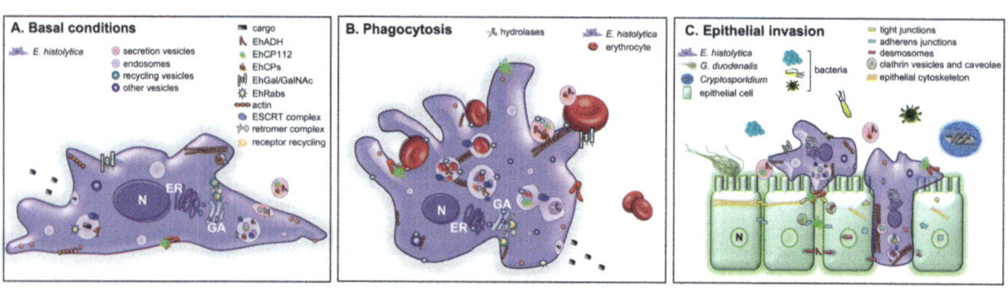

Fig. 20.2 Schematic representation of cellular structures and molecules involved in *E. histolytica* vesicular trafficking. **a) Basal conditions**: constitutive trafficking of parasite molecules within the endomembrane system (ER and GA), such as EhRabs, ESCRT and retromer complexes, as well as surface (EhCP112, EhADH and EhGal/GalNAc lectin) and secreted (e.g. EhCPs) factors that participate during nutrients uptake and metabolic maintenance. **b) Phagocytosis**: transport of molecules (such as ESCRT and retromer complexes, EhRabs, EhCPs and lipids) necessary for phagosomes formation and maturation, during target cells engulfment and degradation, or cargo recycling. **c) Epithelial invasion**: adhesins (EhADH and EhGal/GalNAc) and other surface molecules are transported during host cells recognition and attachment. Then, intercellular junctions (TJs, AJs and DSMs) are opened, and epithelial cells are detached by the action of EhCPs, EhADH and others. In addition to *E. histolytica*, bacteria and other protozoa co-inhabit the gut, presumably affecting its homeostasis. In all these events, actin cytoskeleton plays a crucial role for mobilizing the vesicles. ER, endoplasmic reticulum. GA, Golgi apparatus. N, nucleus

(ii) SNAREs proteins that promote the fusion of membranes to allow the passage of molecules from one compartment to another. (iii) The ESCRT machinery that participate in endocytosis, phagocytosis and MVBs formation. (iv) The retromer, conducting proteins for degradation or recycling. (v) The EhCPs that open the way to reach the epithelia during trophozoites invasion. (vi) The molecules involved in trophozoites adherence to target cells, such as the EhGalGal/NAc lectin and the EhCPADH complex. (vii) The EhADH adhesin that is a scaffold protein involved in adherence, phagocytosis and the epithelial junctional opening.

In conclusion, vesicular trafficking is a cellular event determinant for *E. histolytica* virulence expression. Remarkably, it is fundamental for phagocytosis and motility of trophozoites to reach the epithelia and produce damage to target cells.

Acknowledgments We thank to Tomas Sanchez and Alejandrina Reyes for their technical assistance.

Funding Information This work was supported by the National Council for Science and Technology of Mexico (CONACYT, A1-S-8380 for EO and 284477 for AB).

Disclosure of Interests All authors declare they have no conflict of interest.

References

Avalos-Padilla, Y., Betanzos, A., Javier-Reyna, R., García-Rivera, G., Chávez-Munguía, B., Lagunes-Guillén, A., et al. (2015). EhVps32 is a vacuole-associated protein involved in pinocytosis and phagocytosis of *Entamoeaba histolytica*. *PLoS Pathogens, 11,* e1005079.

Avalos-Padilla, Y., Knorr, R. L., Javier-Reyna, R., García-Rivera, G., Lipowsky, R., Dimova, R., et al. (2018). The Conserved ESCRT-III machinery participates in the phagocytosis of *Entamoeba histolytica*. *Frontiers in Cellular and Infection Microbiology, 8,* 53.

Bañuelos, C., García-Rivera, G., López-Reyes, I., Mendoza, L., González-Robles, A., Herranz, S., et al. (2012). EhADH112 is a Bro1 domain-containing protein involved in the *Entamoeba histolytica* multivesicular bodies pathway. *Journal of Biomedicine and Biotechnology, 2012,* 657942.

Betanzos, A., Bañuelos, C., & Orozco, E. (2019). Host invasion by pathogenic Amoebae: *Epithelial Disruption* by parasite proteins. *Genes (Basel), 10,* 618.

Betanzos, A., Javier-Reyna, R., García-Rivera, G., Bañuelos, C., González-Mariscal, L., Schnoor, M., et al. (2013). The EhCPADH112 complex of *Entamoeba histolytica* interacts with tight junction proteins occludin and claudin-1 to produce epithelial damage. *PLoS ONE, 8,* e65100.

Betanzos, A., Zanatta, D., Bañuelos, C., Hernández-Nava, E., Cuellar, P., & Orozco, E. (2018). Epithelial cells expressing EhADH, An *Entamoeba histolytica* Adhesin, exhibit increased tight junction proteins. *Frontiers in Cellular and Infection Microbiology, 8,* 340.

Biller, L., Matthiesen, J., Kühne, V., Lotter, H., Handal, G., Nozaki, T., et al. (2014). The cell surface proteome of *Entamoeba histolytica*. *Molecular and Cellular Proteomics, 13,* 132–144.

Bolaños, J., Betanzos, A., Javier-Reyna, R., García-Rivera, G., Huerta, M., Pais-Morales, J., et al. (2016). EhNPC1 and EhNPC2 proteins participate in trafficking of exogenous cholesterol in *Entamoeba histolytica* trophozoites: Relevance for phagocytosis. *PLoS Pathogens, 12,* e1006089.

Bracha, R., & Mirelman, D. (1984). Virulence of *Entamoeba histolytica* trophozoites. Effects of bacteria, microaerobic conditions, and metronidazole. *Journal of Experimental Medicine, 160,* 353–368.

Bruchhaus, I., Loftus, B. J., Hall, N., & Tannich, E. (2003). The intestinal protozoan parasite *Entamoeba histolytica* contains 20 cysteine protease genes, of which only a small subset is expressed during in vitro cultivation. *Eukaryotic Cells, 2,* 501–509.

Burgess, S. L., Gilchrist, C. A., Lynn, T. C., & Petri, W. A., Jr. (2017). Parasitic protozoa and interactions with the host intestinal microbiota. *Infection and Immunity, 85,* e00101–e00117.

Burgess, S. L., & Petri, W. A., Jr. (2016). The Intestinal bacterial microbiome and *E. histolytica* infection. *Current Tropical Medicine Reports, 3,* 71–74.

Campsteijn, C., Vietri, M., & Stenmark, H. (2016). Novel ESCRT functions in cell biology: Dpiraling out of control? *Current Opinion in Cell Biology, 41,* 1–8.

Capaldo, C. T., Farkas, A. E., & Nusrat, A. (2014). Epithelial adhesive junctions. F1000Prime Rep. 6.

Castellanos-Castro, S., Cerda-García-Rojas, C. M., Javier-Reyna, R., Pais-Morales, J., Chávez-Munguía, B., & Orozco, E. (2016). Identification of the phospholipid lysobisphosphatidic acid in the protozoan *Entamoeba histolytica*: An active molecule in endocytosis. *Biochemistry and Biophysics Reports, 5,* 224–236.

Chavez-Munguía, B., Espinosa-Cantellano, M., Castanon, G., & Martinez-Palomo, A. (2000). Ultrastructural evidence of smooth endoplasmic reticulum and golgi-like elements in *Entamoeba histolytica* and Entamoeba dispar. *Archives of Medical Research, 31,* S165–S167.

Christy, N. C. V., & Petri, W. A. (2011). Mechanisms of adherence, cytotoxicity and phagocytosis modulate the pathogenesis of *Entamoeba histolytica. Future Microbiology, 6,* 1501–1519.

Constantino-Jonapa, L. A., Hernández-Ramírez, V. I., Osorio-Trujillo, C., & Talamás-Rohana, P. (2018). EhRab21 mobilization during erythrophagocytosis in *Entamoeba histolytica. Microscopy Research and Technique, 81,* 1024–1035.

Cornick, S., & Chadee, K. (2017). *Entamoeba histolytica*: Host parasite interactions at the colonic epithelium. *Tissue Barriers*, 5.

Cornick, S., Moreau, F., & Chadee, K. (2016). *Entamoeba histolytica* Cysteine Proteinase 5 Evokes Mucin Exocytosis from Colonic Goblet Cells via αvβ3 Integrin. *PLoS Pathogens, 12,* e1005579.

Cornick, S., Moreau, F., Gaisano, H. Y., & Chadee, K. (2017). VAMP8 mucin exocytosis attenuates intestinal pathogenesis by *Entamoeba histolytica.* Microbial Cell (Graz, Austria) *4*, 426–427.

Cuellar, P., Hernández-Nava, E., García-Rivera, G., Chávez-Munguía, B., Schnoor, M., Betanzos, A., et al. (2017). *Entamoeba histolytica* EhCP112 dislocates and degrades claudin-1 and claudin-2 at tight junctions of the intestinal epithelium. *Frontiers in Cellular and Infection Microbiology, 7,* 372.

Dacks, J. B., & Doolittle, W. F. (2004). Molecular and phylogenetic characterization of syntaxin genes from parasitic protozoa. *Molecular and Biochemical Parasitology, 136,* 123–136.

Debnath, A., Tashker, J. S., Sajid, M., & McKerrow, J. H. (2007). Transcriptional and secretory responses of *Entamoeba histolytica* to mucins, epithelial cells and bacteria. *International Journal for Parasitology, 37,* 897–906.

Di Genova, B. M., & Tonelli, R. R. (2016). Infection strategies of intestinal parasite pathogens and host cell responses. *Frontiers in Microbiology, 7,* 256.

Elkin, S. R., Lakoduk, A. M., & Schmid, S. L. (2016). Endocytic pathways and endosomal trafficking: A primer. *Wiener Medizinische Wochenschrift, 166,* 196–204.

Emmanuel, M., Nakano, Y. S., Nozaki, T., & Datta, S. (2015). Small GTPase Rab21 mediates fibronectin induced actin reorganization in *Entamoeba histolytica*: Implications in pathogen invasion. *PLoS Pathogens, 11,* e1004666–e1004666.

Espinosa, A., & Paz-y-Miño-C, G. (2019). Discrimination experiments in entamoeba and evidence from other protists suggest pathogenic amebas cooperate with kin to colonize hosts and deter rivals. *The Journal of Eukaryotic Microbiology, 66,* 354–368.

Fernández-López, L. A., Gil-Becerril, K., Galindo-Gómez, S., Estrada-García, T., Ximénez, C., Leon-Coria, A., et al. (2019). *Entamoeba histolytica* interaction with enteropathogenic *Escherichia coli* increases parasite virulence and inflammation in amebiasis. *Infection and Immunity, 87,* e00279–e00319.

Frankel, E. B., & Audhya, A. (2018). ESCRT-dependent cargo sorting at multivesicular endosomes. *Seminars in Cell & Developmental Biology, 74,* 4–10.

Furukawa, A., Nakada-Tsukui, K., & Nozaki, T. (2012). Novel transmembrane receptor involved in phagosome transport of lysozymes and β-hexosaminidase in the enteric protozoan *Entamoeba histolytica. PLoS Pathogens, 8,* e1002539–e1002539.

García-Rivera, G., Rodriguez, M. A., Ocadiz, R., Martinez-Lopez, M. C., Arroyo, R., Gonzalez-Robles, A., et al. (1999). *Entamoeba histolytica*: A novel cysteine protease and an adhesin form the 112 kDa surface protein. *Molecular Microbiology, 33,* 556–568.

Getsios, S., Huen, A. C., & Green, K. J. (2004). Working out the strength and flexibility of desmosomes. *Nature Reviews Molecular Cell Biology, 5,* 271–281.

Ghosh, S. K., Field, J., Frisardi, M., Rosenthal, B., Mai, Z., Rogers, R., et al. (1999). Chitinase secretion by encysting Entamoeba invadens and transfected *Entamoeba histolytica* trophozoites: localization of secretory vesicles, endoplasmic reticulum, and Golgi apparatus. *Infection and Immunity, 67,* 3073–3081.

Gilchrist, C. A., Houpt, E., Trapaidze, N., Fei, Z., Crasta, O., Asgharpour, A., et al. (2006). Impact of intestinal colonization and invasion on the *Entamoeba histolytica* transcriptome. *Molecular and Biochemical Parasitology, 147,* 163–176.

Godbold, G. D., Corbett, K. D., & Mann, B. J. (2002). A Rho-like small GTPase of *Entamoeba histolytica* contains an unusual amino acid residue in a conserved GDP-stabilization region and is not a substrate for C3 exoenzyme. *Experimental Parasitology, 101,* 107–110.

Goplen, M., Lejeune, M., Cornick, S., Moreau, F., & Chadee, K. (2013). Entamoeba histolytica contains an occludin-like protein that can alter colonic epithelial barrier function. *PLoS ONE, 8,* e73339.

Han, J., Pluhackova, K., & Böckmann, R. A. (2017). The multifaceted role of SNARE proteins in membrane fusion. *Frontiers in Physiology, 8,* 5.

Hanadate, Y., Saito-Nakano, Y., Nakada-Tsukui, K., & Nozaki, T. (2016). Endoplasmic reticulum-resident Rab8A GTPase is involved in phagocytosis in the protozoan parasite *Entamoeba histolytica. Cellular Microbiology, 18,* 1358–1373.

Hernández-Nava, E., Cuellar, P., Nava, P., Chávez-Munguía, B., Schnoor, M., Orozco, E., et al. (2017). *Adherens junctions and desmosomes are damaged by Entamoeba histolytica: Participation of EhCPADH complex and EhCP112 protease* (p. 19). USA: Cell Microbiology.

Herrera-Martínez, M., Hernández-Ramírez, V. I., Lagunes-Guillén, A. E., Chávez-Munguía, B., & Talamás-Rohana, P. (2013). Actin, RhoA, and Rab11 participation during encystment in Entamoeba invadens. *BioMed Research International, 2013,* 919345.

Hurley, J. H. (2008). ESCRT complexes and the biogenesis of multivesicular bodies. *Current Opinion in Cell Biology, 20,* 4–11.

Huston, C. D., Boettner, D. R., Miller-Sims, V., & Petri v, A. J. (2003). Apoptotic killing and phagocytosis of host cells by the parasite *Entamoeba histolytica. Infection and Immunity, 71,* 964–972.

Irmer, H., Tillack, M., Biller, L., Handal, G., Leippe, M., Roeder, T., et al. (2009). Major cysteine peptidases of *Entamoeba histolytica* are required for aggregation and digestion of erythrocytes but are dispensable for phagocytosis and cytopathogenicity. *Molecular Microbiology, 72,* 658–667.

Javier-Reyna, R., Hernández-Ramírez, V. I., González-Robles, A., Galván-Mendoza, I., Osorio-Trujillo, C., & Talamás-Rohana, P. (2012). Rab7 and actin cytoskeleton participate during mobilization of β1EHFNR in fibronectin-stimulated *Entamoeba histolytica* trophozoites. *Microscopy Research and Technique, 75,* 285–293.

Javier-Reyna, R., Montaño, S., García-Rivera, G., Rodríguez, M. A., González-Robles, A., & Orozco, E. (2019). EhRabB mobilises the EhCPADH complex through the actin cytoskeleton during phagocytosis of *Entamoeba histolytica. Cellular Microbiology, 21,* e13071.

Juárez-Hernández, L. J., García-Pérez, R. M., Salas-Casas, A., García-Rivera, G., Orozco, E., & Rodríguez, M. A. (2013). *Entamoeba histolytica*: The over expression of a mutated EhRabB protein produces a decrease of in vitro and in vivo virulence. *Experimental Parasitology, 133*, 339–345.

Lauwaet, T., José Oliveira, M., Callewaert, B., De Bruyne, G., Mareel, M., & Leroy, A. (2004). Proteinase inhibitors TPCK and TLCK prevent *Entamoeba histolytica* induced disturbance of tight junctions and microvilli in enteric cell layers in vitro. *International Journal for Parasitology, 34*, 785–794.

Leippe, M. (1997). Amoebapores. *Parasitology Today, 13*, 178–183.

Lejeune, M., Moreau, F., & Chadee, K. (2011). Prostaglandin E2 produced by *Entamoeba histolytica* signals via EP4 receptor and alters claudin-4 to increase ion permeability of tight junctions. *The American Journal of Pathology, 179*, 807–818.

Lidell, M. E., Moncada, D. M., Chadee, K., & Hansson, G. C. (2006). *Entamoeba histolytica* cysteine proteases cleave the MUC2 mucin in its C-terminal domain and dissolve the protective colonic mucus gel. *Proceedings of the National Academy of Sciences of the United States of America, 103*, 9298–9303.

López-Reyes, I., Bañuelos, C., Betanzos, A., & Orozco, E. (2011). A bioinformatical approach to study the endosomal sorting complex required for transport (ESCRT) machinery in protozoan parasites: the Entamoeba histolytica case. In M. Mahdavi (Ed.), *Bioinformatics*, (289–312).Rijeka, Croatia: Intech Open Access Publisher.

López-Reyes, I., García-Rivera, G., Bauelos, C., Herranz, S., Vincent, O., Lpez-Camarillo, C., et al. (2010). Detection of the endosomal sorting complex required for transport in *Entamoeba histolytica* and characterization of the EhVps4 protein. *Journal of Biomedicine and Biotechnology, 2010*, 890674.

Mardu, F., Berhe, B., Tesfay, K., & Negash, H. (2019). Assessment of sanitary condition of services as implication for intestinal parasitic infections among prison inmates: Institutional based cross-sectional study in eastern Tigrai zonal prison, northern Ethiopia, 2018. *BBMC Research Notes, 12*, 393.

Marinets, A., Zhang, T., Guillén, N., Gounon, P., Bohle, B., Vollmann, U., et al. (1997). Protection against invasive amebiasis by a single monoclonal antibody directed against a lipophosphoglycan antigen localized on the surface of *Entamoeba histolytica. Journal of Experimental Medicine, 186*, 1557–1565.

Martinez-Higuera, A., Salas-Casas, A., Calixto-Gálvez, M., Chávez-Munguía, B., Pérez-Ishiwara, D. G., Ximénez, C., et al. (2013). Identification of calcium-transporting ATPases of *Entamoeba histolytica* and cellular localization of the putative SERCA. *Experimental Parasitology, 135*, 79–86.

Martinez-Palomo, A., Gonzalez-Robles, A., Chavez, B., Orozco, E., Fernandez-Castelo, S., & Cervantes, A. (1985). Structural bases of the cytolytic mechanisms of *Entamoeba histolytica. The Journal of Protozoology, 32*, 166–175.

Marumo, K., Nakada-Tsukui, K., Tomii, K., & Nozaki, T. (2014). Ligand heterogeneity of the cysteine protease binding protein family in the parasitic protist *Entamoeba histolytica. International Journal for Parasitology, 44*, 625–635.

Mitra, B. N., Saito-Nakano, Y., Nakada-Tsukui, K., Sato, D., & Nozaki, T. (2007). Rab11B small GTPase regulates secretion of cysteine proteases in the enteric protozoan parasite *Entamoeba histolytica. Cellular Microbiology, 9*, 2112–2125.

Mittal, K., Welter, B. H., & Temesvari, L. A. (2008). *Entamoeba histolytica*: Lipid rafts are involved in adhesion of trophozoites to host extracellular matrix components. *Experimental Parasitology, 120*, 127–134.

Morton, E. R., Lynch, J., Froment, A., Lafosse, S., Heyer, E., Przeworski, M., et al. (2015). Variation in rural African gut microbiota is strongly correlated with colonization by entamoeba and subsistence. *PLoS Genetics, 11*, e1005658–e1005658.

Nakada-Tsukui, K., Okada, H., Mitra, B. N., & Nozaki, T. (2009). Phosphatidylinositol-phosphates mediate cytoskeletal reorganization during phagocytosis via a unique modular protein consisting of RhoGEF/DH and FYVE domains in the parasitic protozoon *Entamoeba histolytica. Cellular Microbiology, 11,* 1471–1491.

Nakada-Tsukui, K., Saito-Nakano, Y., Ali, V., & Nozaki, T. (2005). A retromerlike complex is a novel Rab7 effector that is involved in the transport of the virulence factor cysteine protease in the enteric protozoan parasite *Entamoeba histolytica. Molecular Biology of the Cell, 16,* 5294–5303.

Nozaki, T., & Nakada-Tsukui, K. (2006). Membrane trafficking as a virulence mechanism of the enteric protozoan parasite *Entamoeba histolytica. Journal of Parasitology Research, 98,* 179–183.

Ocádiz, R., Orozco, E., Carrillo, E., Quintas, L. I., Ortega-Lopez, J., Garcia-Perez, R. M., et al. (2005). EhCP112 is an Entamoeba histolytica secreted cysteine protease that may be involved in the parasite-virulence. *Cellular Microbiology, 7,* 221–232.

Perdomo, D., Manich, M., Syan, S., Olivo-Marin, J.-C., Dufour, A. C., & Guillén, N. (2016). Intracellular traffic of the lysine and glutamic acid rich protein KERP1 reveals features of endomembrane organization in *Entamoeba histolytica. Cellular Microbiology, 18,* 1134–1152.

Pereira-Leal, J. B., & Seabra, M. C. (2000). The mammalian Rab family of small GTPases: Definition of family and subfamily sequence motifs suggests a mechanism for functional specificity in the Ras superfamily11Edited by M. *Yaniv The Journal of Molecular Biology, 301,* 1077–1087.

Pertuz Belloso, S., Ostoa Saloma, P., Benitez, I., Soldevila, G., Olivos, A., & García-Zepeda, E. (2004). *Entamoeba histolytica* cysteine protease 2 (EhCP2) modulates leucocyte migration by proteolytic cleavage of chemokines. *Parasite Immunology, 26,* 237–241.

Petri, W. A., Jr., Haque, R., & Mann, B. J. (2002). The bittersweet interface of parasite and host: Lectin-carbohydrate interactions during human invasion by the parasite *Entamoeba histolytica. The Annual Review of Microbiology, 56,* 39–64.

Phillips, B. P., & Gorstein, F. (1966). Effects of different species of bacteria on the pathology of enteric amebiasis in monocontaminated guinea pigs. *The American Journal of Tropical Medicine and Hygiene, 15,* 863–868.

Ralston, K. S., & Petri, W. A., Jr. (2011a). Tissue destruction and invasion by *Entamoeba histolytica. Trends in Parasitology, 27,* 254–263.

Ralston, K. S., & Petri, W. (2011b). The ways of a killer: How does *Entamoeba histolytica* elicit host cell death? *Essays in Biochemistry, 51,* 193–210.

Ravdin, J. I., & Guerrant, R. L. (1981). Role of adherence in cytopathogenic mechanisms of *Entamoeba histolytica.* Study with mammalian tissue culture cells and human erythrocytes. *Journal of Clinical Investigation, 68,* 1305–1313.

Rodríguez, M. A., García-Pérez, R. M., García-Rivera, G., López-Reyes, I., Mendoza, L., Ortiz-Navarrete, V., Orozco, E. (2000). An *Entamoeba histolytica* Rab-like encoding gene and protein: Function and cellular location. *Molecular and Biochemical Parasitology, 108,* 199–206.

Saito-Nakano, Y., Loftus, B. J., Hall, N., & Nozaki, T. (2005). The diversity of Rab GTPases in *Entamoeba histolytica. Experimental Parasitology, 110,* 244–252.

Saito-Nakano, Y., Mitra, B. N., Nakada-Tsukui, K., Sato, D., & Nozaki, T. (2007). Two Rab7 isotypes, EhRab7A and EhRab7B, play distinct roles in biogenesis of lysosomes and phagosomes in the enteric protozoan parasite *Entamoeba histolytica. Cellular Microbiology, 9,* 1796–1808.

Saito-Nakano, Y., Yasuda, T., Nakada-Tsukui, K., Leippe, M., & Nozaki, T. (2004). Rab5-associated vacuoles play a unique role in phagocytosis of the enteric protozoan parasite *Entamoeba histolytica. Journal of Biological Chemistry, 279,* 49497–49507.

Sateriale, A., & Huston, C. D. (2011). A sequential model of host cell killing and phagocytosis by *Entamoeba histolytica. Journal of Parasitology, 2011,* 926706.

Sato, D., Nakada-Tsukui, K., Okada, M., & Nozaki, T. (2006). Two cysteine protease inhibitors, EhICP1 and 2, localized in distinct compartments, negatively regulate secretion in *Entamoeba histolytica. FEBS Letters, 580,* 5306–5312.

Seaman, M. N. J. (2012). The retromer complex - endosomal protein recycling and beyond. *Journal of Cell Science, 125,* 4693–4702.

Serrano-Luna, J., Piña-Vázquez, C., Reyes-López, M., Ortiz-Estrada, G., & De La Garza, M. (2013). Proteases from *Entamoeba* spp. and pathogenic free-living amoebae as virulence factors. *Journal of Tropical Medicine,* 890603.

Srivastava, V. K., Yadav, R., Watanabe, N., Tomar, P., Mukherjee, M., Gourinath, S., et al. (2017). Structural and thermodynamic characterization of metal binding in Vps29 from *Entamoeba histolytica*: Implication in retromer function. *Molecular Microbiology, 106,* 562–581.

Stanley, S. L., Huizenga, H., & Li, E. (1992). Isolation and partial characterization of a surface glycoconjugate of *Entamoeba histolytica*. *Molecular and Biochemical Parasitology, 50,* 127–138.

Takai, Y., Sasaki, T., & Matozaki, T. (2001). Small GTP-binding proteins. *Physiological Reviews, 81,* 153–208.

Thibeaux, R., Ave, P., Bernier, M., Morcelet, M., Frileux, P., Guillen, N., et al. (2014). The parasite *Entamoeba histolytica* exploits the activities of human matrix metalloproteinases to invade colonic tissue. *Nature Communications, 5,* 5142.

Ujang, J. A., Kwan, S. H., Ismail, M. N., Lim, B. H., Noordin, R., & Othman, N. (2016). Proteome analysis of excretory-secretory proteins of *Entamoeba histolytica* HM1:IMSS via LC-ESI-MS/MS and LC-MALDI-TOF/TOF. *Clinical Proteomics, 13,* 1–10.

Varet, H., Shaulov, Y., Sismeiro, O., Trebicz-Geffen, M., Legendre, R., Coppée, J.-Y., et al. (2018). Enteric bacteria boost defences against oxidative stress in *Entamoeba histolytica*. *Scientific Reports, 8,* 9042.

Verma, A. K., Verma, R., Ahuja, V., & Paul, J. (2012). Real-time analysis of gut flora in Entamoeba histolytica infected patients of Northern India. *BMC Microbiology, 12,* 183.

Verma, K., & Datta, S. (2017). The monomeric GTPase Rab35 regulates phagocytic cup formation and phagosomal maturation in *Entamoeba histolytica*. *JJournal of Biological Chemistry, 292,* 4960–4975.

Welter, B. H., Goldston, A. M., & Temesvari, L. A. (2011). Localisation to lipid rafts correlates with increased function of the Gal/GalNAc lectin in the human protozoan parasite, *Entamoeba histolytica*. *International Journal for Parasitology, 41,* 1409–1419.

Welter, B. H., & Temesvari, L. A. (2009). Overexpression of a mutant form of EhRabA, a unique Rab GTPase of *Entamoeba histolytica*, alters endoplasmic reticulum morphology and localization of the Gal/GalNAc adherence lectin. *Eukaryotic Cell, 8,* 1014–1026.

WHO. (2015). *Global health estimates 2015*. Geneva, Switzerland: World Health Organization.

Yavuz, H., Kattan, I., Hernandez, J. M., Hofnagel, O., Witkowska, A., Raunser, S., et al. (2018). Arrest of trans-SNARE zippering uncovers loosely and tightly docked intermediates in membrane fusion. *Journal of Biological Chemistry, 293,* 8645–8655.

Zhang, Z., Duchêne, M., & Stanley, S. L., Jr. (2002). A monoclonal antibody to the amebic lipophosphoglycan-proteophosphoglycan antigens can prevent disease in human intestinal xenografts infected with *Entamoeba histolytica*. *Infection and Immunity, 70,* 5873–5876.

Zhang, Z., Yan, Le, Wang, L., Seydel, K. B., Li, E., Ankri, S., et al. (2000). *Entamoeba histolytica* cysteine proteinases with interleukin-1 beta converting enzyme (ICE) activity cause intestinal inflammation and tissue damage in amoebiasis. *Molecular Microbiology, 37,* 542–548.

Chapter 21
Molecular Characterization of *Entamoeba histolytica* tRNA Genes

N. C. Davhana, P. A. Mbati, and A. Samie

Abstract *Entamoeba histolytica* is a eukaryotic protozoan parasite responsible for amoebiasis. Its genetic diversity throughout the world as well as its impact on the outcome of infection is still a mystery. In the present study, a total of 774 stool samples were tested by microscopy, TechLab enzyme-linked immunosorbent assays (ELISAs) and multiplex PCR for the presence of *Entamoeba histolytica.* For genotype analysis, six different loci (NK, RR, AL, DA and $\mathbf{S}^{\mathbf{TGA}}$-D) of the tRNA genes were amplified by PCR. The genotyping as well as the demographic data were analyzed using the Statistical Package for Social Sciences (SPSS for WINDOWS version 21.0) program. The overall prevalence of *Entamoeba* species varied from 5.4% to 10.5 % and was higher in urban area. The genetic profiling of *E. histolytica* indicated that some strains were specific to each of the two locations while some were common. Profile Number 1 of the NK locus was more prevalent in diarrhea samples and was from Pretoria. Similarly, some profiles were more associated with diarrhea compared to others further indicating that the outcome of the infection by this parasite might be associated with the genotype. The results also indicated the possibility of strains clustering by region. The results obtained in this study confirm that tRNA genes might have a role in the presentation of amoebiasis (symptomatic and asymptomatic infections) depending on the genetic profile of the infecting strain. This genotyping system could also be used to identify the origin of the infection once it has been standardized.

Keywords Diarrheal diseases · *Entamoeba histolytica* · Genotyping · Rural/Urban · South Africa

---

N. C. Davhana · A. Samie (✉)
Department of Microbiology, University of Venda, Private BagX5050, Thohoyandou, South Africa
e-mail: samie.amidou@univen.ac.za

P. A. Mbati
Botho University, Botho Education Park, Kgale, Gaborone, Botswana

© Springer Nature Switzerland AG 2020
N. Guillen (ed.), *Eukaryome Impact on Human Intestine Homeostasis and Mucosal Immunology*, https://doi.org/10.1007/978-3-030-44826-4_21

Abbreviations

tRNA	transfer Ribonucleic Acid
ELISA	Enzyme Linked Immuno-Sorbent Assay
PCR	Polymerase Chain Reaction
NK, RR, AL, DA and S^{TGA}-D	Symbols representing the different tRNA genes
SPSS	Statistical Package for Social Science
STR	Short Tandem Repeat
DNA	Deoxy-ribonucleic acid
OD	Optical density
MEDUNSA	Medical University of South Africa

Introduction

Entamoeba histolytica is a well-known protozoan parasite responsible for the disease called amoebiasis. However, the prevalence of amoebiasis varies significantly and is high in the developing world, especially in areas where sanitation infrastructure and water supply are inadequate (Stark et al. 2008). Currently, what determine the outcome of infection is not well understood as 90 % of the infected individuals do not show any clinical signs while 10 % of the infection results in diarrhea (Samie et al. 2008). Humans are the primary host for *E. histolytica*. Transmission of *E. histolytica* generally occurs through ingestion of cysts from food or water contaminated by feces and unprotected (anal-oral) sex. The hands have been proven to be the common denominator in transferring pathogens from surface, food or animals to humans (Alum et al. 2010).

Currently, it is not clear whether the outcome of the infection by this parasite is associated with the genotype of the organism, but few studies have been conducted to determine the potential role of the parasite genomic features on the outcome of infection (Tawari et al. 2008; Haghighi et al.2002). Studies done in Bangladesh and China have investigated the six pairs of tRNA-linked STR in order to determine if the parasite genotypes are associated with the outcomes of the infection or not (Haghighi et al. 2002; Ali et al. 2005; Jaiswal et al. 2014). Results of these studies have provided evidence that the parasite genome does influence the outcome of infection and have identified few parasite genotypes that are associated with the outcomes of the infection (Jaiswal et al. 2014; Ali et al. 2007; Feng et al. 2012; Das et al. 2014).

Previous studies in the African continent have shown a high prevalence of *E. histolytica* even though there has been overestimation in some cases (Samie et al. 2008; Al-Harthi and Jamjoom 2007). However, very few studies have investigated the implications of the parasite genetic variations on the occurrence of disease. Therefore, the present study, investigated the potential impact of genetic diversity on the outcome of an *E. histolytica* infection.

Materials and Methods

Ethical Considerations

This study was approved by the research and ethics Committee of the University of Venda and MEDUNSA campus of the Limpopo University, South Africa. Permission to use public health facilities to source participants was obtained from the Department of Health and Welfare. The objectives of the study were clearly explained to the participants and they were requested to sign consent forms before the collection of stool samples. The information obtained from the participants was kept confidential.

Sample Collection

A total of 774 stool samples were collected from different clinics in Pretoria, Gauteng province, and Giyani, Limpopo province, South Africa. Both-diarrheal and non-diarrheal stool samples were collected. These samples are those that have been tested by the local laboratories and for routine testing. The stool samples were transported without delay to the University of Venda, Parasitology laboratory in cooler boxes with ice.

Detection of **E. histolytica** *Cysts and Trophozoites in Fecal Specimens by Microscopy*

Iodine wet mounts of fresh unpreserved stool samples were examined microscopically for the detection of *Entamoeba* cysts and trophozoites as described by Parija and Prabhakar (1995). Iodine wet mounts were briefly prepared by adding a portion of each unpreserved stool specimen to a drop of Lugol's iodine on a glass microscope slide and placing a cover slip on the stool suspension. The wet mounts were microscopically examined initially by using a low-power (10×) objective and then using a high-power (40×) objective of a compound light microscope. Unpreserved stool samples were stored at −20 °C until further use for ELISA and DNA extraction.

Antigen Detection of **E. histolytica** *in Stool Samples by ELISA*

Entamoeba histolytica was detected using *E. histolytica* II enzyme-linked Immunosorbent assay (ELISA) antigen detection kit (TechLab, Inc. Blacksburg, VA, USA) as per the manufacturer's instructions. The optical density (OD) was read at 450 nm using Biotek ELx808 reader (Winooski, VT, USA). According to the manufacturer, a sample with an OD suspension greater than 0.170 were considered positive while those that are less than 0.170 were considered negative.

Extraction of Genomic DNA from Stool Samples

The genomic DNA of *E. histolytica* was extracted from stool samples using ZYMO RESEARCH fecal DNA mini Prep kit from Inqaba Biotech. Briefly, 150 mg of stool sample was added to ZR bashing bead lysis tube and 750 μl of the lysis solution was also added and then disrupted at a maximum speed for 5 min. The ZR bashing bead lysis tube was centrifuged at 10,000 rpm for min, the supernatant of the samples was transferred to a Zymo-spin TM IV spin filter in a collection tube. The samples were centrifuged at 7000 rpm for 1 min, thereafter 1200 μl of fecal DNA binding buffer was added to the filtrate in the collection tube. The mixtures were transferred to a Zymo-spin TM IIC column in a collection tube and centrifuged at 10,000 rpm for 1 min. The flow from the collection tube was discarded and 200 μl of DNA pre-wash buffer was added to the Zymo-spin TMIIC column in a new collection tube. The samples were then centrifuged at 10,000 rpm for 1 min and 500 μl of fecal DNA wash buffer was then added to the Zymo-spin TMIIC column and then centrifuged at 10,000 rpm for 1 min. Zymo-spin TMIIC column was transferred to a clean 2 ml microcentrifuge tube and 100 μl of DNA elution buffer was added directly to the column matrix and then centrifuged at 10,000 rpm for 30 s to elute the DNA. The eluted DNA was then transferred to a prepared Zymo-spinTM IV HRC spin filter in a clean 2 ml microcentrifuge tube and then centrifuged at 8000 rpm for 1 min. After centrifugation, the filtered DNA was stored at −20 °C for further analysis.

Molecular Detection of **E. histolytica**

In order to confirm the ELISA positive samples, a multiplex PCR was used to identify *E. histolytica* which will be further used for the molecular characterization of the isolates. The multiplex PCR assay was performed according to a protocol previously described (Hamzah et al. 2006). The multiplex PCR reaction was performed in a total volume of 25 μl. The reaction mixture briefly contained 2.5 μl of 10× PCR buffer with 1.5 μl of $MgCl_2$, 0.5 μl of deoxynucleoside triphosphate mix, 0.2 μl of Taq polymerase, 0.25 μl of BSA, 0.6 μl of the reverse primer (EhR), 12.65 μl of distilled water, 0.6 μl of forward primer (EntaF) (Table 21.1) and 5 μl of the extracted DNA samples. Samples were gently vortexed and then placed in a thermocycler (Thermo-Hybaid, United Kingdom). The conditions for PCR were set at: heat denaturation 1 cycle at 94 °C for 3 min, followed by 30 cycles of 94 °C for 1 min, annealing temperature at 58 °C for 1 min, extension at 72 °C for 1 min, with a final extension at 72 °C

Table 21.1 Primers sequences used for genotyping PCR

Primer name	Sequence (5′–3′)	Annealing (°C)
A-L5	GGATCGATACCCCTCATCTCCA	64
A-L3	CGCATCTTGCGATAGCCGAG	
D-A5	CTGGTTAGTATCCTTCGCCTGT	56
D-A3	GCTACACCCCCATTAACAAT	
N-K5	CGAACGGCTGTTAACCGTTA	55
N-K3	TTCCTAGCTCAGTCGGTAGA	
R-R5	AGCATCAGCCTTCTAAGCTG	55
R-R3	CTTCCGACTGAGCTAACAAG	
STGA-D5	CTCTGGATGCGTAGGTTCAA	58
STGA-D3	GTATCTTCGCCTGTCACGTG	
S-Q5	GTGGTCTAAGGCGTGTGACT	56
S-Q3	GAGATTCTGGTTCTTAGGACCC	

for 7 min. The PCR products were separated by electrophoresis in 1.5 % agarose gel at 100 V for 45 min in Tris-acetate buffer, visualized by UV-transilluminator. A 100 bp DNA ladder was used as a size reference for PCR assay. The forward primer in combination with the appropriate reverse primer generated a 166-bp PCR product with *E. histolytica* DNA.

PCR Genotyping of E. histolytica

The PCR reaction was performed in a total volume of 25 μl. The reaction mixture briefly contain 2.5 μl of 10× PCR buffer with 1.5 μl of $MgCl_2$, 0.5 μl of deoxynucleoside triphosphate (dNTP) mix, 0.2 μl of Taq polymerase, 0.25 μl of BSA, 13.85 μl of distilled water, 5 μl of the extracted DNA samples and 0.6 μl of each of the primers pair (Table 21.1) were used: locus RR (R-R5 and R-R3), locus NK (NK5 and NK3), locus AL (A-L5 and A-L3), locus D-A (D-A5 and D-A3), locus SD (S^{TGA}-D5 and S^{TGA}-D3), and locus SQ (S-Q5 and S-Q3), under the conditions previously described using the primers shown in Table 21.1 (Ali et al. 2005). The PCR products were separated electrophoretically in 1.5 % agarose gels after staining with ethidium bromide and visualized by UV light and photographed.

Statistical Analysis

The results were compiled using an excel spread sheet and edited appropriately (Microsoft office package) and analyzed using the Statistical Package for Social Sciences (SPSS) program, version 17.1 with the fisher chi square test and the difference between two variables was considered significant if the *p* value was less than 0.05.

Results

Demographic Characteristics of the Study Population

From a total of 774 stool samples, males were 391 (50.5 %) and female were 48.6 %. Table 21.2 shows the demographic characteristics of the study population based on the origin, age group, gender and sample consistency. The type of the stool samples was indicated by the physical presentation of the sample at the time of the collection. From the 774 samples, 467 (60.3 %) were loose, 202 (26.1 %) were formed and 104 (13.4 %) were watery. Information on age and gender was only provided for only a small proportion of the participants and showed that the ages between 1 and 25 years contributed the highest percentage of samples 107 (13.8 %), followed by those who were 26–45 years old with 56 (7.2 %) and 46–90 years old with 49 (6.3 %).

Table 21.2 Prevalence of *E. histolytica* according to the origin, gender, age and sample consistency of the participants using ELISA method

Samples characteristics		*E. histolytica* positive	Total	Statistics
Sample origin	Giyani	3 (5.4 %)	56	$\chi^2 = 1.491, P = 0.222$
	Pretoria	75 (10.5 %)	717	
Gender	Male	45 (12 %)	375	$\chi^2 = 2.653, P = 0.103$
	Female	33(8.4 %)	391	
Age group	1–25 years	18 (16.8 %)	107	$\chi^2 = 3.341; P = 0.188$
	26–45 years	11 (21.2 %)	52	
	46–90 years	4 (8.2 %)	49	
Consistency	Formed	12 (5.9 %)	202	$\chi^2 = 5.781; P = 0.056$
	Soft	52 (11.1 %)	467	
	Watery	14 (13.6%)	103	

Detection of Entamoeba spp *by Microscopy and Enzyme-Linked Immunosorbent Assay (ELISA) for the Detection of* E. histolytica *Antigen*

In the present study, 774 samples were examined microscopically for the presence of *Entamoeba* cyst in stool samples and 16.5 % samples were found to be positive. Table 21.2 indicates the prevalence on *E. histolytica* based on age, origin, gender and sample consistency. The highest prevalence of *E. histolytica* was found in Pretoria with 10.5 %, as compared to Giyani which was 5.4 % and the difference was not statistically significant ($\chi^2 = 1.491$; $P = 0.222$). *Entamoeba histolytica* was more common in males (12 %) than in females (8.4 %), but the difference was not statistically significant ($\chi^2 = 2.653$; $P = 0.103$). Most of the participants who were infected were the ages of between 26 and 45 years by 21.2 % followed by those who were in the age group 1–25 years by 16.8 %. The least infected were of the age group 49–90 years by 8.2 %. However, the difference was not statistically significant ($\chi^2 = 3.341$; $P = 0.188$). According to samples consistency, the highest prevalence was found in watery stool samples by 13.6 %, followed by soft with 11.1 % and the least was found with 5.9 %, but the difference was not statistically significant ($\chi^2 = 5.781$; $P = 0.056$).

Molecular Detection of E. histolytica *by PCR method*

The 78 samples that were positive by ELISA were used in the PCR protocol for the detection of *E. histolytica* DNA. From these, 42 samples showed positive for *E. histolytica* small-subunit rRNA gene. Figure 21.1 shows the results obtained from the PCR detection of *E. histolytica* from stool samples.

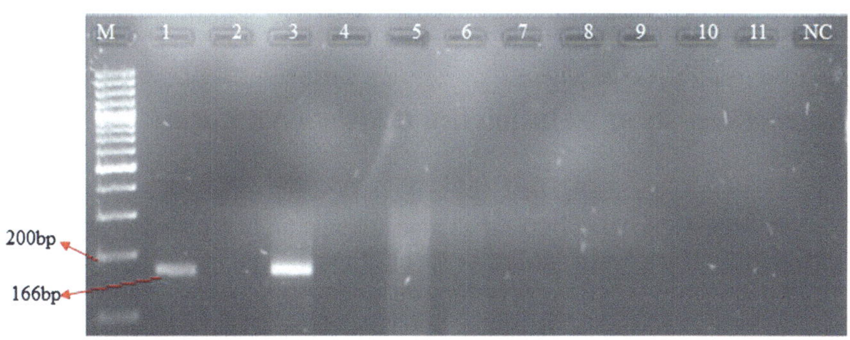

Fig. 21.1 An agarose gel picture showing the PCR results for the detection of *E. histolytica* in the stool samples. M-Molecular marker 100 bp, NC-negative control, 1-11 DNA samples

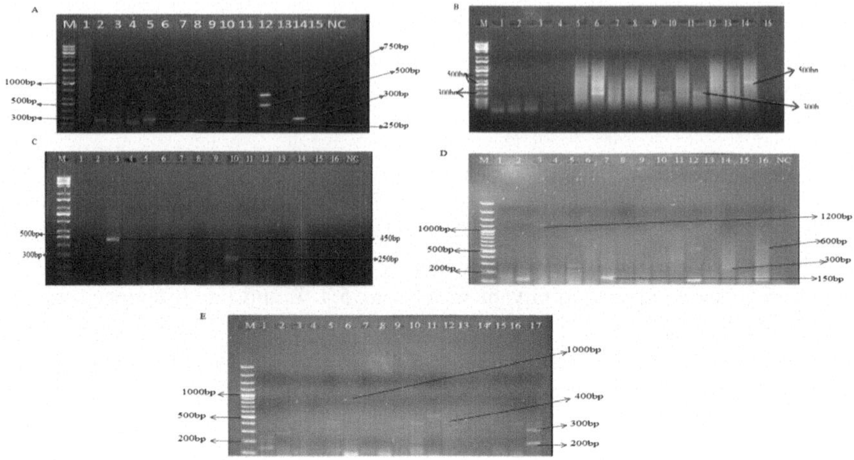

Fig. 21.2 Pictures of the agarose gel electrophoresis showing the profiles of the different loci (A = NK, B = RR, C = AL, D = DA and E = S^{TGA}-D loci). M-Molecular marker 100 bp, NC-negative control, 1-17 DNA samples

Genotyping of E. histolytica *Based on PCR Profile*

In order to investigate the genetic diversity of the *E. histolytica* based on the six tRNA-linked STR loci, 78 DNA samples were used in different PCR reactions targeting the six STR loci. The PCR was unsuccessful in some of the samples for certain loci of the six tRNA-linked STR. Loci S-Q did not show any bands in all tested samples and was excluded from the study. Figure 21.2 shows some examples of the genetic profiles obtained from the samples for the different genes. In the present study, different profiles were observed in different samples of different origins (Giyani and Pretoria).

Genetic Diversity of E. histolytica *Based on NK Locus*

From the 15 samples that amplified for the NK locus, 9 different profiles were obtained as shown in Table 21.3. Of all nine different profiles obtained for the NK locus, profile number 5 was found to be the most common profile compared to other profiles. This profile occurred in five different samples in Giyani and 4 were females and 1 was male. Of these five samples, four samples had loose consistency and one had formed consistency. Therefore this profile was not associated with diarrhea. Another profile that occurred in two samples was profile number 3 and both samples were from Giyani and were both from male participants, and had loose consistency.

Table 21.3 Distribution of the different profiles for the NK locus of *E. histolytic*

	Sample origin		Consistency			Gender		
NK profile	Giyani	Pretoria	Formed	Loose	Watery	Male	Female	Total
1	0	1	0	0	1	1	0	1
2	0	1	0	1	0	0	1	1
3	2	0	0	2	0	2	0	2
4	1	0	1	0	0	0	1	1
5	5	0	1	4	0	1	4	5
6	1	1	0	2	0	2	0	2
7	0	1	0	1	0	1	0	1
8	1	0	1	0	0	0	1	1
9	0	1	0	1	0	1	0	1
Total	10 (66.7 %)	5 (33.3 %)	3 (20 %)	11 (73.3 %)	1 (6.7 %)	8 (53.3 %)	7 (46.7 %)	15 (100 %)

Out of all profile obtained in NK locus, only profile number 1 had watery consistency and was from Pretoria. It is possible that this profile could be associated with diarrhea.

Genetic Diversity of *E. histolytica* Based on the RR Profile

From the 30 samples that amplified for the RR locus, 16 different profiles were obtained as shown in Table 21.4. Of all sixteen different profiles obtained in RR locus, profile number 9 occurred four times in different samples of which three were from Pretoria (two males and one female) and one was from Giyani (female). Of these four samples, the two males from Pretoria had watery consistency. Another profile that occurred in four samples was profile number 11 where two samples were from Giyani (male and female) and another two were from Pretoria (male and female). Of these four samples, three samples had loose consistency and one had formed consistency. These and other profile numbers can be observed in Table 21.4.

Genetic Diversity of **E. histolytica** *Based on AL Locus*

From the 25 samples that amplified for the AL locus, 15 different profiles were obtained as shown in Table 21.5. Of all fifteen different profiles obtained in AL locus, profile number 5 occurred six times in different samples. Out of those six samples observed with profile number 5, three samples were from Pretoria (one male and two females) while three were from Giyani (two males and one female). Of these six samples, only one sample from Giyani had formed consistency and the rest had loose consistency. Another profile that occurred in three samples was profile number 14 of which two samples were from Pretoria (two males) and one was from Giyani (female). Of these three samples, two samples from Pretoria had loose consistency and one from Giyani had formed consistency. Profiles 2, 3 and 8 were observed in two samples each and all of them were from Pretoria (female and male), and only one sample observed in profile number 2 had watery consistency and the rest had loose consistency. Other profiles occurred once in the sample and included profile number 1, 4, 6, 7, 9, 10, 11, 12 and 13. Out of all these profiles, only profiles number 4 and 7 had watery consistency.

Genetic Diversity of **E. histolytica** *Based on the DA Locus*

From the 13 samples that amplified for the DA locus, 6 different profiles were obtained as shown in Table 21.6. Profile number 2 occurred five times in different samples. Out of those five samples observed in profile number 2, four samples were from Pretoria (two males and two females) and one was from Giyani (female).

Table 21.4 Distribution of the different profiles for the RR locus of *E. histolytica*

	Origin		Consistency			Gender		
RR profile	Giyani	Pretoria	Formed	Locse	Watery	Male	Female	Total
1	1	0	0	1	0			1
2	2	0	1	1	0	0	2	2
3	0	2	0	1	1	1	1	2
4	0	1	0	1	0	1	0	1
5	1	0	0	1	0	0	1	1
6	0	1	0	1	0	1	0	1
7	1	2	0	3	0	2	1	3
8	0	1	0	1	0	0	1	1
9	1	3	1	1	2	2	2	4
10	0	3	1	2	0	2	1	3
11	2	2	1	3	0	2	2	4
12	0	1	0	1	0	1	0	1
13	1	0	0	1	0	1	0	1
14	1	0	1	0	0	1	0	1
15	1	2	1	2	0	2	1	3
16	0	1	0	1	0	0	1	1
Total	11 (36.7 %)	19 (63. 3%)	6 (20 %)	21 (70%)	3 (10 %)	16 (55.2 %)	13 (44.8 %)	30 (100 %)

Table 21.5 Distribution of the different profiles for the AL locus of *E. histolytica*

	Origin		Consistency			Gender		
Al profile	Giyani	Pretoria	Formed	Loose	Watery	Male	Female	Total
1	0	1	0	1	0	1	0	1
2	0	2	0	1	1	1	1	2
3	0	2	0	2	0	1	1	2
4	0	1	0	0	1	0	1	1
5	3	3	1	5	0	3	3	6
6	1	0	0	1	0	0	1	1
7	0	1	0	0	1	1	0	1
8	0	2	0	2	0	1	1	2
9	0	1	0	1	0	1	0	1
10	1	0	1	0	0	0	1	1
11	0	1	0	1	0	0	1	1
12	0	1	1	0	0	1	0	1
13	0	1	0	1	0	0	1	1
14	1	2	1	2	0	2	1	3
15	1	0	0	1	0	0	1	1
Total	7 (28 %)	18 (72 %)	4 (16 %)	18 (72 %)	3 (12 %)	12 (48 %)	13 (52 %)	25 (100 %)

Table 21.6 Distribution of the different profiles for the DA locus of *E. histolytica*

	Origin		Consistency			Gender		
DA profile	Giyani	Pretoria	Formed	Loose	Watery	Male	Female	Total
1	0	2	1	1	0	2	0	2
2	1	4	1	4	0	2	3	5
3	1	0	0	1	0			1
4	0	1	0	0	1	0	1	1
5	0	3	0	2	1	3	0	3
6	0	1	0	0	1	1	0	1
Total	2 (15.4%)	11 (84.6%)	2 (15.4%)	8 (61.5%)	3 (23.1%)	8 (66.7%)	4 (33.3%)	13 (100%)

Of these five samples, only one sample from Pretoria had formed consistency and the rest had loose consistency. Profile 5 occurred in three samples, which were from Pretoria and were all males. Two samples had loose consistency and one had watery consistency. Profiles number 4 and 6 were observed in one sample each and were from Pretoria (male and female). Both of them had watery consistency. Another profile that occurred in two samples was profile number 1 and, both samples were from Pretoria (males). Of these two samples, one had loose consistency and another one had formed consistency.

Genetic Diversity of **E. histolytica** *Based on the S^{TGA}-D Locus*

Of the 9 samples that amplified, 6 profiles were generated. Of the six profiles obtained in S^{TGA}-D locus as represented in Table 21.7, profile number 3, 5 and 6 occurred in two different samples each and all samples had loose consistency. Two samples observed in profile number 3 were males while the rest were females. Profiles number 1, 2 and 4 were observed in one sample each and were female but only one sample observed in profile number 4 had watery consistency.

Discussion

The objectives of this study were to determine the prevalence of *E. histolytica* and to determine the molecular characteristics of *E. histolytica* in relation to the occurrence of diarrhea among patients attending a rural primary health care clinic in Giyani and private clinics in the Gauteng region (Pretoria). *Entamoeba histolytica* is the causative agent of one of the most important infectious diseases called amoebiasis that affects mainly tropical and subtropical countries. *Entamoeba histolytica* is the pathogenic species that has been documented to cause diseases in humans and can also cause serious damage to the intestinal and other extra-intestinal organs, such as the brain and liver (Ximenez et al. 2010). Microscopy is still the only tool used for the detection of *E. histolytica* in most developing countries, but cannot distinguish between *E. histolytica* and *E. dispar*. Consequently, it is now believed that epidemiological figures on the disease and its spread are overemphasized since they relied on microscopic identification (Ali et al. 2008).

The prevalence of *E. histolytica* infection varies in different parts of the world. *Entamoeba histolytica* was found to be more prevalent in Pretoria by 10.5 % than in Giyani by 5.4 %. Pretoria is an urban area and one would estimate that the prevalence of *Entamoeba* would be higher in the rural area which in this case in Giyani. This could be due to the fact that the samples from Pretoria were mostly diarrheal samples that have been submitted for routine laboratory analysis. The findings of this study agree with the results of the study done by Al-Shammari et al. (2001), who detected higher rates of infection in urban than rural areas.

Table 21.7 Distribution of the different profile for S^{TGA}-D locus of *E. histolytica*

	Sample origin	Consistency			Gender		
STGA-D profile	Pretoria	Formed	Loose	Watery	Male	Female	Total
1	1	1	0	0	0	1	1
2	1	0	1	0	0	1	1
3	2	0	2	0	2	0	2
4	1	0	0	1	0	1	1
5	2	0	2	0	0	2	2
6	2	0	2	0	0	2	2
Total	9 (100 %)	1 (11.1 %)	7 (77.8 %)	1 (11.1 %)	2 (22.2 %)	7 (77.8 %)	9 (100 %)

In the present study, there was a high prevalence of *E. histolytica* infection in males (12 %) than in females (8.4 %). These results agree with the results of the study done by Ozyurt et al. (2007), in which they demonstrated the high prevalence of *E. histolytica* infection in male with the rate of 67 % as compared to the female with a low rate of 33 %. However, it disagreed with the results of the study done by Ozgumus et al. (2007), in which they reported a high prevalence of *E. histolytica* infection for female with the rate of 64 % and 36 % for male among patients attending health care in Turkey. Also a hospital-based study in Pakistan, observed a high prevalence of *E. histolytica* infection in female which was 31.5 % compared to 19.6 % of male (Ejaz et al. 2011). The present study showed that the participants of age group between 26 and 45 years had higher rate of *E. histolytica* infection as compare with other ages (0–25 and 46–90), although this difference was not statistically significant ($X^2 = 3.341$; $P = 0.188$). This is possibly due to the fact that this age group is active and might be more exposed to the pathogen. *Entamoeba histolytica* infection is a fecal oral disease, thus improper hygiene practice play a major role in the transmission of *E. histolytica*. A study conducted in Vietnam also showed that the transmission of *E. histolytica* was more than three -fold risk increase if hands were not properly washed (Pham Duc et al. 2011). While the results of this study disagree with the results of the study done by Tanyuksel and Petri (Dawood et al. 2002), in which they demonstrated the high prevalence of *E. histolytica* infection at the rate of 66 % among ages of 1–5 years.

Out of the 78 samples found to be positive for *E. histolytica* by ELISA, 42 samples showed positive for *E. histolytica* tRNA gene. The antigen-based ELISA and PCR are two such techniques reported to demonstrate exceptional sensitivities and specificities when compared with microscopy (Tanyuksel and Petri 2003). This is not an indication that PCR is less sensitive than ELISA but the difficulties related to the amplification of tRNA genes that we encountered during our study. Other authors also found that not all samples positive even for PCR using the ssrDNA amplified for the tRNA genes. This could also be an indication of the diversity among the strains (Ali et al. 2014).

The ability to identify strains of *E. histolytica* may lead to insights into the population structure and epidemiology of the organism. Locus S-Q did not show any bands in all tested samples and were excluded from the study. In a previous study (Ali et al. 2005), it was observed that the success rate for PCR amplification was higher for some STRs than others and that nested PCR had a success rate of >99 % across all STRs, irrespective of the sample origin. One of the recommendations of their study was the use of nested PCR when negative results were obtained with single primer pairs or when reagent cost is not a significant consideration. In the present study, all five loci (NK, RR, Al, DA and STGA-D) were amplified in the first PCR round using the forward and reverse primers.

The results obtained in the present study indicated the possibility of strains clustering by region. The genetic profiling of *E. histolytica* indicated that some strains were specific to each of the two locations (Giyani and Pretoria), while some were common. For examples, profile number 3 and 10 of the RR locus; profile number 5 of the AL locus and profile number 2 of the DA locus were more identified in Pretoria than other profile, and profile number 2 of the RR locus and profile number 5 of the AL locus were also more identified in Giyani. These results confirm the diversity of this parasite and suggest that these loci could be useful in geographic segregation of as well as the source tracking of the infections. Furthermore, genetic variants were specifically found in asymptomatic samples while some were found in diarrheal samples. This suggest that some variants might have a role in the presentation of amoebiasis either pathogenic or non-pathogenic (symptomatic or asymptomatic infections). A study done previously (Stanley 2003) confirmed that most of the individuals infected by *E. histolytica* remain asymptomatic while others have clinical signs. Profile Number 1 of the NK locus was more prevalent in diarrhea samples and was from Pretoria. Similarly, some profiles (profile number 10 of the RR locus) were more associated with diarrhea compared to others indicating that the outcome of the infection by this pathogenic parasite might be associated with the genotyping. Diarrheal disease affects people of all ages and is one of the greatest causes of morbidity and mortality throughout the world (Bern and Guerrant 2003).

Conclusion

The present study indicates that infections caused by *E. histolytica* were more prevalent in diarrheal samples even though they were obtained from a metropolitan area (Pretoria) compared to asymptomatic samples from a rural area (Giyani). *Entamoeba histolytica* infection was more prevalent t in males than in females, and in the age group of 26–45 years. According to our results, microscopy is a simple method and it should be combined with other methods such as ELISA and PCR for identification of the species to avoid false and/or insufficient diagnosis and treatment applications. The result obtained in this study indicated that tRNA genes might have a role in the presentation of amoebiasis (symptomatic and asymptomatic infections) depending on the genetic profile of the infecting strain. This genotyping system could also be used to identify the origin of the infection once it has been standardized.

Acknowldgements The authors express their sincere gratitude to the rural primary health care clinics management in Giyani, Limpopo Province and private clinics management in Pretoria, Gauteng Province as well as the patients for their cooperation. This study received financial support from a grant made available to Prof. A Samie through the National Research Foundation.

Competing Interest The authors declare that they have no competing interests.

References

Al-Harthi, S. A., & Jamjoom, M. B. (2007). Enteroparasitic occurrence in stools from residents in Southwestern region of Saudi Arabia before and during Umrah season. *Saudi Medical Journal, 28,* 386–390.

Ali, I. K. M., Clark, C. G., & Petri, W. A., Jr. (2008). Molecular epidemiology of amoebiasis. *Infection, Genetics and Evolution, 8,* 698–707.

Ali, I. K. M., Mondal, U., Roy, S., Haque, R., Petri Jr., W. A., Clark, C.G. (2007) Evidence fora link between parasite genotype and outcome of infection with Entamoeba histolytica.Journal of Clinical Microbiology, *45*, 285–289.

Ali, E., Samie, A., Sandi, E., & Raed, A. O. (2014). Genetic variability of the Serine-Rich *Entamoeba histolytica* protein gene in clinical isolates from the United Arab Emirates. *Tropical Biomedicine, 31*(2), 370–377.

Ali, I. K. M., Zaki, M., & Clark, C. G. (2005). Use of PCR amplification of tRNA gene-linked short tandem repeats for genotyping *Entamoeba histolytica. Journal of Clinical Microbiology, 43,* 5842–5847.

Al-Shammari, S., Khoja, T., El-Khwasky, F., & Gad, A. (2001). Intestinal parasitic diseases in Riyadh Saudi Arabia: Prevalence, sociodemographic and environmental associates. *Tropical Medicine & International Health, 6*(3), 184–189.

Alum, A., Rubino, J. R., & Ijaz, M. K. (2010). The global war against intestinal parasites-should we use holistic approach? *The International Journal of Infectious Diseases (IJID), 14,* e732–e738.

Bern, M. C., & Guerrant, R. L. (2003). The global burden of diarrheal disease, as estimated from studies published between 1992 and 2000. *Bulletin of the World Health Organization (WHO), 81,* 197–204.

Das, K., Mukherjee, A. K., Chowdhury, P., Sehgal, R., & Bhattacharya, M. K. (2014). Multilocus sequence typing system (MLST) reveals a significant association of *Entamoeba histolytica* genetic patterns with disease outcome. *International Journal for Parasitology, 66,* 308–314.

Dawood, K. A., Al-Muhja, R., Al-Zubaidy, F., & Al-Abbas, J. A. (2002). A study of giardia and amoebiasis in maternity and Hospital of Diwania, Kufa. *Journal of Medicine, 5*(1), 167–170.

Ejaz, M., Murtaza, G., Ahmad, M., Khan, S. A., Saqib, Q. N., Asad, M. H. H. B., et al. (2011). Determination of the prevalence of Entamoeba histolytica in human ata private fertilizer company hospital in Pakistan using microscopic technique. *African Journal of Microbiology Research, 5,* 149–152.

Feng, M., Cai, J., Yang, B., Fu, Y., Min, X., Tahibana, H., et al. (2012). Unique short tandem repeat nucleotide sequences in *Entamoeba histolytica* isolates from China. *Parasitology Research, 111,* 1137–1142.

Haghighi, A., Kobayashi, S., Takeuchi, T., Masuda, G., & Nozaki, T. (2002). Remarkable genetic polymorphism among *Entamoeba histolytica* isolates from a limited geographic area. *Journal of Clinical Microbiology, 40,* 4081–4090.

Hamzah, Z., Petmitr, S., Mungthin, M., Leelayoova, S., & Chavalitshewinkoon-Petmitr, P. (2006). Differential detection of *Entamoeba histolytica, Entamoeba dispar*, and *Entamoeba moshkovskii* by a single-round PCR assay. *Journal of Clinical Microbiology, 44,* 3196–3200.

Jaiswal, V., Ghoshal, U., Mittal, B., Dhole, T. N., & Ghoshal, U. C. (2014). Association between allelic variation due to short tandem repeats in tRNA gene of *Entamoba histolytica* and clinical phenotypes of amoebiasis. *Acta Tropica, 133,* 1 7.

Ozgumus, O. B., Celik-Sevim, E., Alpay-Karaoglu, S., Sandallic, C., & Sevim, A. (2007). Molecular characterization of antibiotic resistant *Escherichia coli* strains isolated from tap and spring waters in a Coastal Region in Turkey. *International Journal of Microbiology, 45,* 379–387.

Ozyurt, M., Kurt, O., Yaman, O., Ardic, N., & Haznedaroglu, T. (2007). Evaluation of intestinal parasites in a period of four years in the coprology laboratory of a training hospital. *Turkish Parazirol Derg, 31,* 306–308.

Parija, S. C., & Prabhakar, P. K. (1995). Evaluation of lacto-phenol cotton blue for wet mount preparation of feces. *Journal of Clinical Microbiology, 33,* 1019–1021.

Pham Duc, P., Nguyen-Viet, H., Hattendorf, J., Zinsstag, J., Dac Cam, P., & Odermatt, P. (2011). Risk factors for Entamoeba histolytica infection in an agricultural community in Hanam Province Vietnam. *Parasit Vectors, 4,* 102.

Samie, A., Obi, C. L., Bessong, P. O., Houpt, E., Stroup, S., Njayou, M., et al. (2008). *Entamoeba histolytica*: genetic diversity of African strains based on the polymorphism of the serine-rich protein gene. *Experimental Parasitology, 118,* 354–361.

Stanley, S. L. (2003). Amoebiasis. *Lancet, 361,* 1025–1034.

Stark, D., van Hal, S., Fotedar, R., Butcher, A., Marriott, D., Ellis, J., et al. (2008). Comparison of stool antigen detection kits to PCR for diagnosis of amebiasis. *Journal of Clinical Microbiology, 46,* 1678–1681.

Tanyuksel, M., & Petri, W. A., Jr. (2003). Laboratory diagnosis of amoebiasis. *Clinical Microbiology Reviews, 16,* 713–729.

Tawari, B., Ali, I. K. M., Scott, C., Quail, M. A., Berriman, M., & Hall, N. (2008). Patterns of evolution in the unique tRNA gene arrays of the genus *Entamoeba. Molecular Biology and Evolution, 25,* 187–198.

Ximenez, C., Cerritos, R., & Roja, L. (2010). Human amoebiasis: Breaking the paradigm? *International Journal of Environmental Research and Public Health (IJERPH), 7*(3), 1105–1120.

Chapter 22
Gene Silencing and Overexpression to Study Pathogenicity Factors of *Entamoeba histolytica*

Constantin König and Iris Bruchhaus

Abstract The pathogen *Entamoeba histolytica* can live asymptomatically in the human gut, or it can disrupt the intestinal barrier and induce life-threatening abscesses in different organs, most often in the liver. The molecular framework that enables this invasive, highly pathogenic phenotype is still not well understood. Genes encoding long-time favoured pathogenicity factors like the cysteine peptidases, the surface adherence lectins or the pore forming peptides are also present in non-pathogenic clones of the parasite. Therefore, a key task in amoebic research is to elucidate the mechanisms underlying *E. histolytica* invasion and tissue destruction. A lot of effort has been put into trying to identify molecules that trigger invasion of amoebae to tissues, evasion of the immune response and survival within the unfriendly environment of the host. This also applies to the genetic manipulation of parasites aiming to characterize putative pathogenicity factors. So far, the gene silencing by means of the CRISPR/Cas9 system does not work for *E. histolytica*. However, it has been possible to overexpress genes in *E. histolytica* for quite some time. In addition, a method developed in Upinder Singh's laboratory based on RNAi silencing allows stable silencing of genes. Furthermore, *E. histolytica* clones with different pathogenicity are available, which were derived from the same amoeba isolate. Together with newly developed in vitro infection models, this opens up the possibility of clarifying the complex pathology of an *E. histolytica* infection in the future.

Keywords Transfectants · Overexpression · Gene silencing · Virulence factor

Introduction

The intestinal protozoan *Entamoeba histolytica* is an important human parasite. Recent data clearly indicate that the life-threating amoebic liver abscess (ALA) continues to be a common clinical complication of amoebiasis infection in Asian,

C. König · I. Bruchhaus (✉)
Bernhard Nocht Institut for Tropical Medicine, Hamburg, Germany
e-mail: bruchhaus@bnitm.de

© Springer Nature Switzerland AG 2020

N. Guillen (ed.), *Eukaryome Impact on Human Intestine Homeostasis and Mucosal Immunology*, https://doi.org/10.1007/978-3-030-44826-4_22

African and Latin American countries with 11,300 death in 2013 (Herricks et al. 2017). *E. histolytica* can induce amoebic colitis as well as formation of ALAs in humans. The life cycle of this parasite consists of infectious cysts that survive outside the host and vegetative trophozoites that proliferate in the human gut. In general, trophozoites persist asymptomatically for months or years in the human intestine. However, in 10% of cases, the trophozoites become, under as yet unknown circumstances, invasive and induce extraintestinal amoebiasis. Invasion into the intestinal mucosa can lead to induction of amoebic colitis, whereas dissemination to the liver can result in ALA formation (Blessmann et al. 2003).

Currently, it is discussed that mainly three protein families (Gal/GalNAc lectin, cysteine peptidase and amoebapore family) are essential for ALA formation. The first step of this invasion process is the adhesion of amoebae to target cells via the Gal/GalNAc lectin (Petri and Schnaar 1995). After establishing contact, the amoebae secrete amoebapores, which mediates a contact-dependent lysis of the target cells (Leippe 1997). Important for invasion as well as for the process of tissue degradation are the cysteine peptidases (CPs) of *E. histolytica* (Bruchhhaus 2015) (Fig. 22.1).

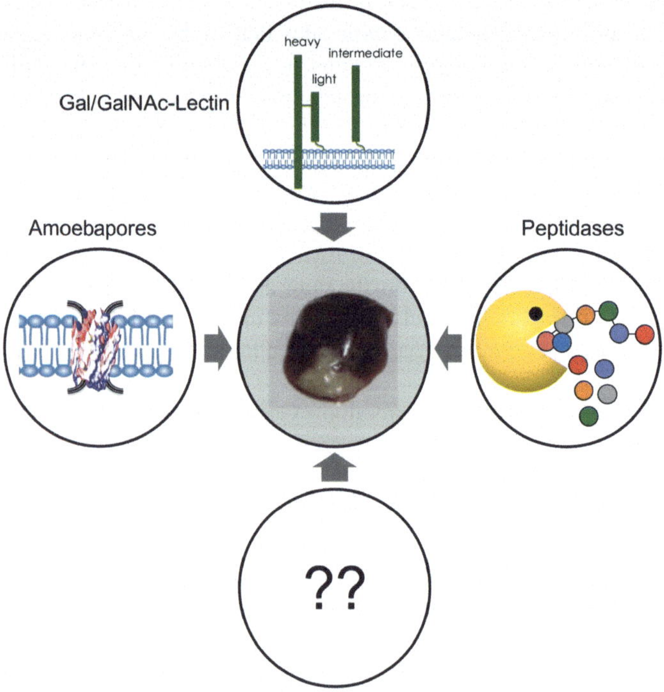

Fig. 22.1 Main pathogenicity factors during amoebic liver abscess formation. The Gal/GalNAc lectins, amoebapores and peptidases are believed to be the main pathogenicity factors of *E. histolytica*. However, it is postulated that *E. histolytica* still has a number of previously unidentified pathogenicity factors

However, most of these molecules are also present in the non-pathogenic species *E. dispar*.

Therefore, a key goal in current amoebic research is to elucidate the mechanisms of *E. histolytica* invasion and tissue destruction. Much effort has been applied trying to identify molecules that trigger invasion of amoebae to tissues, evasion of the host immune response and survival within the unfriendly environment of the host. One promising strategy to analyse the virulence of *E. histolytica* is the genetic manipulation of genes encoding putative pathogenicity factors.

Generation of Overexpression Transfectants for Functional Characterization of *Entamoeba histolytica* Proteins

The generation of overexpression transfectants for *E. histolytica* has been performed mostly with the vector developed by Hamann and colleagues (Hamann et al. 1995; Arhets et al. 1998; Wassmann et al. 1999; Davis et al. 2009; Voigt et al. 1999). Here, the neomycin phosphotransferase-coding sequence flanked by the 5′-untranslated sequence and the 3′-untranslated sequence of an *E. histolytica* actin gene is used for selection. The GOI is flanked by the 5′-untranslated sequence of an *E. histolytica* Gal/GalNAc lectin gene and the 3′-untranslated actin region (Nickel and Tannich 1994; Hamann et al. 1995) (Fig. 22.2a). Other vectors, which are based on very similar structures have also been used (Nozaki et al. 1998, 1999; Katz et al. 2002). Interestingly, ectopic expression does not work in all experiments. For example, successful overexpression of the genes coding for the two cysteine peptidases EhCP-A1 and EhCP-A5 can only be induced if the genes' own promoters are used (Hellberg et al. 2001; Tillack et al. 2006). In addition, it must always be taken into account that mislocalizations can occur. This applies in particular to the expression of fusion proteins. If, for example, EhCP-A2 and EhCP-A5 are expressed fused to GFP, they are stored as insoluble proteins in inclusion bodies in the amoebae (unpublished data). Off-target effects may also occur, but no information is yet available. To regulate overexpression an episomal tetracycline-inducible gene expression system was developed (Ramakrishnan et al. 1997; Hamann et al. 1997). Although the system has been developed for over 20 years, it has hardly been used so far. Overexpression transfectants can also be generated for the reptile pathogen *E. invadens*, which is used as model system to study trophozoite to cyst differentiation (Ehrenkaufer and Singh 2012; Singh et al. 2012).

In the following section, some examples are given that allowed characterizing protein functionally with the help of overexpression transfectants.

Several studies used the overexpression approach to study the importance of CPs for pathogenicity. Through in silico analyses, 35 potential *ehcp* genes could be identified in the genome of *E. histolytica*, which can be divided into three families termed EhCP-A, -B and -C due to their different structures. Only four of the CP encoding genes (*ehcp-a1, -a2, -a5,* and *-a7*) have been shown to be highly expressed in *E.*

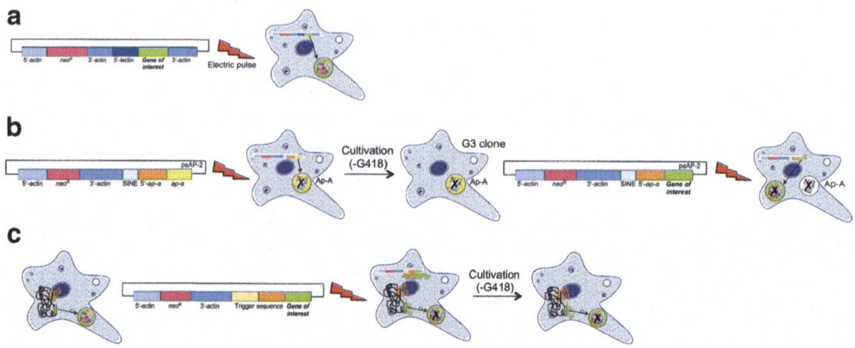

Fig. 22.2 Production of overexpression and silencing transfectants of *E. histolytica*. **a** Overexpression of the gene of interest (GOI) takes place with the help of an expression plasmid containing the neomycin phosphotransferase-coding sequence flanked by the 5′-untranslated sequence and the 3′-untranslated sequence of an *E. histolytica* actin gene. The GOI is flanked by the 5′-untranslated sequence of an *E. histolytica* Gal/GalNAc lectin gene and the 3′-untranslated actin region (Nickel and Tannich 1994; Hamann et al. 1995). **b** Silencing transfectants were produced by using the G3 clone, which must first be produced by transfecting a plasmid containing both the upstream 5′ region of *ehap-a* and the open reading frame of *ehap-a* (psAP-2). After prolonged cultivation without the addition of G418, the trophozoites lose the vector, but *ehap-a* expression is still silenced. To silence a second GOI G3 amoebae must be transfected with the psAP-2 plasmid containing the GOI fused to the 5′-upstream region of *ehap-a* (Bracha et al. 1999, 2003, 2006; Mirelman et al. 2008). **c** Another silencing approach is based on trigger-induced RNA interference (RNAi) gene silencing. This method is based on RNAi using small RNAs, which target genes by sequence-specific silencing. The gene coding region to which large numbers of antisense small RNAs map can serve as a trigger and can be used to silence a gene fused to it (Morf et al. 2013; Pearson et al.,2013; Matthiesen et al. 2019). With this method it was also possible to silence the expression of two genes simultaneously (Khalil et al. 2016)

histolytica under standard axenic culture conditions (Bruchhaus et al. 2003; Tillack et al. 2007; Clark et al. 2007). Overexpression of *ehcp-a1 and ehcp-a2* exclusively increased the activity of the encoded enzyme, whereas overexpression of *ehcp-a5* increased the activity of all major CPs. Furthermore, cytopathic activity, measured by in vitro monolayer disruption, was dramatically increased in *ehcp-a5* overexpression transfectants but showed only a modest increase in *ehcp-a1*- or *ehcp-a2* transfectants. In addition, overexpression of *ehcp-a1* and *ehcp-a2* in *E. histolytica* trophozoites did not augment liver abscess formation in laboratory animals. However, overexpression of *ehcp-a5* led to significant larger ALAs in comparison to the controls. The importance of *ehcp-a5* expression for the pathogenicity of *E. histolytica* trophozoites was further supported as it could be shown that overexpression could compensate reduction of in vivo pathogenicity in parasites that were silenced for the *ap-a* gene coding for the pore forming protein amoebapore A (see below) (Hellberg et al. 2001; Tillack et al. 2006). Also, overexpression of *ehcp-a5* restored the ability of a nonpathogenic cell line to form ALAs, whereas overexpression of *ehcp-a1* and *ehcp-a2* had no influence on virulence. In addition, it was shown that during ALA formation the expression of several *ehcp* genes, which are expressed at low levels under in vitro conditions, is increased in different animal models. Interestingly, overexpression of *ehcp-b8*, *-b9* and *-c13* also restored the pathogenic phenotype of the non-pathogenic clone (Matthiesen et al. 2013).

E. histolytica possesses two CP inhibitors (EhICP1 and EhICP2), which are localized in distinct compartments. With the help of the corresponding overexpression transfectants, it has been suggested that EhCP-A1 accidentally released into the cytosol, is the main target of EhICP1, while EhICP2, in addition to its role in housekeeping processes, may control the proteolytic processing of other hydrolases or performs other tasks different from protease inhibition (Sato et al. 2006; Saric et al. 2012).

Using overexpressing transfectants it could be shown that the *E. histolytica* iron superoxide dismutase (FeSOD) is one of the main components involved in the mechanism of metronidazole resistance (Wassmann et al. 1999). Metronidazole is the drug of choice for the treatment of ALA caused by *E. histolytica*. However, it is possible to generate a certain degree of metronidazole resistance in vitro (Samarawickrema et al. 1997; Wassmann et al. 1999). Overexpression of the FeSOD revealed a significant reduction in susceptibility to metronidazole. This reduction was highest in amoebae simultaneously overexpressing the two enzymatic antioxidants FeSOD and peroxiredoxin. Furthermore, FeSOD overexpressing transfectants quickly adapted to constant exposures of otherwise lethal metronidazole concentrations and metronidazole selection of transfected amoebae favoured retention of the FeSOD-containing plasmid (Wassmann et al. 1999). FeSOD and peroxiredoxin were also identified as two of the molecules present in higher amounts in the pathogenic isolate HM-1:IMSS compared to the non-pathogenic isolate Rahman. Overexpression of peroxiredoxin in Rahman increased the resistance of transfectants to H_2O_2 in vitro and an in vitro infection with these transfectants resulted in a higher degree of intestinal inflammation associated with human colon xenografts. Therefore, the peroxiredoxin may also be involved in amoebic pathogenicity (Davis et al. 2006).

Overexpression of a NAD(H) kinase resulted in an increased tolerance toward H_2O_2 as well as a significantly reduced intracellular concentration of reactive oxygen species. Furthermore, overexpression of a cytosine-5 DNA methyltransferase (Ehmeth) resulted in a pleiotropic phenotype that includes accumulation of multinucleated cells, upregulation of heat shock protein 70 expression and resistance to oxidative stress. Thus, the NAD(H) kinase and Ehmeth are two important molecules that protect amoebae from reactive oxygen species (Jeelani et al. 2013; Fisher et al. 2006).

Using a transcriptome analysis comparing cysts and trophozoites, a Myb DNA binding domain protein (EhMyb-dr) could be identified, which was suspected to be involved in stage conversion (Ehrenkaufer et al. 2007b). With the help of overexpression transfectants, this protein could be characterized as a transcription factor which regulates a set of stage-specific genes (Ehrenkaufer et al. 2009).

E. histolytica contains a homologue (EhPC4) to human positive coactivator 4 (PC4), a DNA-binding protein that facilitates the recruitment of transcription factors. Trophozoites that overexpresses EhPC4 significantly increased their cell proliferation, DNA replication and DNA content. Furthermore, an up-regulation of genes involved in carbohydrate and nucleic acids metabolism, chromosome segregation, and cytokinesis could be identified. These results led to the assumption that EhPC4 may represent a novel regulatory factor of amoebic gene expression (de la Cruz et al. 2014).

Phagocytosis plays an important role for *E. histolytica.* It is important for regular food uptake, but also involved in its pathogenesis, as it is required for killing and evasion of the host immune system (Ralston and Petri 2011; Faust and Guillen 2012). Overexpression of a mutant from of a C2-domain-containing protein kinase (EhC2PK) indicated that this protein is involved in the induction of phagocytic cups (Somlata et al. 2011). Using an overexpression approach RabGTPases, which are GTP-binding proteins were found as being involved in secretory vesicle trafficking events. More than 90 RabGTPases in the genome of *E. histolytica* could be identified by in silico analysis (Saito-Nakano et al. 2005). It was shown, that overexpression of EhRab8A led to increased phagocytosis, while expression of the dominant negative form of EhRab8A led to reduced phagocytosis (Hanadate et al. 2018). In addition, the overexpression of mutant *ehraba* led to reduced adhesion and phagocytosis of erythrocytes and these experiments also implied that EhRabA affect the morphology of secretory organelles and regulates the secretion of proteins in *E. histolytica* (Welter and Temesvari 2009). Overexpression of *ehrab5* enhanced also phagocytosis and was in addition involved in the transport of amoebapore to phagosomes (Saito-Nakano et al. 2004). The overexpression of *ehrab7a* and *ehrab7b* led to an increase in late endosome/lysosome acidification. The overexpression of an *ehrab7b* mutant versions also led to a defect in phagocytosis associated with the impaired formation and disassembly of pre-phagosomal vacuoles (Saito-Nakano et al. 2007). *Ehrab11b* overexpression led to an increase in both intracellular and secreted CP activities. This increase affected all major CPs and correlated with an increase in cytolytic activity. Thus, Rab11B appears to be involved in the transport and secretion of CPs (Mitra et al. 2007).

Generation of *Entamoeba histolytica* Silencing Transfectants for Functional Characterization of Proteins

Another possibility for functional characterisation of proteins is silencing of the expression of the respective genes. Unfortunately, methods like CRISPR/Cas9 or homologous recombination are presently not applicable for *E. histolytica.* So far, two alternative methods have been established that allow gene silencing in *E. histolytica.*

In the first approach trophozoites are used with amoebapore A gene silencing occurring after transfection of amoebae with a hybrid plasmid containing a 5′-upstream region of the *ehap-a* gene. With the help of this so-called G3 clone it is now possible to silence the expression of a second gene (Bracha et al. 1999, 2003, 2006; Mirelman et al. 2008) (Fig. 22.2b). However, this approach has inherent limitations. So far, only one special HM-1: IMSS isolate can be used to generate the G3 clone lacking *ehap-a* expression. Since amoebapore A is one of the pathogenicity factors of *E. histolytica,* the clone G3 cannot induce ALAs and is therefore non-pathogenic (Bracha et al. 1999, 2003, 2006; Mirelman et al. 2008; Zhang et al. 2004; Irmer et al. 2009).

Nevertheless, the functions of several *E. histolytica* proteins were analyzed with this approach, with some examples listed below. EhCP-A5 has been identified as a key virulence factor that induces the release of mucin by goblet cells. Using the silencing approach it has been shown that the binding of EhCP-A5 to αvβ3 receptors of goblet cells can initiate a signalling cascade involving PI3K, PKCδ and MARKKS to promote mucin secretion from goblet cells that are crucial for disease pathogenesis (Cornick et al. 2016). EhCP-A5 also appears to play a crucial role in the penetration of the colonic *lamina propria* during the invasion of the intestinal mucosa. In addition, *E. histolytica* silenced for *ehcp-a5* expression did not induce pro-inflammatory cytokine secretion of the host (Bansal et al. 2009).

The cyclooxygenase of *E. histolytica* (EhCox) was identified as a regulator of CP activity. EhCox stimulates the secretion of IL-8 from mucosal epithelial cells, which leads to the recruitment of neutrophils to the site of infection and exacerbate the infection. EhCox silencing led to a decrease in CP activity as well as in virulence. Therefore it is assumed that the regulation of CP activity by EhCox may be a negative feedback mechanism to reduce proteolytic activity during the process of colonisation (Shahi et al. 2018). Furthermore, by using different silencing transfectants, it could be shown that cell death und cytoskeleton disorganization depend on parasite adhesion via the Gal/GalNAc-lectin and amoebic CP activity (Faust et al. 2011).

Silencing of the gene encoding a leishmanolysin-like metallopeptidase (EhMSP-1) leads to an increase in adhesion and phagocytosis and a reduction in mobility (Teixeira et al. 2012). EhMSP-1 silencing transfectants were no longer able to form dot-like polymerized actin structures after interaction with the human extracellular matrix component fibronectin similar to mammalian invadosome structures (Hasan et al. 2018). Silencing the gene encoding the rhomboid peptidase of *E. histolytica* led to decreased host cell adhesion. The reduction of adhesion was observed only for live cells and not for apoptotic cells. Nevertheless, the influence of the rhomboid peptidase on phagocytosis seems to be independent from its influence on adhesion (Baxt et al. 2010). In a more comprehensive approach where 15 genes were silenced that were more strongly expressed in phagocytic amoebae, 5 genes could be identified in which silencing led to a reduction in the ability to phagocytize apoptotic host cells. Interestingly, the phagocytosis of live cells was not affected (Sateriale et al. 2016).

Trogocytosis has recently been identified as a key step in amoebic cytolysis and invasion (Ralston et al. 2014). Now, with the help of silencing transfectants, a member of the AGC family kinase (EhAGCK1) has been identified that is specifically involved in trogocytosis, while a second member (EhAGCK2) is involved in all actin-dependent endocytic processes (Somlata et al. 2017).

However since the G3 clone used in these silencing approaches is no longer pathogenic due to the inhibition of *ap-a* expression, this method is not suitable to investigate putative pathogenicity factors of *E. histolytica*. Recently a second approach which is based on trigger-induced RNA interference (RNAi) gene silencing and that did not have the limitations was developed (Morf et al. 2013; Pearson et al. 2013). This method is based on RNAi using small RNAs, which target genes by sequence-specific silencing (Zhang et al. 2008). The gene-coding region to which large numbers of antisense small RNAs map can serve as a trigger and can be used to silence a gene fused to it (Morf et al. 2013; Pearson et al. 2013). In all cases, the first 132 bp of the EHI_197520 coding sequence was used as the trigger region. The *ehi_197520* gene has abundant sRNAs and low gene expression in the HM-1:IMSS isolate (Morf et al. 2013). With this method it was also possible to silence the expression of two genes simultaneously (Khalil et al. 2016) (Fig. 22.2c). Furthermore, this approach has also been adapted to use in *E. invadens* (Suresh et al. 2016).

Unfortunately, only a few studies used this approach for characterization of amoeba proteins. Using the silencing approach described by Bracha and colleagues, is was shown that the peptidase EhROM1 is involved in adhesion to and phagocytosis of host cells (Bracha et al. 1999, 2003; Baxt et al. 2010). Based on the novel approach, the influence of EhROM1 on pathogenicity could now also be investigated (Morf et al. 2013). Silencing of EhROM1 led to a reduction in cytotoxicity, hemolytic activity, directed and undirected transwell migration (Rastew et al. 2015). In addition, using the silencing method described above, the molecule EHI_108720 was found to bind to an H_2O_2-regulating motif (HRM) and was characterized as a stress-reactive transcription factor that controls a transcriptional regulatory network associated with oxidative stress (Pearson et al. 2013). Interstingly, the RNAi-based silencing approach failed if genes coding for proteins of the RNA pathway should be targeted including Argonaute genes, RNaseIII, and RNA-dependent RNA polymerase (RdRP) (Pompey et al. 2014).

Comparative Studies Between Pathogenic and Non-pathogenic *Entamoeba histolytica* Isolates

A straight-forward approach to identify molecules involved in ALA formation is the direct and quantitative comparison of pathogenic and non-pathogenic strains at proteome and transcriptome levels. Unfortunately, in most of the previous studies the pathogenic isolate HM-1:IMSS was compared with the non-pathogenic isolate Rahman, although both strains have a different genetic background (MacFarlane and Singh 2006; Davis et al. 2007; Ehrenkaufer et al. 2007b; Davis et al. 2006). Furthermore, the non-pathogenic Rahman strain has several serious defects (e.g. defect in phagocytosis; reduced cytotoxicity, unable to growth in animals, truncated glycan chain of the proteophosphoglycan coating surface). The genomic difference between HM-1: IMSS and Rahman is very low, however DNA fragment duplications were detected (Weedall et al. 2012). In addition, the HM-1: IMSS strain used as a virulent isolate loses its ability to form ALA after prolonged cultivation (Olivos et al. 2005). Comparison of the virulent HM-1: IMSS with the long-term cultivated non-virulent HM-1: IMSS indicated that the contact between trophozoites and the human colon is associated with a unique gene transcription profile required for the development of an invasive phenotype and that the ability to switch to a virulent phenotype has been lost in the Rahman strain (Weber et al. 2016). In addition, most of the comparative transcriptome analyses of the isolates HM-1: IMSS/Rahman were conducted using a microarray approach that did not cover the entire genome of the amoebae (MacFarlane and Singh 2006; Davis et al. 2007; Ehrenkaufer et al. 2007; Davis et al. 2006). Due to these severe limitations it was important to obtain *E. histolytica* isolates with stable phenotypes that were derived from an identical isolate of origin.

Recently, clones of a non-pathogenic cell line (HM-1: IMSS-A) and a pathogenic cell line (HM-1:IMSS-B) were generated and analysed for their ability to induce ALA in two murine models for the disease (Biller et al. 2009, 2010; Meyer et al. 2016). HM-1: IMSS was originally isolated from a colonic biopsy of rectal ulcer from an adult male patient with amoebic dysentery in 1967, Mexico City, Mexico. The monoxenically cultured HM-1: IMSS isolate was passed from Margarita de la Torre to Louis S. Diamond who adapted it to axenic cultivation. Then, this axenically cultivated HM-1: IMSS isolate was transferred to the American Type Culture Collection (ATCC). Cell line A was send to the Bernhard Nocht Institute for Tropical Medicine (BNITM; Hamburg, Germany) in 2001 by Barbara Mann (Charlottesville, University of Virginia) as a batch of cells from the same culture that was used for DNA preparation to sequence the *E. histolytica* genome (Loftus et al. 2005). Cell line B was purchased directly from ATCC in 1991 and has been cultivated at BNITM ever since. Both cell lines were cloned (Fig. 22.3a). All clones derived from HM-1: IMSS-A show no or very small ALA formation in gerbils 7 days post infection. This phenotype is more divergent for clones derived from HM-1: IMSS-B: for 2/3 of the B-clones, clear ALA formation was visible 7 days post infection, whereas for the other 1/3 abscess formation was significantly lower or not present at all (Fig. 22.3b/c). Interestingly, the pathogenic phenotype of the respective B-clones has remained stable over years without further animal passages (Meyer et al. 2016). To identify the factors putatively involved in abscess formation, three clones ($A1^{np}$, $B2^{p}$ and $B8^{np}$) were selected for in-depth transcriptome analyses (Meyer et al. 2016). In total, 76 genes are differentially expressed between clones $A1^{np}$ and $B2^{p}$, and 19 genes are differentially expressed between $B2^{p}$ and $B8^{np}$. Only six genes were similarly regulated between the non-pathogenic clones $A1^{np}$ and $B8^{np}$ and the pathogenic clone $B2^{p}$ (Fig. 22.3d) (Meyer et al. 2016).

To investigate the influence of the proteins encoded by the differentially expressed genes, overexpression transfectants of the most promising candidates were generated. Of 20 genes analysed, ectopic expression of eight reduced pathogenicity. These genes encode for a metallopeptidase (EhMP8-2), C2 domain proteins, alcohol dehydrogenases, and hypothetical proteins. Only one gene, encoding a hypothetical protein (EHI_127670), increased pathogenicity upon overexpression (Meyer et al. 2016).

As mentioned above, another possibility to investigate the influence of differentially expressed genes on pathogenicity is their silencing. Unfortunately, the trigger region *ehi_197520* normally used to silence genes, is expressed in the HM:1-IMSS derived clones $A1^{np}$ $B2^{p}$, and $B8^{np}$ and can therefore not be used to silence genes

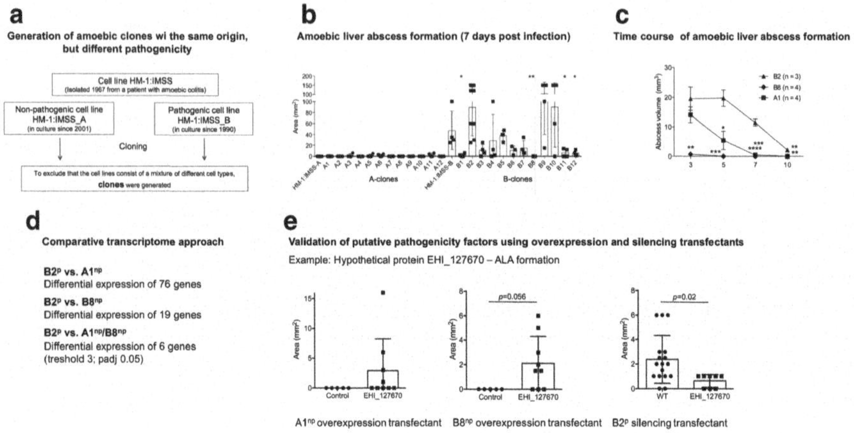

Fig. 22.3 Identification of new pathogenicity factors. One way to identify molecules involved in ALA formation is the direct and quantitative comparison of pathogenic and non-pathogenic strains. At proteome and transcriptome levels. **a** For a direct comparison, cloned amoebae which differ in their pathogenicity but originate from the same isolate are suitable. **b** Analysis of ALA formation after infecting gerbils with various *E. histolytica* clones. **c** Analysis of the time course of ALA formation after infecting gerbils with clones A1^{np}, B2^{p} or B8^{np}. D. Comparative transcriptome approach of in vitro cultivated amoebae of clones A1^{np}, B2^{p} or B8^{np}. Analysis of ALA formation after infecting mice with various *E. histolytica* transfectants. As examples A1^{np} and B8^{np} transfectants silenced for the expression of EHI_127670 and B2^{p} transfectant overexpressing EHI_127670 was shown (Meyer et al. 2016; Matthiesen et al. 2019)

in these amoebae. However, recently, trigger sequences that allow specific silencing in A- and B-clones, respectively have been identified (Matthiesen et al. 2019; Morf et al. 2013). Using the newly identified trigger sequences, the expression of 15 genes differentially expressed between the pathogenic and non-pathogenic clones were silenced. The respective transfectants were analyzed for their ability to induce liver destruction in the murine model for the disease. An influence on pathogenicity could only be shown for two of the genes investigated. Silencing of *ehi_180390* (encoding an AIG1 protein) increased liver pathology induced by a non-pathogenic parental clone. Interestingly, silencing of *ehi_127670*, which was identified as a pathogenicity factor using the overexpression transfectants, decreased the pathogenicity of the pathogenic clone B2^{p}. In addition, in EHI_127670-silencing transfectants a lower activity for EhCP-A1 and EhCP-A5 could be detected, while overexpression did not influence CP activity (Fig. 22.3e) (Matthiesen et al. 2019).

Thus, by combining different experimental approaches and methods such as comparative transcriptome approach, overexpression and gene silencing, it was possible to decipher additional players contributing to the development of *E. histolytica*-induced pathology. Other new techniques, such as 3D cell culture systems and organoids will make it possible to unravel the exact virulence mechanism of *E. histolytica* in the future.

Acknowledgments This work was supported by the Jürgen Manchot Stiftung.

References

Arhets, P., Olivo, J. C., Gounon, P., Sansonetti, P., & Guillen, N. (1998). Virulence and functions of myosin II are inhibited by overexpression of light meromyosin in *Entamoeba histolytica. Molecular Biology of the Cell, 9*(6), 1537–1547.

Bhattacharya, S., & Bhattacharya, A. (2011). A C2 domain protein kinase initiates phagocytosis in the protozoan parasite *Entamoeba histolytica. Nature Communications, 2,* 230.

Bansal, D., Ave, P., Kerneis, S., Frileux, P., Boche, O., Baglin, A. C., et al. (2009). An ex-vivo human intestinal model to study *Entamoeba histolytica* pathogenesis (In Vitro Research Support, Non-U.S. Gov't). *PLOS Neglected Tropical Diseases, 3*(11), e551.

Baxt, L. A., Rastew, E., Bracha, R., Mirelman, D., & Singh, U. (2010). Downregulation of an *Entamoeba histolytica* rhomboid protease reveals roles in regulating parasite adhesion and phagocytosis. *Eukaryotic Cell, 9*(8), 1283–1293. https://doi.org/10.1128/EC.00015-10.

Biller, L., Davis, P. H., Tillack, M., Matthiesen, J., Lotter, H., Stanley, S. L., Jr., et al. (2010). Differences in the transcriptome signatures of two genetically related *Entamoeba histolytica* cell lines derived from the same isolate with different pathogenic properties. *BMC Genomics, 11,* 63.

Biller, L., Schmidt, H., Krause, E., Gelhaus, C., Matthiesen, J., Handal, G., et al. (2009). Comparison of two genetically related *Entamoeba histolytica* cell lines derived from the same isolate with different pathogenic properties. *Proteomics, 9*(17), 4107–4120.

Blessmann, J., Ali, I. K., Nu, P. A., Dinh, B. T., Viet, T. Q., Van, A. L., et al. (2003). Longitudinal study of intestinal *Entamoeba histolytic*a infections in asymptomatic adult carriers. *Journal of Clinical Microbiology, 41*(10), 4745–4750.

Bracha, R., Nuchamowitz, Y., Anbar, M., & Mirelman, D. (2006). Transcriptional silencing of multiple genes in trophozoites of *Entamoeba histolytica. PLoS Pathogens, 2*(5), e48.

Bracha, R., Nuchamowitz, Y., Leippe, M., & Mirelman, D. (1999). Antisense inhibition of amoebapore expression in *Entamoeba histolytica* causes a decrease in amoebic virulence. *Molecular Microbiology, 34*(3), 463–472.

Bracha, R., Nuchamowitz, Y., & Mirelman, D. (2003). Transcriptional silencing of an amoebapore gene in *Entamoeba histolytica*: Molecular analysis and effect on pathogenicity. *Eukaryotic Cell, 2*(2), 295–305.

Bruchhaus, I., Loftus, B. J., Hall, N., & Tannich, E. (2003). The intestinal protozoan parasite *Entamoeba histolytica* contains 20 cysteine protease genes, of which only a small subset Is expressed during in vitro cultivation. *Eukaryotic Cell, 2*(3), 501–509.

Bruchhhaus, I. (2015). Cysteine Peptidases in Pathogenesis. In T. Nozaki & A.Bhattcharya (Eds.), *Amoebiasis—biology and pathogenesis of Entamoeba* (447–458). Springer. ISBN 978-4-431-55199-7.

Clark, C. G., Alsmark, U. C., Tazreiter, M., Saito-Nakano, Y., Ali, V., Marion, S., et al. (2007). Structure and content of the *Entamoeba histolytica* genome. *Advances in Parasitologyol, 65,* 51–190.

Cornick, S., Moreau, F., & Chadee, K. (2016). *Entamoeba histolytica* Cysteine Proteinase 5 Evokes Mucin Exocytosis from Colonic Goblet Cells via alphavbeta3 Integrin. *PLoS Pathogens, 12*(4), e1005579.

Davis, P. H., Chen, M., Zhang, X., Clark, C. G., Townsend, R. R., & Stanley, S. L., Jr. (2009). Proteomic comparison of *Entamoeba histolytica* and Entamoeba dispar and the role of *E. histolytica* alcohol dehydrogenase 3 in virulence. *PLoS Neglected Tropical Diseases, 3*(4), e415.

Davis, P. H., Schulze, J., & Stanley, S. L., Jr. (2007). Transcriptomic comparison of two *Entamoeba histolytica* strains with defined virulence phenotypes identifies new virulence factor candidates and key differences in the expression patterns of cysteine proteases, lectin light chains, and calmodulin. *Molecular and Biochemical Parasitology, 151*(1), 118–128.

Davis, P. H., Zhang, X., Guo, J., Townsend, R. R., & Stanley, S. L., Jr. (2006). Comparative proteomic analysis of two *Entamoeba histolytica* strains with different virulence phenotypes identifies peroxiredoxin as an important component of amoebic virulence. *Molecular Microbiology, 61*(6), 1523–1532.

de la Cruz, O. H., Muniz-Lino, M., Guillen, N., Weber, C., Marchat, L. A., Lopez-Rosas, I., et al. (2014). Proteomic profiling reveals that EhPC4 transcription factor induces cell migration through up-regulation of the 16-kDa actin-binding protein EhABP16 in *Entamoeba histolytica. J Proteomics, 111,* 46–58. https://doi.org/10.1016/j.jprot.2014.03.041.

Ehrenkaufer, G. M., & Singh, U. (2012). Transient and stable transfection in the protozoan parasite Entamoeba invadens. *Molecular and Biochemical Parasitology, 184*(1), 59–62.

Ehrenkaufer, G. M., Eichinger, D. J., & Singh, U. (2007a). Trichostatin A effects on gene expression in the protozoan parasite *Entamoeba histolytica. BMC Genomics, 8,* 216.

Ehrenkaufer, G. M., Haque, R., Hackney, J. A., Eichinger, D. J., & Singh, U. (2007b). Identification of developmentally regulated genes in *Entamoeba histolytica*: Insights into mechanisms of stage conversion in a protozoan parasite. *Cellular Microbiology, 9*(6), 1426–1444.

Ehrenkaufer, G. M., Hackney, J. A., & Singh, U. (2009). A developmentally regulated Myb domain protein regulates expression of a subset of stage-specific genes in *Entamoeba histolytica. Cellular Microbiology, 11*(6), 898–910.

Faust, D. M., & Guillen, N. (2012). Virulence and virulence factors in *Entamoeba histolytica*, the agent of human amoebiasis. *Microbes and Infection, 14*(15), 1428–1441.

Faust, D. M., Marquay Markiewicz, J., Danckaert, A., Soubigou, G., & Guillen, N. (2011). Human liver sinusoidal endothelial cells respond to interaction with *Entamoeba histolytica* by changes in morphology, integrin signalling and cell death. *Cellular Microbiology, 13*(7), 1091–1106.

Fisher, O., Siman-Tov, R., & Ankri, S. (2006). Pleiotropic phenotype in *Entamoeba histolytica* overexpressing DNA methyltransferase (Ehmeth). *Molecular and Biochemical Parasitology, 147*(1), 48–54.

Hamann, L., Buss, H., & Tannich, E. (1997). Tetracycline-controlled gene expression in *Entamoeba histolytica. Molecular and Biochemical Parasitology, 84*(1), 83–91.

Hamann, L., Nickel, R., & Tannich, E. (1995). Transfection and continuous expression of heterologous genes in the protozoan parasite *Entamoeba histolytica. Proceedings of the National Academy of Sciences of the United States of America, 92*(19), 8975–8979.

Hanadate, Y., Saito-Nakano, Y., Nakada-Tsukui, K., & Nozaki, T. (2018). Identification and Characterization of the *Entamoeba Histolytica* Rab8a Binding Protein: A Cdc50 Homolog. *International Journal of Molecular Sciences, 19*(12).

Hasan, M. M., Teixeira, J. E., & Huston, C. D. (2018). Invadosome-Mediated Human Extracellular Matrix Degradation by Entamoeba histolytica. *Infect Immun*, 86(9).

Hellberg, A., Nickel, R., Lotter, H., Tannich, E., & Bruchhaus, I. (2001). Overexpression of cysteine proteinase 2 in *Entamoeba histolytica* or Entamoeba dispar increases amoeba-induced monolayer destruction in vitro but does not augment amoebic liver abscess formation in gerbils. *Cellular Microbiology, 3*(1), 13–20.

Herricks, J. R., Hotez, P. J., Wanga, V., Coffeng, L. E., Haagsma, J. A., Basanez, M. G., et al. (2017). The global burden of disease study 2013: What does it mean for the NTDs? *PLOS Neglected Tropical Diseases, 11*(8), e0005424.

Irmer, H., Tillack, M., Biller, L., Handal, G., Leippe, M., Roeder, T., et al. (2009). Major cysteine peptidases of *Entamoeba histolytica* are required for aggregation and digestion of erythrocytes but are dispensable for phagocytosis and cytopathogenicity. *Molecular Microbiology, 72*(3), 658–667.

Jeelani, G., Husain, A., Sato, D., Soga, T., Suematsu, M., & Nozaki, T. (2013). Biochemical and functional characterization of novel NADH kinase in the enteric protozoan parasite *Entamoeba histolytica. Biochimie, 95*(2), 309–319.

Katz, U., Ankri, S., Stolarsky, T., Nuchamowitz, Y., & Mirelman, D. (2002). *Entamoeba histolytica* expressing a dominant negative N-truncated light subunit of its gal-lectin are less virulent. *Molecular Biology of the Cell, 13*(12), 4256–4265.

Khalil, M. I., Foda, B. M., Suresh, S., & Singh, U. (2016). Technical advances in trigger- induced RNA interference gene silencing in the parasite *Entamoeba histolytica. International Journal for Parasitology, 46*(3), 205–212.

Leippe, M. (1997). Amoebapores. *Parasitology Today, 13*(5), 178–183.

Loftus, B., Anderson, I., Davies, R., Alsmark, U. C., Samuelson, J., Amedeo, P., et al. (2005). The genome of the protist parasite *Entamoeba histolytica. Nature, 433*(7028), 865–868.

MacFarlane, R. C., & Singh, U. (2006). Identification of differentially expressed genes in virulent and nonvirulent Entamoeba species: Potential implications for amebic pathogenesis. *Infection and Immunity, 74*(1), 340–351.

Matthiesen, J., Bar, A. K., Bartels, A. K., Marien, D., Ofori, S., Biller, L., et al. (2013). Overexpression of specific cysteine peptidases confers pathogenicity to a nonpathogenic *Entamoeba histolytica* clone. *mBio, 4*(2), e00072–e00013.

Matthiesen, J., Lender, C., Haferkorn, A., Fehling, H., Meyer, M., Matthies, T., et al. (2019). Trigger-induced RNAi gene silencing to identify pathogenicity factors of *Entamoeba histolytica. The FASEB Journal, 33*(2), 1658–1668.

Meyer, M., Fehling, H., Matthiesen, J., Lorenzen, S., Schuldt, K., Bernin, H., et al. (2016). Overexpression of differentially expressed genes identified in non-pathogenic and pathogenic *Entamoeba histolytica* clones allow identification of new pathogenicity factors involved in amoebic liver abscess formation. *PLoS Pathogens, 12*(8), e1005853.

Mirelman, D., Anbar, M., & Bracha, R. (2008). Epigenetic transcriptional gene silencing in *Entamoeba histolytica. IUBMB Life, 60*(9), 598–604.

Mitra, B. N., Saito-Nakano, Y., Nakada-Tsukui, K., Sato, D., & Nozaki, T. (2007). Rab11B small GTPase regulates secretion of cysteine proteases in the enteric protozoan parasite *Entamoeba histolytica. Cellular Microbiology, 9*(9), 2112–2125.

Morf, L., Pearson, R. J., Wang, A. S., & Singh, U. (2013). Robust gene silencing mediated by antisense small RNAs in the pathogenic protist *Entamoeba histolytica. Nucleic Acids Research, 41*(20), 9424–9437.

Nakada-Tsukui, K., & Nakada-Tsukui, T. (2017). AGC family kinase 1 participates in trogocytosis but not in phagocytosis in *Entamoeba histolytica. Nature Communications, 8*(1), 101.

Nickel, R., & Tannich, E. (1994). Transfection and transient expression of chloramphenicol acetyltransferase gene in the protozoan parasite *Entamoeba histolytica. Proceedings of the National Academy of Sciences of the United States of America, 91*(15), 7095–7098.

Nozaki, T., Asai, T., Kobayashi, S., Ikegami, F., Noji, M., Saito, K., et al. (1998). Molecular cloning and characterization of the genes encoding two isoforms of cysteine synthase in the enteric protozoan parasite *Entamoeba histolytica. Molecular and Biochemical Parasitology, 97*(1–2), 33–44.

Nozaki, T., Asai, T., Sanchez, L. B., Kobayashi, S., Nakazawa, M., & Takeuchi, T. (1999). Characterization of the gene encoding serine acetyltransferase, a regulated enzyme of cysteine biosynthesis from the protist parasites *Entamoeba histolytica* and Entamoeba dispar. Regulation and possible function of the cysteine biosynthetic pathway in Entamoeba. *Journal of Biological Chemistry, 274*(45), 32445–32452.

Olivos, A., Ramos, E., Nequiz, M., Barba, C., Tello, E., Castanon, G., et al. (2005). *Entamoeba histolytica*: Mechanism of decrease of virulence of axenic cultures maintained for prolonged periods. *Experimental Parasitology, 110*(3), 309–312.

Pearson, R. J., Morf, L., & Singh, U. (2013). Regulation of H2O2 stress-responsive genes through a novel transcription factor in the protozoan pathogen *Entamoeba histolytica. Journal of Biological Chemistry, 288*(6), 4462–4474.

Petri, W. A., Jr., & Schnaar, R. L. (1995). Purification and characterization of galactose- and N-acetylgalactosamine-specific adhesin lectin of *Entamoeba histolytica. Methods in Enzymology, 253,* 98–104.

Pompey, J. M., Morf, L., & Singh, U. (2014). RNAi pathway genes are resistant to small RNA mediated gene silencing in the protozoan parasite *Entamoeba histolytica*. *PLoS ONE, 9*(9), e106477.

Ralston, K. S., & Petri, W. A., Jr. (2011). Tissue destruction and invasion by *Entamoeba histolytica*. *Trends in Parasitology, 27*(6), 254–263.

Ralston, K. S., Solga, M. D., Mackey-Lawrence, N. M., Somlata, Bhattacharya, A., & Petri, W. A., Jr. (2014). Trogocytosis by *Entamoeba histolytica* contributes to cell killing and tissue invasion. *Nature, 508*(7497), 526–530.

Ramakrishnan, G., Vines, R. R., Mann, B. J., & Petri, W. A., Jr. (1997). A tetracycline-inducible gene expression system in *Entamoeba histolytica*. *Molecular and Biochemical Parasitology, 84*(1), 93–100.

Rastew, E., Morf, L., & Singh, U. (2015). *Entamoeba histolytica* rhomboid protease 1 has a role in migration and motility as validated by two independent genetic approaches. *Experimental Parasitology, 154*, 33–42.

Saito-Nakano, Y., Loftus, B. J., Hall, N., & Nozaki, T. (2005). The diversity of Rab GTPases in *Entamoeba histolytica*. *Experimental Parasitology, 110*(3), 244–252.

Saito-Nakano, Y., Mitra, B. N., Nakada-Tsukui, K., Sato, D., & Nozaki, T. (2007). Two Rab7 isotypes, EhRab7A and EhRab7B, play distinct roles in biogenesis of lysosomes and phagosomes in the enteric protozoan parasite *Entamoeba histolytica*. *Cellular Microbiology, 9*(7), 1796–1808.

Saito-Nakano, Y., Yasuda, T., Nakada-Tsukui, K., Leippe, M., & Nozaki, T. (2004). Rab5- associated vacuoles play a unique role in phagocytosis of the enteric protozoan parasite *Entamoeba histolytica*. *Journal of Biological Chemistry, 279*(47), 49497–49507.

Samarawickrema, N. A., Brown, D. M., Upcroft, J. A., Thammapalerd, N., & Upcroft, P. (1997). Involvement of superoxide dismutase and pyruvate:ferredoxin oxidoreductase in mechanisms of metronidazole resistance in *Entamoeba histolytica*. *Journal of Antimicrobial Chemotherapy, 40*(6), 833–840.

Saric, M., Irmer, H., Eckert, D., Bar, A. K., Bruchhaus, I., & Scholze, H. (2012). The cysteine protease inhibitors EhICP1 and EhICP2 perform different tasks in the regulation of endogenous protease activity in trophozoites of *Entamoeba histolytica*. *Protist, 163*(1), 116–128.

Sateriale, A., Miller, P., & Huston, C. D. (2016). Knockdown of five genes encoding uncharacterized proteins inhibits *Entamoeba histolytica* phagocytosis of dead host cells. *Infection and Immunity, 84*(4), 1045–1053.

Sato, D., Nakada-Tsukui, K., Okada, M., & Nozaki, T. (2006). Two cysteine protease inhibitors, EhICP1 and 2, localized in distinct compartments, negatively regulate secretion in *Entamoeba histolytica*. *FEBS Letters, 580*(22), 5306–5312.

Shahi, P., Moreau, F., & Chadee, K. (2018). *Entamoeba histolytica* Cyclooxygenase-like protein regulates cysteine protease expression and virulence. *Frontiers in Cellular and Infection Microbiology, 8*, 447.

Singh, N., Ojha, S., Bhattacharya, A., & Bhattacharya, S. (2012). Stable transfection and continuous expression of heterologous genes in Entamoeba invadens. *Molecular and Biochemical Parasitology, 184*(1), 9–12.

Suresh, S., Ehrenkaufer, G., Zhang, H., & Singh, U. (2016). Development of RNA Interference Trigger-Mediated Gene Silencing in Entamoeba invadens. *Infection and Immunity, 84*(4), 964–975.

Teixeira, J. E., Sateriale, A., Bessoff, K. E., & Huston, C. D. (2012). Control of *Entamoeba histolytica* adherence involves metallosurface protease 1, an M8 family surface metalloprotease with homology to leishmanolysin. *Infection and Immunity, 80*(6), 2165–2176.

Tillack, M., Biller, L., Irmer, H., Freitas, M., Gomes, M. A., Tannich, E., et al. (2007). The *Entamoeba histolytica* genome: Primary structure and expression of proteolytic enzymes. *BMC Genomics, 8*, 170.

Tillack, M., Nowak, N., Lotter, H., Bracha, R., Mirelman, D., Tannich, E., et al. (2006). Increased expression of the major cysteine proteinases by stable episomal transfection underlines the important role of EhCP5 for the pathogenicity of *Entamoeba histolytica*. *Molecular and Biochemical Parasitology, 149*(1), 58–64.

Voigt, H., Olivo, J. C., Sansonetti, P., & Guillen, N. (1999). Myosin IB from *Entamoeba histolytica* is involved in phagocytosis of human erythrocytes. *Journal of Cell Science, 112*(Pt 8), 1191–1201.

Wassmann, C., Hellberg, A., Tannich, E., & Bruchhaus, I. (1999). Metronidazole resistance in the protozoan parasite *Entamoeba histolytica* is associated with increased expression of iron-containing superoxide dismutase and peroxiredoxin and decreased expression of ferredoxin 1 and flavin reductase. *Journal of Biological Chemistry, 274*(37), 26051–26056.

Weber, C., Koutero, M., Dillies, M. A., Varet, H., Lopez-Camarillo, C., Coppee, J. Y., et al. (2016). Extensive transcriptome analysis correlates the plasticity of *Entamoeba histolytica* pathogenesis to rapid phenotype changes depending on the environment. *Scientific Reports, 6,* 35852.

Weedall, G. D., Clark, C. G., Koldkjaer, P., Kay, S., Bruchhaus, I., Tannich, E., et al. (2012). Genomic diversity of the human intestinal parasite *Entamoeba histolytica* (Research Support, N.I.H., Extramural Research Support, Non-U.S. Gov't). *Genome biology, 13*(5), R38. https://doi.org/10.1186/gb-2012-13-5-r38.

Welter, B. H., & Temesvari, L. A. (2009). Overexpression of a mutant form of EhRabA, a unique Rab GTPase of *Entamoeba histolytica*, alters endoplasmic reticulum morphology and localization of the Gal/GalNAc adherence lectin. *Eukaryotic Cells, 8*(7), 1014–1026. https://doi.org/10.1128/EC.00030-09.

Zhang, H., Ehrenkaufer, G. M., Pompey, J. M., Hackney, J. A., & Singh, U. (2008). Small RNAs with 5′-polyphosphate termini associate with a Piwi-related protein and regulate gene expression in the single-celled eukaryote *Entamoeba histolytica*. *PLOS Pathogens, 4*(11), e1000219. http://www.ncbi.nlm.nih.gov/entrez/query.fcgi?cmd=Retrieve&db=PubMed&dopt=Citation&list_uids=19043551.

Zhang, X., Zhang, Z., Alexander, D., Bracha, R., Mirelman, D., & Stanley, S. L., Jr. (2004). Expression of amoebapores is required for full expression of *Entamoeba histolytica* virulence in amebic liver abscess but is not necessary for the induction of inflammation or tissue damage in amebic colitis. *Infection and Immunity, 72*(2), 678–683. http://www.ncbi.nlm.nih.gov/entrez/query.fcgi?cmd=Retrieve&db=PubMed&dopt=Citation&list_uids=14742508.

Chapter 23
Polyadenylation Machineries in Intestinal Parasites: Latest Advances in the Protozoan Parasite *Entamoeba histolytica*

Juan David Ospina-Villa, Brisna Joana Tovar-Ayona, Nancy Guillen, Esther Ramírez-Moreno, César López-Camarillo, and Laurence A. Marchat

Abstract In eukaryotic cells, nuclear cleavage and polyadenylation of mRNA precursors (pre-mRNA) generate functional protein encoding transcripts that can be exported to the cytoplasm and translated. Nevertheless, in protozoan parasites that cause intestinal infections in humans, the current knowledge on mRNA 3′-end formation is limited. We performed a genomic survey in *Entamoeba histolytica*, *Giardia lamblia*, and *Cryptosporidium parvum* databases and predicted that polyadenylation machineries are generally well conserved in these pathogens. Notably, most parasites have the 25 kDa subunit of the heterotetrameric CFIm (CFIm25), the 77 kDa subunit of the heterohexameric CPSF (CPSF73) and the poly(A) polymerase (PAP), which are essential proteins for poly(A) site selection, RNA cleavage and poly(A) tail synthesis. However, several differences exist that may be useful to develop new methods to control these pathogens. Taking *E. histolytica* as a study model, we focused on the EhCFIm25 protein, because it is the unique subunit of CFIm in this pathogen, whereas active CFIm in humans is formed by two 25 kDa subunits interacting with two larger subunits. Human and parasite CFIm25 proteins only share 32% identity but they have a similar three-dimensional folding. Interestingly, trophozoites loose virulence and are induced to death when EhCFIm25 expression is silenced, which denotes the relevance of this protein for *E. histolytica* control. To assess this hypothesis, we obtained RNA aptamers that specifically recognize EhCFIm25 by using the SELEX (systematic evolution of ligands by exponential enrichment) procedure and

J. D. Ospina-Villa
Independent Researcher, Medellin, Colombia

B. J. Tovar-Ayona · E. Ramírez-Moreno · L. A. Marchat (✉)
Posgrados en Biomedicina Molecular y en Biotecnología, ENMH, Instituto Politécnico Nacional, Ciudad de México, Mexico
e-mail: lmarchat@gmail.com; lmarchat@ipn.mx

N. Guillen
Unité d'Analyse d'Images Biologiques, Institut Pasteur, Paris, France

C. López-Camarillo
Posgrado en Ciencias Genómicas, Universidad Autónoma de la Ciudad de México, Ciudad de México, Mexico

© Springer Nature Switzerland AG 2020
N. Guillen (ed.), *Eukaryome Impact on Human Intestine Homeostasis and Mucosal Immunology*, https://doi.org/10.1007/978-3-030-44826-4_23

showed that their ingestion by phagocytosis dramatically reduces trophozoites proliferation. Moreover, RNA-protein binding and molecular modeling assays allowed us to identify that the GUUG motif is the binding site of EhCFIm25, while it is the UGUA sequence for the human protein. All these observations led us to propose that aptamers targeting specific parasite proteins, alone or in combination with the conventional treatment, could represent a new tool for controlling the development of amoebiasis and other challenging parasitic diseases.

Keywords Aptamers · Cleavage/polyadenylation · *Entamoeba histolytica* · 3′-UTR

Abbreviations

3′-UTR	3′-Untranslated region
APA	Alternative polyadenylation
CF	Cleavage factor
CPSF	Cleavage and polyadenylation specificity factor
CstF	Cleavage stimulation factor
ncRNAs	Non-coding RNAs
PABPN1	Poly(A) binding protein
PAP	Poly(A) polymerase
pre-mRNA	mRNA precursors
SELEX	Systematic evolution of ligands by exponential enrichment

Introduction

The regulation of gene expression is an essential process for all living organisms. This process allows the transformation of the information contained in the DNA into functional proteins necessary for cell development and functioning. In eukaryotes, nuclear DNA is transcribed into pre-mRNA that are processed by three coordinated processes: capping, which consists of the addition of the structure called cap at the 5′-end, splicing, which consists of the removal of introns, and cleavage/polyadenylation of the 3′-untranslated region (3′-UTR). This complex process involves the selection of the polyadenylation site, the cleavage of pre-mRNA 3′-end and the addition of a variable number of adenosine residues depending on the cell needs. The poly(A) tail favors the stability of mRNAs, facilitates their export to the cytoplasm and enhances translation efficiency (Jalkanen et al. 2014). Almost all eukaryotic mRNA and many non-coding RNAs (ncRNAs) are polyadenylated (Chen et al. 2017).

Polyadenylation in Humans

In humans, the cleavage/polyadenylation process is carried out by a large machinery that is composed of about 80 proteins, of which, at least 20 form the four main complexes that recognize target sequences in the 3′-UTR region of pre-mRNA (Fig. 23.1). The key specific targets and protein complexes for this process are:

- The Poly(A) signal (AAUAAA) which is recognized by the cleavage and polyadenylation specificity factor (CPSF).
- The U/GU-rich sequence which is recognized by the cleavage stimulation factor (CstF).
- The Poly(A) site that usually corresponds to the dinucleotide CA, is determined by Cleavage factors I and II (CFIm and CFIIm), which are recruited to this site to complete the formation of the complex and promote the cleavage of the 3′-end by CPSF73.

Finally, the synthesis of the poly(A) tail is carried out by means of the poly(A) polymerase (PAP) with the cooperation of the poly(A) binding protein (PABPN1). Additional factors, such as symplekin, PP1, and RBBP6 among others, contribute to regulate this processing event (Xiang et al. 2014; Neve et al. 2017). Among this large set of factors, three proteins are specifically important because of their function in the cleavage/polyadenylation process and poly(A) tail formation: CFIm25, CPSF73 and PAP.

The 25 kDa subunit of the heterotetrameric CFIm complex is an essential protein that interacts directly with UGUA RNA motif (Yang et al. 2011). The importance of this protein has been described very accurately in the literature, since it is essential

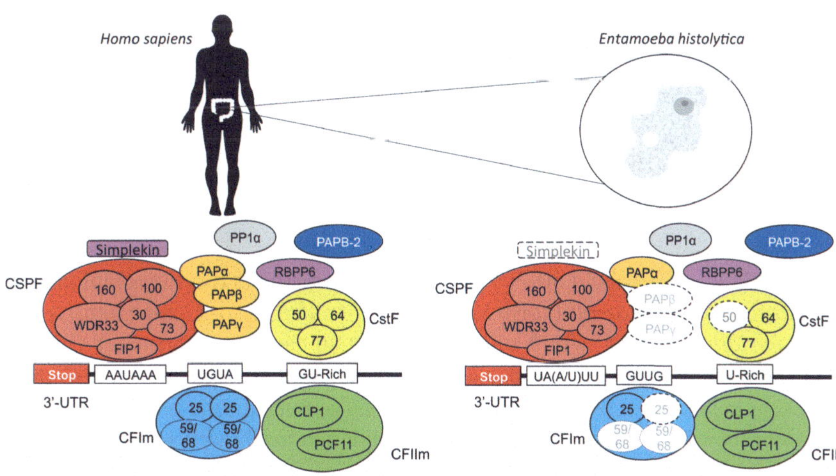

Fig. 23.1 Schematic representation of polyadenylation machineries in *Homo sapiens* and *Entamoeba histolytica*. Icon made by Freepik from www.flaticon.com

for the recruitment of cleavage/polyadenylation factors to pre-mRNA 3′-UTR; it is also necessary for the cleavage of mRNA and polyadenylation reactions (Brown and Gilmartin 2003). In addition, it is involved in the selection of alternative polyadenylation sites (Kubo et al. 2006). Many of the eukaryotic genes contain more than one polyadenylation site, which leads to the generation of different mRNA isoforms of the same gene (Tian et al. 2005); this phenomenon is known as alternative polyadenylation (APA). Depending on the selected polyadenylation site, some mRNA isoforms may or may not have important regulatory sites for translation efficiency, localization, and stability (Gilmartin 2005). Moreover, 3′-UTR also contains target sequences for miRNAs (Vasudevan et al. 2007). Hence the correct selection of poly(A) sites can affect gene expression and defects in APA have been related to several diseases, including cancer (Jafari Najaf Abadi et al. 2019).

The 73 kDa subunit is one of six polypeptides that constitute the CPSF complex along with CPSF160, CPSF100, CPSF30, Fip1 and WDR33. CPSF30 and WDR33 are the subunits that recognize directly AAUAAA polyadenylation signal in RNA (Chan et al. 2014; Sun et al. 2017). CPSF100, CPSF73 and symplekin form the core cleavage complex (CCC) (Dominski et al. 2005). CPSF73 is an essential protein for the cleavage/polyadenylation process because it is the endonuclease responsible for catalysing the cleavage reaction at the poly(A) site previously determined by CFIm25 (Ryan et al. 2004).

The Poly(A) polymerase controls the synthesis of the poly(A) tail. At least six isoforms of canonical PAPα are generated by alternative splicing, but PAP II is generally the main isoform (Raabe et al. 1991). After pre-mRNA cleavage, PAP starts the slow polymerization of adenosine residues; the nascent poly(A) tail is rapidly protected by interaction with the PABN1 and CPSF is recruited, which speeds up the addition of adenosine residues (Kühn et al. 2009). When the poly(A) tail reaches ~ 250 residues, the formation of a spherical structure that involves the poly(A) tail coated by PABPN1 and the multi-functional protein nucleophosmin (NPM1) seems to govern the termination of polyadenylation (Sagawa et al. 2011).

Polyadenylation in *Giardia lamblia* and *Cryptosporidium parvum*

To date, little is known about mRNA poly(A) tail formation in two important intestinal parasites, *Giardia lamblia* (also known as *G. duodenalis* or *G. intestinalis*) and *Cryptosporidium parvum*, which represent important causes of diarrhea in tropical and developing regions, and can also affect certain populations in developing countries (Davis et al. 2002). By screening the parasites genome databases, we found that *G. lamblia* seems to have a very small polyadenylation machinery with only six out of the 21 most relevant human factors used as queries; in contrast, this machinery is almost complete in *C. parvum* (Ospina-Villa et al. 2020).

Entamoeba histolytica as a Study Model

Entamoeba histolytica is the causative agent of intestinal amebiasis that affects 10% of the world's population (Silva et al. 2014). Each year, it affects 50 million people and causes approximately 100,000 deaths, which makes it the third parasite responsible for deaths worldwide after malaria and schistosomiasis (Pineda and Perdomo 2017). In 2005, our group reported the polyadenylation machinery of *E. histolytica* from the screening of the first version of the parasite genome. We found that both the proteins and the RNA motifs in 3′-UTR of pre-mRNAs are quite conserved between the pathogen and its human host, although some differences exist that may be used to identify new molecular targets for parasite control (López-Camarillo et al. 2005; Zamorano et al. 2008) (Fig. 23.1). Several proteins, such as EhPAP (García-Vivas et al. 2005), have been characterized. Notably, EhCFIm25 is the only subunit of the CFIm complex, whereas active CFIm in humans is formed by two 25 kDa subunits interacting with two larger subunits. Moreover, although proteins have similar three-dimensional structures, there are marked differences between amino acid sequences and proteins display sequence identity of only 32% (Pezet-Valdez et al. 2013). Interestingly, EhCFIm25 gene silencing produces multinucleated trophozoites, with deficient mobility and erythrophagocytosis capacity, and a reduced proliferation, prompting us to hypothesize that polyadenylation factors may be relevant targets for *E. histolytica* control (Ospina-Villa et al. 2017) as it has been reported for other protozoan parasites (Hendriks et al. 2003; Sidik et al. 2016; Palencia et al. 2017; Sonoiki et al. 2017).

To assess this hypothesis, we obtained RNA aptamers that specifically recognize EhCFIm25 by using the SELEX (systematic evolution of ligands by exponential enrichment) procedure. Aptamers sequencing, RNA-protein binding and molecular modeling assays allowed us to identify that the GUUG motif is the binding site of EhCFIm25, while it is the UGUA sequence for the human protein. Moreover, aptamers ingestion by phagocytosis severely reduces trophozoites proliferation (Ospina-Villa et al. 2018). All these observations led us to propose that aptamers targeting specific parasite proteins, alone or in combination with the conventional treatment, could represent a new tool for controlling the development of amoebiasis and other challenging parasitic diseases.

Conclusion

The cleavage/polyadenylation process represents an important pathway for the regulation of gene expression that is worth being explored for the identification of potential biochemical targets in intestinal parasites that affect human health. Particularly, our results obtained in *E. histolytica* indicate that key factors such as CFIm25, but also CPSF73 and PAP, should also be evaluated in *G. lamblia*, and *C. parvum* as an alternative in the treatment or diagnosis of these infections. Moreover, they promote the use of aptamers as a new tool for the recognition of parasite proteins.

Acknowledgments This work was supported by ECOS/ANUIES/SEP/CONACyT (France-Mexico), SEP/CONACyT and SIP/IPN (Mexico).

Conflicts of Interest The authors declare no conflict of interest. The founding sponsors had no role in the design of the experiments and the writing of the manuscript.

References

Brown, K. M., & Gilmartin, G. M. (2003). A mechanism for the regulation of pre-mRNA 3′ processing by human cleavage factor Im. *Molecular Cell, 12,* 1467–1476.

Chan, S. L., Huppertz, I., Yao, C., Weng, L., Moresco, J. J., Yates, J. R., et al. (2014). CPSF30 and Wdr33 directly bind to AAUAAA in mammalian mRNA 3′ processing. *Genes & Development, 28,* 2370–2380.

Chen, W., Jia, Q., Song, Y., Fu, H., Wei, G., & Ni, T. (2017). Alternative polyadenylation: Methods, findings, and impacts methods and findings of alternative polyadenylation. *Genomics, Proteomics and Bioinformatics, 15,* 287–300.

Davis, A. N., Haque, R., & Petri, W. A. J. (2002). Update on protozoan parasites of the intestine. *Current Opinion in Gastroenterology, 18,* 10–14.

Dominski, Z., Yang, X., Purdy, M., Wagner, E. J., & Marzluff, W. F. (2005). A CPSF-73 homologue is required for cell cycle progression but not cell growth and interacts with a protein having features of CPSF-100. *Molecular and Cellular Biology, 25,* 1489–1500.

García-Vivas, J., López-Camarillo, C., Azuara-Liceaga, E., Orozco, E., & Marchat, L. A. (2005). *Entamoeba histolytica*: Cloning and expression of the poly(A) polymerase EhPAP. *Experimental Parasitology, 110,* 226–232.

Gilmartin, G. M. (2005). Eukaryotic mRNA 3′ processing: A common means to different ends. *Genes & Development, 19,* 2517–2521.

Hendriks, E. F., Abdul-Razak, A., & Matthews, K. R. (2003). tbCPSF30 depletion by RNA interference disrupts polycistronic RNA processing in *Trypanosoma brucei*. *Journal of Biological Chemistry, 278,* 26870–26878.

Jafari Najaf Abadi, M. H., Shafabakhsh, R., Asemi, Z., Mirzaei, H. R., Sahebnasagh, R., Mirzei, H., et al. (2019). CFIm25 and alternative polyadenylation: Conflicting roles in cancer. *Cancer Letters, 459,* 112–121.

Jalkanen, A. L., Coleman, S. J., & Wilusz, J. (2014). Determinants and implications of mRNA poly(A) tail size–does this protein make my tail look big? *Seminars in Cell & Developmental Biology, 34,* 24–32.

Kubo, T., Wada, T., Yamaguchi, Y., Shimizu, A., & Handa, H. (2006). Knock-down of 25 kDa subunit of cleavage factor Im in Hela cells alters alternative polyadenylation within 3′-UTRs. *Nucleic Acids Research, 34,* 6264–6271.

Kühn, U., Gündel, M., Knoth, A., Kerwitz, Y., Rüdel, S., & Wahle, E. (2009). Poly(A) tail length is controlled by the nuclear Poly(A)-binding protein regulating the interaction between Poly(A) polymerase and the cleavage and polyadenylation specificity factor. *Journal of Biological Chemistry, 284,* 22803–22814.

López-Camarillo, C., Orozco, E., & Marchat, L. A. (2005). *Entamoeba histolytica*: Comparative genomics of the pre-mRNA 3′ end processing machinery. *Experimental Parasitology, 110,* 184–190.

Neve, J., Patel, R., Wang, Z., Louey, A., & Furger, A. M. (2017). Cleavage and polyadenylation: Ending the message expands gene regulation. *RNA Biology, 14,* 865–890.

Ospina-Villa, J. D., Dufour, A., Weber, C., Ramirez-Moreno, E., Zamorano-Carrillo, A., Guillen, N., et al. (2018). Targeting the polyadenylation factor EhCFIm25 with RNA aptamers controls survival in *Entamoeba histolytica. Scientific Reports 8.*

Ospina-Villa, J. D., Guillén, N., Lopez-Camarillo, C., Soto-Sanchez, J., Ramirez-Moreno, E., Garcia-Vazquez, R., et al. (2017). Silencing the cleavage factor CFIm25 as a new strategy to control *Entamoeba histolytica* parasite. *Journal of Microbiology, 55,* 783–791.

Ospina-Villa, J. D., Tovar-Ayona, B. J., López-Camarillo, C., Soto-Sánchez, J., Ramírez-Moreno, Castañón-Sánchez, C. A., Marchat, L. A. (2020). mRNA Polyadenylation Machineries in Intestinal Protozoan Parasites. *Journal of Eukaryotic Microbiology.*

Palencia, A., Bougdour, A., Brenier-Pinchart, M. P., Touquet, B., Bertini, R. L., Sensi, C., Gay, G., et al. (2017). Targeting Toxoplasma gondii CPSF3 as a new approach to control toxoplasmosis. EMBO Molecular Medicine, 1–10.

Pezet-Valdez, M., Fernández-Retana, J., Ospina-Villa, J. D., Ramírez-Moreno, M. E., Orozco, E., Charcas-López, S., et al. (2013). The 25 kDa subunit of cleavage factor Im Is a RNA-binding protein that interacts with the poly(A) polymerase in *Entamoeba histolytica. PLoS ONE, 8,* e67977.

Pineda, E., & Perdomo, D. (2017) *Entamoeba histolytica* under oxidative stress: What countermeasure mechanisms are in place? *6*, 44.

Raabe, T., Bollum, F. J., & Manley, J. L. (1991). Primary structure and expression of bovine poly(A) polymerase. *Nature, 353,* 229–234.

Ryan, K., Calvo, O., & Manley. J. L. (2004). Evidence that polyadenylation factor CPSF-73 is the mRNA 3′ processing endonuclease. *10*, 565–573.

Sagawa, F., Ibrahim, H., Morrison, A. L., Wilusz, C. J., & Wilusz, J. (2011). Nucleophosmin deposition during mRNA 3′ end processing influences poly(A) tail length. *The EMBO Journal, 30,* 3994–4005.

Sidik, S. M., Huet, D., Ganesan, S. M., Huynh, M., Wang, T., Nasamu, A. S., et al. (2016). A genome-wide CRISPR screen in toxoplasma identifies essential Apicomplexan Genes. *Cell, 166,* 1423–1435.

Silva, M. T. N., Santana, J. V., Bragagnoli, G., da Marinho, A. M., & Malagueño, N. E. (2014). Prevalence of *Entamoeba histolytica*/Entamoeba dispar in the city of Campina Grande, in northeastern Brazil. *Journal of the São Paulo Institute of Tropical Medicine, 56,* 451–454.

Sonoiki, E., Ng, C. L., Lee, M. C. S., Guo, D., Zhang, Y. K., Zhou, Y., et al. (2017). A potent antimalarial benzoxaborole targets a Plasmodium falciparum cleavage and polyadenylation specificity factor homologue. *Nature Communications, 8,* 14574.

Sun, Y., Zhang, Y., Hamilton, K., Manley, J. L., Shi, Y., Walz, T., et al. (2017). Molecular basis for the recognition of the human AAUAAA polyadenylation signal. *Proceedings of the National Academy of Sciences of the United States of America, 115,* E1419–E1428.

Tian, B., Hu, J., Zhang, H., & Lutz, C. S. (2005). A large-scale analysis of mRNA polyadenylation of human and mouse genes. *Nucleic Acids Research, 33,* 201–212.

Vasudevan, S., Tong, Y., & Steitz, J. A. (2007). Switching from repression to activation: microRNAs can up-regulate translation. *Science, 318,* 1931–1934.

Xiang, K., Tong, L., & Manley, J. L. (2014). Delineating the Structural Blueprint of the Pre-mRNA 3′-End Processing Machinery. *Molecular and Cellular Biology, 34,* 1894–1910.

Yang, Q., Gilmartin, G. M., & Doublie, S. (2011). The structure of human cleavage factor I(m) hints at functions beyond UGUA-specific RNA binding: A role in alternative polyadenylation and a potential link to 5′ capping and splicing. *RNA Biology, 8,* 748–753.

Zamorano, A., López-Camarillo, C., Orozco, E., Weber, C., Guillen, N., & Marchat, L. A. (2008). In silico analysis of EST and genomic sequences allowed the prediction of cis-regulatory elements for *Entamoeba histolytica* mRNA polyadenylation. *Computational Biology and Chemistry, 32,* 256–263.

Chapter 24
Probiotics as Anti-*Giardia* Defenders: Overview on Putative Control Mechanisms

Jana Alazzaz, Soraya Chaouch, Anne-Sophie Boucard, Luis G. Bermudez-Humaran, and Isabelle Florent

Abstract *Giardia intestinalis* is a protist intestinal parasite responsible for giardiasis, a disease whose impact is recognized in public health. After ingestion of *Giardia* cysts from either contaminated food or water, the trophozoite proliferative form, responsible for pathogenic effects, develops in the proximal small intestine of the host where it coexists with gut microbiota. Several studies have revealed the importance of this gut ecosystem and/or some probiotic bacteria in providing protection against *G. intestinalis* infections through partially known mechanisms (Travers et al. Journal of Parasitology Research, 2011). In the last years, our team has shown, using biological and biochemical approaches, that some probiotic strains of *Lactobacillus*, in particular *L. johnsonii* La1 and *L. gasseri* CNCM-I 4884, display anti-*Giardia* effects both in vitro and in vivo (Travers et al. Frontiers in Microbiology 2016; Allain et al. Frontiers in Microbiology 8:2707, 2018a, Frontiers in Microbiology 9:98, b). Our investigations have demonstrated that the supernatant of these strains contains Bile-Salt-Hydrolase (BSH)-like activities mediating toxic effects on *Giardia*. This effect is not directly, but by converting non-toxic components of bile (conjugated bile salts) into bile salts deconjugates proved to be highly toxic to the parasite. These anti-*Giardia* effects could be mimicked in vitro by treating *Giardia* cultures with either commercially available BSH bacterial enzymes (Travers et al. 2016) or two *L.*

J. Alazzaz (✉) · S. Chaouch · I. Florent (✉)
UMR7245, Molécules de Communication et Adaptation des Microorganismes, Muséum National d'Histoire Naturelle, Centre National de la Recherche Scientifique, CP52, 57 Rue Cuvier, 75005 Paris, France
e-mail: jana.alazzaz@mnhn.fr

I. Florent
e-mail: isabelle.florent@mnhn.fr

S. Chaouch
e-mail: soraya.chaouch@mnhn.fr

A.-S. Boucard · L. G. Bermudez-Humaran
INRAE, Micalis Institute, INRAE, AgroParisTech, Université Paris-Saclay, 78352 Jouy-En-Josas, France
e-mail: anne-sophie.boucard@inrae.fr

L. G. Bermudez-Humaran
e-mail: luis.bermudez@inrae.fr

© Springer Nature Switzerland AG 2020

N. Guillen (ed.), *Eukaryome Impact on Human Intestine Homeostasis and Mucosal Immunology*, https://doi.org/10.1007/978-3-030-44826-4_24

johnsonii La1 BSH enzymes produced and purified from recombinant *Escherichia coli* strains (Allain et al. 2018a), in the presence of bile, or even directly with some deconjugated bile salts (Travers et al. 2016). Currently, we are focusing on understanding the mechanism of action (MoA) of toxic metabolites generated by these BSH activities on the parasite itself using imaging and RNA sequencing methods in order to explore the changes in gene expressions in *Giardia.* Altogether, these data pave the way for new approaches for the treatment of this widespread neglected infectious disease.

Keywords *Giardia intestinalis* · *Lactobacillus* · Bile salt hydrolase · Deconjugated bile salts · Anti-*Giardia* activity

Abbreviations

BSH	Bile Salt Hydrolase
C	Cholic acid
CDC	Chenodeoxycholic acid
CNCM	Collection Nationale de Cultures de Microorganismes (National Collection of Microrganisms Cultures)
DAPI	4′,6-diamidino-2-phenylindole
DC	Deoxycholic acid
FAO	Food and Agriculture Organization
FCS	Fetal Calf Serum
FDA	Food and Drug Administration
GC	Glyco-cholic acid
GCDC	Glyco-chenodeoxycholic acid
GDC	Glyco-deoxycholic acid
GRAS	Generally Recognized As Safe
IC50	Inhibitory Concentration yielding 50% inhibition
LAB	Lactic Acid Bacteria
MoA	Mechanism of Action
TC	Tauro-cholic acid
TCDC	Tauro-chenodeoxycholic acid
TDC	Tauro-deoxycholic acid
WGA	Wheat-Germ Agglutinin
WHO	World Health Organization

Giardia

Giardia intestinalis (also called *G. lamblia* or *G. duodenalis*) is a parasitic protozoa belonging to the order Diplomonadida of the super-group Excavata (Adl et al. 2019). It is responsible for giardiasis, an acute or chronic intestinal disorder, characterized by malabsorption, diarrhea, weight loss, dehydration, and abdominal pain in humans and a variety of vertebrates (Cotton et al. 2011). Giardiasis is a public health issue mainly in developing countries but also in developed countries, since outbreaks have been associated with drinking water contaminations with a low infectious dose (10 cysts) resulting from runoffs of contaminated soils by rainfalls, agricultural practices, and sewage treatment plant dysfunctions (Mons et al. 2009; Baldursson and Karanis 2004; Rendtorff 1954). Widely distributed in the environment as resistant cysts, *G. intestinalis* infects many mammals including humans by fecal-oral transmission. Following ingestion, cysts differentiate during their gastrointestinal transit into the motile and replicative flagellated form known as trophozoites (responsible for the pathogenic effects), before their release with the host feces into the environment as infective cysts (Ankarklev et al. 2010). Trophozoite forms proliferate in the host gut lumen, where they transiently adhere to the gut epithelium and coexist with the gut microbiota (Allain et al. 2017). Recent data show that this microbiota and/or some probiotic strains can protect hosts against *Giardia* infections, but the protective mechanisms involved in these effects are poorly understood (Travers et al. 2011; Burgess et al. 2017). *Giardia* belongs to a complex of species currently composed of eight different genotypes (called "assemblages") with variable host and host range specificities. Moreover, assemblages A and B display a wide host diversity infecting both human and animals and are thus considered zoonotic, contrary to other assemblages which have reduced or even specific host tropisms (Cacciò et al. 2018). Thus, assemblages C and D are mainly observed in dogs, assemblage F in cats and assemblage G in rodents (Cacciò et al. 2018).

Fighting against giardiasis is actually possible by using anti-infectious molecules such as metronidazole, tinidazole and benzimidazoles (Leitsch et al. 2011). However, treatments based on these drugs have their limits due to the emergence of strains resistant to these compounds, that are now becoming general for most infectious agents (Kirk et al. 2010). In light of these limitations, the development of new "therapeutic" strategies does also concern the intestinal parasite *G. intestinalis* itself, for which the use of probiotics, for the prevention or the treatment of giardiasis, is becoming an active emerging field, although the molecular mechanisms involved remain poorly described (Travers et al. 2011; Vitetta et al. 2016).

Probiotics

The last update on the definition of probiotics by the Food and Agriculture Organization/World Health Organization (FAO/WHO), dating back to October 2013, states: "*live microorganisms that, when administrated in adequate amounts, confer a health benefit to the host*" (Hill et al. 2014). An ideal probiotic, which is always defined at a strain level, should be able to positively modulate host intestinal microbiota, stabilizing resident microorganisms and restraining colonization by pathogens (Bakhtiar et al. 2013). Moreover, as beneficial microorganisms, probiotics should respond to a list of criteria (Ouwehand and Salminen 1992), detailed below under the prism of more recent data. Notably, once identified and selected, probiotic strains of interest should be characterized for their MoA since probiotic properties are also dependent on conditions of use and doses (Bakhtiar et al. 2013). Nowadays, an increasing number and diversity of commercial probiotic strains are available on market. The commercial and those under studies mostly target bacterial infectious diseases, however, little is known for fighting against parasitic and viral illness (Berrilli et al. 2012; Liévin-Le Moal and Servin 2014; Zare Mirzaei et al. 2018; Kiousi et al. 2019).

Lactic Acid Bacteria (LAB) are commonly used as probiotics as some strains are Generally Recognized As Safe (GRAS) for humans, according to the US Food and Drug Administration (FDA) and fulfill criteria of the Qualified Presumption of Safety (QPS) according to the European Food Safety Authority (EFSA). Among LAB, lactobacilli and bifidobacteria have been extensively studied by scientists and industrials for their potential as probiotics (Azad et al. 2018). Probiotic strains belonging to *Lactobacillus* genus have been until now the focus of our work on *Giardia*. *Lactobacillus* spp. are non-sporulating facultative anaerobic or microaerophilic Gram-positive bacteria (Holzapfel et al. 2001). Their fermentative metabolism is characterized by the production of lactic acid that has been implicated in lactobacilli ability to inhibit intestinal pathogens development, these latter ones being mostly documented as bacterial pathogens (Vandenbergh 1993). In addition to lactic acid production, the inhibitory effect of *Lactobacillus* spp. on these bacterial pathogens relies on the production of antimicrobial peptides (bacteriocins), the competition for mucosal site adhesion and nutrients consumption and also the modulation of the immune system (Figueroa-González et al. 2011). Probiotics ingestion has been also suggested to modulate the gut microbiota composition (Isolauri et al. 2012). For the screening procedure, as the FAO/WHO has recommended, selected probiotic strains should provide (i) resistance to gastric acidity and bile salts, (ii) adhesive properties to mucus and intestinal epithelial cells and finally (iii) anti-microbial and antagonism activities against potentially pathogenic microorganisms (Markowiak et al. 2017). *Lactobacillus* spp. consumption present no risk of mortality to humans, and side effects following their administration are scarce (0.05%–0.4% of cases) (Gasser 1994). Thus, *Lactobacillus* spp. have a well-reported history of safety and are awarded GRAS by the FDA (Sorokulova et al. 2008).

Characterization of *Giardia*-Probiotic Crosstalk

Since probiotic microorganisms provide health benefits to their hosts through the protection against pathogens and the modulation of both innate and adaptive immunity at local and systemic levels (Cebra 1999; Haller et al. 2000; Isolauri et al. 2001), trials have been exerted aiming at exploring whether these organisms could also be used to treat *Giardia* infections. Certainly, colonization of the intestine by *Giardia* strongly depends on the intestinal microbiota and its susceptibility (Singer and Nash 2000; Torres et al. 2000). In addition, *Giardia* infection, in its turn, may exert changes in the composition of the host microbiota and its diversity (Burgess et al. 2017). Since lactobacilli are one of the most common bacteria in the human duodenum (Mitsuoka 1992), several studies have focused on the ability of *Lactobacillus* probiotic strains to shield the host from the detrimental effects mediated by *Giardia* infections (Pérez et al. 2001; Humen et al. 2005; Goyal et al. 2013; Shukla et al. 2008, 2019; Vivancos et al. 2018). It was shown that *Lactobacillus casei* MTCC 1423 strain is effective in eliminating *Giardia* in mice (Shukla and Sidhu 2011). Moreover, in 2001, Perez and collaborators found that the culture supernatant of *L. johnsonii* La1 was able to control *G. lamblia* (*intestinalis*) growth in vitro, an effect that was strain-dependent since six other strains of *Lactobacillus acidophilus* (tested in parallel), did not show any noticeable effect on the parasite (Pérez et al. 2001). This effect was confirmed in vivo using a gerbil model that evaded *Giardia* colonization when treated with *L. johnsonii* La1 administrated by gavage, in addition to a reinforcement of the host immune response against the parasite (Humen et al. 2005). Based on available literature, *L. johnsonii* La1 appeared as a good model of choice, to study the molecular crosstalk between *Giardia* and probiotic bacteria, with, in addition, well-known genomes for both partners (Pridmore et al. 2004; Morrison et al. 2007; Franzén et al. 2009). Other non-lactobacilli probiotics have demonstrated anti-giardial effects such as *Enterococcus faecium* SF68, and Slab51 (Benyacoub et al. 2005; Perrucci et al. 2019). Beyond bacteria, it must be noted that trials have used yet other probiotics microorganisms including yeasts, as for example *Saccharomyces boulardii*, that have shown promising results in the protection against giardiasis with a decreased number of parasite cysts in feces from patients treated with a combination of *S. boulardii* and metronidazole *versus* patients treated only with metronidazole (Besirbellioglu et al. 2006). Quite recently, this observation was further supported by another study showing that *S. boulardii* could be also used as a co-adjuvant in giardiasis treatment since it shows a reduction in intestinal damages caused by *Giardia* with an approximate reduction of 70% of the parasite load in vivo model of infected gerbil mice (Ribeiro et al. 2018).

We have been interested in studying these probiotic-parasite interactions for prophylactic and/or therapeutic purposes, applied to *G. intestinalis,* and focused our initial interests on the probiotic strain *L. johnsonii* La1, based on the promising results cited above (Pérez et al. 2001; Humen et al. 2005; Pridmore et al. 2004; Morrison et al. 2007; Franzén et al. 2009). By combining biological, biochemical and metabolomic approaches, we have discovered that the MoA of *L. johnsonii* La1 against *G. intestinalis* is partially indirect, and involves Bile Salt Hydrolase (BSH) type enzyme activities produced by this bacterium, which provoke the death of the parasites by converting bile components (identified as being conjugated bile salts, non-toxic to *Giardia*) into toxic compounds (identified as deconjugated bile salts) (Travers et al. 2016). The deleterious action of deconjugated bile salts on parasites growth in vitro was then directly confirmed (Travers et al. 2016), and the recombinant BSH enzymes of this probiotic strain, produced in *E. coli* (currently 2 of the 3 encoded by its genome: BSH-47 and BSH-56), also allowed to block the parasite proliferation in vitro (parasites in culture) in the presence of bile and in vivo in a murine model of giardiasis: newborn mice, line OF1 (Allain et al. 2018a). Several in vitro tests reflected the potential benefit provided by BSH enzymes although displaying different substrate specificities: indeed, BSH-47 and BSH-56 (from *L. johnsonii* La1), have distinct substrate specificities—BSH-56 mainly hydrolyzing Tauro-conjugates but also Glyco-conjugates, whereas BSH-47 hydrolyzes mostly Tauro-conjugates—both being active in vitro in a dose-dependent manner and in vivo (BSH-47) (Allain et al. 2018a). These BSH effects, potentially distinct, are important to know since it is well established that bile composition differs dramatically according to the host (Farthing et al. 1985; Aguiar Vallim et al. 2013). In parallel, we tested 29 lactobacilli strains, isolated from a variety of environments, for their "anti-*Giardia*" and "BSH activity" properties in vitro (Allain et al. 2018b). These studies allowed establishing (1) a positive correlation between these two properties, making it possible to propose a screen of potentially anti-*Giardia* strains based on their BSH activities (Allain et al. 2018b). In addition, (2) it allowed to discover the novel *L. gasseri* CNCM I-4884 probiotic strain, that proved to be as active as *L. johnsonii* La1 strain in vitro but much more active in vivo, in the murine model of the newborn infant mouse OF1 (Allain et al. 2018b). Indeed, while *Giardia* trophozoite loads were reduced in infected mice by gavages with either probiotics, *Giardia* cyst loads were significantly more highly reduced by using *L. gasseri* CNCM I-4884 gavages as compared to using *L. johnsonii* La1 gavages (Allain et al. 2018b). The mechanism responsible for this higher activity of *L. gasseri* CNCM I-4884 compared to *L. johnsonii* La1 at controlling *Giardia* in vivo is not yet established, but we are currently investigating it (Fig. 24.1).

To our knowledge, the sum of these studies, that lead to the discovery of a possible MoA of probiotics on the development of *G. intestinalis* parasite involving BSH activities of lactobacilli (Travers et al. 2016; Allain et al. 2018a, b), remain rather unique in the emerging field of *Giardia*-probiotics cross-talk exploration. It would be obviously interesting to also test the specificity of this control mechanism of lactobacilli against the different assemblages of *G. intestinalis*. Hence, to go deeper, it is interesting to understand what is happening at the parasite level, especially the MoA of toxic metabolites that are generated by active BSH or present in lactobacilli supernatants, in both in vitro and in vivo models.

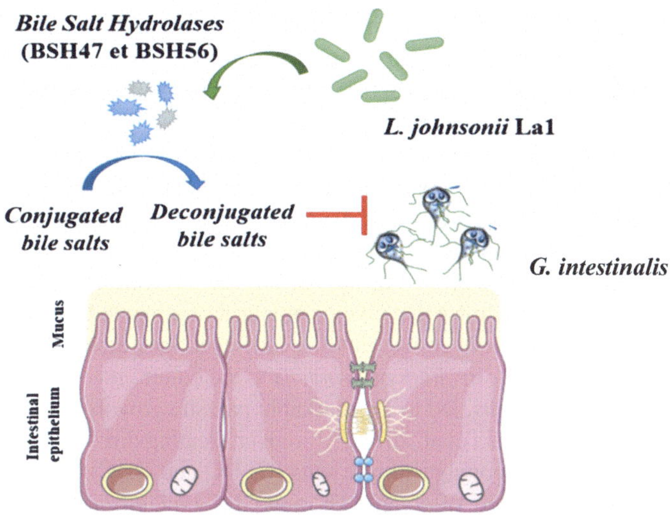

Fig. 24.1 Our data suggest a possible MoA by which the probiotic strain of *Lactobacillus johnsonii* La1, by secreting/releasing BSH-like enzymes in an environment where bile is present and abundant, can fight the *Giardia* parasite through the conversion of conjugated bile salts (non-toxic to *Giardia*) to deconjugated bile salts (toxic to *Giardia*)

The Direct Effect of Bile Active Compounds on the *Giardia* Parasite

Giardia culture medium has been historically supplemented with bile in order to promote the parasite in vitro growth; indeed, parasites have been documented to use bile lipids as metabolites or source of phospholipids for membrane biosynthesis (Farthing et al. 1985; Halliday et al. 1995; Das et al. 1997). However, in the presence of high concentrations of bile and bile salts, growth reduction rate is observed (Farthing et al. 1983, 1985; Gillin 1987; Gillin et al. 1989). Which mechanisms happen behind these observations, i.e., which of the bile components are beneficial/detrimental to *Giardia* remains incompletely explored, as only a few of these bile components have been tested on *Giardia* growth and survival (Farthing et al. 1983, 1985; Gillin 1987; Gillin et al. 1989). However, this remains challenging due to the complexity of bile composition, both in terms of metabolites and of their respective concentrations, depending on their biological sources (Aguiar Vallim et al. 2013; Farthing et al. 1985). Interestingly, intestine bile salts have been shown to be potent antimicrobial agents, involved in innate defenses (Sung et al. 1993; Itoh et al. 1999; Begley et al. 2005). Sannasiddappa and collaborators have found that bile salts exert an antibacterial effect on *Staphylococcus aureus* (Sannasiddappa et al. 2017); however, the potential antiparasitic activity exerted by bile salts on *G. intestinalis* is still poorly understood. Their exploration might however hold potential as recently, bile salts have been indeed considered as therapeutic agents (Donker et al. 2019). Since our studies have clearly pointed towards modifications of bile composition, mediated by lactobacilli BSH enzymes, as important drivers of *G. intestinalis* development in vitro and in vivo (Travers et al. 2016; Allain et al. 2018a, b), a logical follow-up of this discovery was to focus on these bile salts.

Experiments have been conducted to study individually a series of pure tauro- and glyco-conjugated bile salts (tauro- and glyco-cholic (TC, GC), tauro- and glyco-deoxycholic (TDC, GDC), tauro- and glyco-chenodeoxycholic (TCDC, GCDC) acids) as well as their deconjugated counterparts: cholic-acid (C), deoxycholic acid (DC) and chenodeoxycholic (CDC) acids, on *Giardia* (Travers et al. 2016; Allain et al. 2018a). Results have shown that glycine or taurine conjugated bile salts have no toxic impact on *G. intestinalis* growth in vitro. However, the addition of a recombinant BSH enzyme from *Clostridium perfringens* to the culture medium, in presence of these conjugated bile salts, led to a remarkable parasite growth inhibition within the 24 h of the assay (Travers et al. 2016). Moreover, treating directly *G. intestinalis* trophozoites with pure deconjugated bile salts (C, DC and CDC) have also shown a toxic dose-dependent effect of notably DC and CDC on *Giardia* growth (IC50 of 132 μM and 147 μM respectively), which was not observed by using cholic acid (IC50 > 400 μM) or, as mentioned above, conjugated bile salts (Travers et al. 2016). Interestingly, the killing effect of deconjugated bile salts on *Giardia* have been correlated with their hydrophobicity properties, no inhibition being observed with the most hydrophilic salt cholate contrary to the more hydrophobic salts, deoxycholate and chenodeoxycholate (Travers et al. 2016). This could explain also the non-toxic effect of conjugated bile salts since deconjugated bile salts are more hydrophobic than their conjugated counterparts (Ridlon et al. 2016). However and very importantly, the IC50 values of these active deconjugated bile salts are much lower than their critical micellar concentrations, which means that the parasite killing effect is not simply related to their surfactant properties (Critical micellar concentrations of C ~ 14 mM, DC ~ 1.4 mM and CDC > 7 mM, based on the manufacturers). Based on these findings, comparative studies of deconjugated bile salts effects on *Giardia*, have been designed using cholic acid as negative control. Microscopic observations of treated and untreated *Giardia* cells using scanning electron microscopy, revealed altered morphology and plasma membrane disruptions under both recombinant bile-salts BSH treatment in presence of bile (Allain et al. 2018a), and pure DC treatment (0.1 g/L i.e. 232 μM) in absence of bile (Allain et al. 2018a; Fig. 24.2), compared to controls.

Current Investigations on the Toxic Effects of Bile Salts

Although numerous studies have converged towards the existence of a beneficial effect provided by different types of probiotics to control giardiasis, few have provided hints on the molecular MoA behind these effects (Amer et al. 2014; Allain et al. 2018a, b). Currently, our studies aim to determine the mechanism of killing of these *Lactobacillus* defenders on *Giardia* parasite itself since no data in this concern

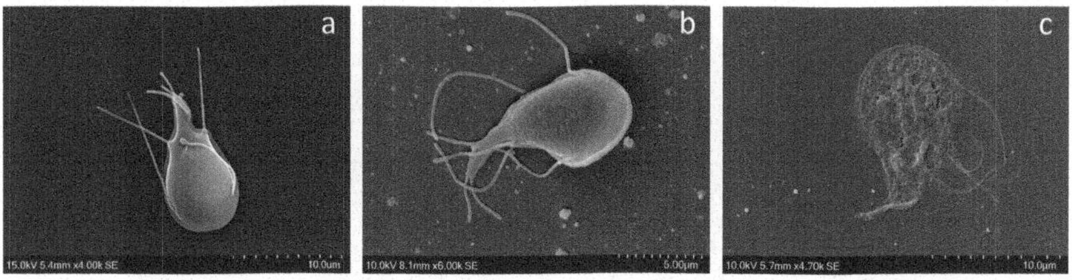

Fig. 24.2 Morphological alterations following in vitro treatments of *G. intestinalis* WB6 strain trophozoites by deoxycholic (DC) acid. **a** Modified TYI-S-33 medium control (+10% FCS) without bile. **b** Modified TYI-S-33 medium control (+10% FCS) with bile (bovine bile 0.6 g/L) showing the characteristic pear-shaped of *G. intestinalis* WB6 strain trophozoites in vitro. **c** Modified TYI-S-33 medium control (+10% FCS) with DC (0.1 g/L) displays alterations and a disruption of plasma membrane exposing cell interior. Scale bar = 5 μm (**b**) or 10 μm (**a**, **c**). See also Allain et al. (2018a)

have been published. Our hypothesis was that this detrimental effect—mediated by some deconjugated bile salts—could be due to their ability to disrupt essential membrane functions and bioenergetic processes, as shown for *Staphylococcus aureus* and *Clostridium difficile* (Thanissery et al. 2017; Sannasiddappa et al. 2017).

The main primary bile acids, produced by the human liver, are cholic acid and chenodeoxycholic acid, mostly conjugated to taurine and glycine (Donkers et al. 2019). Under intestinal microbiota deconjugation and by dehydroxylation at C7, primary bile acids are converted into secondary bile acids resulting in deoxycholic acid and lithocholic acid, respectively. In the intestine, bile salts are considered as digestive molecules helping in lipid digestion as well as important innate defenses and potent antibacterial agents due to their soap-like character (Sung et al. 1993; Itoh et al. 1999; Begley et al. 2005). Bile salts are known to inhibit bioenergetic processes by intracellular acidification, dissipation of the proton motive force, and induction of DNA damage and protein denaturation (Kurdi et al. 2006; Merritt and Donaldson 2009). Interestingly, Pérez and collaborators suggested that extracellular factors of *L. johnsonii* La1 arrest the in vitro growth of *G. intestinalis* at the G1 phase indicating that this bacterium may directly affect parasite replication (Pérez et al. 2001). Note that the suspected bacterial metabolite (< 1 kDa) (Pérez et al. 2001) is of a different nature than the BSH (> 30 kDa) we have identified. Thus, the interaction between *Giardia* and human bile salts is an important factor in its ability to colonize the host intestine. Despite this importance, the MoA of these bile salts on *Giardia* is not fully understood. Several experiments established in our laboratory, using fluorescent microscopy, have shown that *Giardia* trophozoites treated for 22 h with DC (0.1 g/L) and CDC (0.1 g/L) turn into roundish cells with some disassembled flagella, compared to untreated and C-treated conditions (Fig. 24.3). These results recall those recently obtained by Sievers and collaborators on a different biological model, the bacterium *C. difficile,* in which less flagellated cells were observed in the presence of DC and CDC but no influence on flagella was induced by C, cholic acid (Sievers et al. 2019). These findings also recall morphological observations accompanying *Giardia* encystation program, experimentally induced

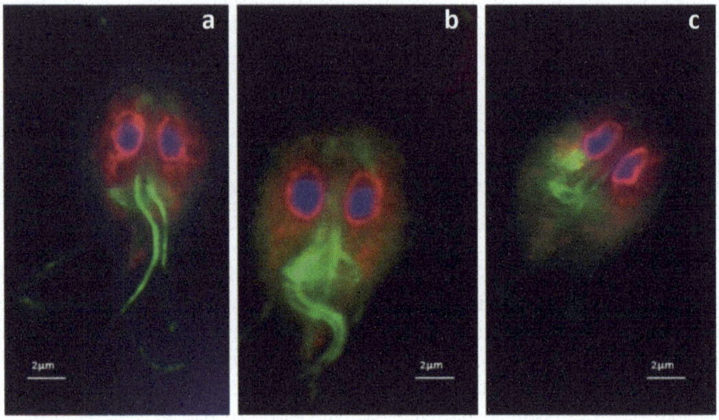

Fig. 24.3 Fluorescence microscopy images of *Giardia intestinalis* trophozoites labeled with DAPI (blue), anti-Tubulin antibody (green, Woods et al. 1989) and Wheat-Germ Agglutinin (WGA, red, Ratner et al. 2008) after 22 h of culture in vitro. **a** Modified TYI-S-33 medium control (+10% FCS, no bile), **b** Modified TYI-S-33 medium control (+10% FCS, no bile), with cholic acid (Sigma, 0.1 g/L), **c** Modified TYI-S-33 medium control (+10% FCS, no bile), with deoxycholic acid (Sigma, 0.1 g/L). Scale bar = 2 μm (**a**, **b**, **c**)

in vitro by modifying bile (and bile components?) supply as well as pH in the culture conditions (Lujan and Svärd 2011). We have described phenotypic *Giardia* growth alterations upon (1) *L. johnsonii* La1 supernatant culture challenge in the presence of bile, (2) rBSH supplementation in the presence of bile and (3) pure deconjugated bile salts treatments (Travers et al. 2016; Allain et al. 2018a, b), but no information on the adaptation of gene expression patterns is available so far in these various conditions. As a mean to get further insight into the molecular mechanisms involved, we have designed to use a transcriptomic approach (RNA sequencing) to explore the gradation of these responses, when *Giardia* cells are exposed to these three challenging but progressively simplified conditions.

As a starting point, we logically choose to focus on the least complex configuration: (3), i.e. pure deconjugated bile salts treatments, using DC (0.1 g/L) in vitro culture. Preliminary results, from triplicated cultures in presence and absence of DC sampled at T0, and after 7 and 16 h of treatments, have shown that there is a part of the parasite transcriptome that is altered due to the deconjugated bile salt treatment. Some genes, documented to be involved in encystation (Einarsson et al. 2016) seems to be altered in their transcriptomic profile, suggesting that *Giardia* cells in presence of DC have a tendency to encystation. This response is also accompanied with a deregulated cell cycle and translational phase. In addition, some morphological alterations of *Giardia* trophozoites under DC treatment (Fig. 24.3) could be paralleled with some changes at transcriptomic level. Genes encoding proteins participating in cytoskeleton components were modified after 16 h exposure to DC, in comparison to control, untreated, conditions.

Certainly, studying the "simple" treatment of *Giardia* parasites with deconjugated bile salts individually might not be sufficient to accurately reflect the more complex experimental situation (Sannasiddappa et al. 2017), but it forms a simple model to explore individual bile acids activities on *Giardia*. Next, the higher levels of experimental complexity will be progressively explored as mentioned above such as (1) the combination of recombinant BSH enzymes in presence of bile, and (2) *Lactobacillus* culture supernatants in presence of bile, for their effects on *Giardia*. In both conditions, mixtures of deconjugated bile salts of several types (Cholic acid, deoxycholic acid, chenodeoxycholic acids and others) are expected to be produced, the combined effect of which on *Giardia* remains to be explored. Moreover, in the latter condition (probiotic supernatants), additional active principles may potentially be also present as previously demonstrated (Travers et al. 2016), which may further influence the biological response on the parasite.

Finally, in order to carefully understand how probiotics exert their defence on *Giardia* in vivo, it will be necessary to investigate the molecular mechanisms exerted on *Giardia* parasite following either probiotic gavages of murine models of giardiasis, or, ultimately, patients treatments by these probiotics.

Acknowledgements We deeply thank the Region Ile de France-DIM (One Health) for funding this work, notably post-doctoral grant to JA. We also thank: Evelyne DUVERNOIS-BERTHET and Loïc PONGER (CSB-AVIV department, MNHN) for assistance to RNA sequencing experiments/analysis; Geraldine TOUTIRAIS (TME, Plateau Technique de Microscopie Électronique et de Microanalyses du MNHN, Paris) for SEM imaging and Marc GEZE and Cyril WILLIG (CEMIM, MNHN) for fluorescence imaging.

References

Adl, S. M., Bass, D., Lane, C. E., Lukeš, J., Schoch, C. L., Smirnov, A., … & Cárdenas, P. (2019). Revisions to the classification, nomenclature, and diversity of eukaryotes. *Journal of Eukaryotic Microbiology*, *66*(1), 4–119.

Allain, T., Amat, C. B., Motta, J. P., Manko, A., & Buret, A. G. (2017). Interactions of *Giardia* sp. with the intestinal barrier: Epithelium, mucus, and microbiota. *Tissue Barriers*, *5*(1), e1274354.

Allain, T., Chaouch, S., Thomas, M., Travers, M.-A., Vallée, I., Langella, P., et al. (2018b). Bile salt hydrolase activities: A novel target to screen Anti-*Giardia Lactobacilli*? *Frontiers in Microbiology, 9,* 89.

Allain, T., Chaouch, S., Thomas, M., Vallée, I., Buret, A. G., Langella, P., et al. (2018a). Bile-salt-hydrolases from the probiotic Strain *Lactobacillus johnsonii* La1 mediate anti-Giardial activity in vitro and in vivo. *Frontiers in Microbiology, 8,* 2707.

Amer, E. I., Mossallam, S. F., & Mahrous, H. (2014). Therapeutic enhancement of newly derived bacteriocins against *Giardia lamblia*. *Experimental Parasitology*, *146*, 52–63.

Ankarklev, J., Jerlström-Hultqvist, J., Ringqvist, E., Troell, K., & Svärd, S. G. (2010). Behind the smile: Cell biology and disease mechanisms of *Giardia* species. *Nature Reviews Microbiology, 8*(6), 413–422.

Azad, M., Kalam, A., Sarker, M., Li, T., & Yin, J. (2018). Probiotic species in the modulation of gut microbiota: An overview. *BioMed Research International*.

Bakhtiar, S. M., LeBlanc, J. G., Salvucci, E., Ali, A., Martin, R., Langella, P., … & Azevedo, V. (2013). Implications of the human microbiome in inflammatory bowel diseases. *FEMS Microbiology Letters*, *342*(1), 10–17.

Baldursson, S., & Karanis, P. (2011). Waterborne transmission of protozoan parasites: Review of worldwide outbreaks–an update 2004–2010. *Water Research, 45*(20), 6603–6614.

Begley, M., Gahan, C. G., & Hill, C. (2005). The interaction between bacteria and bile. *FEMS Microbiology Reviews, 29*(4), 625–651.

Benyacoub, J., Pérez, P. F., Rochat, F., Saudan, K. Y., Reuteler, G., Antille, N., et al. (2005). *Enterococcus faecium* SF68 enhances the immune response to *Giardia intestinalis* in mice. *The Journal of Nutrition, 135*(5), 1171–1176.

Berrilli, F., Di Cave, D., Cavallero, S., & D'Amelio, S. (2012). Interactions between parasites and microbial communities in the human gut. *Frontiers in Cellular and Infection Microbiology, 2*, 141.

Besirbellioglu, B. A., Ulcay, A., Can, M., Erdem, H., Tanyuksel, M., Avci, I. Y., ... & Pahsa, A. (2006). *Saccharomyces boulardii* and infection due to *Giardia lamblia*. *Scandinavian Journal of Infectious Diseases, 38*(6–7), 479–481.

Burgess, S. L., Gilchrist, C. A., Lynn, T. C., & Petri, W. A. (2017). Parasitic protozoa and interactions with the host intestinal microbiota. *Infection and Immunity, 85*(8), e00101-17.

Cacciò, S. M., Lalle, M., & Svärd, S. G. (2018). Host Specificity in the *Giardia intestinalis* Species Complex. *Infection, Genetics and Evolution: Journal of Molecular Epidemiology and Evolutionary Genetics in Infectious Diseases, 66*, 335–345.

Cebra, J. J. (1999). Influences of microbiota on intestinal immune system development. *The American Journal of Clinical Nutrition, 69*, 1046S–1051S.

Cotton, J. A., Beatty, J. K., & Buret, A. G. (2011). Host parasite interactions and pathophysiology in *Giardia* infections. *International Journal for Parasitology, 41*(9), 925–933.

Das, S., Schteingart, C. D., Hofmann, A. F., Reiner, D. S., Aley, S. B., & Gillin, F. D. (1997). *Giardia lamblia*: Evidence for carrier-mediated uptake and release of conjugated bile acids. *Experimental Parasitology, 87*(2), 133–141.

de Aguiar Vallim, T. Q., Tarling, E. J., & Edwards, P. A. (2013, May). Pleiotropic roles of bile acids in metabolism. *Cell Metabolism, 17*(5), 650–669.

Donkers, J. M., Abbing, R. L. R., & Van de Graaf, S. F. (2019). Developments in bile salt based therapies: A critical overview. *Biochemical Pharmacology, 161*, 1–13.

Einarsson, E., Troell, K., Hoeppner, M. P., Grabherr, M., Ribacke, U., & Svärd, S. G. (2016). Coordinated changes in gene expression throughout encystation of *Giardia intestinalis*. *PLoS Neglected Tropical Diseases, 10*(3), e0004571.

Farthing, M. J., Keusch, G. T., & Carey, M. C. (1985). Effects of bile and bile salts on growth and membrane lipid uptake by *Giardia lamblia*. Possible implications for pathogenesis of intestinal disease. *The Journal of Clinical Investigation, 76*(5), 1727–1732.

Farthing, M. J., Varon, S. R., & Keusch, G. T. (1983). Mammalian bile promotes growth of *Giardia lamblia* in axenic culture. *Transactions of the Royal Society of Tropical Medicine and Hygiene, 77*(4), 467–469.

Figueroa-González, I., Quijano, G., Ramírez, G., & Cruz-Guerrero, A. (2011). Probiotics and Prebiotics-perspectives and challenges. *Journal of the Science of Food and Agriculture, 91*(8), 1341–1348.

Franzén, O., Jerlström-Hultqvist, J., Castro, E., Sherwood, E., Ankarklev, J., Reiner, D. S., et al. (2009). Draft genome sequencing of *Giardia intestinalis* assemblage B isolate GS: Is human giardiasis caused by two different species? *PLoS Pathogens, 5*(8), e1000560.

Gasser, F. (1994). Safety of lactic acid bacteria and their occurrence in human clinical infections. *Bulletin de l'Institut Pasteur (France)*.

Gillin, F. D. (1987). *Giardia lamblia*: The role of conjugated and unconjugated bile salts in killing by human milk. *Experimental Parasitology, 63*(1), 74–83.

Gillin, F. D., Boucher, S. E., Rossi, S. S., & Reiner, D. S. (1989). *Giardia lamblia*: The roles of bile, lactic acid, and pH in the completion of the life cycle in vitro. *Experimental Parasitology, 69*(2), 164–174.

Goyal, N., Rishi, P., & Shukla, G. (2013). *Lactobacillus rhamnosus* GG antagonizes *Giardia intestinalis* induced oxidative stress and intestinal disaccharidases: An experimental study. *World Journal of Microbiology & Biotechnology, 29*(6), 1049–1057.

Haller, D., Blum, S., Bode, C., Hammes, W. P., & Schiffrin, E. J. (2000). Activation of human peripheral blood mononuclear cells by nonpathogenic bacteria in vitro: Evidence of NK cells as primary targets. *Infection and Immunity, 68,* 752–759.

Halliday, C. E., Inge, P. M., & Farthing, M. J. (1995). Characterization of bile salt uptake by *Giardia lamblia. International Journal for Parasitology, 25*(9), 1089–1097.

Hill, C., Guarner, F., Reid, G., Gibson, G. R., Merenstein, D. J., Pot, B., … & Calder, P. C. (2014). Expert consensus document: The International Scientific Association for Probiotics and Prebiotics consensus statement on the scope and appropriate use of the term probiotic. *Nature Reviews Gastroenterology & Hepatology, 11*(8), 506–514.

Holzapfel, W. H., Haberer, P., Geisen, R., Björkroth, J., & Schillinger, U. (2001). Taxonomy and important features of probiotic microorganisms in food and nutrition. *The American Journal of Clinical Nutrition, 73*(2 Suppl), 365S–373S.

Humen, M. A., De Antoni, G. L., Benyacoub, J., Costas, M. E., Cardozo, M. I., Kozubsky, L., … & Pérez, P. F. (2005). *Lactobacillus johnsonii* La1 antagonizes *Giardia intestinalis* in vivo. *Infection and Immunity, 73*(2), 1265–1269.

Isolauri, E., Rautava, S., & Salminen, S. (2012). Probiotics in the development and treatment of allergic disease. *Gastroenterology Clinics of North America, 41*(4), 747–762.

Isolauri, E., Sutas, Y., Kankaanpaa, P., Arvilommi, H., & Salminen, S. (2001). Probiotics: effects on immunity. *The American Journal of Clinical Nutrition, 73,* 444S–450S.

Itoh, M., Wada, K., Tan, S., Kitano, Y., Kai, J., & Makino, I. (1999). Antibacterial action of bile acids against *Helicobacter pylori* and changes in its ultrastructural morphology: Effect of unconjugated dihydroxy bile acid. *Journal of Gastroenterology, 34*(5), 571–576.

Kiousi, D. E., Karapetsas, A., Karolidou, K., Panayiotidis, M. I., Pappa, A., & Galanis, A. (2019). Probiotics in extraintestinal diseases: Current trends and new directions. *Nutrients, 11*(4), 788.

Kirk, M. D., Pires, S. M., Black, R. E., Caipo, M., Crump, J. A., Devleesschauwer, B., … & Hall, A. J. (2015). World Health Organization estimates of the global and regional disease burden of 22 foodborne bacterial, protozoal, and viral diseases, 2010: a data synthesis. *PLoS medicine, 12*(12), e1001921.

Kurdi, P., Kawanishi, K., Mizutani, K., & Yokota, A. (2006). Mechanism of growth inhibition by free bile acids in *Lactobacilli* and *Bifidobacteria. Journal of Bacteriology, 188*(5), 1979–1986.

Leitsch, D., Burgess, A. G., Dunn, L. A., Krauer, K. G., Tan, K., Duchêne, M., et al. (2011). Pyruvate: ferredoxin oxidoreductase and thioredoxin reductase are involved in 5-nitroimidazole activation while flavin metabolism is linked to 5-nitroimidazole resistance in *Giardia lamblia. The Journal of Antimicrobial Chemotherapy, 66*(8), 1756–1765.

Liévin-Le Moal, V., & Servin, A. L. (2014). Anti-infective activities of *Lactobacillus* strains in the human intestinal microbiota: From probiotics to gastrointestinal anti-infectious biotherapeutic agents. *Clinical microbiology reviews, 27*(2), 167–199.

Lujan D. H., & Svärd S. (2011). *Giardia*: a model organism. ISBN 978-3-7091-0198-8, Springer-Verlag Wien.

Markowiak, P., & Śliżewska, K. (2017). Effects of probiotics, prebiotics, and synbiotics on human health. *Nutrients, 9*(9), 1021.

Merritt, M. E., & Donaldson, J. R. (2009). Effect of bile salts on the DNA and membrane integrity of enteric bacteria. *Journal of Medical Microbiology, 58*(Pt 12), 1533–1541.

Mirzaei, E. Z., Lashani, E., & Davoodabadi, A. (2018). Antimicrobial properties of lactic acid bacteria isolated from traditional yogurt and milk against*Shigella* strains. *GMS Hygiene and Infection Control, 13.*

Mitsuoka, T. (1992). The human gastrointestinal tract. In B. J. B. Wood (Ed.), *The lactic acid bacteria in health and disease* (Vol. 1, pp. 69–114). London, United Kingdom: Elsevier Applied Science.

Mons, C., Dumètre, A., Gosselin, S., Galliot, C., & Moulin, L. (2009). Monitoring of *Cryptosporidium* and *Giardia* river contamination in Paris area. *Water Research, 43*(1), 211–217.

Morrison, H. G., McArthur, A. G., Gillin, F. D., Aley, S. B., Adam, R. D., Olsen, G. J., … & Davids, B. J. (2007). Genomic minimalism in the early diverging intestinal parasite *Giardia lamblia*. *Science, 317*(5846), 1921–1926.

Ouwehand, A. C., & Salminen, S. J. (1998). Adhesion inhibitory activity of β-lactoglobulin isolated from infant formulae. *Acta Paediatrica, 87*(5), 491–493.

Pérez, P. F., Minnaard, J., Rouvet, M., Knabenhans, C., Brassart, D., De Antoni, G. L., et al. (2001). Inhibition of *Giardia intestinalis* by extracellular factors from *Lactobacilli*: An in vitro study. *Applied and Environmental Microbiology, 67*(11), 5037–5042.

Perrucci, S., Fichi, G., Ricci, E., Galosi, L., Lalle, M., & Rossi, G. (2019). In vitro and ex vivo evaluation of the anti-*Giardia intestinalis* activity of the supernatant of Slab51 (SivoMixx). *PLoS ONE, 14*(3), e0213385.

Pridmore, R. D., Berger, B., Desiere, F., Vilanova, D., Barretto, C., Pittet, A. C., … & Mollet, B. (2004). The genome sequence of the probiotic intestinal bacterium *Lactobacillus johnsonii* NCC 533. *Proceedings of the National Academy of Sciences, 101*(8), 2512–2517.

Ratner, D. M., Cui, J., Steffen, M., Moore, L. L., Robbins, P. W., & Samuelson, J. (2008). Changes in the N-glycome, glycoproteins with Asn-linked glycans, of *Giardia lamblia* with differentiation from trophozoites to cysts. *Eukaryotic Cell, 7*(11), 1930–1940.

Rendtorff, R. C. (1954). The experimental transmission of human intestinal protozoan parasites. II. *Giardia lamblia* cysts given in capsules. *American Journal of Hygiene, 59*(2), 209–220.

Ribeiro, M. R. S., Oliveira, D. R., Oliveira, F. M. S., Caliari, M. V., Martins, F. S., Nicoli, J. R., … & Gomes, M. A. (2018). Effect of probiotic *Saccharomyces boulardii* in experimental giardiasis. *Beneficial Microbes, 9*(5), 789–797.

Ridlon, J. M., Harris, S. C., Bhowmik, S., Kang, D. J., & Hylemon, P. B. (2016). Consequences of bile salt biotransformations by intestinal bacteria. *Gut microbes, 7*(1), 22–39.

Sannasiddappa, T. H., Lund, P. A., & Clarke, S. R. (2017). In vitro antibacterial activity of unconjugated and conjugated bile salts on *Staphylococcus aureus*. *Frontiers in Microbiology, 8*, 1581.

Shukla, G., Devi, P., & Sehgal, R. (2008). Effect of *Lactobacillus casei* as a probiotic on modulation of Giardiasis. *Digestive Diseases and Sciences, 53*(10), 2671–2679.

Shukla, G., Kamboj, S., & Sharma, B. (2019). Comparative Analysis of Antigiardial Potential of Heat Inactivated and Probiotic Protein of Probiotic*Lactobacillus rhamnosus* GG in Murine Giardiasis. *Probiotics and Antimicrobial Proteins*, 1–9.

Shukla, G., & Sidhu, R. K. (2011). *Lactobacillus casei* as a probiotic in malnourished *Giardia lamblia*-infected mice: a biochemical and histopathological study. *Canadian Journal of Microbiology, 57*(2), 127–135.

Sievers, S., Metzendorf, N. G., Dittmann, S., Troitzsch, D., Gast, V., Tröger, S. M., … & Riedel, K. (2019). Differential view on the bile acids stress response of *Clostridioides difficile*. *Frontiers in Microbiology, 10*, 258.

Singer, S. M., & Nash, T. E. (2000). The role of normal flora in *Giardia lamblia* infections in mice. *The Journal of Infectious Diseases, 181*, 1510–1512.

Sorokulova, I. B., Pinchuk, I. V., Denayrolles, M., Osipova, I. G., Huang, J. M., Cutting, S. M., et al. (2008). The safety of two bacillus probiotic strains for human use. *Digestive Diseases and Sciences, 53*(4), 954–963.

Sung, J. Y., Shaffer, E. A., & Costerton, J. W. (1993). Antibacterial activity of bile salts against common biliary pathogens. Effects of hydrophobicity of the molecule and in the presence of phospholipids. *Digestive Diseases and Sciences, 38*(11), 2104–2112.

Thanissery, R., Winston, J. A., & Theriot, C. (2017). Inhibition of spore germination, growth, and toxin activity of clinically relevant *C. difficile* strains by gut microbiota derived secondary bile acids. *Anaerobe, 45*, 86–100.

Torres, M. F., Uetanabaro, A. P., Costa, A. F., Alves, C. A., Farias, L. M., Bambirra, E. A., et al. (2000). Influence of bacteria from the duodenal microbiota of patient with symptomatic giardiasis on the pathogenicity of *Giardia intestinalis* in gnotoxenic mice. *Journal of Medical Microbiology, 49,* 209–215.

Travers, M. A., Florent, I., Kohl, L., & Grellier, P. (2011). Probiotics for the control of parasites: An overview. *Journal of Parasitology Research, 2011,* 610769.

Travers, M. A., Sow, C., Zirah, S., Deregnaucourt, C., Chaouch, S., Queiroz, R. M., … & Grellier, P. (2016). Deconjugated bile salts produced by extracellular bile-salt hydrolase-like activities from the probiotic *Lactobacillus johnsonii* La1 inhibit *Giardia duodenalis* in vitro growth. *Frontiers in Microbiology*, *7,* 1453.

Vandenbergh, P. A. (1993). Lactic acid bacteria, their metabolic products and interference with microbial growth. *FEMS Microbiology Reviews, 12*(1–3), 221–237.

Vitetta, L., Saltzman, E. T., Nikov, T., Ibrahim, I., & Hall, S. (2016). Modulating the gut micro-environment in the treatment of intestinal parasites. *Journal of Clinical Medicine*, *5*(11), 102.

Vivancos, V., González-Alvarez, I., Bermejo, M., & Gonzalez-Alvarez, M. (2018). Giardiasis: Characteristics, pathogenesis and new insights about treatment. *Current Topics in Medicinal Chemistry, 18*(15), 1287–1303.

Woods, A., Sherwin, T., Sasse, R., MacRae, T. H., Baines, A. J., & Gull, K. (1989). Definition of individual components within the cytoskeleton of *Trypanosoma brucei* by a library of monoclonal antibodies. *Journal of Cell Science, 93*(1989), 491–500.

Chapter 25
A Protocol to Quantify Cellular Morphodynamics: From Cell Labelling to Automatic Image Analysis

Maria Manich, Aleix Boquet-Pujadas, Stéphane Dallongeville, Nancy Guillen, and Jean-Christophe Olivo-Marin

Abstract Cellular morphodynamics can be used as markers for many physiological and pathological processes. This protocol provides a step-by-step guide to identify variations in motility and morphology within (or across) cell populations using non-invasive live imaging and reproducible image analysis techniques such as segmentation and tracking. Detailed instructions cover all the way from cell culturing and labelling to automatic image and statistical analyses, including the definition of multiple descriptors that characterise the shape and movement of cells in a quantitative manner. All methods are available as free open-source software and illustrated by video tutorials.

Keywords Cell biology · Image analysis · Software · Segmentation · Tracking · Shape descriptors

---

Electronic supplementary material The online version of this chapter (https://doi.org/10.1007/978-3-030-44826-4_25) contains supplementary material, which is available to authorized users.

M. Manich · A. Boquet-Pujadas · S. Dallongeville · N. Guillen · J.-C. Olivo-Marin (✉)
Institut Pasteur, Bioimage Analysis Unit, 75015 Paris, France
e-mail: jcolivo@pasteur.fr

M. Manich · A. Boquet-Pujadas · S. Dallongeville · J.-C. Olivo-Marin
Centre National de La Recherche Scientifique, CNRS, UMR3691 Paris, France

A. Boquet-Pujadas
Sorbonne Université, Paris, France

N. Guillen
Centre National de La Recherche Scientifique, CNRS, ERL9195 Paris, France

© Springer Nature Switzerland AG 2020
N. Guillen (ed.), *Eukaryome Impact on Human Intestine Homeostasis and Mucosal Immunology*, https://doi.org/10.1007/978-3-030-44826-4_25

Introduction

Advances in microscopy techniques and fluorescent probes have long being helping the scientific community determine the importance of cell movement and deformation in multiple biological processes. However, many studies remain qualitative, i.e. differences in shape or motility are assessed visually, adding subjectivity to potential biological conclusions. Conversely, using image analysis to assign numerical values to both shape and movement does not only guarantee the reproducibility of the conclusions but also opens the door to statistical analyses that allow classifying cell populations and phenotyping. Accordingly, we present a step-by-step manual that shows how to quantify cellular morphodynamics in a non-invasive and reproducible way using only confocal microscopy and fluorescent markers.

The present protocol details both biological and computational experiments. We first describe the necessary biological techniques, namely culturing the cells and fluorescently labelling the cytoplasm; next, we comment on how to perform non-invasive imaging using a confocal microscope; and, finally, we provide a ready-to-use image analysis workflow that goes all the way from raw images to biological conclusions in a reproducible manner. More specifically, we present automatic tools for cell segmentation and tracking that are freely accessible as modules in the Icy platform; as well as multiple descriptors that quantify cell shape and movement from the resulting contours and tracks. These descriptors serve as a basis from which to perform statistical tests and assess any possible correlation between morphodynamical variables. All the key steps of the protocol are available as video tutorials and are exemplified using a population of *Entamoeba histolytica*, a highly motile parasite that migrates through diverse human tissues, including the intestine and the liver.

Results

Wet-Lab Protocol: Culturing Cells and Acquiring Images

To quantify movement and deformation using image analysis (see dry-lab protocol), it is paramount to image the cells non-invasively (physiological relevance) and in good spatiotemporal resolution (easier analysis). To meet these two criteria, we label the cytoplasm with a fluorescent dye and use a spinning-disk confocal microscope.

Cell Culture and Staining

Trophozoites of the *Entamoeba histolytica* strain HM1: IMSS were grown overnight at 37 °C in TYI-S-33 medium (Diamond et al. 1978). Medium was then replaced by incomplete TYI-S-33 medium (serum/vitamines-free) (TYI). Cells were labeled with Cell Tracker™ Red CMTPX, a fluorescent dye that is well suited for monitoring cell movement and displacements (Petropolis et al. 2014). The dye has low cytotoxicity, does not affect cell viability nor proliferation, and its fluorescence was stable during the entire imaging process, allowing us to track cellular movements with a red excitation/emission spectra (577/602 nm). In this case, we have used the fluorescent dye that emits in the red spectra because *Entamoeba histolytica* autofluoresces at 488 nm. Since the forthcoming image analysis methods are based on accurate cell segmentation, in this paper we used non-confluent cultures. Other image analysis tools are required to deal with confluent cell cultures, but they are not the focus of this paper.

Cells were incubated for 45 min at 37 °C, and then washed with TYI pre-warmed at 37 °C by reversing the tube and simply discarding the medium. No centrifugation is required because amoebas are adherent cells and remain attached to the glass tube during the process. Trophozoites were gently suspended in pre-warmed TYI by shaking the tube and then seeded on 35 mm glass-bottomed imaging Ibidi dishes, obtaining an estimate of 5×10^3 cells.

Microscopy Experiment

Images were taken with a spinning disk confocal microscope (25 × objective) inside an incubator at 37 °C to keep the parasites at a physiological temperature where they are specially motile. Indeed, at these temperature *E. histolytica* parasites can move at up to 1 μm/s in 2D culture conditions (Dufour et al. 2015). Fortunately, with the spinning disk microscope, images can be acquired at very high frame rates with minimal illumination and photo-bleaching of the living samples.

Videos were recorded for four minutes at an imaging rate of one frame per second (i.e. a total of 240 frames) and at a pixel size of 0.48 μm. The fields of view were taken to be of around 512 × 512 pixels, corresponding to $246 \times 246\ \mu m^2$, which typically contained around 2–6 cells. The z position was set at a height of around 2 μm from the glass.

Both pixel size and frame rate are necessary for the posterior image analysis, for example to obtain the speed in real units, and therefore need to be stored. They are typically stored automatically in the metadata of the image files by the software associated with the microscope, but we recommend to double-check that this is indeed the case. In our case, all images were acquired with the Volocity 3D image analysis software (Perkin Elmer, USA) and the files and their associated metadata stored in the mvd2 format.

There are no potential dangers involved in the experiments, neither because of laser beams nor of parasite pathogenicity. However, a P2-class laboratory is needed to handle the living trophozoites. The protocol was set up according to the guidelines provided by the Safety Authorities and the Image Microscopy Facility platform of Institut Pasteur.

Dry-Lab Protocol: Analysing Images

The motility of a cell population can be studied quantitatively using image analysis. In this context, each individual cell in a video sequence is first singled out of the background in a process called segmentation. Cell segmentation not only allows to delimit the borders of the cells present in an image, but also to calculate their centroid and thus to track the displacement of the cells over time. On the one hand, digitising the contours of the cell opens the door to characterising the cell shape with descriptors such as roundness; on the other hand, the time tracks contain information on the movement of the cell such as its speed or the straightness of its trajectory, which shed light on the reasons behind its migration (random, directed chemotaxis, etc.). Therefore, these data enable a rich quantification of both cell morphology and motility that ideally translates into cell phenotyping when complemented with an extensive statistical and correlation analysis.

The three main steps (segment, track and statistical assessment) are visited in detail in respective Sects. "Hierarchical K-Means", "Active Contours", "Cell Tracking with Track Manager" and are automatised by bioimaging softwares such as Icy (de Chaumont et al. 2012; Wiesmann et al. 2015), which we address immediately in Sect. 25.2.2. All steps are available as video tutorials.

Bioimage Analysis Software

To quantify cell motility, we present Icy, a free and open-source platform for bioimage analysis that provides multiple resources to visualize, annotate and quantify bioimaging data (http://icy.bioimageanalysis.org).

Icy provides a user-friendly approach to new and classical image analysis techniques alike: filtering, segmentation, tracking.... They are all available under different modules called plug-ins who all share the same graphical interface. Examples of segmentation plug-ins are Thresholding, Active Contours, Parametric Snakes, Potts Segmentation, Spot Detector (Olivo-Marin 2002) and HK-Means (Dufour et al. 2008); whereas plug-ins such as Spot Tracking, Track Manager and Kymograph Tracker provide different approaches to tracking. In this protocol, we will focus on HK-Means (Sect. "Hierarchical K-Means"), Active Contours (Sect. "Active Contours"), and Track Manager (Sect. "Cell Tracking with Track Manager") in order to provide a step-by-step guide on how to analyse cell shape and motility.

The graphical interface integrates 2D and 3D visualisation resources, as well as a series of tools to easily crop and cut through time series, z-stacks or multi-channel sequences. Also intuitive is the management of so-called Regions Of Interest (ROIs), i.e. delimited areas of the image that are of special interest and that might want to be analysed aside, for example a cell segmented from the image. In the Icy platform, ROIs are superimposed over the original image and can be manipulated as independent objects on which common operations such as "copy/paste" (ctrl + c/v) or "delete" can be applied, allowing to easily combine analyses performed on different channels or sequences. ROIs are deeply integrated into Icy so that any analysis or quantification from them is automatic and straightforward. For instance, cell segmentation results are represented as ROIs from which multiple descriptors such as area, mean fluorescence intensity or roundness can be directly accessed in the ROI menu and exported into Excel files for further data analysis.

As a last remark, in its most recent version 2.0, Icy has introduced a new image handling engine that allows working with big video sequences, be it either because they are long or because they were taken at very high spatial and/or temporal resolution. The idea behind the new engine is that only a portion of the image sequence is loaded into the local RAM memory, while the rest is stored on hard-disk at the price of longer processing times.

We have used the sequence called 25 × 40 to illustrate this protocol (Movie 1) over its several steps. The Time Stamp Overlay plugin (Tutorial 1) can be used to stamp the elapsed time onto the video.

Cell Segmentation with Hierarchical K-Means and Active Contours

Hierarchical K-Means

Hierarchical K-Means (HK-means or HKM here) is a segmentation method based on a K-Means clustering of the image histogram, i.e. an algorithm that divides the different intensity values in the image into groups according to a similarity measure. Ideally, these groups correspond to the different cells and to the background. However, the K-means algorithm requires the number of groups to be specified in advance. To tackle this problem, a hierarchical strategy is introduced. In this way, the algorithm attempts to find the ideal number of groups using a bottom-up approach. This process can be helped if the user specifies a value for the expected minimum and maximum size of the cells.

HKM is a fundamental tool in image processing; it is one of the go-to algorithms if the user wants to segment cells in a quick, ready-to-use and quasi-automatic manner. And precisely because of its hierarchical clustering approach, it performs better than classical clustering and thresholding algorithms. However, HKM suffers from three main drawbacks. First, it has difficulty telling apart cells that are in contact with each other. Second, big intensity heterogeneities inside the cell might trigger multiple detections. And third, since the resulting segmentations are groups of pixels rather than polygonal contours, some accuracy may be lost when computing shape descriptors.

A step-by-step guide to the HKM plug-in in Icy can be found in Fig. 25.1 and is accompanied by Tutorial 2.

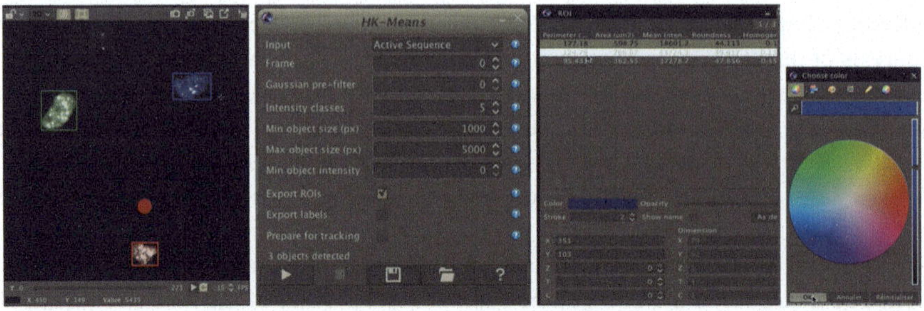

Fig. 25.1 Cell segmentation with HK-Means and ROI color selection. HK-Means can be opened from the detection tab or from the search bar. The first step is to choose which frames we are interested in segmenting (here we select frame 0, but 'ALL' is also a possibility). We have to specify the number of intensity classes (see the text for an explanation on HK-Means) in which the histogram is to be split. That is the number of different intercellular intensities (e.g. if all cells are the same intensity, two classes are enough). When in doubt, we recommend choosing higher values, but we also remark that it comes at a computational price. To aid the segmentation of the image into the different classes the user can also input an expected minimum and maximum size for the cells so as to eliminate possible groups that are respectively too small (e.g. debris) or too big to possibly be a cell (e.g. cell clusters). Notice that these sizes are required in pixels, to have a rough idea of the cell size in pixels draw a ROI around the cell and check its size ("Interior" descriptor in the ROI tab at the right-hand side). Finally, applying the Gaussian pre-filter can help improve the segmentation of noisy images. Since the segmentation output are ROIs, we can obtain any descriptor directly from the ROI tab. Here we show the perimeter, the area, the mean intensity, the roundness and the homogeneity inside the ROIs, but many more shape descriptors can be selected using the "gear" button. ROI colors can be chosen (see Tutorial 2)

Active Contours

Active Contours (Zimmer et al. 2002) (AC) are well adapted to study cell morphodynamics; they provide accurate cell contours and are capable of segmenting cells that are in close contact, as well as cells with inner heterogeneities. However, in contrast to the more classical segmentation methods, AC need to be initialised. The user has to specify an approximative initial contour (ROI) around the cell so that the algorithm can pick up on it. This initialization can be done manually, by drawing the ROI over the object of interest, or automatically, using other segmentation tools (e.g. the above-described HKM). The initial ROI contours are then refined by the AC method, which slowly deforms the contour. In this way, the contour is progressively fitted to the cell shape in an attempt not only to separate the image into multiple intensities, but also to find the edges of the cells in the gradient of the image. When the segmentation spans a whole video sequence, the ROI resulting from segmenting a given frame can be used as an initial ROI for the following frame (see '*track objects over time*' in the AC plug-in). Therefore, if the image acquisition is relatively fast, initialising the ROIs at the very first frame is enough to segment the entire sequence.

In summary, whereas HKM is fast and does not need to be initialised, it is most performant when image quality is good and cells are well separated; otherwise AC take over at the price of initialisation and speed. In fact, we remark that a good

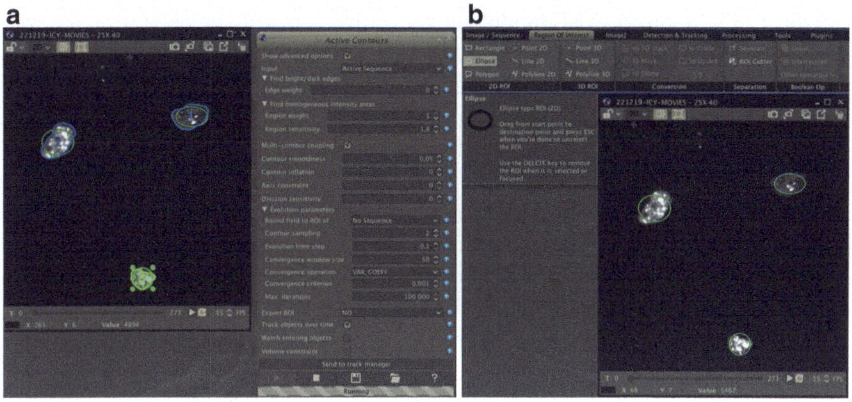

Fig. 25.2 Manual drawing of approximative ROIs and automatic cell segmentation with Active Contours. **a** The first step before running Active Contours is to draw approximative ROIs around each of the cells; we do it manually here (green ellipses), but one could use HK-Means to initialize the method automatically. **b** Active Contours can be opened from the detection tab or from the search bar. In order to segment and track all of the time frames make sure to activate "tracks objects over time". Perhaps the two most important parameters of the plug-in are the edge and region weights, which control the balance between the importance of (1) the big intensity differences that are expected at the border of the cell and (2) the homogeneous intensity that is expected inside the cell as opposed to that of the outside. In addition, "contour inflation" might help compensate for a lack of contrast between cell and background by adding an artificial expansion rate. On the other hand, the set of evolution parameters are more technical but can help speed up the process and/or make the final contours more accurate. All parameter settings can be readily stored and loaded using the save icon on the bottom. The results of the segmentation are also presented as ROIs (see text and Tutorial 3)

approach is to combine the two; that is to use HKM (only) on the first frame to automatically set the initial ROIs required by AC. However, in this protocol we have found it more pedagogic to set the initial contours manually.

A step by step guide to the AC plug-in in Icy can be found in Fig. 25.2 and is accompanied by Tutorials 2, 3 and 4. Movie 2 shows the segmented cells with Active Contours.

Cell Tracking with Track Manager

Using either of the segmentation plug-ins on a video sequence results in a time-series of ROIs that can potentially be linked together to generate the track of a cell, i.e. to draw the path that the cell followed. At our spatiotemporal resolution, it suffices to associate a ROI at a given time point with the closest ROI at the following time point to accurately track cells; more precisely, it is the centroids of the successive ROIs that are concatenated into a cell track. However, more advanced tracking tools such as Multiple Hypothesis Tracking (Chenouard et al. 2013) become necessary for high-speed particle tracking. In either case, the resulting tracks can be analyzed with the Track Manager (TM) module in Icy, which is readily invoked from the very segmentation plug-ins using the '*send to track manager*' button.

TM displays the resulting tracks directly overlayed on the original sequence. The tracks can also be analyzed through an accessible interface, for instance to investigate motility parameters such as cell speed or mean squared displacement (MSD) and compare them between populations or correlate them with other descriptors, for example of cell morphology (see below). All these quantifications tasks are done through so called Track Processors (TPs). Each TP has a specific function: from filtering unwanted tracks, to quantifying movement, passing by a myriad of display functionality such as color-coding the tracks ("TP Color"). In this protocol, we have used several TPs. Briefly, (i) "Motion Profiler" computes multiple motion descriptors such as the average speed or the linearity/persistence of the tracks; (ii) "Instant Speed" displays the speed of the cell as a time curve; whereas (iii) "ROI Statistics" (ROIS) displays time curves of several shape descriptors as is described in Sect. "Statistical Tests with R".

A step-by-step guide to the TM plug-in in Icy can be found in Figs. 25.3, 25.4 and 25.5 and is accompanied by Tutorial 5. Movie 3 shows the segmented cells with Active Contours and their centroid tracks.

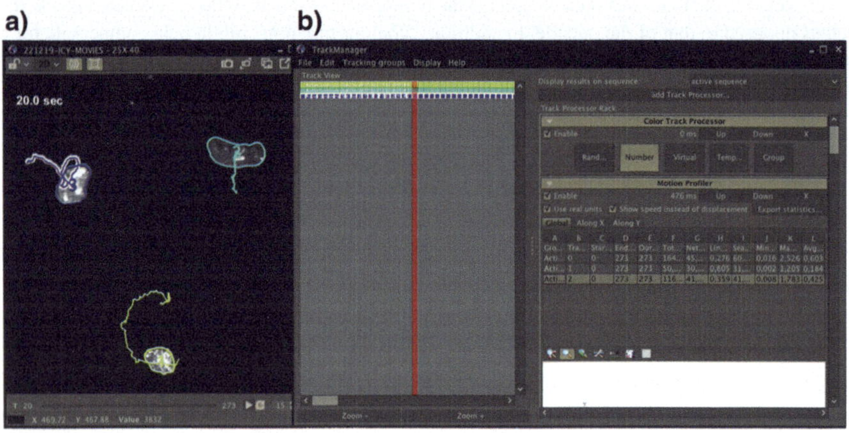

Fig. 25.3 Cell track analysis with Track Manager and Track Processors. a Track Manager can be opened from the tabs or directly from most segmentation plug-ins (e.g. HK-Means or Active Contours, see respective Figs. 25.1, 25.2 and Tutorial 4). The tracks for each of the cells are automatically overlaid on the video sequence in the corresponding colors. **b** Each track is a separate entity and can be filtered or quantified by adding Track Processors. Here we are displaying the Color and Instant Speed Track Processors, but many others are available (e.g. see Figs. 25.4, 25.5 and 25.6). The red vertical bar displays the current time point, and can be dragged to navigate the time sequence. Tracks can be saved into an.xml file

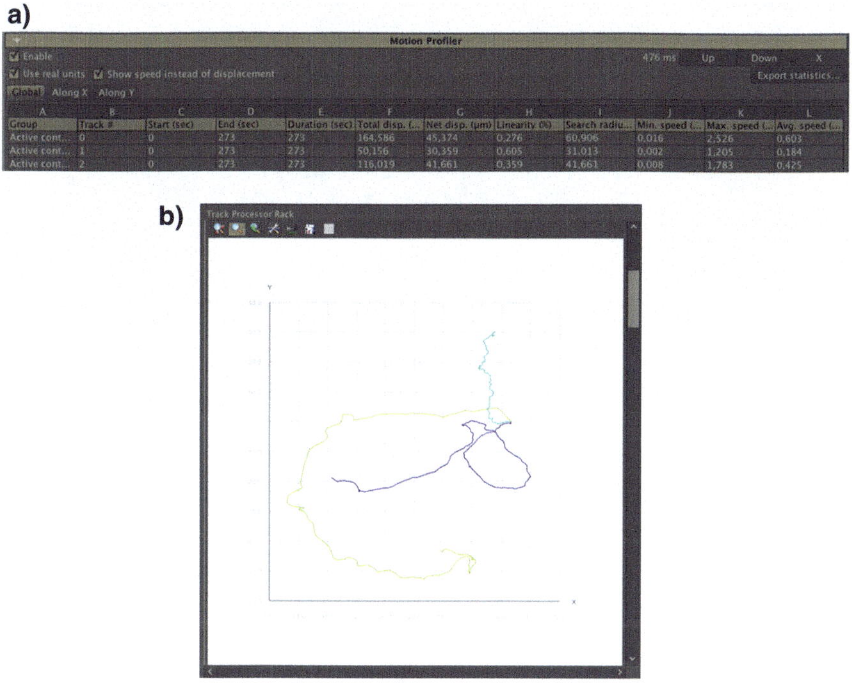

Fig. 25.4 Descriptors and visualisation of cell tracks with the Motion Profiler Track Processor in Track Manager. a By selecting Motion Profiler from the track processors in Track Manager, we obtain multiple descriptors of the three tracks. For example, we can see typical minimum, maximum and average speed values; and we can also quantify how straight the cells are moving with the measures of linearity and search radius. In addition, the processor can take the metadata into account to offer the values in real units. All these results can be exported to an Excel file. **b** Motion Profiler also provides a graphical representation of all the tracks from a common origin, from where we can visually assess whether motion is random or directed

Morphological Descriptors and Statistical Tests with ROI Statistics and R

Cell Descriptors with ROI Statistics

Different cell populations might be characterised by different morphologies. Given a time sequence of already segmented cells in the form of ROIs (e.g. with AC), the ROI Statistics (ROIS) processor in TM provides a wide range of geometrical properties that describe the shape of each ROI. Together with a posterior statistical analysis, these descriptors may help tell apart different populations or be used for phenotyping. Many such descriptors are available in ROIS; in this study, we only consider the fol-

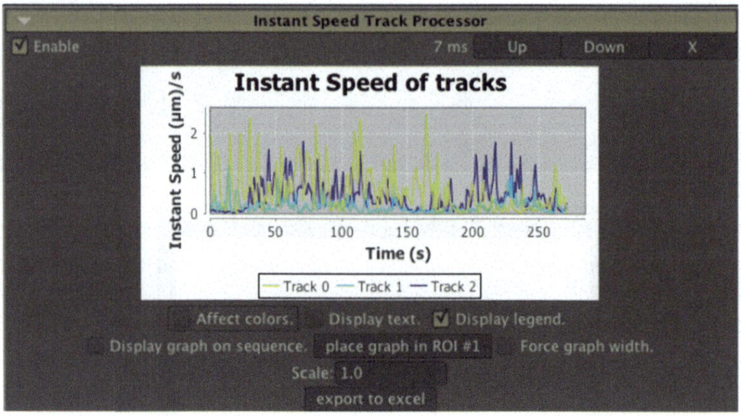

Fig. 25.5 Time curves of cell speed with the Instant Speed Track Processor in Track Manager. By selecting Instant Speed from the track processors in Track Manager, we obtain the time curves of the speed at each time point along each of the cell tracks. These data can be used to explore the cycles of acceleration and deceleration of the cells for example. The curves can be exported to an Excel file

lowing: area (μm^2), perimeter (μm), roundness (%), mean intensity values (a.u.) and homogeneity (a.u.). "Perimeter" measures the perimeter of the ROI in micrometers (here the scaling information is extracted automatically from the metadata). Equivalently, "Area" measures the ROI area in micrometers squared. "Mean intensity" averages the intensity values inside the ROI, whereas "Homogeneity" highlights the internal variations of the intensity distribution within the cell. Lastly, "Roundness" is a measure of how similar to a circle the ROI is. These data are displayed directly in Icy, but can also be exported to an Excel file (Table 25.1) for further analysis, for example to perform statistical tests that assess the correlation between each of the descriptors. For instance, we study the correlations between the temporal mean of all these parameters and the Speed ($\mu m/s$) resulting from TM. Alternative shape descriptors can be extracted by rewriting the cell shape in different mathematical basis such as Fourier (2D) or Spherical Harmonics (3D); these work well to separate populations, but often lack biological interpretability (Ducroz et al. 2012).

A step-by-step guide to the TM plug-in in Icy can be found in Fig. 25.6.

Statistical Tests with R

In order to assess whether any trend or correlation exists between the extracted descriptors we perform a visual pairwise comparison educated with Spearman's rank correlation coefficient. So-called Spearman's "rho" attempts to quantify the monotonicity of the relationship between a pair of variables, irrespectively of its linearity. The coefficient spans the interval $[-1, 1]$, where the extremes correspond to perfectly monotonic functions, respectively decreasing or increasing (i.e. functions

Table 25.1 Cell shape descriptors and speed for the 42 cells of the experiment. (25 × 40 illustrates the protocol, values are means)

Name	Area (µm2)	Homogeneity (561)	Intensity (561)	Perimeter (µm)	Roundness (%)	Speed (µm/s)
25 × 54	316.08	0.08	24,203	75.58	40.05	1.16
	2407.56	0.12	19,189	196.18	59.33	0.08
25 × 49	2779.13	0.13	24,801	208.46	59.11	0.24
	308.78	0.15	15,070	69.42	48.16	1.21
25 × 47	771.28	0.13	16,320	119.90	34.82	0.17
	411.95	0.13	21,341	81.62	49.3	0.43
25 × 44	760.11	0.10	25,160	114.79	45.55	0.40
	684.88	0.10	25,544	100.88	68.20	0.12
	545.58	0.11	21,754	90.64	58.77	0.97
	689.35	0.10	23,625	108.82	48.97	0.25
	658.80	0.10	23,308	109.13	46.66	0.49
	929.79	0.10	23,857	119.42	61.02	0.16
25 × 40	713.25	0.18	42,758	109.02	54.33	0.42
	438.46	0.32	46,123	88.07	42.48	0.60
	773.17	0.15	17,600	118.27	36.11	0.18
25 × 38	362.51	0.11	18,333	73.27	59.32	1.05
	1045.82	0.10	22,956	132.80	37.29	0.10
	814.01	0.10	21,460	113.92	54.76	0.33
25 × 36	1093.23	0.10	21,429	135.41	52.52	0.14
	525.07	0.08	27,100	92.00	50.12	0.31
	2713.55	0.10	36,252	233.99	29.84	0.10
	600.94	0.11	18,237	95.02	60.56	0.61
25 × 35	329.77	0.1	22,166	73.21	47.95	1.10
	630.77	0.1	24,413	101.25	51.82	1.37
	681.54	0.1	21,612	107.25	48.32	0.14
	466.57	0.1	33,879	85.38	53.27	0.23
	617.05	0.11	19,998	104.93	41.30	0.22
25 × 34	656.30	0.34	49,554	108.70	43.97	0.32
	663.35	0.21	10,132	103.12	47.39	0.29
25 × 33	207.05	0.07	24,868	56.05	52.87	0.62
	595.10	0.13	16,759	97.52	52.17	0.86
25 × 32	523.20	0.13	16,121	94.08	45.30	0.54
	653.94	0.11	19,977	105.08	41.45	0.14
25 × 30	1084.28	0.11	23,447	130.04	30.10	0.15

(continued)

Table 25.1 (continued)

Name	Area	Homogeneity	Intensity	Perimeter	Roundness	Speed
	(μm2)	(561)	(561)	(μm)	(%)	(μm/s)
	887.42	0.09	23,499	128.19	39.44	0.24
	738.61	0.12	17,261	107.66	51.55	0.30
	791.83	0.08	26,903	115.63	47.52	0.15
	515.45	0.10	20,549	85.64	70.17	0.12
25 × 29	430.99	0.09	23,945	90.56	35.33	0.54
	532.69	0.12	17,155	93.70	49.06	0.47
	437.21	0.12	16,415	85.08	47.43	0.37
	799.27	0.10	22,078	118.07	42.83	0.64

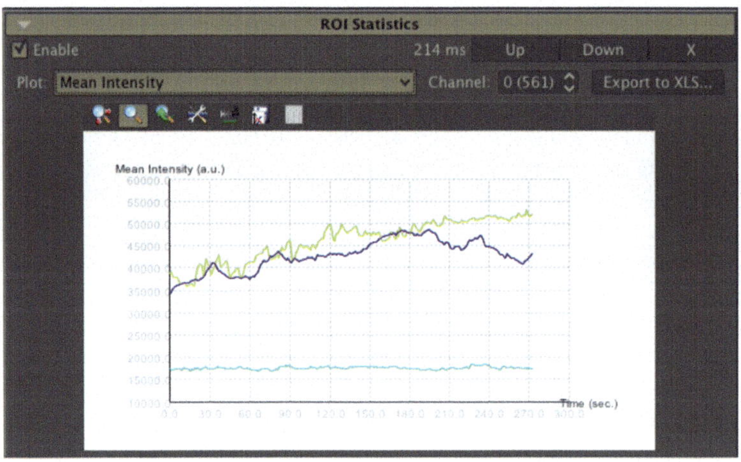

Fig. 25.6 Cell shape descriptors with the ROI Statistics Track Processor in Track Manager. By selecting ROI Statistics from the track processors in Track Manager, we obtain different descriptors of cell shape (perimeter, roundness, etc.) for each time point along a cell track. Here are presented the fluorescence average intensity values inside the ROIs. The data can be exported to an Excel file

that always go down, or up, without fluctuations); and 0 indicates a lack of correlation. Precisely, the p-value associated with the coefficient results from testing whether this coefficient is significantly different from 0.

Statistical analysis software can directly read the output values exported from Icy. Here, we use a short R routine that can automatically generate the pairwise graphics showing possible trends, as well as the correlation values and their corresponding *p*-values (Fig. 25.7). This program uses some functionality from the ggplot2 library. While it is not the aim of the paper to provide in-depth statistical insight, we remark that it is important to check whether your data satisfies all the assumptions made during the statistical analysis. For illustrative purposes, in Fig. 25.7 generated by

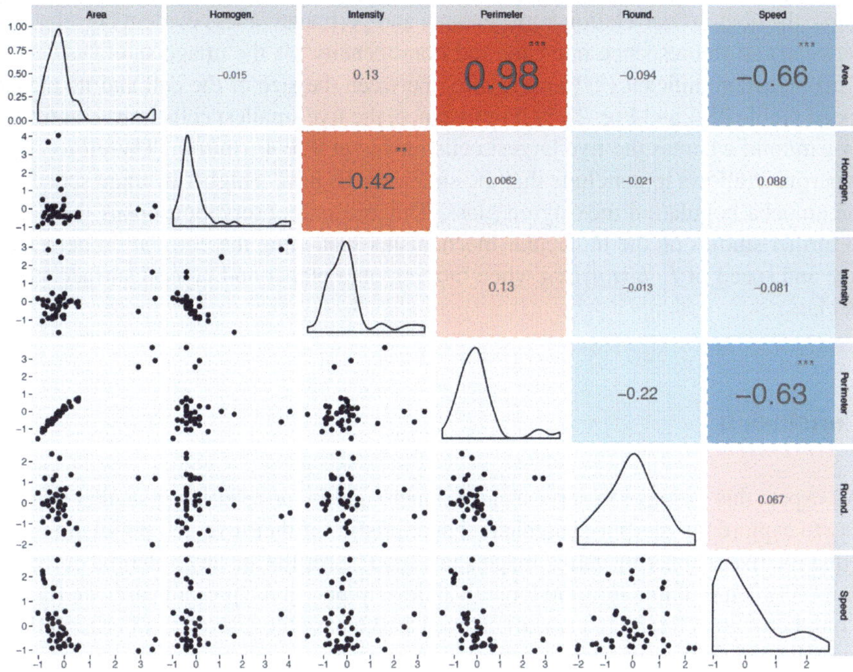

Fig. 25.7 Statistical analysis of cell descriptors shows a significant correlation between cell speed and size. The figure is a matrix quantifying the correlation between pairs of descriptors for n = 42 cells from 13 different movies. On the upper triangular side, Spearman's correlation values resulting from descriptor pairs are displayed accompanied by their significance in the form of stars. The values are also displayed using a color gradient (red positive, blue negative) to facilitate the analysis. On the lower triangular side, we plot descriptor pairs on a normalised scale to show any possible trend. The diagonal contains univariate density plots

the R routine, we display univariate descriptor density plots (diagonal) and pairwise descriptor plots (lower diagonal), but directly compute the pairwise correlation coefficients and their associated tests (upper diagonal) with no prior analysis.

Example to Illustrate the Proposed Protocol for Image Analysis

During the in vitro growth of *E. histolytica*, it is common to observe diverse phenotypes regarding the size of the cells, their mobility, the heterogeneity of fluorescence during labeling, etc. We wondered whether the protocol proposed here could help us identify any correlations between these phenotypes. After acquisition of videomicroscopies of *E. histolytica* seeded on glass, the image analysis was performed on n = 42 cells from 13 different video sequences. The data highlights several relationships: the obvious correlation between area and perimeter, a less evident correlation between cell fluorescence intensity and homogeneity (as the image saturates), and a strong and significant (***) correlation between the size of the cell and its mean speed (Table 25.1 and Fig. 25.7). For instance, the five smallest cells moved at 62 ± 14 μm/min, whereas the five largest cells moved at 9 ± 4 μm/min. Therefore, this experiment allows to conclude that the smaller cells have a higher average speed in the amoeba population moving on glass. This original observation opens the door to further studies on the molecular mechanisms sustaining the correlation between size and speed of *E. histolytica* when moving on a planar and neutral surface such as glass.

Conclusion

We expect this protocol to serve as a beginner's guide for cell biologists that would like to capture the morphodynamical characteristics of their live cell populations in a quantitative manner by using image analysis. The results are any potential correlations between multiple morphodynamical descriptors (in the present case, we found a link between cell size and speed), as well as the possible discovery of criteria that can tell apart subpopulations of cells.

Materials and Basic Methods

Biological Materials

- Trophozoites of Entamoeba *histolytica* strain HM1:IMSS growing in TYI-S33 media (Diamond et al. 1978).
- Cell Tracker™ Red CMTPX (ThermoFisher, catalog number C34554, final concentration 2.5 μM). Before use, suspend the dessicated dye (50 μg) in 8.33 μl of DMSO to obtain a 10 mM stock solution. An intermediate dilution (1/200) has to be prepared to avoid aggregates of DMSO and Cell Tracker in the media.
- 35 mm high glass-bottom Ibidi dish (catalog number 81158, Ibidi, France).

Equipment

- Microbiological safety station with laminar flow to manipulate the cells; wearing a blouse and gloves is mandatory during the experimental steps.
- Spinning disk confocal microscope (UltraVIEW VoX, Perkin Elmer, USA; excitation: 561 nm; objective: 25×; temperature control set to 37 °C).

Softwares

- Volocity (Perkin Elmer, USA) to perform imaging.
- Icy (Institut Pasteur, France) to perform image analysis.

Summary of the Protocol

Procedure—The protocol can be summarised as a general workflow (Fig. 25.8) in the following steps: culture the cells and label them with a fluorescent cytoplasm dye; image the cells with a temporal resolution that is appropriate to the cell movement. Save the video sequences on the hard-disk; open the Icy software and allocate RAM according to the expected image size, open the sequence, and double-check the metadata; draw initial ROIs over the cells and run the Active Contours plug-in; send the resulting segmentation to the Track Manager, use the different track processors to analyse cell movement and shape and export them to Excel; perform statistical and correlation tests on the data, for example using R.

Timing—Cell labelling and preparation takes between one and two hours. Live imaging only involves setting up the sample on the microscope and taking multiple video sequences of around 240 frames (i.e. around 4 min). Segmentation and tracking takes a fraction of a second per frame. Statistical analysis takes well under an hour.

Troubleshooting—1. Check that the Java version in your computer is compatible with Icy. 2. From within the preferences tab in Icy assign RAM memory to the software according to the potential size of your images. 3. Check that your temporal resolution is adequate: if there are too many frames per second compared to the speed of the cells, remove frames in constant intervals in order to lift some computational burden. 4. All stages of the quantification can be saved in their corresponding formats. For example, image sequences can be saved in.tif, whereas ROIs and tracks are saved in.xml. This guarantees complete reproducibility, as slightly different ROIs can result in slightly different segmentations.

Data availability—All data presented in this protocol (files as.tif,.xml,.avi,.mov) and tutorials are available online (Manich 2020) so that any potential user can reproduce the results by following the protocol.

Acknowledgements We acknowledge Marion Louveaux for advice on image analysis reproducibility and data online availability. We are grateful to the technical unit of BioImagerie Photonique of Institut Pasteur for their help with microscopy experiments. A.B.P. is part of the Pasteur-Paris University (PPU) International Ph.D. Program.

Funding This project has received funding from the European Union's Horizon 2020 research and innovation program under the Marie Skłodowska-Curie grant agreement No 665807, the Institut Carnot Pasteur Microbes & Sante (ANR 16 CARN0023-01), the Labex IBEID (ANR-10-LABX-62-IBEID), France-BioImaging infrastructure (ANR-10-INBS-04) and the program PIA INCEPTION (ANR-16-CONV-0005).

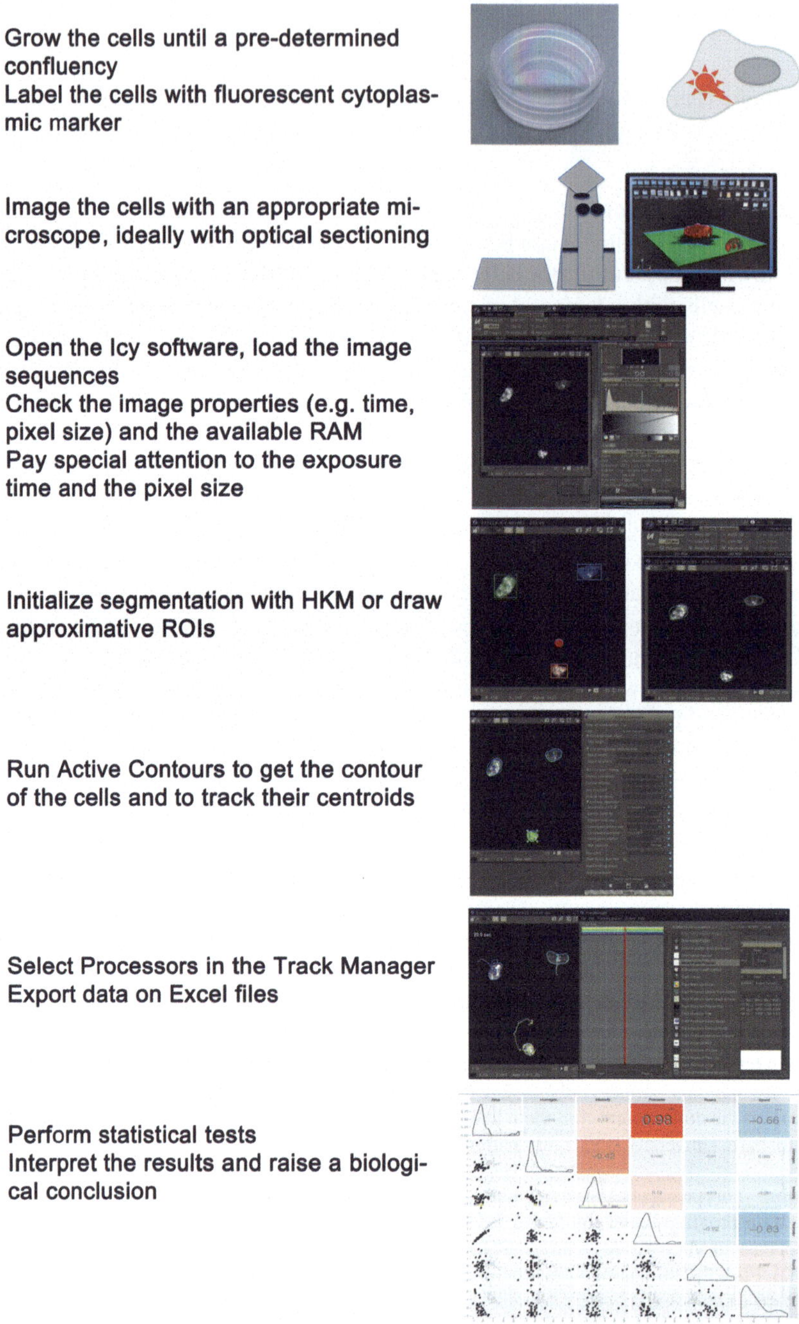

Fig. 25.8 Summary of the protocol. See text for a complete description of the protocole

References

Chenouard, N., Bloch, I., & Olivo-Marin, J. C. (2013). Multiple hypothesis tracking for cluttered biological image sequences. *IEEE Transactions on Pattern Analysis and Machine Intelligence, 35*(11), 2736–3750. https://doi.org/10.1109/TPAMI.2013.97.

De Chaumont, F., Dallongeville, S., Chenouard, N., Hervé, N., Pop, S., Provoost, T., ... & Lagache, T. (2012). Icy: an open bioimage informatics platform for extended reproducible research. *Nature methods*, *9*(7), 690.

Diamond, L. S., Harlow, D. R., & Cunnick, C. C. (1978). A new medium for the axenic cultivation of *Entamoeba histolytica* and other *Entamoeba*. *Transactions of the Royal Society of Tropical Medicine and Hygiene, 72*(4), 431–432. https://doi.org/10.1016/0035-9203(78)90144-X.

Ducroz, C., Olivo-Marin, J. C., & Dufour, A. (2012). Characterization of cell shape and deformation in 3d using spherical harmonics. In *2012 9th IEEE International Symposium on Biomedical Imaging (ISBI)* (pp. 848–851). IEEE. Barcelona. https://doi.org/10.1109/isbi.2012.6235681.

Dufour, A., Meas-Yedid, V., Grassart, A., & Olivo-Marin, J. C. (2008, December). Automated quantification of cell endocytosis using active contours and wavelets. In *2008 19th International Conference on Pattern Recognition* (pp. 1–4). IEEE, Tampa, FL. https://doi.org/10.1109/icpr.2008.4761748.

Dufour, A. C., Olivo-Marin, J. C., & Guillén, N. (2015). Amoeboid movement in protozoan pathogens. *Seminars in Cell & Developmental Biology, 46,* 128–134. https://doi.org/10.1016/j.semcdb.2015.10.010.

Manich, M. (2020). Online data of "A protocol to quantify cellular morphodynamics: from cell labelling to automatic image analysis" (Version http://icy.bioimageanalysis.org). Zenodo. https://doi.org/10.5281/zenodo.3594363.

Olivo-Marin, J. C. (2002). Extraction of spots in biological images using multiscale products. *Pattern Recognition, 35,* 1989–1996. https://doi.org/10.1016/S0031-3203(01)00127-3.

Petropolis, D. B., Faust, D. M., Deep Jhingan, G., & Guillen, N. (2014). A New Human 3D-Liver Model Unravels the Role of Galectins in Liver Infection by the Parasite *Entamoeba histolytica. PLoS Pathogens, 10*(9), e1004381. https://doi.org/10.1371/journal.ppat.1004381.

Wiesmann, V., Franz, D., Held, C., Munzenmayer, C., Palmisano, R., & Wittenberg, T. (2015). Review of free software tools for image analysis of fluorescence cell micrographs. *Journal of Microscopy, 257,* 39–53. https://doi.org/10.1111/jmi.12184.

Zimmer, C., Labruyere, E., Meas-Yedid, V., Guillen, N., & Olivo-Marin, J. C. (2002). Segmentation and tracking of migrating cells in videomicroscopy with parametric active contours: a tool for cell-based drug testing. *IEEE Transactions on Medical Imaging, 21*(10), 1212–1221. https://doi.org/10.1109/TMI.2002.806292.

The manufacturer's authorised representative in the EU is Springer Nature Customer Service Centre GmbH, Europaplatz 3, 69115 Heidelberg, Germany. If you have any concerns regarding our products, please contact ProductSafety@springernature.com

Printed and bound by CPI Group (UK) Ltd, Croydon, CR0 4YY
23/09/2026
02221532-0001